NUCLEAR POWER PERFORMANCE AND SAFETY

VOLUME 4
SAFETY TECHNOLOGY

The following States are Members of the International Atomic Energy Agency:

AFGHANISTAN
ALBANIA
ALGERIA
ARGENTINA
AUSTRALIA
AUSTRIA
BANGLADESH
BELGIUM
BOLIVIA
BRAZIL
BULGARIA
BURMA
BYELORUSSIAN SOVIET SOCIALIST REPUBLIC
CAMEROON
CANADA
CHILE
CHINA
COLOMBIA
COSTA RICA
COTE D'IVOIRE
CUBA
CYPRUS
CZECHOSLOVAKIA
DEMOCRATIC KAMPUCHEA
DEMOCRATIC PEOPLE'S REPUBLIC OF KOREA
DENMARK
DOMINICAN REPUBLIC
ECUADOR
EGYPT
EL SALVADOR
ETHIOPIA
FINLAND
FRANCE
GABON
GERMAN DEMOCRATIC REPUBLIC
GERMANY, FEDERAL REPUBLIC OF
GHANA
GREECE
GUATEMALA
HAITI
HOLY SEE
HUNGARY
ICELAND
INDIA
INDONESIA
IRAN, ISLAMIC REPUBLIC OF
IRAQ
IRELAND
ISRAEL
ITALY
JAMAICA
JAPAN
JORDAN
KENYA
KOREA, REPUBLIC OF
KUWAIT
LEBANON
LIBERIA
LIBYAN ARAB JAMAHIRIYA
LIECHTENSTEIN
LUXEMBOURG
MADAGASCAR
MALAYSIA
MALI
MAURITIUS
MEXICO
MONACO
MONGOLIA
MOROCCO
NAMIBIA
NETHERLANDS
NEW ZEALAND
NICARAGUA
NIGER
NIGERIA
NORWAY
PAKISTAN
PANAMA
PARAGUAY
PERU
PHILIPPINES
POLAND
PORTUGAL
QATAR
ROMANIA
SAUDI ARABIA
SENEGAL
SIERRA LEONE
SINGAPORE
SOUTH AFRICA
SPAIN
SRI LANKA
SUDAN
SWEDEN
SWITZERLAND
SYRIAN ARAB REPUBLIC
THAILAND
TUNISIA
TURKEY
UGANDA
UKRAINIAN SOVIET SOCIALIST REPUBLIC
UNION OF SOVIET SOCIALIST REPUBLICS
UNITED ARAB EMIRATES
UNITED KINGDOM OF GREAT BRITAIN AND NORTHERN IRELAND
UNITED REPUBLIC OF TANZANIA
UNITED STATES OF AMERICA
URUGUAY
VENEZUELA
VIET NAM
YUGOSLAVIA
ZAIRE
ZAMBIA
ZIMBABWE

The Agency's Statute was approved on 23 October 1956 by the Conference on the Statute of the IAEA held at United Nations Headquarters, New York; it entered into force on 29 July 1957. The Headquarters of the Agency are situated in Vienna. Its principal objective is "to accelerate and enlarge the contribution of atomic energy to peace, health and prosperity throughout the world".

Printed by the IAEA in Austria
June 1988

PROCEEDINGS SERIES

NUCLEAR POWER PERFORMANCE AND SAFETY

VOLUME 4
SAFETY TECHNOLOGY

PROCEEDINGS OF AN INTERNATIONAL CONFERENCE
ON NUCLEAR POWER PERFORMANCE AND SAFETY
ORGANIZED BY THE
INTERNATIONAL ATOMIC ENERGY AGENCY
AND HELD IN
VIENNA, 28 SEPTEMBER – 2 OCTOBER 1987

In six volumes

INTERNATIONAL ATOMIC ENERGY AGENCY
VIENNA, 1988

NUCLEAR POWER PERFORMANCE AND SAFETY, VOL. 4
IAEA, VIENNA, 1988
STI/PUB/761
ISBN 92-0-050388-8

FOREWORD

The International Conference on Nuclear Power Performance and Safety, organized by the International Atomic Energy Agency, was held at the Austria Centre Vienna (ACV) in Vienna, Austria, from 28 September to 2 October 1987.

The objective of the Conference was to promote an exchange of worldwide information on the current trends in the performance and safety of nuclear power and its fuel cycle, and to take a forward look at the expectations and objectives for the 1990s. This objective was accomplished through presentation and discussion of about 200 papers at the Conference.

Almost 500 participants and observers from 40 countries and 12 organizations discussed three major questions which were posed as the focus for this Conference:

(1) What are the current trends and major issues with regard to performance and safety of nuclear power, the nuclear fuel cycle and radioactive waste management?

(2) What steps are being taken or need to be taken to resolve outstanding issues in order to improve the performance of nuclear power with assured safety?

(3) What performance objectives and achievements can be anticipated for the 1990s?

The trend of nuclear power plant performance is demonstrated by the improvement which has been achieved since the Conference on Nuclear Power and its Fuel Cycle in 1977. At that time, the average energy availability factor for the 137 nuclear power reactor units reported to the IAEA Power Reactor Information System (PRIS) was only 64.7%. By 1986, the most recent year for which complete data are available, that figure had improved to 70.4% for the 288 nuclear power reactor units listed in PRIS. It is even more significant that 55% of these nuclear power units are operating with an energy availability of 75% or better. Indeed, since 1984 some 40% of the units have been consistently reporting an energy availability of more than 80%.

These data clearly show that there are lessons which can be learned through means such as this Conference, by improving the information exchange among all concerned and highlighting the standards of performance which can be achieved.

Essentially all countries with nuclear power plants have carefully analysed the Chernobyl accident in relation to the safety of their own nuclear power plants and concluded that this type of accident would not happen in other types of nuclear power reactors. Nevertheless, the issues of nuclear power safety were of high interest during the Conference, demonstrated by the fact that more than 70 papers on safety were presented and discussed.

The discussion of severe accidents has received much attention during the last year and this was reflected in a number of papers. An overview assessment of the source term discussion was presented by the International Nuclear Safety Advisory Group (INSAG); this concluded, among other things, that accident management is a fruitful path to reduce risk. Another important goal must be to protect the containment function. Both these aspects were discussed in a number of papers.

Many papers in the Conference discussed aspects of safety goals on the basis of probabilistic methods. Apart from quantitative methods for improving the precision of probability and consequence predictions, there are some fundamental points of terminology and criteria needing clarification and agreement. The papers provided some insight into the problems, but not solutions for them. The way in which various relevant parameters should be combined and limited for individual and societal (collective) risk such as mortality, low dose effects, land use, capital loss, etc., is still very much debated. Therefore, decisions on design bases will continue for some time to be made on the basis of deterministic engineering judgement supported by quantitative probabilistic analyses.

The communication of safety information to the public was also addressed in the Conference. Factual and objective information routinely made available is one prerequisite for clarifying to the public the risks of nuclear power. It is clear that not only must nuclear power plants be safe but also that their safety must be made evident to the public.

Considering that the nuclear fuel cycle is an inseparable part of nuclear power performance and safety, it is not surprising that about one third of the papers presented during the Conference dealt with this subject.

Notwithstanding the positive experience in the operation of water cooled nuclear power plants with uranium oxide fuel and the good performance of existing schemes of fuel utilization, research and development continue on improvements in reactor fuel design, performance, reliability and utilization, with the aim to make nuclear energy more competitive. Assessments reported during the Conference show that advanced fuel management schemes (including extended burnup, low leakage loading patterns, fuel assembly reconstitution, etc.) can reduce uranium consumption for light water reactors by about 20% and enrichment requirements by 10 to 14%.

The back end of the nuclear fuel cycle is currently of concern to the nuclear community, not only because of its impact on uranium requirements (reprocessing and recycling can reduce natural uranium requirements by up to 40%), but also because of its political, environmental, socio-economic and technical aspects.

In the back end of the nuclear fuel cycle there is a clear trend to a rapprochement of positions, and countries which in previous years were very strong in supporting the so-called once-through fuel cycle now take a more flexible policy which does not in principle exclude the possibility of accepting the reprocessing strategy in the future.

Policy decisions for waste management have already been taken in many countries and the 1990s should be a period of demonstration and implementation of these policies. As illustrated by data presented from a number of countries, many years of experience in radioactive waste management have been achieved and the technology exists to implement the national plans and policies that have been developed.

The establishment of criteria, the development of safety performance methodology and site investigation work are key activities essential to the successful selection, characterization and construction of geological repositories for the final disposal of radioactive waste. Considerable work has been done in these areas over the last ten years and will continue into the 1990s. However, countries that are considering geological disposal for high level waste now recognize the need for relating the technical aspects to public understanding and acceptance of the concept and decision making activities. The real challenge for the 1990s in waste disposal will be successfully to integrate technological activities within a process which responds to institutional and public concern.

The International Atomic Energy Agency wishes to record its sincere thanks to the Federal Government of Austria and to the authorities of the City of Vienna for the facilities made available and the substantial support provided. These contributed greatly to the smooth running and success of the Conference.

EDITORIAL NOTE

The Proceedings have been edited by the editorial staff of the IAEA to the extent considered necessary for the reader's assistance. The views expressed remain, however, the responsibility of the named authors or participants. In addition, the views are not necessarily those of the governments of the nominating Member States or of the nominating organizations.

The use of particular designations of countries or territories does not imply any judgement by the publisher, the IAEA, as to the legal status of such countries or territories, of their authorities and institutions or of the delimitation of their boundaries.

The mention of names of specific companies or products (whether or not indicated as registered) does not imply any intention to infringe proprietary rights, nor should it be construed as an endorsement or recommendation on the part of the IAEA.

The authors are responsible for having obtained the necessary permission for the IAEA to reproduce, translate or use material from sources already protected by copyrights.

Material prepared by authors who are in contractual relation with governments is copyrighted by the IAEA, as publisher, only to the extent permitted by the appropriate national regulations.

SUMMARY OF COMPLETE CONTENTS

CONTENTS OF VOLUME 4

NUCLEAR SAFETY EXPERIENCE AND SAFETY ASSESSMENT TRENDS (Technical Sessions 3.1A and 3.1B)

SAFETY AND SEVERE ACCIDENTS (Technical Sessions 3.2A and 3.2B)

SAFETY IMPROVEMENTS IN DESIGN AND OPERATION
(Technical Sessions 3.3A and 3.3B)

POSTER PRESENTATIONS

NUCLEAR SAFETY EXPERIENCE AND SAFETY ASSESSMENT TRENDS

Technical Sessions 3.1A and 3.1B

Chairman (Technical Session 3.1A)

H. RABOLD
German Democratic Republic

Chairman (Technical Session 3.1B)

Wan-Yong CHON
Republic of Korea

APPLICATION OF RADIATION PROTECTION OPTIMIZATION PRINCIPLES TO POTENTIAL EXPOSURES FROM ACCIDENTS

D. BENINSON
Comisión Nacional de Energía Atómica,
Buenos Aires, Argentina

Abstract

APPLICATION OF RADIATION PROTECTION OPTIMIZATION PRINCIPLES TO POTENTIAL EXPOSURES FROM ACCIDENTS.

The optimization requirement of radiation protection is that all doses should be kept 'as low as reasonably achievable' (ALARA), economic and social factors being taken into account. This requirement consists in selecting protection options that restrict the detriment from a given practice or radiation source to a value such that further reductions would be considered less significant than the additional efforts required to achieve such reductions. Differential cost-benefit optimization has already become established as a technique in radiation protection and alternative optimization techniques are being explored. A present development in radiation protection is expected to influence the optimization approach. This development is the extension of the principles of the dose limitation system to the case of 'probabilistic exposures', i.e. exposures that would result from random disruptive events. In the case of 'probabilistic exposures', the use of the expectation values (implicit in cost-benefit optimization) would be invalid in the case of low probability-high consequence situations. Furthermore, even the use of collective doses would be precluded if high doses, beyond the range of stochastic linearity, are possible. It seems, therefore, interesting to examine other optimization techniques and the basic tools of decision analysis as possible general procedures for optimization, applicable both to normal situations and to 'probabilistic exposures'. The paper discusses the application of classical decision theory using probabilistically defined utility functions, in a method that seems to be applicable to both exposure situations. For 'normal' exposures, the cost-benefit optimization is shown to be a special case of the general procedure. In the case of optimization of safety features for potential disruptive events, consequences could include components which are very difficult to express in a single quantity. However, ranking them by preference and assigning utilities could be carried out by the decision maker, ensuring consistent decisions. The paper discusses such utility assignment in the case of mortality as a consequence of accidental exposures.

1. INTRODUCTION

The strong emphasis placed by modern radiation protection on optimization is related to the fact that protection is not taken to be absolute. For radiation protection purposes it is assumed that there is a proportional relationship between dose and the probability of incurring a stochastic effect, within the relevant range of doses. A consequence of this assumption is that radiation sources, in spite of the protection efforts involved, imply some residual 'detriment' (mathematical expectation of harmful stochastic effects in exposed individuals).

The optimization requirement of radiation protection is that all doses should be kept 'as low as reasonably achievable' (ALARA), economic and social factors being taken into account. This requirement consists in selecting protection options that restrict the detriment from a given practice or radiation source to a value such that further reductions would be considered less significant than the additional efforts required to achieve such reductions.

Optimization of protection, therefore, implies the reasonable selection of the best option among a set of protection options. Procedures for this decision making range from simple common sense and good practice to the use of elaborated decision analysis aids.

The more sophisticated optimization techniques tend to be more applicable to the design stage of protection systems rather than to the conduct of operations. Multi-criteria methods compare the different options by pairs, taking into account all predetermined criteria, to decide if one option is better than another, iterating the process to the final decision. The value of these methods is that they take account of many criteria and can deal with qualitative aspects. However, they do not generally lead to a complete ranking of options.

Other methods, called aggregative methods, instead of comparing options in pairs, attempt to combine the values of the re-established decision criteria of each option into single values. The most widely used aggregative methods are based on utility functions.

It should be recognized that all optimization methods require a number of assumptions and social judgements. In some cases the assumptions may be implicit and in others explicit. The use of decision analysis techniques usually makes the assumptions more explicit, and provides a coherent approach which could lead to consistent decisions in the trade-offs involved in nuclear safety and radiation protection requirements.

An approach applicable both to optimization of protection for 'probabilistic' exposures from random disruptive events and 'normal' exposures must rest on some common characteristic of these two types of exposure. An important common characteristic is the random nature of the consequences.

In the case of 'normal situations' the consequences of the residual exposures at a given protection level are stochastic effects. The actual number of stochastic effects would always be a random variable distributed in binomial distributions. Probabilities would be a function of the distribution of the effective dose equivalents, and the mathematical expectation would be proportional to the collective dose.

In the case of exposures that would result from random disruptive events, a second order stochasticity would apply. Several events could occur with different probabilities, and each would cause a distribution of doses and thus an uncertain number of stochastic effects plus some non-stochastic effects if the resulting doses are high enough. Non-health detriments are also possible in this case, such as land use invalidation, cost and impact of required countermeasures, etc.

As in all optimization methods, the framework presented in this paper results in the expression of the 'preference' of the decision maker; its only purpose is to provide a mechanism for coherent preference when many decisions are involved.

2. DECISION ANALYSIS APPLIED TO RADIATION PROTECTION AND SAFETY

Provided that individual risk limitation is satisfactory, both for 'normal' and 'probabilistic' exposures [1, 2], the requirement of optimization implies the examination of available protection options which would reduce the risks but would also involve social costs.

Both protection for normal cases and safety features to account for probabilistic exposures would involve decisions in a situation where the actual consequences are uncertain, but where probabilities can be assigned to the possible out-comes. A typical approach for decisions in such situations can be found in the use of 'utility' functions probabilistically based: optimization would then involve minimizing (or maximizing) some aggregate of probabilities. This, a classical approach which can be found in decision analysis text books, is summarized in the following paragraphs, adapted to the case of radiation protection [3].

The first step in the procedure is to identify a set of options that would be effective in reducing risks, preferably from the beginning of the design stage. Where experience exists of operating similar installations or systems, the identification of other options to the one used will be facilitated by the operating experience, including information on maintenance, repairs, modifications, radiation exposures and costs. The set of options should preferably include only the practicable options to avoid unnecessary cost of analysis.

These protection options may be listed as O_i (O_1, O_2,... O_m), and must be analysed in connection with the possible occurrences of random events E_j (E_1, E_2,... E_n), for example the number of stochastic effects caused by the source in the case of 'normal' exposures, the probabilities of which may be denoted by P_j (P_1, P_2,... P_n). Obviously $\Sigma_j P_j = 1$, if the non-occurrence of the possible events is also listed as an event and if the occurrence of one event precludes the others.

Each protection option O_i involves a given social 'cost', including monetary costs, restrictions and inconveniences for the design, in order to reduce the probability of the possible events $E_1 \ldots E_n$. In order to define the consequences, one can assume that a particular event E_j actually occurs in spite of the risk-reducing efforts implied by O_i. All uncertainty is removed from the problem and a definite consequence is associated with the selected protective option. Since the protective option O_i not only has influenced the probability of occurrence of event E_j but also involves a given cost (component of the consequence), the 'consequence' will depend on both O_i and E_j. All the components of this consequence C_{ij} are not necessarily amenable to aggregation into a single quantity.

Having established the set of protection options, the random events against which protection is planned, with their probabilities of occurrence and set of consequences, the next stage is to formalize the expression of preference of the decision maker regarding these consequences.

Each of the defined consequences may consist in a list of components. However, it is conceptually possible to rank all the possible consequences as defined above (i.e. C_{ij} for all values of i and j) in order, from the best to the worst, according to the *preferences* of the decision maker. The ranking must be coherent, in the sense that if a consequence C_1 is preferred to C_2 and C_2 is preferred to C_3, then C_1 must also be preferred to C_3. It is conceivable that two consequences may be equally desirable (or undesirable).

In order to assign a probabilistically based utility value to each consequence, it is possible to design a conceptual random game with only two possible outcomes: a good consequence C_g (at least as good as the best in the ranking) or alternatively a bad consequence C_b (at least as bad as the worst in the ranking). The probability of obtaining C_g in a trial of the game is U, and obviously the probability of obtaining C_b is $1 - U$. A further characteristic of the conceptual game is that the value of U can be selected at will before the trial.

For each consequence C_{ij} of the ranking, there will be a value of U at which the *actual occurrence* of C_{ij} would be judged by the decision maker to be *equal in preference* (equally desirable or undesirable) to entering in a trial of the conceptual game. This particular value of U, $U(C_{ij})$, is the utility of consequence C_{ij}. As defined, utility can be used in probabilistic expressions because, under the equal preference concept, it is equivalent to the conditional probability of obtaining the good outcome given both the adoption of protective option O_i and the actual occurrence of the event E_j:

$$U(C_{ij}) = P(C_g | O_i \text{ and } E_j) \tag{1}$$

Implementing a protection option O_i influences the probabilities of occurrence of the random events E_j. As the events E_j exclude each other, the conditional probability given O_i of obtaining the good consequence (or more exactly situations equivalent in preference to the good consequence) is

$$P(C_g | O_i) = \sum_j U(C_{ij}) \, P_{ij} \tag{2}$$

where P_{ij} is the probability of event E_j if protective option O_i has been implemented.

Optimization of protection involves the selection of the option that maximizes the above probability. Optimization of protection should therefore be achieved with an option O_i such that

$$\sum_{j} U(C_{ij})\, P_{ij} = \text{maximum} \qquad (3)$$

There is an apparent element of arbitrariness in such an optimization procedure, because there is indetermination in selecting the good and bad consequences of the preference random game. However, provided that they are chosen outside the range of all actual possible consequences, it can readily be shown that the determination of the optimized protection option is independent of the actual values selected.

3. PROBABILISTIC DECISION ANALYSIS IS CONSISTENT WITH OPTIMIZATION IN NORMAL SITUATIONS

Assigning probabilistic utilities is basically the result of social valuations intrinsic in the concept of preference. It should be noted that judgements of this type are inherent in all procedures of radiation protection optimization. For example, in the cost-benefit optimization for normal situations, social valuations are involved in selecting the value of α, the monetary value assigned to the unit of collective dose [4].

As the collective dose in normal situations is proportional to the mathematical expectation of consequence (expected number of stochastic effects), the selection of a value of α implies the indirect assignment of a monetary cost per statistical life, $v = \alpha/R$, where R is the risk of stochastic effects per unit effective dose equivalent. The selection of a value of α, therefore, implies a judgement of equal preference for consequences of either one more statistical death or one extra expenditure of a value v.

It is easy to show that the decision analysis approach presented in this paper is consistent with the usual cost-benefit optimization used for 'normal' exposure situations.

If decision analysis is applied to optimize protection for these 'normal' situations, using Eq. (3), utilities must be assigned to consequences, involving at least the number of stochastic effects (0, 1, 2,...n) and the cost of the protection option O_i, namely χ_i. The probabilities P_{ij} of the different consequences are a function of the distribution of doses resulting from protection option O_i, the probability of incurring in a stochastic effect being linear with the effective dose equivalent.

If the assignment of utilities is linear with the actual number of stochastic effects, which is quite reasonable in view of the low individual probabilities involved, in the range of interest, and if it is judged possible to aggregate this part of the consequence with the cost of the protection option as χ_i/v, then

Eq. (3) can easily be manipulated to give the objective function of cost-benefit optimization:

$$\chi_i + \alpha S_i = \min.$$

provided no other preference criteria are taken to apply.

4. OPTIMIZATION OF PROTECTION FOR DISRUPTIVE EVENTS

Actual consequences of disruptive events could include components which are very difficult to express in a single quantity (number of non-stochastic and stochastic effects, land contamination, cost of the protective option, etc.). However, ranking by preference and utility assignment can be carried out by the decision maker, ensuring consistent decisions, which is the main purpose of decision aids.

It is possible at present to elaborate somewhat more regarding utility functions for one type of component of the consequences, namely the loss of life. For prospective random accidental events, the risk to an individual depends on the probability of the event and on the resulting dose, and therefore the conditional probability of death, given the event. For radiation effects one deals, in this case, with effects of second order stochasticity. For example, the probability of dying (risk) as a result of a given potential disruptive event can be expressed as $P = P_i P_2$, where P is the 'risk', P_1 is the probability of occurrence of the event which would result in a given dose, and P_2 is the conditional probability of death given that dose. For P_2, for radiation protection purposes, the probability of stochastic effects (namely fatal cancer and serious genetic effects) is taken to be of the order of 10^{-2} per sievert of effective dose equivalent [1]. On the other hand, the probability of non-stochastic effects, above the threshold, increases from zero to one over a usually small range of dose. For example, for mortality due to the acute radiation syndrome resulting from whole body irradiation, the probability at less than 2 Gy is practically zero, while it is nearly one above 6 Gy.

Acute non-stochastic death can be dealt with together with or separated from stochastic late death. This is a first decision expressing the preference of the decision maker. Regarding the number of lives lost, the utility function could take different forms depending again on preference.

However, it is likely that decreases of utility with increases of number of lives lost will be judged to depend on the actual numbers involved. For example, accidents killing 1 or 100 individuals would be considered of quite different magnitudes, while accidents killing 10 001 or 10 100 individuals would be judged quite similar, although the absolute difference is the same in both cases.

Utility in these cases could be taken to be an increasing function of the total number of individuals not affected by the event U(N − n), where N is the total number of individuals potentially at risk and n is the number killed in the event. It should be noted that the selection of 'good' and 'bad' consequences for the conceptual game for assigning utilities would imply n = 0 and n = N respectively.

If one assumes a constant risk aversion of the decision maker over the entire range of values, the valuation of the magnitude of accidents on a relative basis as exemplified above (which implies a logarithmic nature for the 'disutily' (1 − U) leads to a utility function of the type

$$U(N - n) = \frac{1 - e^{-k(N-n)}}{1 - e^{-kN}}$$

where k is a constant, which could be assessed from actual judgements of the decision maker.

An interesting feature of the above utility function is that it leads to linearly decreasing utilities in the case of very small values of n compared to N, consistent with the assumptions of cost-benefit optimization of 'normal' exposure situations, where individual risk is minute. It also indicates that high individual risk, approaching one, would lead to very small utility, consistent with the different social judgements regarding a statistical death or an identified one.

A final comment is necessary. Many prospective decision makers could perhaps conclude that optimization of protection by decision analysis techniques really boils down to the advice: 'Select the protection option you like; anyway it is a matter of your preference'. This is only partially true. It is a matter of *coherent* preference, sometimes (but not always) provided by experience alone, used together with the assessed probabilities of events which may occur in spite of all risk reducing efforts.

REFERENCES

[1] INTERNATIONAL COMMISSION ON RADIOLOGICAL PROTECTION, Recommendations of the International Commission on Radiological Protection, ICRP Publication 26, Pergamon Press, Oxford and New York (1983).

[2] BENINSON, D., IAEA-CN-48/36, these Proceedings, Vol. 3.

[3] BENINSON, D., "Optimization of radiation protection as a special case of decision theory", Optimization of Radiation Protection (Proc. Symp. Vienna, 1986), IAEA, Vienna (1986) 45–50.

[4] INTERNATIONAL COMMISSION ON RADIOLOGICAL PROTECTION, Cost-Benefit Analysis in the Optimization of Radiological Protection, ICRP Publication 37, Pergamon Press, Oxford and New York (1983).

SAFETY GOALS AND DOSE-FREQUENCY CRITERIA TO CONTROL INDIVIDUAL RISK

P.A. GOTTSCHALK, W. WERNER, K. KOEBERLEIN
Gesellschaft für Reaktorsicherheit GmbH,
Cologne, Federal Republic of Germany

Abstract

SAFETY GOALS AND DOSE-FREQUENCY CRITERIA TO CONTROL INDIVIDUAL RISK.

Recently, the International Commission on Radiological Protection and advisory groups of the Organisation for Economic Co-operation and Development and the International Atomic Energy Agency have published probabilistic safety criteria for nuclear installations. A common feature is that these criteria have been formulated as dose-frequency curves which limit individual risk. The objective is to extend the established radiation protection principles to accident situations where there is a potential for an exposure, but not a certainty. The paper points out some difficulties regarding the practical application of these approaches.

Introduction

Probabilistic risk assessment techniques have matured significantly during the past decade. Probabilistic approaches to nuclear safety issues are more widely accepted than some years ago. Insights gained from risk studies have influenced system design and licensing in many instances. Risk studies have received increased attention by the public. In this situation there is a growing need for a reference scale to assess the results of risk studies in quantitative terms.

It is common understanding that the relative values of frequencies and potential consequences of accidents and above all their interrelations are the most important results of risk studies. They provide a measure to evaluate and possibly improve the safety concept. Such interrelations represent risk-profiles.

An important question is whether a particular risk-profile depicts an adequate and balanced level of safety. For this purpose a yardstick is needed.

Recent initiatives taken by ICRP and advisory groups of OECD and IAEA strive for the establishment of a radiological reference scale. The scale is formulated as a criterion curve. The curve attempts to control individual risk over the whole spectrum of events by placing limits on the probability of annual doses to members of the critical group. The curve is formulated as a FARMER limit line.

A successful application of the criterion curve is based on two premises: First, the occurrence frequencies of the events and their corresponding individual doses must be provided with sufficient confidence, and second, the results of such probabilistic evaluations must be readily comparable with the criterion curve.

In the following some difficulties regarding practical applications of the criterion curve approach are pointed out. We'll confine ourselves to the above mentioned second aspect.

The Criterion Curve of ICRP-46

ICRP has recommended in its publication ICRP-46 /1/ a criterion which is setting a limit to the probability of annual dose to members of the critical group. The criterion curve is shown in Fig. 1.

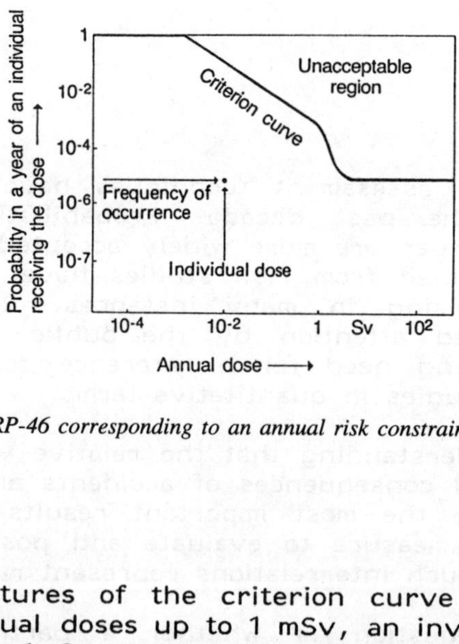

FIG. 1. Criterion of ICRP-46 corresponding to an annual risk constraint of 10^{-5} from all events.

The relevant features of the criterion curve are: probability limit of 1 for annual doses up to 1 mSv, an inverse proportionality region up to 1 Sv, a non-proportional region for doses in the range 1 Sv up to about 5 Sv, and a constant probability for doses which are larger than about 5 Sv.

The critical group is defined as being representative of those individuals in the population who are expected to receive the highest dose equivalent. When such a group cannot be defined, a hypothetical group or a representative individual should be considered who, due to location and time, would receive the highest dose.

The ICRP-46 criterion is to be applied to an individual affected by the cumulative risk resulting from all events and all nuclear installations. For a single nuclear installation - for instance a solid-waste disposal - some fraction of the risk limit is to be selected as a source specific upper bound.

According to ICRP-46 the radiological risk should be judged on the basis of point values and single, discrete release categories with specific provision to limit the cumulative frequencies of particular doses in the respective dose spectrum.

The curve in the dose-frequency diagram has been chosen with the objective that each point on the curve represents an annual (mortality) risk of 10^{-5}, regardless of the types of serious radiation health effects.

The curve should be used as a basic decison tool for checking whether a nuclear installation is unacceptable. Events which might create radiation exposures should be selected, their probabilities of occurrences should be assessed and the consequent exposures of individuals of the critical group should be evaluated. Finally, the point representing the probability of occurrence of the initial event and all other environmental conditions and the corresponding maximum dose should be plotted.

The criterion curve belongs to the so-called "Farmer curves".

Similar concepts have been published by OECD /2/, CNEA /3/ and IAEA /4/.

The Farmer Approach

Since Farmer /5/ has proposed his criterion considerable confusion arose about the correct interpretation of the Farmer type limit-lines. Attempts to evaluate overall risk for a given Farmer limit-line by integrating the area below the line has led to controversy /6-11/. Individual mortality risk (per year) has been evaluated between 1.4×10^{-6} and 2.9×10^{-4}, based on the same Farmer line and using identical assumptions in converting exposure into mortality risk. The different values are traceable to different interpretations of the Farmer line.

The problem will be made clear if one ponders on the widely adopted interpretation that the placing of the limit-line in Fig. 1, for example, is intended to reflect an annual frequency of 10^{-2} for a dose of 0.1 Sv. However, it is only possible to assign finite values to the frequencies that a dose will be within finite intervals. The probability that a dose will be exactly 10^{-2} (or any other numbers) is zero. One must indicate the intervals on the abscissa to which a given ordinate refers.

A formal mathematical definition of the Farmer line has been proposed in ref. /12/. Here, the original Farmer line has been interpreted as a frequency density distribution for release in an infinitesimal interval. Since the Farmer line has been presented on a log-scale, Farmer interpreted his limit line as the frequency density function on a logarithmic scale which differs from the previous interpretation by ln 10 = 2.303. Other experts have, furthermore, interpreted the Farmer line as applying to median frequencies for a decade interval of exposures (in suitable units) or even as a complementary distribution function. Obviously, these interpretations are of great significance for the overall risk and, moreover, for the concept of equal risk associated with different values on the abscissa if the limit line takes the form of a straight line with constant slope.

Another problem has been pointed out by Okrent /6/ and has been stressed by Cox et al. /11/:

The Farmer limit line sets acceptability criteria to individual sequences or sequence classes. The total amount of sequences, however, is not limited. Such an approach may lead to unacceptable overall risk despite adequately low risk posed by each individual sequence or sequence class. In addition, concentration in some sequence categories may cause an unbalanced safety concept. Also, a specific event sequence might easily be split into two or more sequences in order to fulfill the Farmer criterion for each of the partial sequences without being in compliance with regard to their cumulative frequency.

A typical example is shown in Fig. 2. The figure depicts the results of a probabilistic study of a reprocessing plant. Shown are the collective doses to the general public caused by different accidents.

Apparently, it is very difficult to judge the adequacy of the design using Farmer-type assessment criteria.

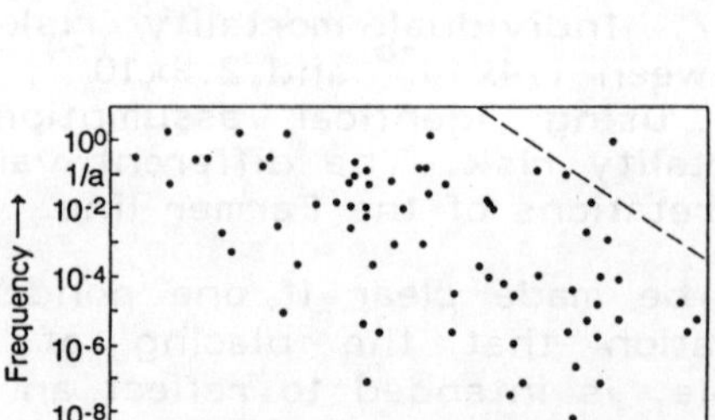

FIG. 2. Farmer diagram for accidents in a reprocessing plant.

The 'Worst Case'

Within the ICRP-46 approach highest doses should be plotted.

A characteristic of theoretical risk studies is that, for all accident scenarios, which cannot be excluded by basic principles, nonzero probabilities are found. Given a scenario with certain consequences, another scenario can always be constructed with larger magnitudes of consequences.

Moreover, results of risk studies should be viewed very cautiously when very remote scenarios and very low probabilities are involved. Such results are very often artefacts of the analysis and affected by very large uncertainties.

Thus, attempts to assess and evaluate risk on the basis of "worst cases" suffer from important limitations:

They focus on the weakest point of risk study results, which are most heavily affected by inaccuracies, they emphasize a concept which is mathematically and technically poorly defined, and, moreover, they provide no guidance with respect to the high priority principle in nuclear safety and risk management, namely the avoidance of damages by putting emphasis on preventive measures.

The CCFD Approach

In WASH-1400, /13/, in the German Risk Study /14/ and in current fullscope risk studies the frequencies of accidents are shown as a function of the magnitude of their consequences. The state-of-the-art method for presenting the results is to use complementary cumulative frequency distributions (CCFD). The CCFD is a graph showing the (analytically calculated) frequency (per year) of the totality of events causing damage greater than a given value.

In cases where the magnitude of the consequences and also their frequencies may vary over a wide range, presenting the results of risk studies in form of CCFDs is adequate for several reasons:

- through this form of presentation, the whole range of frequencies and consequences can be easily comprehended,
- the CCFD is obtained by integrating probability density functions; therefore, its use avoids discussions about appropriate lengths of intervals for calculation of interval probabilities.
- the corresponding expectation values and the particular "worst cases" are also known or can be derived from the CCFD.

In close spirit with the CCFD-presentation of risk study results the reference scale should be a CCFD-criterion curve, too.

In fact, a possible CCFD approach to limit lines of risk has been formulated by Cox et al. /11/. These authors suggested a hypothetical CCFD as alternative to the Farmer line with appropriately chosen parametric form to meet all requirements of the regulator or decision maker.

An example of a dose-frequency curve in form of a CCFD and possible applications for the assessments of accidents can be found in ref. /15/. This curve has been constructed similar to the ideas of ref. /11/ and uses radiological principles laid down in the Radiological Protection Ordinance of the Federal Republic of Germany.

Problems Regarding Practical Application of the ICRP-46 Criterion Curve

Dose-distributions

Regarding practical application, the ICRP-46 criterion curve creates difficulties. As mentioned, the state-of-the-art method for presenting and comparing the results of risk studies is to use CCFD's. However, the ICRP-46 criterion curve has to be compared with points which represent the probabilities of occurrence of events and the corresponding doses.

Typical results of individual dose calculation for accidental releases are shown in Figures 3 und 4. The figures show the probability distributions of the expected effective dose (70 years dose commitment) for a most exposed individual at unfavourable distance from the power plant (625 m). The source term has been kept fixed. It represents a postulated core melt accident and early containment failure. The probability distributions are caused by the different weather conditions. Each weather condition has been statistically weighted according to its occurrence frequency. Both, the CCFD and the interval probabilities in different dose intervals are shown.

The following can be concluded: First, interval probabilities do not render a clear picture, their use will create ambiguities. Second, the magnitudes of individual doses vary by orders of magnitude. Thus, a single dose value which corresponds to the accident and which might be used as a representative for assessing the respective radiological risk cannot be defined. Any selection of a distinctive point value is arbitrary and renders the comparison with the ICRP-46 criterion doubtfull.

The decision concerning acceptance or non-acceptance of a specific facility or radiation source depends on the arbitrary selection of the dose point values.

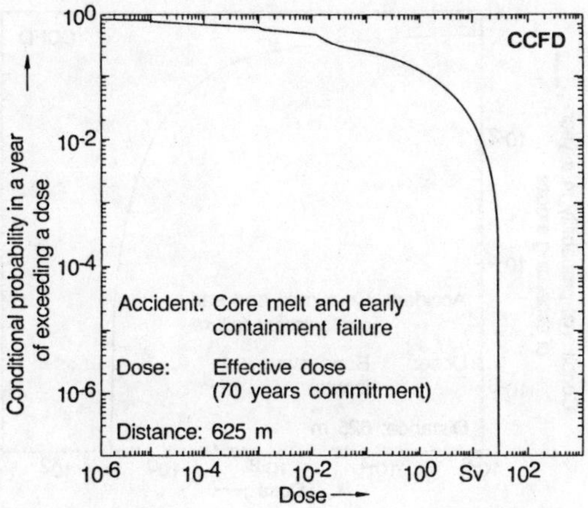

FIG. 3. Complementary cumulative frequency distribution of dose commitments (individual doses of a reference person at a particularly unfavourable distance from the emitting source) caused by core melt and early containment failure. The expected effective dose is shown (70 years dose commitment) including all relevant organs or body tissues.

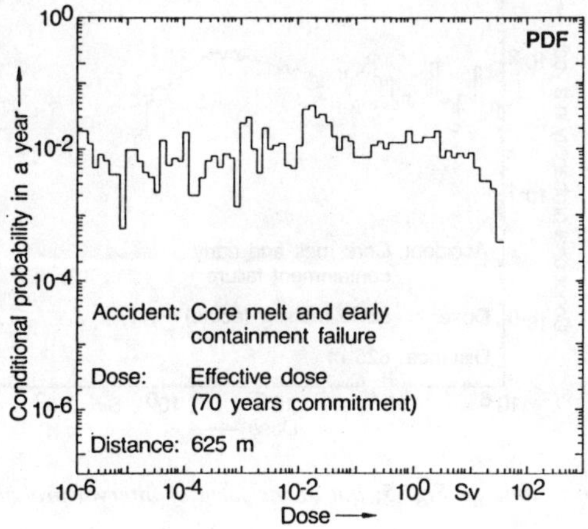

FIG. 4. Dose commitments in the form of interval probabilities.

Early and latent radiation health effects

Moreover, in risk studies early and latent radiation health effects are distinguished and, accordingly, probability distributions for different types of doses are calculated which are used as the appropriate radiological quantities for the different kinds of health effects.

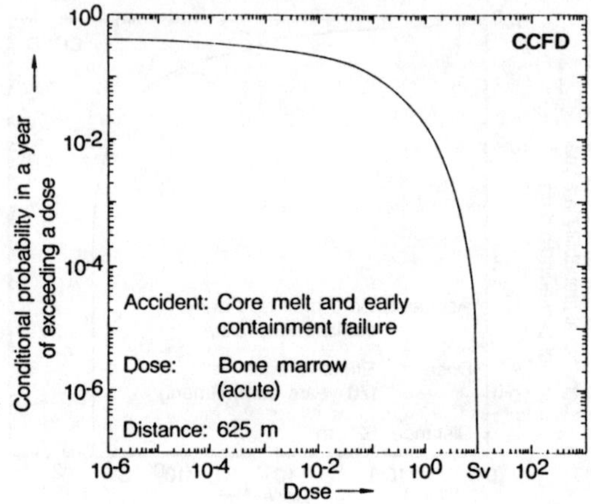

FIG. 5. Same as Fig. 3, but for the acute radiation dose absorbed by red bone marrow.

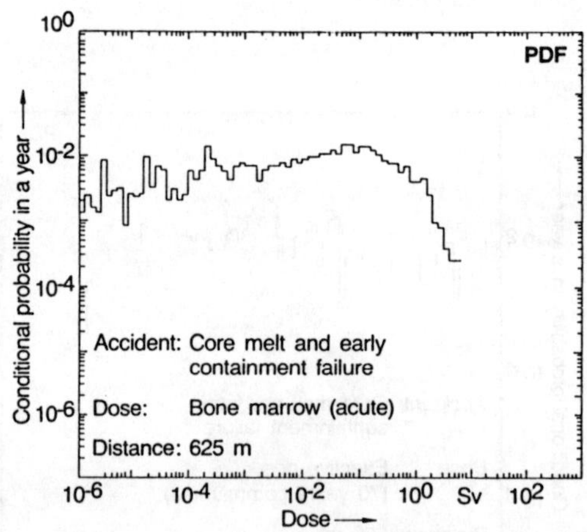

FIG. 6. Same as Fig. 5, but in the form of interval probabilities.

Figures 5 and 6 show the results of individual dose calculations for the same source terms as in Figures 3 and 4, but for the acute radiation dose absorbed by red bone marrow. This might be considered as a representative for individual early fatality risk. Again, the results are shown both in CCFD form and as interval probabilities.

Obviously, the expected long-term effective dose may differ significantly from the acute dose. In particular, exposure scenarios are possible where the acute dose is less than the threshold for non-stochastic injuries, however, the long-term dose might be quite large (about 10 Sv), so that life-threatening is caused by the late cancer risk only. Such scenarios are difficult to assess by means of the ICRP-46 criterion curve.

A correct superposition has to be made of the probability that different kinds of detrimental health effects will occur to an exposed individual. Moreover, separate dose-frequency criterion curves are needed for the control of each of the different kinds of risk.

Construction Principles for Dose-Frequency Criteria

1. Definition of the (individual) risk:

   If several events E_i contribute to the particular kind of detriment k and to the dose interval $\Delta D = (D, D+dD)$, the differential risk contribution from event E_i to interval ΔD is written as

$$dR_k = f^i(D)P_k(D)dD$$

   with the frequency and conditional probability densities $f^i(D)$ and $P_k(D)$, respectively, for dose D and detriment k at a given dose D.
   The total risk from all events for a fixed kind of detriment k is

$$R_k = 1 - \prod_i (1 - R_k^i)$$

   with

$$R_k^i = \int_0^\infty f^i(D)P_k(D)dD$$

However, this only applies to events E_i which are independent of each other.

2. Superposition of different kinds of detrimental health effects

In the case of mutually exclusive detrimental effects, the probabilities of detriment have to be superposed correctly. In addition, they have, in general, to be correlated with various dose concepts (e.g. effective dose, bone marrow dose, etc.).

This superposition may be written as follows:

$$R_M = 1 - (1 - R_F) \cdot (1 - R_S)$$

with: R_M = total fatality

R_F = early fatality

R_S = late fatality

and $$R_S = (1 - R_F) \cdot (1 - (1 - R_1)(1 - R_2)),$$

if the acute fatality and two different kinds of late detriment (e.g. fatality from stochastic and from non-stochastic detriment) are considered.

In general, the probabilities of detriment, R_F, R_1 and R_2, refer to different kinds of doses D_F, D_1 and D_2 and their respective frequencies.

If only late fatality from stochastic effects is considered, the conditional probability for late fatality (under the condition of no early lethal health effects) can be related to the organ-specific dose commitments and organ-specific risk-coefficients as follows

$$R_1 = 1 - \prod_{og} (1 - \alpha^{og} \times D^{og}_{late})$$

Then, an effective long-term dose commitment relevant for only somatic latent health effects and accounting for the reduction of latent health risk by the competing acute mortality risk can be calculated according to

$$D^{eff}_{late} = \alpha^{-1}(1-R_F)R_1 = \alpha^{-1}R_s$$

Here, α denotes the effective risk coefficient of ICRP-30 /16/

$$\alpha = 1.25 \times 10^{-4} \text{ fat/rem}$$

3. Construction of CCFD criterion curves /11, 15/

Starting point is the density function g(s). Total risk, R_o (e.g. the expectation value), the expected total occurrence frequency, h, and the complementary cumulative frequency H (at the value S=s) are given by[1]

$$R_0 = \langle S \rangle = \int_0^\infty s g(s) ds$$

$$h = \int_0^\infty g(s) ds$$

$$H(S > s) = \int_s^\infty g(s') ds'$$

Here, S is e.g. the individual mortality risk, $0 \leqq S \leqq 1$.

Let us specialize to the case where g(s) is given by the following functional type

$$g(s) = \begin{cases} k_0 s^{-1/2} & 0 < s < s_0 \\ k_1 s^{-1} & s_0 \leq s \leq s_1 \\ k_2 s^{-\nu} & s_1 < s \end{cases}$$

with four parameters, k_o, k_1, k_2 and ν, ($\nu > 1$). Then

$$h = \int_0^1 g(s) ds + k_2 (\nu - 1)^{-1}$$

$$R_0 = \int_0^1 g(s) s ds + k_2 (\nu - 1)^{-1}$$

[1] H and h refer to the occurrence frequencies of the events which are taken into account for calculating the individual risk.

The coefficients k_o, k_1, k_2 and ν are found by the conditions

$$k_1 s_1^{-1} = k_2 s_1^{-1} \qquad 2$$

$$\left.\begin{aligned} \int_0^{s_0} g(s)sds &= \int_{s_0}^{s_1} g(s)sds + \int_{s_1}^{1} g(s)sds + k_2(\nu-1)^{-1} \\ \int_{s_0}^{s_1} g(s)sds &= \int_{s_1}^{1} g(s)sds + k_2(\nu-1)^{-1} \end{aligned}\right\} \qquad 3$$

and

$$R_0 = \langle S \rangle = \int_0^1 sg(s)ds + k_2(\nu-1)^{-1}$$

The last requirement is due to the condition that the total risk associated with the criterion curve should not exceed the predefined level R_o.

The transformation from a mortality-frequency CCFD into a dose-frequency CCFD is straight forward for a given dose-mortality relationship

$$s = s(D), \quad ds = \frac{ds}{dD} dD .$$

For stochastic late effects the linear-relationship might be used

$$s_{late} = \alpha D_{late}^{eff} \quad \text{with } \alpha = 10^{-4}/\text{rem}.$$

For acute health effects a sigmoidale relationship which includes a threshold is more appropriate.

In cases where acute and late effects have to be considered the competition of both effects must be accounted for by (see previous chapter)

$$D_{late}^{eff} = \alpha^{-1}(1-R_F)R_1$$

---

[2] Continuous matching.

[3] Equal contribution of risk in different ranges ("balanced safety concept").

Fig. 7 shows a dose-frequency criterion in CCFD-form which has been constructed according to the above mentioned principles /15/. The placement of the curve is dictated by the requirement that the corresponding total individual risk R_o should not exceed the cancer risk resulting

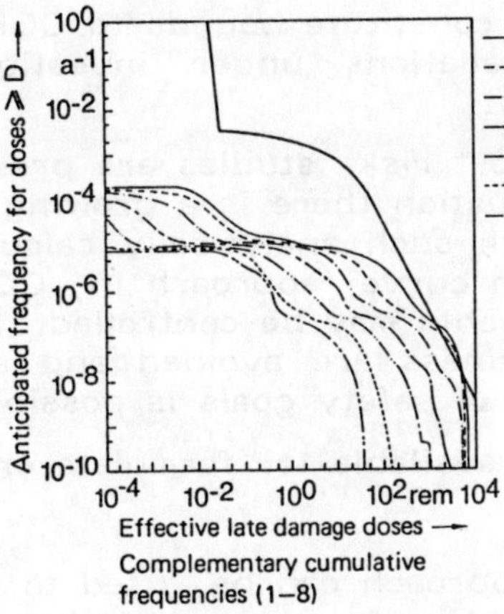

FIG. 7. Complementary cumulative frequency distribution (CCFD) of committed effective dose-equivalents relevant for stochastic health effects as a function of the distance of a reference person from the emitting source. The competing acute mortality risk has been taken into account (see text). Cumulative distributions are shown where all release categories (LOCA and core melt accidents) have been added up. The CCFD criterion curve corresponds to a total annual risk of about 3 × 10⁻⁶ fat/a (≙ 0.3 mSv).

from the permissible operational dose commitments in the FRG, e.g. $3 \cdot 10^{-4}$ Sv/a (about 10^{-6} fat/a). The curve concerns the total late fatality risk. Also shown are the results of individual dose calculations for a reference person at different distances from the plant. The graphs show the complementary cumulative frequency distributions of the effective long-term dose commitments relevant for somatic latent health effects (including all relevant organs or body tissues and accounting for the competing acute mortality risk). Here, the source terms and frequencies of the German Risk Study (Phase A) have been used.

Conclusions

Frequencies and magnitudes of possible consequences of accidents in nuclear installations vary by orders of magnitude. This variety introduces a new element into the assessment of risk.

Risk oriented numerical safety criteria which are defined as point estimates suffer from a tendency to suppres patterns, ranges and relative ranking of safety issues.

Safety criteria in form of criterion curves attempt to control risk with specific recognition of the relationship between frequencies and consequences.

Criterion curves should constitute bounds to CCFD curves describing the risks of installations under investigation for several reasons:

- The results of most risk studies are presented in form of CCFDs. In this situation there is a growing need for a reference scale to rank such analytically calculated risk profiles. By a criterion curve approach in CCFD-form the entire spectrum of events may be controlled. In addition, ambiguity and arbitrariness are avoided and a mathematically concise formulation of safety goals is possible.
- Methodologies are available to formulate criterion curves in CCFD form.

The criterion curve approach can be used to formulate safety (risk) goals in terms of dose-frequency relationship.

Particular in severe accidental situations it should be distinguished between acute and late effects of radiation.

For the control of each of the different kinds of risk, a separate consequence - frequency criterion curve may be important. No single dosimetric quantity appears to be appropriate in formulating the risk limitation.

The shape of the reference CCFD and the actual variables in which the reference CCFDs are expressed should be carefully examined. They should directly reflect the objectives of the regulatory bodies or the decision makers. The results of full scope risk studies should be compared with the proposed criterion curves before such criteria are established. Also, the impact of plant modifications and their risk ranking in the light of proposed criterion curves would deepen the understanding of these risk criteria and would help to confirm whether they reflect their intended objectives.

Because of the Farmer-type approach and the simplified dose concept, the ICRP-46 concept has strong limitations. It is questionable whether the ICRP-46 criterion curve can be used as a sound probabilistic reference scale.

It is felt, that there is an important need for exchange of technical information among those advocating various frequency-consequence criteria.

References

/1/ Annals of the ICRP, Radiation Protection Principles for the Disposal of Solid Radioactive Waste, ICRP Publication No. 46, Pergamon Press, 1986

/2/ Nuclear Energy Agency, Long term radiation protection objectives for radioactive waste disposal Report of a Group of Experts OECD/NEA, Paris, 1984

/3/ Comision Nacional de Energia Atomica, Buenos Aires. Consejo Asesor para el Licenciamiento de Instalaciones Nucleares. Criterios radiológicos relativos a accidentas. CNEA, 1979. 2 p Norma Calin no. 3.1.3

/4/ IAEA, The Application of Radiation Protection Principles to Sources of Potential Exposure: Towards a Unified Approach to Radiation Safety Draft prepared by an Advisory Group Meeting (Jan. 1987), Vienna 87

/5/ F.R. Farmer, Nucl. Safety 8 (1967) 539

/6/ D. Okrent, Nucl. Technol. 27 (1975) 304

/7/ M. Meleis, R.C. Erdmann, Nucl. Safety 13 (1972) 22

/8/ H.J. Otway, R.C. Erdmann, Nucl. Eng. Design 13 (1970) 365

/9/ F.R. Farmer, Nucl. Safety 13 (1972) 362

/10/ R.C. Erdmann, Nucl. Safety 13 (1972) 365

/11/ D.C. Cox, P. Baybutt, Nucl. Technol. 57 (1982) 320

/12/ J.R. Beattle, G.D. Bell, J.E. Edwards UKAEA Report AHSB (S) R 159 (1969)

/13/ Nuclear Regulatory Commission Reactor Safety Study: An Assessment of Accident Risks in U.S. Commercial Nuclear Power Plants, NRC Report WASH-1400 (NUREG-75/014), NTIS, Oct. 1975

/14/ Gesellschaft für Reaktorsicherheit; Deutsche Risikostudie Kernkraftwerke, Eine Untersuchung zu dem durch Störfälle in Kernkraftwerken verursachten Risiko, Verlag TÜV Rheinland, Köln, 1979; engl. translations A.W. Barsell and I.B. Wall: German Risk Study-Main Report, A Study of the Risk due to Accidents in Nuclear Power Plants, Report, EPRI-NP-1804-SR, April 1981

/15/ Schriftenreihe Reaktorsicherheit und Strahlenschutz, BMU, Konzept zur Unterstützung der Sicherheitsbeurteilung von Kernkraftwerken mit probabilistischen Methoden P.A. Gottschalk et al., BMU-1986-133 (ISSN 0724-3316)

/16/ International Commission on Radiological Protection (ICRP) Report 30, 1979

PROBABILISTIC SAFETY CRITERIA FOR NUCLEAR POWER PLANTS

B. EDMONDSON
Central Electricity Generating Board,
London, United Kingdom

F. NIEHAUS
Division of Nuclear Safety,
International Atomic Energy Agency,
Vienna

Abstract

PROBABILISTIC SAFETY CRITERIA FOR NUCLEAR POWER PLANTS.

With the increasing use of probabilistic safety assessment many countries have developed probabilistic criteria for nuclear safety to complement deterministic criteria. However their status and use vary. One group of countries is already using them (or is close to a final decision to do so) for safety decisions including design, licensing and operation. Another group of countries is at present debating the basic concepts and their implications. The third group of countries is monitoring the developments for possible use. Based on qualitative objectives a number of quantitative criteria have been proposed. There is a wide range of attitudes. For example, there is a contrast between the reactor safety practitioners' views and those expressed by the International Commission on Radiological Protection (ICRP) Publication 46 and practitioners in the radwaste field. The latter impose individual risk limits, limits in the sense that they must not be exceeded. This position was arrived at via the established risk levels for individual continuous exposure at the ICRP limits. They also take the view that other objectives should be subsumed in an optimization process which takes account of all relevant factors, including costs, societal risk, amenity detriment, etc. Reactor practitioners do not accept the concept of individual risk *limits*. Rather, they prefer to see probabilistic safety criteria as design guidelines, objectives, assessment principles. Within the community of reactor safety practitioners itself, there is a variety of attitudes to such criteria depending on local circumstances. In some countries strenuous efforts are being made to develop the concept to the point where it can ease the burden of rule making requirements. In other countries it is thought that the value of the concept is limited by the likelihood that the burden to demonstrate compliance will be very severe except at the safety function/system level. The paper summarizes and reviews used or proposed probabilistic safety criteria.

1. INTRODUCTION

Recent experience in the nuclear industry, and also with chemical facilities, has demonstrated the need to pay more attention to severe accidents. Safety of industrial installations is so far mainly assured by a deterministic approach which consists of a set of fundamental principles for siting, design, construction, operation, maintenance, accident management and emergency preparedness. These fundamental principles are implemented by specific engineering principles.

Experience shows that in spite of all precautions taken, accidents can happen. Often these accidents occur due to a combination of design weaknesses, human failures and maintenance errors. The tool to analyse and evaluate the potential for severe accidents is quantitative safety assessment, in particular Probabilistic Safety Analysis (PSA). In recent years this methodology has matured to a stage where in many countries it has become a standard tool complementing the deterministic approach towards safety.

In addition to the many qualitative conclusions which can be drawn, PSAs also provide quantitative results which in turn need quantitative criteria to allow for an assessment. Many countries have therefore developed probabilistic criteria for nuclear safety to complement deterministic criteria. However their status and use varies. One group of countries is already using them (or is close to a final decision to do so) for safety decisions. Another group of countries is at present debating the basic concepts and their implications and makes use of such criteria on a trial basis. The third group of countries is monitoring the developments and the experience gained for possible use.

In discussing the various approaches to Probabilistic Safety Criteria (PSC), it is important to note the present use of PSA. PSA is used as a decision aiding tool in nearly all areas of nuclear safety including design, licensing and operation. Emphasis on one or the other area is necessarily reflected in the various national approaches to PSC. Recently there is a trend to use PSA in particular for operational safety. In this regard PSA results provide a more logical basis for derivation of consistent Technical Specifications including optimization of maintenance schedules and limiting conditions for operation. Also Level-1 PSAs can be stored on personal computers in the control room as an aid to the operator. This development is reflected in the present discussion which is focusing on PSC at the level of safety functions/systems.

This paper summarizes and reviews PSC which are in use or have been proposed. It is based on material presented at two IAEA Technical Committee Meetings [1,2].

2. QUALITATIVE PROBABILISTIC SAFETY CRITERIA

Deterministic safety principles rule out severe consequences for events representative of design basis accidents and provide for a design that is tolerant to events that moderately exceed the design envelope. Even though all reasonably practicable steps are taken to prevent anticipated operational occurrences from turning into accidents, accidents may nevertheless occur and have indeed occurred. The probabilistic approach analyses for all initiating faults the probability that safety systems will fail to operate correctly. It is thus possible to estimate total risk including probabilities and consequences. PSC put constraints on the total risk and thus have to deal with the old question of what is a tolerable risk. No general answer can be provided since "acceptable risk is the risk of the best alternative" [3] including all technical, economic and social aspects of a facility. Thus PSC provide screening criteria for which risks are definitely "unacceptable" and/or which are not worthwhile pursuing further in any detail. In order to derive any numerical values, qualitative objectives have been put forward, which are sufficiently general to meet largely with common approval. The final decision on what is tolerable, of course, has to be based on a political process.

The following qualitative PSC for severe accidents have been proposed at the level of public health and safety [see e.g. 4,5]:

- individuals should bear no significant additional risk;
- society should bear no significant additional risk;
- risks from nuclear power should be not greater than risks from viable competing technologies;
- risks from accidents should be no greater than risks from normal operation.

In addition qualitative PSC for severe accidents have been proposed at the plant level [see e.g. 2,4,6].

- It should be reasonably assured that no severe coremelt accident would occur in a population of plants during their lifetime.
- No single category of accident sequences should dominate the probability of severe core damage.
- No single safety system should contribute significantly to the probability of core damage.

It should be noted that the criteria at the level of public health and the core melt criterion establish first principles from which quantitative PSC can be derived. The last two of the plant level criteria are aimed at a balanced design based on present practice.

TABLE 1. LEVELS OF PROBABILISTIC SAFETY CRITERIA (PSC) AND REQUIRED LEVEL OF PROBABILISTIC SAFETY ANALYSIS (PSA) TO DETERMINE COMPLIANCE

	Level of Probabilistic Safety Criteria	Level of Probabilistic Safety Analysis
Plant	Safety components Safety systems Safety functions Initiating events	Level 0 = reliability studies
	Accident sequences Core melt	Level 1
	Containment performance Large release of radioactive materials	Level 2
Public health	Individual risk Societal risk Cost-benefit (effectiveness)	Level 3

3. QUANTITATIVE PROBABILISTIC SAFETY CRITERIA

Turning these generalities into quantitative PSC is a topic of widespread activity, the nature of which is obviously conditioned by the purpose conceived for such criteria in specific cases. In turn, this is related to specific regulatory procedures. The kinds of quantitative PSC that are being explored are summarized in Table 1 as they relate to the level of PSA which would be required to show compliance.

There is a wide range of attitudes. For example, there is a contrast between the reactor safety practitioner's views and those expressed by ICRP-46 [7] and by the practitioners in the radwaste field. The latter impose individual risk limits, limits in the sense that they must not be exceeded. This position was arrived at via the established risk levels for individual continuous exposure at the ICRP limits. They also take the view that other objectives should be subsumed in an optimization process which takes account of all relevant factors, including costs, societal risk, amenity detriment, etc. Reactor practitioners do not accept the concept of individual risk limits with reference to accident situations. Rather, they prefer to see PSC as design guidelines, objectives, assessment principles.

Within the community of reactor safety practitioners itself, there is a variety of attitudes to such criteria depending on local circumstances. In some countries strenuous efforts are being made to develop the concept to the point where it can ease the burden of rulemaking requirements. In other countries it is thought that the value of the concept is limited by the likelihood that the burden to demonstrate compliance will be very severe except at the safety function/system level.

Intermediate between these positions are a number of countries which see value in the concept, strictly as guidance to plant designers or as principles against which designs might be assessed.

The following gives examples representative of the level of PSC proposed and indicates the spread of numbers being discussed.

3.1. PSA at the level of public health

Most efforts at this level concentrate on individual risk. Though many suggestions have been made regarding societal risk, only one country is actually using such a societal PSC [8] and one country is using cancer fatality risk [4]. Possible criteria for cost-benefit analysis have not yet been implemented in any country. In defining PSC at this level the different health effects have to be considered. They can be either

- kept separate and treated separately
- aggregated by weighting factors
- defined in terms of dose, in the case of radiation risk

or one or more of the effects can be selected as representing the risk.

3.1.1. Individual risk

Based on qualitative PSC at the level of public health discussed above, some countries have adopted or are discussing quantitative PSC for individual risk typically in the range of less than 10^{-6} per reactor year (or per industrial facility in general).

Such a PSC can be expressed in various ways:

(a) Probability of health effect per reactor year

In comparing any numbers great care has to be taken how the health effect is defined (prompt, latent, or both) how the individual is defined (e.g. average individual) and how the location of the individual (e.g. average location within a certain distance) is specified. The example in [4] states that "the risk to an average individual in the vicinity of a nuclear power plant of prompt fatalities that might result from reactor accidents should not exceed one-tenth of one percent (0.1 percent) of the sum of prompt fatality risks resulting from other accidents to which members of the U.S. population are generally exposed. The average individual in the vicinity of the plant is defined as the average individual biologically (in terms of age and other risk factors) and locationally who resides within a mile from the plant site boundary. This means that the average individual is found by accumulating the estimated individual risks and dividing by the number of individuals residing in the vicinity of the plant."

This translates to a PSC for risk of individual fatality of 5×10^{-7}/year.

Ref. [8] gives a range of risk of individual fatality of 10^{-6} to 10^{-8} per plant year where risk reduction is desired and applies to all industrial plants. Higher risk than 10^{-6} per plant year is unacceptable and lower than 10^{-8} per plant year is defined as acceptable. These numbers are based on 1 percent of the risk of the lowest risk group in society (children between 10 to 14 years of age having a risk of death of 10^{-4} per year) and apply to a person residing 24 hours a day at a certain location. They include latent effects (see also Fig. 1).

(b) Probability of exposure levels

Such an individual risk criterion has the advantage that it keeps the probability of exposure of an individual separate from the probability of death given that exposure. An example in the form of a "boundary line" is given in Fig. 2, where each point on the criterion curve is equivalent to a risk of 10^{-7} per reactor year [6].

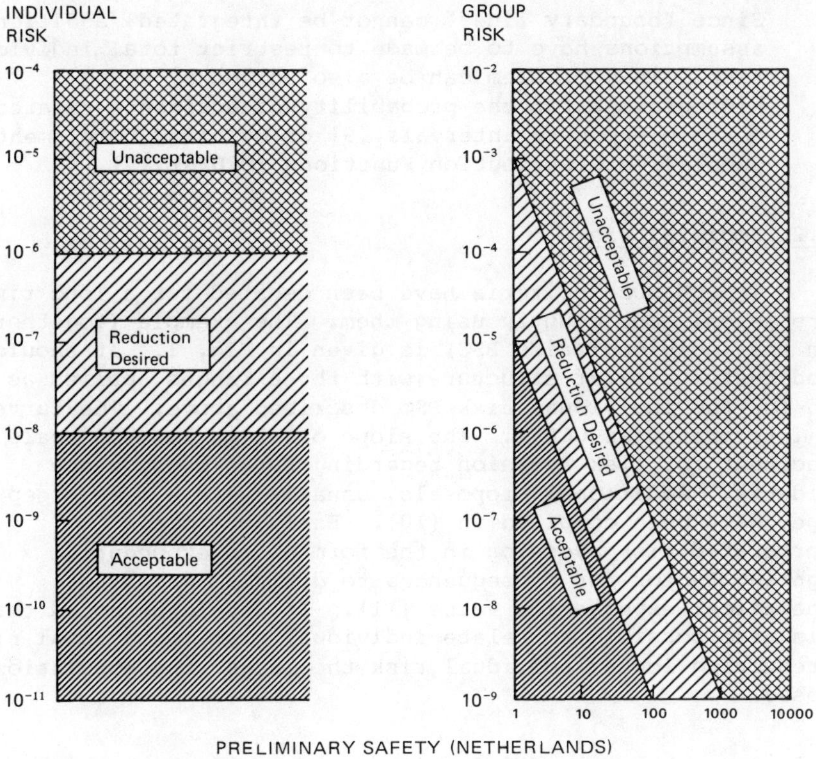

FIG. 1. Criterion for individual risk and for societal prompt fatality risk.

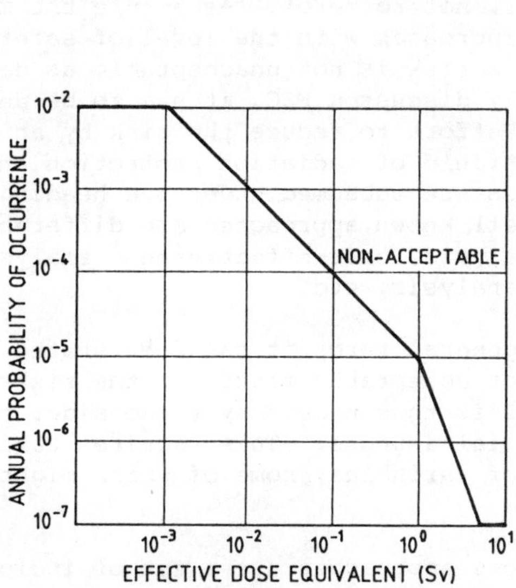

FIG. 2. Criterion curve.

Since "boundary lines" cannot be integrated, additional assumptions have to be made to restrict total individual risk. This problem can be also solved by using histogrammes for the probability of accidents related to certain exposure intervals [9] or by using Complementary Cumulative Distribution Functions (CCDFs) [5].

3.1.2. Societal risk

Though such criteria have been debated for a long time, there is only one country using them. The example (together with the individual risk PSC) is given in Fig. 1. It should be noted that a similar structure with three ranges applies as used for the individual risk PSC described above. The curves are interpreted as CCDFs. The slope of -2 on log-log scale is to account for risk aversion regarding large consequence accidents. Additional proposals, usually with a less steep slope, have been compiled in [10]. Early proposals incorporated risk aversion in the form of an exponential weighting factor for consequences to derive a weighted mathematical expectation value [11]. It should be noted that it is also possible to relate individual risk to societal risk by requiring lower individual risk the larger the population exposed.

3.1.3. Approaches to optimization

It is always possible to reduce a risk at additional expense (however, not to zero). As a rule the marginal cost of risk reduction increases with the level of safety achieved [12]. Thus, if a risk is not unacceptable as determined by one of the previously discussed PSC, it has to be decided whether it is worth the effort to reduce the risk by an incremental amount. In the field of radiation protection any such decision aiding techniques are subsumed under the heading of optimization. Well known approaches are differential cost-benefit analysis, cost-effectiveness analysis, multi-criteria analysis, etc.

In more general terms it has been stated in the introduction that acceptable risks are the risks of the best alternative. It is thus necessary to consider technical economic and social aspects. This requires taking into account a large number of variables, some of which might be difficult to quantify.

So far some approaches use a set of indicators which are representative of certain categories of risk. An approach, which considers core melt probability, PSC for early and latent health effects, US $ 1000 per person-rem[1/] as a surrogate for off-site impact, and on-site costs, is given in [13].

3.2 Level-1 and Level-2 PSC

Logically the highest plant level PSC, and at the same time one which is providing a close link to PSC at the level of public health, would be to provide constraints for the probability of major releases of radioactive materials. An example is given in [4] in form of a general performance guideline which has been proposed for further examination:

"Consistent with the traditional defence-in-depth approach and the accident mitigation philosophy requiring reliable performance of containment systems, the overall mean frequency of a large release of radioactive materials to the environment from a reactor accident should be less than 1 in 1,000.000 per year of reactor operation."

Other practices or proposals which have been reviewed in [2] are basically consistent with such an approach.

For instance it has been proposed that core melt (severe core damage) frequency should be less than 10^{-5} per reactor year and that for all plant damage states leading to coremelt the likelihood of leading to containment failure should be less than 10^{-1} per event on average. In order to achieve a balanced design some proposals include the requirement that specific categories of accident sequences (which could be grouped by classes of initiating events) should not contribute more than 10 percent to the total frequency of coremelt or releases of radioactive materials. There is general agreement that such PSC should use the mean value (rather than the median or the mode) since it is a better representation of the upper tail of the uncertainty distribution.

In [14] the phrase "significant releases of radioactive materials" has been interpreted as those leading to one or more early fatalities. Thus this would be also consistent with a less than 10^{-6} probability of death for the worst exposed individual as discussed above [8] or a smaller value if an average individual is defined [4]. In spite of this general agreement on numerical target values in countries using or considering such PSC, it should be noted that it will be necessary to reach a better consensus on how categories of accident sequences are defined, which initiating events are

1/ 1 rem = 1.00 x 10^{-2} Sv

considered (especially which external events), and how core melt and major releases of radioactive materials are specified.

3.3. PSC for safety functions/systems

Recent discussions concentrate on the development of PSC for safety functions/systems. At this level a direct measure of overall safety (e.g. risk to the public or probability of core melt) is not explicit in the PSC. A set of PSC at the safety function and/or system level aims at providing guidance for designers and operators of plants as well as regulators in a number of areas, for example [2]:

(a) During the preoperational phase including design, siting and licensing: The aim of having specific probabilistic criteria against which to design safety functions and/or systems should lead to a more efficient way for designing safe and economic plants and possibly shorten the time required for licensing.

(b) During the operating phase: A set of such PSC would provide a rational framework for the following purposes:

- To monitor the safety performance of the plant, i.e. to ensure continuous compliance with the design objectives and continuity of safety practice into the operational phase of systems for which a PSC existed at the design phase. This monitoring will provide a way of verification of the predicted values and hence an indirect way for validating techniques, models, databases, etc.

- To harmonize Technical Specifications (TecSpecs), and in particular limiting conditions of operation and surveillance frequencies which can be set more rationally when their impact on the performance of the corresponding systems is assessed as the latter is measured by appropriate probabilistic indices.

- To verify the effectiveness of backfitting measures and to identify the most effective ways for further backfitting system hardware and/or operating procedures.

Various ways have been used or proposed for deriving PSC for safety functions/systems [2]:

(a) It would be logical to relate them to higher level PSC. This might be achieved in a "top-down" approach by disaggregating a set of higher level PSC into a set of

TABLE 2. EXAMPLES OF PROBABILISTIC SAFETY CRITERIA AT THE SAFETY FUNCTION LEVEL

Example 1

System function	System considered	Unavailability (mean)
Coping with small leaks in the primary circuits	Emergency cooling systems	4×10^{-3}
	Emergency cooling systems plus operational cooling systems	3×10^{-4}
Coping with emergency power case	Emergency power systems	2×10^{-5}
	Emergency power plus operational backup systems	5×10^{-6}

Example 2

Safety function	Unreliability
Making the reactor subcritical	10^{-5}
Isolation of the containment	10^{-3}
Supply of feedwater when the off-site power or the main feedwater supply is lost	10^{-4}
Operation of emergency core cooling in the case of a small reactor coolant leak	10^{-4}
Rapid reactor pressure reduction and long term pool cooling (BWR)	10^{-4}

PSC for safety functions or systems. However, this approach is not as straightforward as it may seem at first glance, since the complex interdependencies of functions or systems make it difficult to find an optimum set of such PSCs [15]. This situation is aggravated by the fact that uncertainties have to be considered. Thus, a combination of such an approach with a "bottom-up" approach seems to be advisable where possible sets of PSC at the lower level are compared on the bases of their implied performance at the higher level.

(b) They might be derived by comparison with similar designs which have been thoroughly analysed by deterministic and probabilistic means and which have been accepted as "safe". Such an approach could be useful when it is required to evaluate limited design changes or changes in operational procedures.

(c) Rather than using PSC as such, it has also been suggested that a principle safety goal is to achieve a balanced design, ensuring through application of PSA no undue contribution to severe accidents by any single safety system or function.

In any case it is of utmost importance to specify carefully the boundaries of safety functions or systems, in particular which support systems are included. Examples of such PSC which are in preliminary use are given in Table 2 [16,2]. In Example 2, a 95 percent confidence concept has been adopted.

4. UNCERTAINTIES AND COMPLIANCE

Precise definition of how to deal with the uncertainties of PSA results is the single most important problem in establishing PSC which are of practical use. Suggestions made include use of confidence intervals [16] or compliance indices which could be derived using fuzzy set theory [17]. A pragmatic approach which needs to be better explored in the future could be based on a combination of the probabilistic and the deterministic approach. It could have the following features:

(a) Standardization of PSA as will be suggested in the PSA guidelines under development by the IAEA, including standardized methods for uncertainty and sensitivity analysis.

(b) Description of what is technically leading to the tails of the uncertainty distributions exceeding predefined PSC (see also [4]).

(c) Deterministic approach to exclude or design against the tails of these distributions.

The first point still needs much development work. the last two points still need to be further explored since they also pose serious mathematical and conceptual problems.

5. CONCLUSIONS

With the increasing use of PSA it is also necessary to develop quantitative PSC which allow evaluation of their results. Many countries are using or discussing PSC at various levels. Level-3 criteria are appealing because of their overall approach, though, they pose the difficult problem of demonstrating compliance. Level-1 and Level-2 criteria seem to be most consistently developed at the present time. Problems of demonstrating compliance are less, though large uncertainties are still involved in particular in Level-2 analyses. Development work is concentrating on Level-0 criteria since they provide a pragmatic approach to nuclear safety:

- They reflect the state-of-the-art in development of modern PSA which tend to be structured so as to give the reliabilities of systems in performing specific functions. They can therefore be calibrated by the large variety of quantitative PSA-results now available.
- They address issues which are technically complex and which cannot be easily evaluated by other means. They will help to prevent lack of balance in system design.
- They are effective for broad classes of accident sequences and thus have an overall impact on nuclear safety.
- They focus on two main principles in reactor safety technology, namely 'defence-in-depth' and the avoidance of damage by putting emphasis on preventive measures.
- It is simpler to demonstrate compliance, and the reliability of safety functions/systems can be partially verified during testing and operation.

However, the burden of dealing with uncertainties is shifted towards deriving such PSC.

It should be noted that no matter at which level PSC are derived, lower level criteria always imply higher level criteria and vice versa. It thus seems that PSC are needed as a consistent set at all levels in the form of objectives, targets or goals, to assist safety decisions in design, licensing, and operation.

REFERENCES

[1] INTERNATIONAL ATOMIC ENERGY AGENCY, 1987. Status, Experience, and Future Prospects for the Development of Probabilistic Safety Criteria. Report on a Technical Committee Meeting, 27-31 January 1986, Vienna, Austria.

[2] INTERNATIONAL ATOMIC ENERGY AGENCY, 1987. Probabilistic Safety Criteria at the Safety Function/System Level.

Report on a Technical Committee Meeting, 26-30 January 1987, Vienna, Austria.

[3] KEENEY, R., Decision Criteria for Siting Complex Industrial Facilities. Paper presented at the IIASA Task Force Meeting on Technological Risk in Modern Society, International Institute for Applied Systems Analysis, March 18-20, 1987, Laxenburg, Austria.

[4] U.S. NUCLEAR REGULATORY COMMISSION, 1986. Safety Goals for the Operations of Nuclear Power Plants; Policy Statement. 10 CFR PART 50. USNRC, Washington, D.C.

[5] GOTTSCHALK, P., 1986. Approach to PSA Methods and Probabilistic Safety Criteria in Licensing of NPP in the Federal Republic of Germany. Paper presented at the Technical Committee Meeting "Status, Experience, and Future Prospects for the Development of Probabilistic Safety Criteria ", Vienna, 27-31 January 1986.

[6] GONZALEZ, A., The Regulatory Use of Probabilistic Safety Analysis in Argentina, (Proc. Int. Meeting on Thermal Nuclear Reactor Safety, Chicago, Aug. 29 - Sept. 2, 1982, NUREG-0027, Vol. 1).

[7] INTERNATIONAL COMMISSION ON RADIOLOGICAL PROTECTION, 1985. Radiation Protection Principles for the Disposal of Solid Radioactive Waste. ICRP Publication 46, Annals of the ICRP, Vol. 15, No. 4. Pergamon Press, Oxford, New York, Frankfurt.

[8] VERSTEEG, M.F., 1986. External Safety Policy in the Netherlands: An Approach to Risk Management. Paper presented at the Technical Committee Meeting "Status, Experience, and Future Prospects for the Development of Probabilistic Safety Criteria ", Vienna, 27-31 January 1986.

[9] SNELL, V.G., 1986. Probabilistic Safety Assessment Goals in Canada. Paper presented at the Technical Committee Meeting "Status, Experience, and Future Prospects for the Development of Probabilistic Safety Criteria ", Vienna, 27-31 January 1986.

[10] HIGSON, D.J., 1986. Some Questions on the Development of Probabilistic Risk Criteria. Paper presented at the Technical Committee Meeting "Status, Experience, and Future Prospects for the Development of Probabilistic Safety Criteria ", Vienna, 27-31 January 1986.

[11] GRIESMEYER, J.M., OKRENT, D., 1980. Risk Management and Decision Rules for Light Water Reactors. In: An Approach to Quantitative Safety Goals for Nuclear Power Plants. NUREG-0739, U.S. Nuclear Regulatory Commission, Washington, D.C.

[12] INTERNATIONAL ATOMIC ENERGY AGENCY, 1986. Introduction to Cost-effectiveness Analysis of Risk Reduction Measures in Energy Systems. IAEA-TECDOC-383.

[13] SNIEZEK, J.H., 1986. An Integrated Safety Goal Concept. Paper presented at the Technical Committee Meeting "Status, Experience, and Future Prospects for the Development of Probabilistic Safety Criteria ", Vienna, 27-31 January 1986.

[14] U.S. NUCLEAR REGULATORY COMMISSION, 1987. Reactor Risk Reference Document; Main report/draft for comment, NUREG-1150, Feb. 1987.

[15] PAPAZOGLOU, I.A., FITZPATRICK, R., BARI, R., 1987. Can an Advanced LWR Design Meet a 10^{-5}/ry Core Melt Criterion?" (Proc. Int. ANS/ENS Topical Meeting on Thermal Reactor Safety, San Diego, California, February 1987).

[16] VIROLAINEN, R., 1987. Use of Reliability and Risk Standards as Bases for Acceptance in Licensing and Regulation of Nuclear Power Plants. Paper presented at the Technical Committee Meeting "Status, Experience, and Future Prospects for the Development of Probabilistic Safety Criteria ", Vienna, 27-31 January 1986.

[17] UNWIN, S.D., HAYNS, M.R., 1985. Rational Quantitative Safety Goals, (Proc. ANS/ENS International Topical Meeting on Probabilistic Safety Methods and Applications, Vol. 1, Sess. 2 (Safety Goals), 11-1 to 11-10, San Francisco, California, February 24 - March 1, 1985).

ACTIVITIES WITHIN THE EUROPEAN COMMUNITY RELATED TO SAFETY GOALS FOR NUCLEAR POWER PLANTS

S.A. HARBISON
Health and Safety Executive,
HM Nuclear Installation Inspectorate,
Bootle, United Kingdom

G.N. KELLY
Central Electricity Generating Board,
London, United Kingdom

E.V. GILBY
Gilby Associates,
Knutsford, United Kingdom

F. LANGE, P.A. GOTTSCHALK
Gesellschaft für Reaktorsicherheit mbH,
Cologne, Federal Republic of Germany

B.J. TOLLEY
Commission of the European Communities,
Brussels

Abstract

ACTIVITIES WITHIN THE EUROPEAN COMMUNITY RELATED TO SAFETY GOALS FOR NUCLEAR POWER PLANTS.

In February 1982, after consultation with its Safety of Light Water Reactors Working Group No. 1 ''Methodology, Criteria, Codes and Standards'', the Commission of the European Communities appointed a special Task Force to review existing safety objectives for nuclear power plants and to report on the degree of coherence between the various approaches. The Group considered all relevant safety objectives which existed within Community Member countries as well as the Safety Goals proposed by the United States Nuclear Regulatory Commission and presented its general conclusions and perspectives in its first Status Report in May 1983. The paper gives a brief overview of the more important observations in the first Status Report and reports on progress made by the Task Force since then. In particular it summarizes the preliminary results of a benchmark exercise organised by the Group to elucidate the degree of coherence between the dose-frequency targets used in the different Member States. Eight organizations from seven countries participated in the exercise, which was organized in a form that was as unambiguous as possible to facilitate the investigation of differences in the dose predictions of the various participants. The exercise was also structured in such a way that comparisons could readily be made at intermediate stages in the calculation. The comparison of the various results is currently at a provisional stage but a number of significant differences have already emerged for particular releases in particular release conditions. These differences mainly result from the different assumptions used by the various participants in relation to the conditions (e.g. distance, meteorological conditions, etc.) under which the dose is calculated. The paper also discusses the role of individual dose-frequency targets in relation to other targets. It presents some general conclusions about safety topics which the Task Force has identified as being important in developing a systematic approach to safety objectives.

1. The Task Force's First Status Report and subsequent developments

When the Task Force was set up in February 1982 it was mandated to consider on a technical basis the need, possibilities and limitations of overall safety objectives for nuclear power plants, taking stock of various trial-approaches being discussed in EC-member states and abroad. It was also to examine the merits and implications of the various approaches and to determine the degree of coherence and the divergences in their possible applications, and to prepare a Status Report.

1.1 Major observations in first Status Report

In its first Status Report in 1983 [1], the Task Force recognised that there are a variety of reasons for safety objectives which in part represent the different responsibilities of the groups involved. However, these different safety objectives should at least be consistent and complementary. While recognising that the specification of quantified safety objectives is difficult the Task Force considered it was an important activity because

(i) it establishes an appropriate framework for discussions between regulatory authorities and designers, constructors and operators; and

(ii) it provides a rational basis for a coherent and consistent approach to the safety of nuclear power plants.

The Task Force found it convenient to categorise safety goals under the following five general headings:

1. Reliability figures for safety systems in connection with frequency values of specific accident conditions.

2. Release values in connection with frequency values of specific or overall accident conditions.

3. Individual risk values.

4. Societal risk values.

5. Comparative risk values.

The first and second types of safety goals are normally expressed in straightforward engineering values. The risk values used in the third and fourth types may be expressed in various ways. One approach is the use of the radiation dose received by an individual (or the integrated radiation dose received by the population) per year per event as a surrogate for risk. By using dose-effect relationships these dose values may also be directly expressed in terms of risk as the probability of prompt or late morbidity or fatality.

1.2 Relating risk-based objectives to engineering targets

As far as appreciating the overall risk from the plant is concerned, the most readily understood safety objectives are those expressed in terms of individual and societal risk. Since the time of the first Status Report there has been increasing agreement between Member countries that an individual fatality risk level of no more than 10^{-5} per year should be achievable in new plant and that efforts should be made to achieve a target of 10^{-6} per year. However, such statements about individual risk targets need to be qualified by a realistic assessment of the uncertainties and incompleteness of the probabilistic risk assessments (PRAs) that are used to verify compliance with them. The setting and demonstration of compliance with societal risk targets is subject to even greater uncertainty.

Although general forms of safety objectives based on risk are very useful for communicating with the public and decision makers they do not easily lend themselves for use in the engineering of nuclear power plants. In its first Status Report the Task Force concluded that a safety goal based solely on release (or dose) limits versus probability is insufficient to guarantee that adequate attention is given to ensuring that accidents are prevented, as far as reasonably achievable. They felt that it is necessary to define additional criteria or targets for the reliability of engineered safety systems and also possibly to limit the frequency of severe plant damage. It was, however, recognised that such targets should be consistent with the release-frequency goals and also reflect actual operating experience, e.g. as regards the recorded reliability of safety provisions. During the past 3-4 years it has been a major concern of the Task Force to establish whether reliability targets for safety systems can be developed logically from risk targets.

1.3 Core coolability targets

One particular aspect of the above studies concerned the interrelationship between a numerical value for a core coolability or core melt target and the off-site dose or risk which might be expected to be the basis of a licensing approach. The obvious conclusion reached by the Task Force was that while the use of a core melt or core degradation criterion can be of value in specific areas the dominant contributors to the frequency of core melt/degradation are not necessarily those which would dominate offsite dose and risk. Nevertheless some Member countries have indeed proposed frequency targets in the range of 10^{-5} to 10^{-6} per year as a useful method of ensuring an adequate standard of accident prevention. This recognises the fact that a major contribution to safety derives from ensuring that a significant loss of core coolability is prevented by core design provisions.

Another aspect recognised by the Task Force was that small differences in both design and operational procedures can have major effects on both the absolute magnitude of the core melt frequency and on the relative importance of the various accident sequences. Thus the adoption of a core melt criterion does not explicitly give detailed guidance on design and operational aspects although it does provide a target against which there can be checks to ensure general compatibility.

1.4 Containment performance guidelines

The subject of developing containment performance guidelines has also received continuing attention. This has to be related to the general consideration that the objective is to prevent uncontrolled release conditional upon a major loss of core coolability. It should also be recognised that the performance of containment involves not only the integrity of the structure itself but also that of the associated engineered safeguard systems. One basis is that of a combination of a core melt criterion and a containment guideline to provide a basis for prevention and mitigation compatible with some overall risk requirement. In one Member country, Italy, the concept of developing a containment performance guideline has received major attention and specific values have been suggested for discussion [2]. More recently papers from a number of countries at a CEC seminar on LWR Severe Accident Studies [3] showed a convergence of views in a number of Member countries on both the extent to which containment could be expected to be effective and on the best method of achieving numerical performance limits. In the latter instance a significant increase was noted in the interest in the use of filtered vented containment concepts.

The Task Force plans to continue to monitor on-going work in this field and particularly to review the trends in terms of containment design and performance including associated engineered safeguard systems. In particular this will indicate whether there is sufficient agreement between Member countries about the principles of containment design and its required performance under accident conditions to allow criteria related to such performance to be adopted in conjunction with the reliability of major safety functions.

1.5 The role of siting rules and emergency procedures

The Task Force has also recognised the importance of siting rules and emergency procedures in demonstrating compliance with quantified safety objectives but is aware that these effects have not been studied in a complete and systematic way in relation to their potential effect on overall safety objectives. It intends to continue to evaluate the extent to which a safety objective based on individual risk in combination with siting criteria and emergency response plans can provide adequate control of risk.

1.6 The application of Cost Benefit Analysis

The Task Force has considered the possible use of Cost Benefit Analysis (CBA) in relation to the development of safety objectives. Having reviewed the methodology and use of CBA the Task Force reached the following conclusions:

i) Further improvements to the design of modern nuclear power stations are unlikely to be justified on cost benefit grounds if consideration is limited to the expected direct monetary costs of off-site consequences.

ii) The inclusion of on-site economic effects (e.g. loss of capital investment, replacement generation costs) in the calculation of benefit from averting accidents can significantly affect the balance of cost and benefit. However, the possible incorporation of such effects in safety objectives remains a controversial matter.

iii) A framework exists for the application of CBA to aid decisions on the provision of design changes to reduce the frequency or consequences of accidents. The use of expected risks and costs in this context is, however, deficient in several respects; further progress needs to be made on the appropriate parameters to be included in the analysis and the weight to be given to each (e.g. consequences occurring at different levels of probability) before CBA is likely to find widespread and agreed use in this area.

iv) There are a number of decisions to which CBA can make a contribution and many of the uncertainties appear capable of resolution by the adoption of appropriate rules by the decision making authority.

1.7 Risk Aversion

Another topic considered by the Task Force was risk aversion, which relates to the basic argument that a single event involving many casualties is less acceptable than a series of events with small numbers of casualties even when the total societal impact is the same. There also appears to be an additional aversion to certain types of risk (e.g. from nuclear power) even when they are equal in magnitude to other (conventional) risks. The basic need for the inclusion of a factor to reflect risk aversion in considering different types of safety objectives remains a controversial issue. An approach based strictly on scientific and technical terms would be compatible with the non-inclusion of a risk aversion factor and would have the advantage of facilitating comparisons between potentially hazardous activities of various kinds. Socio-political factors complicate such a simple approach and inevitably the inclusion of appropriate factors increases the complexity of debate. Clearly there are many factors which are potentially relevant including (i) voluntary vs involuntary risk, (ii) group vs individual risk, (iii) immediate vs deferred hazards, (iv) unfamiliar vs natural or established risks, (v) catastrophic vs regularly occurring effects, etc. The aggregation of these factors can result in a large degree of aversion to nuclear power plants. While no specific recommendations are made the Task Force has noted developments in Member countries and plans to base its future work on the concept that some degree of risk aversion to large accidents may need to be included in the future bases of their continuing review of safety objectives.

2. Regulatory Dose Benchmark Exercise

2.1 Introduction

One specific observation of the Task Force in its first Status Report was that there appeared to be a considerable degree of coherence between the dose-frequency targets used in different Member countries but they warned that the coherence might be more apparent than real because of the different models and assumptions used in the different countries. The reasons for this could be

- differences in the status of these dose-frequency targets in the regulatory framework of the different countries,
- differences in the models and assumptions adopted in the various countries to estimate doses from design basis accidents within a regulatory framework,
- differences in the classification of accidents, and in the estimation of the frequency and source terms of radionuclides released to the environment.

To enable a more definite intercomparison and to identify the origins of possible differences the Task Force has co-ordinated a benchmark exercise on dose estimation for design basis accident conditions of nuclear power plants in a regulatory context. The following EC countries and institutions have participated in the benchmark exercise:

- Belgium : TRACTEBEL
- France : CEA, Fontenay-aux-Roses
- Federal Republic of Germany : GRS
- Italy : ENEA-DISP
- Netherlands : TNO
- United Kingdom : CEGB and NII

In addition, STUK from Finland has participated in the exercise.

2.2 Contents and structure of the benchmark exercise

Differences in regulatory dose estimation for accidental releases of radionuclides to the atmosphere can originate from differences in modelling, in the values assigned to model parameters and in the protocols used in the different calculational steps starting with the release to the environment and ending with potential doses from different pathways to a critical individual. The more important of these steps are:

- modelling of release conditions,
- atmospheric transport for a variety of weather conditions,

- calculation of time integrated air concentrations of released radionuclides and of dry, and possibly wet, deposition levels at various distances,
- the protocol (e.g. choice of distance, meteorological conditions, deposition velocity, etc) adopted in selecting air concentrations and deposition levels for regulatory calculations,
- calculations of dose estimates for a number of exposure pathways such as external radiation from the cloud and deposited material, inhalation and ingestion.
- the protocol adopted in the summation of doses over possible pathways, the identification of the critical organ and the specification of the habits of critical groups,
- the extent to which remedial measures (e.g. food bans) are taken into account.

The benchmark exercise was structured in such a way that the origin of differences in the dose predictions of the various participants could be identified. This was achieved by comparing results at various intermediate steps in the calculation. A principal feature of the benchmark exercise was to compare estimates of dose per unit release of a number of potentially significant radionuclides and for two releases (source terms) having radionuclide compositions typical of faults postulated within the design basis for PWRs. A wide range of different release conditions (i.e. duration, height, releases into building wakes and from point sources, physicochemical form of the nuclide, etc) were analysed in each case. In addition to providing in tabular form detailed information concerning the data used, each participant was asked to supply additional information on the models and protocols adopted for regulatory dose calculations.

2.3 Preliminary information from the benchmark exercise

A preliminary analysis of the information and calculations by the eight organisations participating in the benchmark exercise was presented at the PSA'87 Conference in Zurich [4], and it indicated the following important findings:-

- In all participating countries the straight line Gaussian plume model is used for atmospheric dispersion.

- With the exception of Belgium and France all participants use the Pasquill classification scheme for atmospheric stability classes. In general the condition leading to the highest dose is adopted for regulatory purposes.

- The distances at which regulatory doses are calculated for members of the public range between a minimum of 100m and a maximum of 1000m.

- With one exception all participants use dose factors for inhalation and for ingestion based on models whose origins are to be found in ICRP publications 26 and 30, supplemented by consideration of other age groups. Belgium uses dose factors based on ICRP publication 2. The Federal Republic of Germany uses dose factors based on ICRP 30 which will become official shortly.

- All participants calculate the potential doses from cloudshine and inhalation.

- With the exception of Belgium all participants calculate the external dose from gamma-emitters deposited on the ground but the range of integration times is large, lying between 1 day and 50 years.

- Two countries (the Netherlands and Finland) do not currently include the ingestion pathway in the summated dose for regulatory purposes, and Belgium calculates only the pasture-cow-milk pathway for the radiologically important radionuclide I-131. The integration times (period during which consumption of contaminated food is assumed) for the ingestion pathway are 2 months (summer period assumed) for France and essentially 50 years in all other cases although the difference in dose consequent upon these two assumptions is far less than might be implied by the difference in the two integration periods, at least if I-131 is the most relevant nuclide.

- There are some important variations among participants about which exposure pathways are summed when calculating doses for comparison with targets. This makes the comparison of results within the benchmark exercise very complicated.

- There are some differences in the countermeasures assumed to be taken in estimating doses for regulatory purposes.

2.4 Some preliminary conclusions

The benchmark exercise has shown that some differences exist among participants in the total doses calculated for the two source terms and for single radionuclides in a regulatory context. The structuring of the exercise allows the origin of coherences and differences to be traced, and that analysis is continuing. However, the differences that have been identified so far should not be interpreted as necessarily representing differences in nuclear safety standards. Any real differences in such standards can only be determined when all aspects of the regulatory approach have been taken into account. As a further step towards understanding the full implications of the various regulatory approaches in Member countries the Task Force has initiated a second benchmark exercise to compare the models and assumptions used for regulatory purposes in calculating the source term released to the environment for design basis accidents. The participants will present results for the successfully terminated LOCA and it is anticipated that preliminary results will be available later this year or early next year.

3. Concluding Remarks

The Task Force recognises that its work is by no means complete even though significant progress has been made in understanding the basis of existing safety objectives in Member countries. It is anticipated that the second benchmark exercise will add considerably to the valuable insights already emerging from the first exercise. However, the Task Force recognises that any comprehensive safety objectives for nuclear power plants must take account of accidents more severe than the design basis accidents which form the basis of the present benchmark exercises. In theory there should be no real difference in the treatment of design basis accidents and beyond design basis accidents, from the point of view of safety goals, in that the totality of risk is the fundamental basis of comparisons between the various sizes of accidents. However, this simple view is complicated by a number of factors, including the conservative assumptions which are generally applied to the analysis of design basis accidents for regulatory purposes compared with the more "realistic" assumptions applied to beyond design basis accidents, and the possible need to incorporate some measure of risk aversion for potential accidents with large off-site consequences. The Task Force will continue to assess ongoing work, both within its Member countries and in other international groups, on how to incorporate more severe accidents into safety goals for nuclear power plants.

ACKNOWLEDGEMENT

The authors wish to acknowledge the contribution made by the other members of the Task Force and other members of the EC Secretariat to the work described in this paper. The views expressed, however, are those of the authors and are not necessarily the views of their national organisations.

REFERENCES

[1] TASK FORCE ON SAFETY OBJECTIVES OF THE COMMISSION OF THE EUROPEAN COMMUNITIES, Status Report on Safety Goals/Objectives, Ref. No. XII/476/83, CEC, Brussels (1983).

[2] ZAFFIRO, C., ''Probabilistic assessment of the containment performance during severe accidents in PWRs'', Studies of Severe Accidents in LWRs (Proc. CEC Sem. Brussels, 1986), EUR-11019, CEC, Brussels (1987).

[3] COMMISSION OF THE EUROPEAN COMMUNITIES, Studies of Severe Accidents in LWRs, ibid.

[4] LANGE, F., et al., ''Safety objectives for nuclear power plants in terms of dose-frequency targets; a comparison exercise performed by the Commission of the European Communities on dose assessment within a licensing framework'' (Proc. Conf. PSA '87, Zurich, 1987), Vol. 3, TÜV Rheinland, Cologne (1987) 1066–1076.

METHODOLOGIE DE L'ANALYSE DU FACTEUR HUMAIN

M. GRIFFON-FOUCO
Service de la production thermique,
Electricité de France,
Paris, France

Abstract-Résumé

METHODOLOGY OF HUMAN FACTOR ANALYSIS.

The paper describes the manner in which the Heat Production Department of Electricité de France analyses the human factors in nuclear power plants. After describing the teams and structures set up to deal with this subject, the paper emphasizes two types of methods which are used, most often in complementary fashion: (1) an a posteriori analysis, which consists in studying the events which have taken place at nuclear power plants and in seeking the deep-seated causes so as to prevent their reoccurrence in future; (2) an a priori analysis, which consists in analysing a work situation and in detecting all its potential failure factors so as to prevent their resulting once again in dysfunctions of the facility. To illustrate these two types of analysis, two examples are given: first, a study of the telephonic communications between operators in one plant (in which the a posteriori and a priori analysis are developed) and, second, a study of stress in a plant (in which only the a priori analysis is used).

METHODOLOGIE DE L'ANALYSE DU FACTEUR HUMAIN.

Ce mémoire décrit la façon dont le Service de la production thermique d'Electricité de France analyse les facteurs humains dans les centrales nucléaires. Après une description des équipes ou structures mises en place pour traiter ce thème à Electricité de France, l'accent est mis sur deux types de méthodes qui sont le plus souvent utilisées de façon complémentaire: 1) une analyse a posteriori, qui consiste à étudier les événements survenus dans les centrales et à en rechercher les causes profondes de façon à prévenir leur répétition dans l'avenir; 2) une analyse a priori, qui consiste à analyser une situation de travail et à y détecter tous les facteurs potentiels de défaillance de façon à éviter qu'ils ne conduisent un jour à des dysfonctionnements de l'installation. Afin d'illustrer ces deux types d'analyse, deux exemples sont donnés: d'une part, une étude des communications téléphoniques entre opérateurs dans une centrale (les analyses a posteriori et a priori y sont développées), et, d'autre part, une étude du stress dans une centrale (seule l'analyse a priori est utilisée pour cette étude).

1. INTRODUCTION

Il devient banal de dire que le facteur humain joue un rôle non négligeable sur la performance et la sécurité des centrales nucléaires. En effet, les résultats sont là:

1) Les défaillances humaines interviennent dans presque la moitié des arrêts d'urgence et ceci est vrai non seulement pour la France mais aussi pour la plupart des pays équipés de centrales nucléaires.

2) Les accidents de TMI et de Tchernobyl nous rappellent combien l'intervention inadaptée de l'homme peut entacher profondément la sûreté des installations.

Dès 1982, Electricité de France (EDF) a mis en place une véritable structure d'analyse du facteur humain. Au départ, les études et actions ont essentiellement porté sur l'amélioration de l'interface homme–machine (salles de commande, procédures, automatisation de certaines tâches, etc.). De plus en plus, les analyses s'orientent vers la définition de nouvelles pratiques de travail (manière de communiquer entre opérateurs, relations entre différents services, relèves de quart, etc.) ou vers des études plus fondamentales (phénomènes de stress, de routine).

Nous présentons successivement dans ce mémoire:
– les structures d'analyse du facteur humain mises en place à EDF;
– les méthodes d'analyse et de prise en compte du facteur humain;
– des exemples illustrant ces méthodes d'analyse.

2. LES STRUCTURES D'ANALYSE DU FACTEUR HUMAIN A EDF

2.1. Au Service de la production thermique (SPT)

En 1982, conscient de l'acuité du problème posé par les défaillances humaines dans l'exploitation des centrales nucléaires, le Service de la production thermique, responsable de la sûreté et de l'exploitation de ces centrales, a décidé de se doter d'une structure spécialisée dans la prise en compte du facteur humain.

L'équipe créée à cette époque est aujourd'hui constituée de six personnes, dont quatre spécialistes des facteurs humains (ergonomes et psychologues) et deux anciens exploitants de centrale (un ancien chef de quart et un ingénieur).

La multidisciplinarité de cette équipe lui confère une incontestable richesse d'approche des problèmes. Cependant, nous connaissons tous des cas où des équipes multidisciplinaires se sont délitées rapidement du fait d'écarts trop importants dans les modes d'appréhension des sujets d'étude.

En fait, une telle équipe ne peut fonctionner qu'à la condition d'avoir des objectifs communs exigeant une étroite collaboration entre ses membres et créant ainsi une certaine communauté d'esprit.

C'est pourquoi le rattachement à un service de production est une condition primordiale de réussite: l'analyse du facteur humain exige un contact direct avec les exploitants et une confrontation permanente avec la réalité, ceci pour éviter de «lancer des études pour des études» et n'admettre au contraire que les études dérivées d'événements réellement survenus en exploitation.

2.2. A la Direction des études et recherches

Jusqu'en 1986, la Direction des études et recherches possédait des spécialistes des facteurs humains disséminés dans différentes équipes. En novembre 1986, un regroupement a été effectué au sein d'un Département d'études de sûreté et de fiabilité. L'équipe «facteurs humains» de ce département est composée d'environ huit personnes qui sont pour la plupart des spécialistes des facteurs humains (ergonomes, psychologues).

Une collaboration très étroite s'est établie entre cette équipe et celle du SPT: en effet, cette dernière est parfois saturée par le travail en temps réel du fait de l'analyse au jour le jour des événements survenus en centrale; elle n'a donc pas toujours la possibilité de mener des études de réflexion à long terme et les sous-traite à l'équipe de la Direction des études et recherches. Celle-ci, composée essentiellement de spécialistes devrait pouvoir apporter une plus-value appréciable dans les années à venir en nous permettant, dans un premier temps, d'approfondir notre connaissance du comportement humain en exploitation pour, dans un deuxième temps, en déduire des actions d'amélioration des performances humaines.

Notons que le Département d'analyse de sûreté du Commissariat à l'énergie atomique possède également un Laboratoire d'étude du facteur humain, composé d'environ six personnes, avec lequel les équipes d'EDF ont de fréquents contacts, certaines études pouvant même être menées en commun.

3. LES METHODES D'ANALYSE ET DE PRISE EN COMPTE DU FACTEUR HUMAIN

Quels que soient les thèmes traités, la méthodologie d'analyse du facteur humain reste sensiblement la même et combine deux types d'investigations:

– une analyse a posteriori qui consiste à étudier des événements survenus dans les centrales et à en rechercher les causes profondes de façon à prévenir leur répétition dans l'avenir;

– une analyse a priori qui consiste à analyser une situation de travail et à y détecter tous les facteurs potentiels de défaillance de façon à éviter qu'ils ne conduisent un jour à des dysfonctionnements de l'installation.

3.1. L'analyse a posteriori

Nous avons élaboré des outils d'investigation des événements survenus en centrale, reposant sur une méthodologie d'analyse à trois niveaux:

1) *Une analyse statistique simplifiée des défaillances humaines* avec une grille d'investigation comportant environ dix rubriques, parmi lesquelles on trouve

le personnel concerné par la défaillance, le type d'activité, la localisation, les mécanismes de la défaillance et ses causes profondes.

2) *Une analyse statistique approfondie des défaillances humaines* avec une grille d'investigation comportant environ 70 rubriques. Cette grille a été établie en collaboration avec l'Institute of Nuclear Power Operations (Programme HPES: «Human Performance Evaluation System») et reprend la logique de l'analyse simplifiée précédente en approfondissant les différents thèmes. Notons cependant deux différences avec l'approche précédente:

– la découverte de l'événement (qui le découvre? où? quand? comment?) est particulièrement fouillée;

– l'analyse de la situation ne consiste pas uniquement à rechercher les causes de la défaillance mais aussi à en faire une description objective sans idées préconçues: par exemple on s'attache à décrire le type de communication – téléphonique, face à face, etc. – pour ne rechercher qu'ensuite la relation de causalité entre la défaillance et le facteur «communications».

Il est en effet possible que des caractéristiques de la situation de travail qui ne sont pas identifiées ponctuellement comme étant des causes de défaillances apparaissent, seules ou combinées à d'autres, comme génératrices de défaillances (un exemple a priori farfelu: les opérateurs ayant un an d'ancienneté dans le poste et travaillant en quart de nuit seraient prédisposés à commettre des erreurs).

3) *Une analyse clinique approfondie d'incident,* qui repose sur une méthode logique de reconstitution de la chronologie de l'incident, de construction d'un arbre des défaillances ayant généré l'incident et enfin d'élaboration d'un arbre des causes pour chaque défaillance humaine.

Notons que les deux premiers niveaux d'analyse sont utilisés systématiquement pour tous les incidents significatifs ayant pour origine une composante humaine. Le premier niveau, l'analyse statistique simplifiée, est utilisé comme détecteur de tendance puisqu'un bilan de l'ensemble des défaillances humaines ayant conduit à des incidents significatifs est effectué chaque année, ce qui permet de voir l'évolution des défaillances et de leurs causes. Ce niveau d'analyse est trop simplifié pour nous permettre de définir de véritables mesures préventives. Par contre, le second niveau, c'est-à-dire l'analyse statistique approfondie, a pour rôle de détecter des facteurs ou des configurations de facteurs pouvant générer des défaillances et de définir ainsi des actions préventives appropriées.

Le troisième niveau, l'analyse clinique approfondie, a un statut complètement différent puisqu'il est utilisé pour un nombre d'événements (une dizaine chaque année) dont la complexité et la richesse justifient des investigations plus approfondies. Ce type d'analyse a en particulier l'avantage de ne pas détruire la structure de l'incident et les chaînages entre causes, objectif parfois difficile à réaliser avec une analyse statistique qui a tendance à morceler les événements.

Pour conclure ce paragraphe sur l'analyse a posteriori, un point est à souligner: le même type de méthodologie peut être utilisé pour les événements et les défaillances humaines recensés sur simulateurs.

3.2. L'analyse a priori

Il est beaucoup plus difficile d'élaborer une méthodologie d'analyse a priori des situations de travail. La méthode utilisée dépend essentiellement des thèmes abordés: on n'étudie pas de la même manière l'aménagement ergonomique d'une salle de commande et le contenu des communications dans une centrale, quoique l'analyse du travail sous-jacente utilise souvent les mêmes mécanismes d'investigation.

Compte tenu de cela, nous préférons illustrer cette méthodologie d'analyse à l'aide de deux exemples concrets qui sont:

– l'étude des communications dans une centrale;
– la recherche des facteurs de stress.

4. UN PREMIER EXEMPLE D'ILLUSTRATION: L'ANALYSE DES COMMUNICATIONS EN CENTRALE NUCLEAIRE

4.1. L'analyse a posteriori

Cette méthode repose à l'origine sur une analyse a posteriori des événements. On a en effet constaté que des défaillances de communication pouvaient générer des événements touchant aussi bien à la disponibilité des installations qu'à leur sûreté ou à la sécurité du personnel.

L'analyse statistique simplifiée fait par exemple apparaître comme premier facteur de défaillance humaine le facteur «organisation», dans lequel sont inclus les défauts de communication: ceci représente 43% des défaillances humaines ayant conduit à des incidents significatifs.

L'analyse statistique approfondie, qui permet de fouiller les causes des défaillances humaines, fait apparaître la cause «communication» comme la toute première: elle représente à elle seule 19% des défaillances recensées.

Enfin, l'analyse ponctuelle d'incidents nous montre qu'il existe une dépendance étroite entre l'activité de communication et le maintien des critères de sûreté, disponibilité et sécurité:

– *pour la sûreté:* à la centrale du Bugey, des difficultés de communication téléphonique entre la salle de commande et le local ont obligé l'opérateur chargé des actions en local à faire trois allers-retours ce qui, en retardant le démarrage de l'installation, a entraîné une basse température du circuit primaire qui, combiné à d'autres facteurs, a conduit à une injection de sécurité, donc à une action de sauvegarde sur l'installation;

– *pour la disponibilité:* à la centrale de Cruas, une transmission d'information insuffisamment explicite entre deux opérateurs («faire un appoint» à l'alternateur sans préciser s'il s'agissait d'un appoint en air ou en hydrogène) a conduit à l'explosion de l'alternateur due à un mélange air–hydrogène; la centrale a été indisponible pendant un an;

– *pour la sécurité du personnel:* à la centrale de Cruas, un chef de quart est resté enfermé pendant plusieurs heures dans la station de pompage, à la suite du blocage de la porte d'accès; l'absence de moyen de communication avec l'extérieur ne lui a pas permis de signaler sa présence dans ce local.

Tous ces éléments viennent confirmer l'existence d'un problème de communication dans nos centrales, d'où l'idée d'engager une analyse a priori sur ce thème. L'analyse s'est concrétisée au travers de deux enquêtes de types différents:

1) La première enquête a consisté à faire un bilan des moyens de communication existants afin d'étudier leur adaptation aux besoins des exploitants qui sont de trois sortes:
– rechercher une personne,
– échanger des messages parlés,
– échanger des documents écrits.

2) La deuxième enquête a consisté à analyser finement les flux de communication dans une centrale pour en évaluer la quantité et le contenu compte tenu des moyens à mettre en œuvre.

Nous nous limiterons dans ce mémoire à la présentation de la deuxième enquête, car elle illustre mieux l'analyse a priori.

4.2. L'analyse a priori: analyse des flux de communication dans une centrale

L'analyse a porté sur les communications téléphoniques car ce sont les données les plus faciles à recueillir. Sous réserve de garantie d'anonymat, les équipes de conduite ont accepté de se prêter à l'expérience.

L'enregistrement des communications téléphoniques a été réalisé pendant deux semaines dans la centrale: une semaine en arrêt de tranche et une semaine en fonctionnement normal. Les postes enregistrés sont ceux utilisés par les équipes de conduite.

L'analyse a été à la fois quantitative (flux des messages) et qualitative (contenu des messages).

4.2.1. Analyse quantitative des flux

Il apparaît que l'évolution des messages échangés suit celle de l'activité des tranches.

a) Sur 24 heures:

On constate une augmentation de l'activité téléphonique lorsque le personnel qui travaille en horaire discontinu est présent: deux pics d'activité se situent respectivement aux alentours de 10 h et de 15 h 30. Par ailleurs, on note un ralentissement nocturne, plus marqué après minuit.

b) Sur la semaine:

Il est naturel de trouver une baisse d'activité téléphonique le week-end. On pourrait s'attendre à trouver un pic d'activité le lundi matin, car les opérateurs de la salle de commande disent être très perturbés par les nombreuses personnes qui viennent s'informer des événements du week-end et des prévisions pour la semaine à venir. Il s'avère que ces communications se font face-à-face car on ne constate pas d'augmentation de l'activité téléphonique. Par contre, l'augmentation est nette en fin de semaine, le vendredi. Ceci peut être attribué au désir de solder les actions avant le week-end, ce qui occasionne un plus grand nombre de coordinations.

c) Selon les locaux:

Le bureau de consignation ou bureau du chef de quart et de l'adjoint est le principal lieu de coordination entre les agents de conduite et les autres intervenants Cette coordination s'exerce au cours de la préparation et à la fin des actions. Ceci se retrouve au niveau des communications puisqu'il y a quatre pics, en début et en en fin de chaque demi-journée: 8 h, 12 h, 14 h, 17 h.

La salle de commande est le pivot de circulation des informations. Elle gère la coordination pendant le déroulement des actions. En effet, la quantité de messages échangés augmente linéairement jusqu'à un maximum en milieu de demi-journée, 9 h 30 et 15 h, principalement le matin: il peut y avoir jusqu'à 23 messages échangés en une heure.

Le pupitre du BAN (Bâtiment des auxiliaires nucléaires dont l'accès – en zone contrôlée – est soumis à des contraintes sévères: les opérateurs chargés du BAN y passent en général toute la durée du quart) est le poste le plus actif la nuit. En effet, l'activité y est presque aussi intense de nuit que de jour. On note une nette diminution au moment des relèves à cause de l'accès en zone contrôlée.

4.2.2. Analyse du contenu des messages

Le contenu des messages a été analysé selon leur fonction (recherche de personnes, ordre, etc.), selon les interlocuteurs, en fonction du respect de certains principes de communication et enfin selon le vocabulaire employé.

a) Fonction des messages:

Pour 20%, l'activité téléphonique reçue en salle de commande consiste en demandes de recherche de personnes. Deux raisons à cela:

- la salle de commande représente un lieu nécessaire de convergence des agents;
- l'appel par haut-parleur est émis depuis la salle de commande.

La fonction «carrefour d'information» de la salle de commande est confirmée par la répartition entre messages émis et reçus: alors que dans le BAN et au bureau de consignation la répartition est à peu près équivalente entre messages émis et reçus, par contre pour la salle de commande, 70% des messages proviennent de l'extérieur contre 30% seulement émis par la salle de commande. Notons également que 30% des messages échangés sur les postes de la salle de commande sont initiés par une question et qu'il y a une augmentation du nombre de ces questions lors des périodes précédant les relèves de quart.

b) Les interlocuteurs:

Les communications du BAN sont centrées sur l'équipe de conduite: 97% des messages se font au sein de l'équipe. En salle de commande, les échanges au sein de l'équipe ne représentent plus que 60% des messages et au bureau de consignation 50%. La salle de commande et le bureau de consignation sont donc également des lieux d'échanges avec les autres intervenants de la centrale.

c) Respect de certains principes de base pour «bien communiquer»:

Nous nous limiterons dans ce mémoire à deux exemples:

– l'indication du réacteur concerné par la communication: ce principe devrait surtout être respecté dans le BAN car, celui-ci étant commun à deux réacteurs, les possibilités de confusion sont importantes; or, le numéro du r-réacteur n'est spécifié explicitement que dans 33% des cas;

– la présentation du contexte général du message: l'explication de l'objectif et du cadre général des actions qu'on demande d'engager est un gage de fiabilité dans la mesure où elle est souvent le meilleur moyen de lever les éventuels doutes et incompréhensions, par exemple la confusion entre deux circuits; or, nous avons noté que, dans 30% des cas, on formule directement un ordre sans autre explication et on raccroche.

d) Vocabulaire utilisé:

Une analyse détaillée a permis de faire une typologie des actions à travers les messages échangés. Une vue partielle de cette typologie est représentée sous forme de tableau (tableau I).

4.2.3. Conclusion de l'analyse des communications

Cette analyse a priori des flux de communications, combinée à l'analyse des moyens de communication existants dans les centrales, a permis à EDF de dégager deux enseignements principaux:

1) Premier enseignement: l'amélioration des outils de communication

– un système de radiocommunication va être mis à la disposition des équipes de conduite et permettre la recherche de personnes, l'échange de messages parlés et la sauvegarde de l'agent isolé (alarme manuelle d'urgence ou alarme automatique fonctionnant au bout d'une minute d'immobilité de l'opérateur);

– un système de messagerie écrite est en cours d'étude; il devrait permettre d'établir une liaison entre la salle de commande et des boîtes aux lettres réparties

TABLEAU I. VUE PARTIELLE DE LA TYPOLOGIE

Types d'action	Objets des actions	Actions
Actionnement de matériels et de systèmes	Les matériels	Démarrer Arrêter Régler Ouvrir Fermer Manœuvrer Etablir/couper un courant
	Les circuits et portions de circuit	Ligner Remettre en configuration
	Les circulations de fluides liquides ou gazeux	Faire/arrêter • un appoint • un transfert • une circulation

sur l'installation; il aurait pour but de décharger les opérateurs des salles de commande des tâches d'information sur l'état de l'installation, de réduire les délais d'acheminement et d'améliorer la fiabilité des messages, certains ordres pouvant ainsi être confirmés par écrit.

2) Deuxième enseignement: l'amélioration des pratiques de communication

Un manuel de communication a été établi. Il comporte deux parties, l'une concernant les principes à respecter (par exemple, dire où l'on est, répéter le message, etc.), l'autre concernant le vocabulaire à utiliser ou à ne pas utiliser. Une sensibilisation et une formation des exploitants aux règles énoncées dans ce manuel se fait en introduisant certains principes de base dans des séances de formation des équipes d'exploitation.

5. UN DEUXIEME EXEMPLE: L'ETUDE DU STRESS EN CENTRALE NUCLEAIRE

5.1. Présentation de l'étude

Dans ce cas, l'analyse a posteriori est très réduite car il est difficile d'attribuer un incident ou une défaillance humaine à un stress excessif du personnel concerné.

La décision d'analyser le stress en centrale nucléaire a donc été prise indépendamment des événements recensés dans nos centrales: elle a été suscitée par des réflexions d'opérateurs parlant du «trou noir» ressenti dans certaines situations perturbées de l'installation.

L'analyse a priori a donc été le principal mode d'approche de ce thème. L'analyse a débuté par une analyse bibliographique permettant de définir les situations de stress: elles sont classiquement définies comme le produit de deux facteurs constitutifs, d'une part le fait que la maîtrise de la situation comporte des enjeux élevés, et d'autre part l'existence d'une incertitude (subjective et/ou objective) quant à la possibilité pour l'individu d'avoir ou de conserver la maîtrise de la situation.

L'analyse s'est ensuite poursuivie dans une centrale nucléaire où une enquête menée par questionnaire s'est adressée à l'ensemble des agents de la centrale. Le questionnaire a été construit à partir de données recueillies préalablement au cours d'observations et d'entretiens. Il tient compte de la diversité des tâches exécutées et des situation rencontrées par les différentes catégories d'agents de la centrale. Le taux de réponse a été de 2/3 et varie peu selon les catégories d'agents. Le grand nombre de catégories professionnelles a conduit à regrouper certaines d'entre elles et à se limiter à huit «groupes professionnels»: les chefs de quart et les ingénieurs travaillant en 3 X 8, les adjoints chefs de quart, les chefs de bloc, les rondiers, les automaticiens et électriciens, les mécaniciens et chaudronniers, les chimistes et le service technique, les cadres et les personnels administratifs.

Les résultats de l'enquête peuvent être exposés selon trois axes:

1) Un premier ensemble de questions a pour objet d'établir une évaluation quantifiée des enjeux auxquels sont confrontés les différents groupes professionnels dans leurs situations de travail respectives, enjeux estimés par les agents eux-mêmes. Quels sont les risques pour soi-même, leur nature et leur fréquence? Quels sont les risques pour le fonctionnement de la centrale?

2) Un deuxième ensemble de questions est consacré au recensement des occasions où les agents peuvent se trouver dans l'incertitude sur le fait de détenir ou de conserver la maîtrise de la situation. Certaines circonstances de l'exécution des tâches sont-elles repérables comme spécialement génératrices de difficultés et d'incertitude sur la bonne fin des opérations assignées à chaque catégorie d'agents?

3) La troisième série de questions vise à apprécier l'importance de ces facteurs de stress potentiels à travers une évaluation de leur impact sur les individus. Dans quelle mesure observe-t-on chez les agents d'une centrale nucléaire un certain nombre d'éléments psychosomatiques classiques, généralement relevés dans les groupes souvent confrontés à des situations de stress? Quelle est la fréquence de ces répercussions pour chacune des catégories d'agents? Peut-on en vérifier le lien avec les situations d'enjeux et d'incertitude auxquelles sont confrontés ces agents?

5.2. Résultats de l'étude

Nous nous limiterons dans ce mémoire à la présentation d'un exemple illustrant chaque ensemble de questions.

5.2.1. Les enjeux du travail dans une centrale nucléaire: les risques pour soi et pour le système

Chacun des six groupes professionnels de la maintenance et de la conduite a été interrogé sur ses scores moyens d'exposition aux risques suivants: brûlures par fuite de vapeur, risque azote, intoxication par produits chimiques, explosion, irradiation, contamination, électrocution, chute de personne, chute de matériel, enfermement dans un local.

Il ressort un fait majeur: les rondiers disent être fortement exposés aux différents risques. Leurs estimations les placent en tête pour tous les types de risques. Pour sept types de risques (brûlures, azote, intoxication, irradiation, contamination, électrocution, chute de matériel), plus de 50% d'entre eux considèrent que leur fréquence d'exposition est égale ou supérieure à «plusieurs fois par semaine».

Le second groupe qui se détache est celui des adjoints chefs de quart; il se distingue nettement des autres groupes sur deux types de risques: brûlures et électrocution.

Vient ensuite le groupe des automaticiens et électriciens, qui occupe le troisième rang pour les risques chute de matériel, chute de personne et électrocution.

Ces résultats montrent que les personnes estimant être les plus exposées aux risques sont effectivement celles qui se déplacent beaucoup sur l'installation et qui interviennent couramment sur des systèmes en fonctionnement.

5.2.2. Les incertitudes sur le fait de détenir la maîtrise des situations en fonction des circonstances

Il a été demandé aux groupes professionnels travaillant en 3 × 8 s'ils se sentaient plus susceptibles de commettre des erreurs pendant le quart du matin, de l'après-midi ou de la nuit.

Les réponses des différents groupes concernés ont la même distribution: ils s'estiment plus faillibles pendant la nuit et plus le matin que l'après-midi. Une interprétation physiologique de ces réponses consisterait à dire que celles-ci illustrent le rythme circadien des individus, avec une température minimale la nuit (4 h du matin) associée à des performances minimales et une température maximale l'après-midi (16 h) associée à des performances maximales.

Cependant, une autre interprétation vient se superposer à celle-ci pour expliquer la différence entre le matin et l'après-midi; l'équipe de quart du matin arrive à 6 h et repart à 13 h alors que les équipes de jour arrivent à 7 h 30: elle est donc pendant très peu de temps seule sur l'installation et chacun sait que la multiplication des intervenants fragilise la fiabilité de chacun. Par contre l'après-midi, l'équipe de quart est seule de 17 h à 21 h.

5.2.3. Les incidences des facteurs de stress sur les agents

On rejoint ici une des approches les plus utilisées dans les études de stress au travail, mais le plus souvent séparée d'une analyse fine des situations de travail.

L'exposition répétée à des situations de stress est classiquement considérée comme susceptible d'entraîner à terme des effets pathologiques, en particulier sur le plan de la nutrition et du sommeil. Les réponses à ces questions sur le trouble du sommeil et de l'appétit montrent que les groupes qui s'estiment être les plus touchés sont ceux qui travaillent en 3 X 8. En effet, les troubles qui peuvent survenir en ces domaines procèdent généralement d'une causalité complexe et peuvent rarement être imputés à un facteur unique.

5.3. Conclusion de l'étude

Une étude de corrélation a été faite entre les scores de facteurs de stress (exposition aux risques personnels, risque de commettre différentes erreurs) et le score de troubles psychosomatiques. On peut considérer qu'il y a une relation entre la fréquence et la confrontation aux facteurs de stress, tels qu'ils ont été inventoriés au moyen du questionnaire, et les troubles de santé déclarés par les agents.

Par ailleurs, on a noté que les groupes témoignant d'un niveau plus important de troubles de santé sont aussi soumis aux contraintes du travail posté. Et l'on sait que la pratique des horaires postés peut également avoir des répercussions sur la santé. Le facteur à incriminer ne serait-il pas alors le travail posté plutôt que les sources de stress spécifiques au travail en centrale nucléaire?

Comme on peut le constater, cette étude se termine par un grand nombre d'interrogations et d'incertitudes. Son statut est tout à fait particulier par rapport à d'autres investigations. Nous sommes encore au stade de la recherche, peut-être parce que les événements réels, c'est-à-dire l'analyse a posteriori, ne nous permettent pas de déceler des axes prioritaires pour l'analyse a priori.

6. CONCLUSION GENERALE

Les exemples donnés dans ce mémoire ont été présentés pour illustrer notre méthodologie d'analyse a priori et a posteriori des facteurs humains. Ils peuvent cependant donner une image quelque peu déformée des actions menées par EDF dans ce domaine. En effet, les études présentées sont essentiellement des constats et ne proposent pas d'actions véritablement concrètes. Ceci vaut surtout pour l'étude sur le stress.

Il nous semble donc important de conclure ce mémoire en rappelant que la prise en compte des facteurs humains en exploitation consiste le plus souvent en actions concrètes, du type de celles engagées par EDF suite à l'analyse a posteriori des incidents et des défaillances humaines survenus sur l'ensemble des centrales: automatisation de certaines tâches, amélioration ergonomique des salles de commande REP 900 MW, refonte des procédures de conduite incidentelles et accidentelles, meilleur repérage des matériels et des locaux, amélioration de la coordination entre équipes de conduite et de maintenance, etc.

QUANTIFICATION OF HUMAN PERFORMANCE IN NUCLEAR TASKS

A.M. METWALLY
Faculty of Engineering,
Alexandria University,
Alexandria

F.A. MOHAMMED, A.A. OMAR
Atomic Energy Authority,
Nuclear Regulatory and Safety Centre,
Cairo
Egypt

Abstract

QUANTIFICATION OF HUMAN PERFORMANCE IN NUCLEAR TASKS.

The paper discusses how performance shaping factors (PSFs) can be quantitatively assessed. PSFs are those factors that affect human reliability in situations such as power plant operation, maintenance and testing. The study suggests the classification of the PSFs into qualitative and quantitative groups. Multi-attribute utility theory is then used to represent the PSFs as utility functions which need to be optimized. The Simplex method is suggested for use in optimizing the utility functions expressing the PSFs. The quantitative group of PSFs is formulated in a general analytical form. A trial was made to find the optimum characterization of some members in the quantitative group of PSFs and their effect on operator performance.

1. INTRODUCTION

Performance shaping factors (PSFs) [1] are those factors that affect human reliability in situations such as power plant operation, maintenance and testing. They control the amount of variability in human performance in a specific task in such a manner as to make it highly reliable, highly unreliable or at some level in between. PSFs interact in a very complex manner that makes it difficult to develop analytical expressions describing the relative influence of these factors on human performance.

Swain [2] suggested the classification of PSFs into three major sets:

(1) Extra-individual PSFs; these include the situational characteristics, task and equipment characteristics and job instructions.
(2) Intra-individual PSFs; these include individual factors.
(3) A set of PSFs that forms a bridge between the two previous sets and includes both psychological and physiological stresses.

The approach suggested in this paper classifies the PSFs into two major sets:

(1) A set that includes those factors that can be expressed qualitatively such as procedure required, organizational structure, perceptual requirements, motor requirements, the control display relationship, anticipatory requirements, task criticality, skill and experience.
(2) A set that includes those factors that can be expressed quantitatively.

This paper is concerned with the second set of factors, which may include temperature, noise level, work hours, rewards, radiation level, etc. Each factor's effect on human performance can be represented by a three region curve.

The mathematical model of operator performance can be derived using Taylor's model. The Simplex method is suggested to obtain the minimum weights of parameters that optimize operator performance.

2. QUANTITATIVE REPRESENTATION OF PSFs

It is suggested that the effect of the members of the quantitative set listed in Table I on human performance be put in a general mathematical form [3, 4] characterized by three regions:

In the first region $V_0 e^{\sigma X}, \ 0 < X < T_1$
In the second region $V, \ T_1 < X < T_1 + T_2$
In the third region $V \exp\{-\mu(x - (T_1 + T_2)\}$
$T_1 + T_2 < X < T_1 + T_2 + T_3$

TABLE I. PERFORMANCE SHAPING FACTORS

(1)	Temperature and humidity
(2)	Air quality
(3)	Noise and vibration
(4)	Degree of general clarity
(5)	Work hours/work breaks
(6)	Action by supervisors, co-workers, union representatives and regulatory personnel
(7)	Rewards, recognition, benefits
(8)	Level of radiation
(9)	Atmospheric pressure extreme
(10)	Degree of insufficient oxygen.

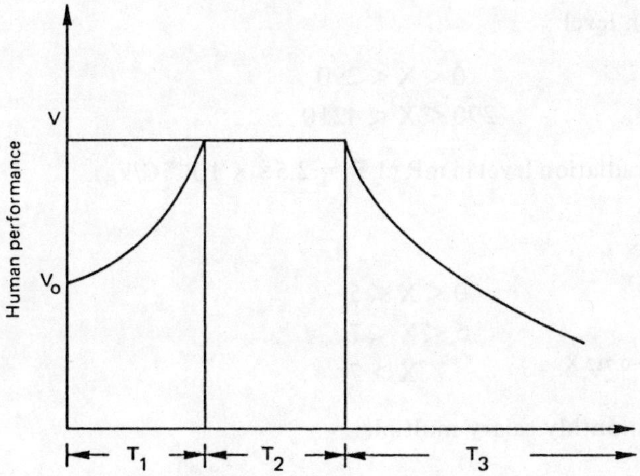

FIG. 1. Graphical representation of the effect of certain PSFs on human performance

Each factor has a curve that satisfies one or more of these three regions. This curve is illustrated in Fig. 1. The values of the constants included in the mathematical form are determined either by judgement or from literature on human factors. As an illustration, the following are some analytical equations showing the effects of some quantitative PSFs.

1. Temperature:

   $0.3\,e^{0.046\,X}$ $\quad 0 < X < 25$
   $2\,e^{-0.046\,X}$ $\quad X > 25$

   X is the temperature in °C.

2. Working hours:

   $0.7\,e^{0.178\,X}$ $\quad 0 < X < 2$
   1 $\quad 2 > X < 6$
   $2.91\,e^{-0.178\,X}$ $\quad 6 < X < 8$

   X is the number of working hours.

3. Noise level:

   $0.5\,e^{0.0116\,X}$ $\quad 20 < X < 60$
   1 $\quad 60 < X < 70$
   $2.252\,e^{-0.0116\,X}$ $\quad 70 < X < 120$

   X is the noise level in dB.

4. Radiation level:

1	$0 < X < 290$
$e^{-0.0004\,X}$	$290 < X < 4210$

X is the radiation level in mR (1 R = 2.58×10^{-4} C/kg).

5. Rewards:

$0.5\,e^{0.157\,X}$	$0 < X < 5$
1	$5 < X < 7$
$147.08\,e^{-0.713\,X}$	$X > 7$

X is the monthly salary multiple.

6. Degree of clarity:

The performance increases linearly up to $X = 1$ and then reaches a plateau.

X is a fraction <1 representing the degree of clarity.

7. Air quality:

Same as degree of clarity:

8. Air pressure:

$0.1\,e^{2.878\,X}$	$0 < X < 0.8$
1	$0.3 < X < 1.2$
$31.614\,e^{-2.878\,X}$	$X > 1.2$

X is the air pressure in atmosphere.

3. PSFs AS UTILITY FUNCTIONS

Utility theory and utility functions can be utilized to evaluate the weight of each member in a set of quantitative PSFs. By the weight we mean the significance of certain PSFs in controlling human reliability. The present study suggests the use of a general form of utility functions as composed of two groups:

Group I: includes all PSFs that can be expressed quantitatively. The utility of the group will be designated by U_I.

Group II: includes all PSFs that can be expressed qualitatively. The utility of this group will be designated by U_{II}. According to the suggested formalism, the general form of utility function is

$$U = U_I + U_{II} \tag{1}$$

The elements of Group I can be weighted according to the effect of each element on the human reliability using the concept of utility functions. So U_I can be written as:

$$U_I = \sum_{i=1}^{N} A_i U_i(x) \tag{2}$$

Where for an element (PSF) X_i, A_i and U_i represent its weighting and utility respectively.

4. GENERALIZATION OF TAYLOR'S MODEL

In studying improvement in human performance of complex tasks, it has been found that the exponential model proposed by Taylor [5] can be applied. The model can be formulated as:

$$Y_m(t) = Y_c + Y_f(1 - e^{-t/\tau}) \tag{3}$$

where:

$Y_m(t)$ is human performance at time t
Y_e is initial performance at the commencement of a certain task
Y_f is final performance
t is practice time
τ is learning time which is constant for a particular task.

The graphical illustration of this model is shown in Fig. 2.

According to the suggested classification of the PSFs and Taylor's model, one can expand the definition of Y_c as follows:

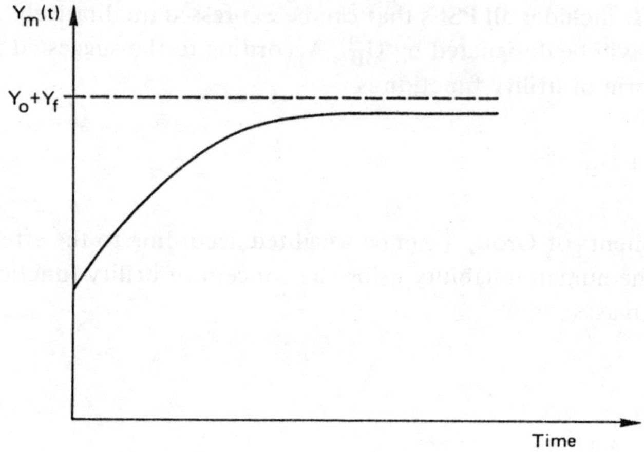

FIG. 2. Graphical illustration of human performance.

Since training is accomplished on a nuclear simulator having nearly the same environmental conditions as in the NPP, the set of the time independent PSFs such as humidity, noise level, etc. (Group I PSFs) controls the value of Y_c. Hence Y_c takes a new definition as it is the human performance determined by Group I PSFs.

Group II PSFs will manifest themselves for controlling the time dependent part of Eq. (1).

Combining Eqs (1) and (3), one can write

$$U = U_1 + A(e^{1-t/\tau})$$

Where:

U is human performance expressed as utility function
UI is human performance control led by Group I PSFs
A is utility of Group II PSFs at $t = \tau$
t is time of practice
τ is time constant of learning.

5. SIMPLEX METHOD TO OPTIMIZE PSFs

The Simplex method [6] is one of the powerful optimization methods in linear programming as applied in solving problems which are entirely linear.

In this study, Simplex methods can be utilized to find the optimum values of A_i, $i = 1,2, -N$ Eq. (2).

It should be noticed that, depending on the number of relevant quantitative PSFs, one must use the same number of Eqs which have the same form as Eqs (2).

To explain how to use these equations, one should assume a value of U_I and determine the values of U_i (x_i) from the corresponding analytical form of each PSF. Finally the Simplex method is applied with the governing equation:

$$U_i = \sum_{i=1}^{N} C\, A_i$$

to find the optimum values of A_i, where C = constant. The significance of this evaluation is to rank the relevant quantitative PSFs according to their effect on human performance.

Another approach in using the Simplex method is to assign different values by judgement for A_i and optimize for U_i; the optimum value of U_i can then be used for the ith quantitative PSF.

REMARKS

This paper concentrates on developing the model based on the generalization of the Taylor model. The human performance is considered a utility function consisting of two utilities U_I: one corresponding to the time independent part of the Taylor model representing the quantitative PSFs and the other representing the qualitative PSFs and corresponding to the qualitative part. The paper emphasizes the first group and suggests that the Simplex method optimizes the weights of the utilities and then optimizes the utilities, indicating the effect of each factor on human performance.

REFERENCES

[1] SWAIN, A.D., GUTTMAN, H.E., in Handbook of Human Reliability Analysis with Emphasis on Nuclear Power Plant Application, Rep. NUREG/CR-1287, Sandia Natl Labs, Albuquerque, NM (1983).

[2] SWAIN, A.D., A Method for Performing a Human Factor Reliability Analysis, Rep. SCR-685, Sandia Natl Labs, Albuquerque, NM (1963).

[3] DARWISH, N.H., Analysis and Modelling of Human Performance in Nuclear Power Plants, MSc Thesis, Alexandria Univ., Egypt (1987) 60–80.

[4] METWALLY, A.M., Modelling and Analysis of Human Performance in NPPs, PhD Thesis, ISV, USA (1982).

[5] HUSSEINY, A.A., SABRI, Z.A., DANOFSKY, R.A., Qualification of operation experience for reliability analysis, Trans. Am. Nucl. Soc. **21** (1976).

[6] HAMDY, A.T., Operations Research: An Introduction, Macmillan, London and New York (1987) 59–80.

NEED FOR A SYSTEMATIC APPROACH TO SAFETY: PROBABILISTIC STUDIES FOR NUCLEAR POWER PLANTS IN ITALY

M. NOBILE
Production and Transmission Division,
Ente Nazionale per l'Energia Elettrica

M. PENNAZZI, S. SERRA
Construction Division,
Ente Nazionale per l'Energia Elettrica

Rome, Italy

Abstract

NEED FOR A SYSTEMATIC APPROACH TO SAFETY: PROBABILISTIC STUDIES FOR NUCLEAR POWER PLANTS IN ITALY.

The paper gives an overview of the probabilistic safety studies performed in Ente Nazionale per l'Energia Elettrica's NPPs now being constructed or designed (Alto Lazio BWR and Progetto Unifacato Nucleare (PUN) PWR). This activity proved quite useful in ascertaining compliance with the predetermined safety objectives, and as a tool for identification of the main corrective actions required. The approach was then used to evaluate the safety level of the plants in operation (Latina Magnox, Caorso BWR, Trino PWR). The experience gained from these plants is almost complete and provides a few useful indications on the possible use of this tool as a source of additional information that is not available from other sources. Every NPP has its own frequency of core damage that depends on the peculiar aspects of the station under examination. A comparison of the results from different plants should be interpreted taking into consideration the completeness of the analysis and the methodology used.

1. ALTO LAZIO PROBABILISTIC SAFETY STUDY

1.1. Introduction

An extensive probabilistic study of the safety related systems was performed for the Alto Lazio nuclear plant, a two unit BWR/6 plant now being constructed about 100 km northwest of Rome, Italy.

The analysis, known as the Alto Lazio Station Reliability Analysis (ALSRA), was performed by a team consisting of engineers from Ente Nazionale per l'Energia Elettrica (ENEL, utility), General Electric (GE, nuclear system vendor), Ansaldo Impianti (AI, architect engineer), and Comitato Nazionale per la Ricerca e per lo Sviluppo dell'Energia Nucleare e delle Energie Alternative (ENEA) (safety authority).

Extensive consultations on the design of the systems, the performance of special accident analyses with multiple failures and verification of the documents issued were held by the design engineers of AI, ENEL and GE. The total effort amounted to more than 250 man-months, with contributions from more than one hundred engineers.

Although the ALSRA is similar in technology to the probabilistic risk assessment (PRA) studies which have been performed on other plants, there are several differences, decided by ENEL, for better use of the analysis as a support for the licensing process and plant operation and as an independent functional design verification. In particular, the final quantitative goal of the ALSRA analysis was the evaluation of the frequency of exceeding the licensing acceptance limits for the safety functions, such as core cooling, containment cooling and dose control function. In the first phase of the study (1984) licensing models were used to find the minimum number of necessary mitigating systems [1, 2]. Furthermore, in 1986 ENEL performed a probabilistic safety study (PSS/AL) with realistic bases.

1.2. Analysis purposes and methodology

The purposes of the ALSRA study were:

- To satisfy the ENEA requirement for the Alto Lazio construction permit for a reliability analysis of the safety systems;
- To provide a plant reliability model to obtain more rational and unique technical specifications for Alto Lazio;
- To identify critical areas in the design which need correction or improvement from the safety or plant availability point of view;
- To make decisions on proposed design modifications and to choose the optimal solution among the alternatives;
- To support the resolution of licensing issues or of problems which might occur during the nuclear test and power operation;
- To provide a better understanding of the integrated response of the plant to transients and accidents, even in the event of multiple failures.

As the first step of the analysis [3], comprehensive failure modes and effect analyses (FMEAs) were performed for all safety systems and for additional systems important to plant availability. The FMEA task provided detailed knowledge of the system, the basis for the construction of the system fault tree, the single failure criterion verification at system level, the review of the design from the point of view of the surveillance test feasibility, the basis for the operating personnel to decide whether the system is to be declared inoperable, following the failure of a particular component, and the basis for the construction of a trouble shooting chart.

After the FMEA task and the thermohydraulic analysis to determine the minimum configuration of the core cooling system, the tasks performed were the development of the event scenarios and of the interdependence matrices, the fault tree construction, the functional fault tree analysis by the SETS/SEP computer codes and the event tree analysis.

The design was close to completion so that a precise and detailed analysis was possible, and the plant construction status was such that it was still possible to correct the discrepancies and to incorporate significant reliability improvements without any plant construction delay.

In the second phase, on the basis of the ALSRA study the PSS/AL was performed according to realistic ground rules, such as:

- The acceptance and the success criteria were defined in a realistic way;
- The timing for operator recovery action was taken into account;
- The credit for non-safety grade systems was considered.

1.3. Results

The ALSRA was concluded in 1984 with the publication of 23 reports containing the details of qualitative and quantitative analysis.

As a consequence of this analysis many recommendations for modifications in the design were implemented to improve plant safety and availability or to correct design inconsistencies. The most significant design changes were made on the control building and control room ventilating systems [4]. The preliminary analysis found that the failure of one of the above systems during normal operation was the major contributor to the plant risk of exceeding the licensing criteria for the core cooling and containment cooling functions. The proposed modifications in the design were based on the reliability models, which allowed a quick decision on their implementation. The prompt response minimized the construction delay and cost.

The quantitative analysis results for the frequency of exceeding the accident licensing criteria for the core cooling show a value of around 6 E-06/year [5]. The results of the PSS/AL second phase [6] show a core damage frequency around 2E-06/year and the contribution from LOCAs in this realistic analysis decreases by more than one order of magnitude in comparison with the ALSRA results.

These results show that Alto Lazio has an excellent reliability of the network of the mitigating systems and of their auxiliaries to respond to the transients and accident events.

2. PUN PWR PROBABILISTIC SAFETY STUDY

2.1. Introduction

A requirement of the Italian energy plan is that a detailed safety evaluation be performed on the preliminary approved standard PWR plant conceptual design 'Progetto Unificato Nucleare' (PUN). Therefore, in December 1982 ENEL decided to perform a level 1 PRA on the preliminary approved PUN design. The primary goal for the safety of the PUN design was attained by means of an assessment of the core damage frequency (CDF); the second goal was to identify the system's weak points.

For this purpose, ENEL, AI and Westinghouse carried out and completed in 1984 a Level 1 PRA [7] for a total effort of 250 man-months. A revision of the PUN PSS performed by the Italian Safety Authority pointed out that the model gives a measure of the core damage probability that is representative of the safety level of the standard plant. Furthermore, the Italian Safety Authority required the PUN PSS to be interactive with the design (living PRA). Therefore, a third goal is to provide a living PRA model which can be used during the plant design and later during operation to optimize design features. The experience gained in performing the study revealed that, through careful application of the probabilistic technique, the study can be used as a tool to develop an optimum plant design in terms of safety and cost.

2.2. Probabilistic safety assessment (PSA) design evaluation process as a living tool

After developing the conceptual design, the intent of the PSA [8] design evaluation process was directed towards the identification of potential improvements to the conceptual design which would significantly improve public safety. These design improvements can be implemented in the conceptual design before beginning construction, thus reducing plant modification costs. Then, alternative design options can be evaluated using probabilistic techniques, and a simplified design can be implemented to reduce capital cost without sacrificing public health and safety.

The overall use of the PUN PSS starts with the development of the conceptual design based on traditional deterministic safety criteria. The conceptual design can then be evaluated using PRA techniques to produce a baseline risk.

After quantification of the CDF, engineering recommendations on the design were proposed to achieve the best value of CDF. These recommendations stemmed from the sensitivity studies carried out previously. At the conclusion of the individual sensitivity study, those factors in the plant design configuration with the strongest impact were combined to determine the overall impact on CDF. The PSS model and methodology had to be flexible to accommodate design changes and to be automated to the maximum possible extent in order to minimize the manpower required for an analysis of potential design changes. To meet this requirement, the PSS became an automated living tool which could be quickly updated with little effort. The living PSS model will be used to evaluate design changes and operating procedures implemented during the construction, testing and operational phases of the plant. In addition, the PSS model will be used to evaluate regulatory requirements that may be imposed during the life of the plant.

With this in mind, the PSS was developed on a modular basis in which each module represents a particular individualized analysis such as initiating events, system fault tree models, event tree models and support state model. These modules were then analysed and combined using integrated PRA microcomputer codes which enable the risk assessment engineers to evaluate any changes by using automated techniques. Having the entire model computerized and readily available for re-analysis helps to fulfil the living PSS requirement [9].

2.3. Results and further developments

The first phase of the PSS was concluded in July 1984 and the mean value of CDF was calculated to be 4.4 E-05/year for internal events and loss of off-site power. The outcome of the study was generally favourable for the design, showing a good balance of plant systems; no single accident sequence accounted for more than 10% of the total CDF.

However, this final result did not comply with the Safety Authority criterion of CDF ($\leqslant$ 1.0 E-05/year) and, therefore, some engineering recommendations were proposed [5]. Since no single system modification could decrease the CDF significantly, ENEL decided to study a set of reasonable design changes. The basic criterion was that any proposal should not deviate from the proven and mature PWR technology. Taking into account also the relevant conservatism included in the PSS methodology and hypothesis, and the advantages that could be gained from more sophisticated studies, any design modification impacting the CDF more than 5% was considered (risk reduction effectiveness). A series of sensitivity analyses were performed in order to identify pertinent investigation areas; they covered the PSS model, operator's actions, initiating event frequencies and possible alternative design configurations. Fifteen alternative design configurations were examined on the basis of one design modification at a time, and then by evaluating the contribution of each modification as a per cent reduction in CDF of the PSS base case. The modifications that have been recommended to the designer are related to the following systems:

- Safety injection and recirculation;
- Residual heat removal;
- Service water and component cooling;
- Vital AC/DC electrical;
- Backup protection.

On the basis of all the preceding modifications a new combined sensitivity for the PSS was calculated, and the result showed a reduction in CDF to 8.0 E-06/year. Moreover, the severe accident analysis showed that in order to further reduce the probabilities and the consequences to the environment of these beyond design basis events, other systems should be modified (e.g. emergency seal injection, electric power) [10]. Therefore, ENEL decided to proceed to an updated PSS taking into consideration all the modifications to the PUN design made from July 1983 to July 1987 that will be incorporated in the PSS (new base case) using the living PRA tool. Some parts of the model will be improved (e.g. recovery action, support state), and some conservatism will be removed (e.g. success criteria); area events (fire, flooding, etc.) and uncertainty analysis will be incorporated.

3. LATINA PROBABILISTIC SAFETY STUDY

3.1. Introduction

The Latina Nuclear Power Station comprises a single gas cooled graphite moderated reactor of the Magnox type. It is located 80 km south of Rome and has operated successfully for over 20 years. As part of a continuing safety review a plant specific PSS was performed by the National Nuclear Corporation and ENEL [5, 11] in 1986. The overall objective of the study was to provide an evaluation of the safety status of the existing protection systems and to ascertain the possible benefit deriving from certain planned modifications already defined by other deterministic studies. The effort amounted to around 100 man-months with contributions from 20 engineers.

3.2. Methodology

The approach of the study and the methodology followed are in line with Level 1 of the 'PRA Procedures Guide' [8]. However, as this guide has particular relevance to LWRs, account has also been taken of PSA methods used for gas cooled reactors in the United Kingdom and also of factors relevant to the Magnox reactor.

The study concentrates on fault conditions that could arise from plant failures. Some qualitative consideration, however, is given to internal hazards such as fire, hot gas release, etc., by way of a review of the design. The assessment requires a definition of core damage in order to categorize the fault sequences that might occur.

The transient analysis for the fault studies performed in recent years specifically for Latina by the Central Electricy Generating Board (CEGB, United Kingdom) was used to this aim. In fact, the limiting transients determined by these analyses have been considered as defining a reference case for which no fuel damage occurs or other safety constraints would be exceeded. The assessment is, therefore, directed to estimating the frequency of fault sequences outside these limiting transients.

This approach is conservative, as the limiting transients analysis includes allowances for uncertainties in the performance of the plant or in the modelled physical properties, which are pessimistic in a best estimate probabilistic study. It is therefore expected that the fault frequencies estimated in the study are conservative in respect of the possibility of serious damage to the reactor core.

The best component failure database estimate was developed for the study. As a database relating only to Latina operating experience was considered insufficient to establish adequate confidence for all the relevant components, a wider basis for the selection of suitable data was considered. The approach followed was to use the NCSR Databank, operated by UKAEA, supplemented with data from the CEGB and Latina operating experience and with other data relative to UK gas cooled reactors, when judged applicable.

3.3. Results

As an important feature of Magnox reactors is the length of time available for the safe initiation of many of the protection systems following a reactor trip, the reliability of operation personnel to ensure satisfactory safe conditions was established. The analyses performed conclude that the influence of human errors on the final figures is very limited. This conclusion is based upon several factors, of which the main ones are:

- The slow response of the graphite moderated gas cooled reactor in post-trip conditions;
- The straightforward thermal-hydraulics governing the reactor response, which in turn influences the operator reliability;
- The simplicity of the plant systems and the small number of controls and indications, allowing easier diagnosis and action;
- The close contact between the operators and the plant, allowing some fault symptoms to be more easily and quickly identified;
- The familiarity of the operators with manual operation of the plant system.

The analysis showed that the possibility to cool the core through natural circulation (pressurized conditions) is an important factor that greatly influences the final results. In this context a figure of around 4.0 E-05/year for CDF has been obtained.

Further analyses show that limited plant modifications made essentially to improve the reliability of the secondary side, this figure can be reduced to 1.5 E-05/year aligning the Latina CDF to that of the other Italian plants.

4. CAORSO PROBABILISTIC SAFETY STUDY

4.1. Introduction

The PSS of Caorso BWR/4 – Mark II containment, in operation since 1981 – (PSS/C) was initiated in August 1986 as a joint effort by ENEL and NUS Corporation. The primary motivation of the study was used to provide a nuclear safety perspective of the plant from the probabilistic viewpoint in response to the public concern arising from the Chernobyl accident. Though accelerated by the occurrence of this accident, the study represents a continuation of ENEL and ENEA's policy to subject all nuclear plants in Italy to a probabilistic analysis of vulnerability. The total effort amounted to around 50 man-months with contributions from 20 engineers.

The main objective of the study was a thorough assessment of the severe CDF associated with internal events. An equally important objective was to provide a characterization of the dominant accident sequences that could leave the reactor core in a condition considered vulnerable to severe core damage. Another objective was to obtain insights to contributors to the public risk by grouping the dominant accident sequences with similar accident phenomenology, and construct and quantify containment event trees.

To analyse the potential release under a severe accident, an Individual Plant Evaluation on Source Term with Industry Degraded Core Rulemaking Programme methodology was conducted by Fauske Associates Inc. and ENEL using the MAAP3 code.

4.2. Methodology

The initial intent was an assessment of the Caorso CDF by performing comparisons with and modification of PRAs of similar plants. However, it rapidly devolved into a plant specific study. In fact, the unique configuration of the support systems as well as the Caorso operating experience were analysed in detail. The classical methodology of event tree, fault tree with SETS/SEP computer codes has been applied. There are, however, some features utilized in the present study which are worth mentioning. The human cognitive reliability correlation was used to assess the probability of human errors. The existence of this correlation enables the user to predict time dependent human response probability in terms of key engineering and operation parameters such as type of behaviour, available response time, and performance shaping factors like stress, experience and quality of man–machine interface. New Caorso symptom–based emergency procedures were the basis for the study.

Of significant assistance in performing the work was the use of the micro-computer workstation, that is, an integrated software package that allows the analyst to prepare, store, retrieve, evaluate and modify the fault and event trees [12].

4.3. Results and further developments

The estimated severe CDF is around 6.5 E-06/year and lies within the range of 'good' plants for contemporary BWRs for which PRAs have been performed.

This is primarily due to good design of the balance of plant and supporting systems (e.g. presence of a motor driven feedwater pump and four diesel generators) as well as the front line systems, to a reliable power grid and to the implementation of the new symptom oriented emergency procedures.

Another preliminary result has proved that the Caorso design has a substantial margin beyond the design basis and the MAAP analysis shows that this provides extensive time for recovery of the essential core cooling and containment functions.

Some design options are still under evaluation which could attain the goal of low environmental releases also in severe accident conditions.

5. CONCLUSIONS

The PSA tool should be a useful means to understand the level of safety of an NPP and will be the rationale for a systematic approach to safety. However, trying to directly transfer the numerical results of a probabilistic assessment from one plant to another could jeopardize the credibility of the tool. Every NPP has its own frequency of core damage and of serious releases that depend on the peculiar aspects of the station under examination, provided the completeness of the analysis and the methodology used are pursued on the same basic rules. On the other hand even two plants of the same design can obtain significantly different results because of their different operating histories , as they affect the generic failure data extracted from the databanks and the initiating events frequency.

All the work for the Italian plants is still in progress; apart from the developments mentioned previously, the Trino Enrico Fermi operating plant will also be evaluated. However, the experience gained from the studies described in this paper provides the following insights which differentiate between plants under construction or being designed and plants in operation.

For plants under construction or being designed:

- Early application of PSA techniques to identify weak points of the design allows for inexpensive system modifications (no dramatic changes are needed) that not only reduce the CDF, but can also be used to reduce the frequency of serious releases.
- A major side benefit of evaluating the design has been the integration of the designers and safety engineering teams. This aspect will be enhanced by introducing at least a reliability analysis on the designers' work.
- Application of PSA techniques to the total plant design allows the risk effectiveness determination to be considered as part of the final decision.

For plants in operation:

- The evaluation of the safety level through PSA allows identification by the utility of the weak aspects (operator, maintenance, tests, etc.), especially during operation of the decay heat removal systems.
- All process systems and not only the safety related equipment provide a capability to cope with severe accidents.
- Support system configuration as well as front line systems are important for the final results and justify the differences between plants.

For all the plants the possibility of enhancing the understanding of accident sequences, systems and components so that appropriate attention can be paid to these areas during design, construction and operation phases is consistent with maintaining the concept of defence in depth, since the PSA tool gives the overall view of the plant behaviour in accident conditions.

REFERENCES

[1] CAVICCHIA, V., PENNAZZI, M., TOCCAFONDI, G., VEZEY, E.E., "Reliability analysis of the Alto Lazio nuclear station safety systems using licensing criteria", Proc. Top. Mtg on Probabilistic Safety Methods and Applications, San Francisco, CA, 1985, EPRI, Palo Alto, CA (1985) 168.

[2] CAVICCHIA, V., PENNAZZI, M., TOCCAFONDI, G., "Probabilistic analysis of exceeding licensing of the Alto Lazio nuclear power plant: an alternative to PRA", Proc. Reliability '85 Conf. Birmingham, England, 1985.

[3] CAVICCHIA, V., PENNAZZI, M., TOCCAFONDI, G., "Reliability analysis methodology in Alto Lazio nuclear power plant", Proc. 3rd International Conf. on Nuclear Technology Transfer, Madrid, 1985.

[4] CAVICCHIA, V., PENNAZZI, M., VALLERO, P., "Control room ventilation design improvement following Alto Lazio station reliability analysis", Proc. 5th Eurodata Conf. Heidelberg, 1986.

[5] CAVICCHIA, V., NOBILE, M., SERRA, S., "Overview of the probabilistic safety studies for nuclear power plants in Italy. Points of view and prospects for the utilization of the probabilistic methodology", Workshop on Probabilistic Safety Assessment as an Aid to Nuclear Plant Management, Brighton, 1986.

[6] CAVICCHIA, V., TOCCAFONDI, G., "Comparison of two probabilistic safety studies, one using licensing criteria and the other realistic criteria", Proc. Int. Top. Conf. on Probabilistic Safety Assessment and Risk Management, Zurich, 1987.

[7] RICHARDSON, D.C., RUSSINO, G., VALENTINI, V., "Use of a probabilistic safety study in the design of the Italian reference PWR", Proc. Int. Top. Mtg on Probabilistic Safety Methods and Applications, San Francisco, CA, 1985, EPRI, Palo Alto, CA (1985) 123.

[8] UNITED STATES NUCLEAR REGULATORY COMMISSION, A Guide to the Performance of Probabilistic Risk Assessments, Rep. NUREG/CR-2300, Ch. 9, USNRC, Washington, DC (1983).

[9] SANCAKTAR, S., SHARP, D.R., SERRA, S., "Conversion of the PUN probabilistic safety study into a living PRA format", Proc. Int. Top. Conf. on Probabilistic Safety Assessment and Risk Management, Zurich, 1987.

[10] BARBUCCI, P., et al., ''Beyond design basis accidents for the Italian standard nuclear power plant: response studies and provisions for prevention and mitigation'', these Proceedings, Vol. 4.

[11] INGHAM, E.L., NOBILE, M., ''Application of PSA to the long-term safety review of Latina NPP'', Proc. Int. Top. Conf. on Probabilistic Safety Assessment and Risk Management, Zurich, 1987.

[12] CAVICCHIA, V., LUKIC, Y.D., ''Experience using a PC based workstation in a level 1 PRA effort'', ibid.

UNITED KINGDOM TRENDS IN SAFETY ASSESSMENT

P.D. JENKINS
Generation Development and Construction Division,
Central Electricity Generating Board,
Gloucester

J.R. HARRISON
Health and Safety Department,
Central Electricity Generating Board,
London

United Kingdom

Abstract

UNITED KINGDOM TRENDS IN SAFETY ASSESSMENT.

Some years ago the electricity utilities in the United Kingdom documented their safety criteria, and the guidelines to be used as targets for the design of future nuclear power stations. These criteria and guidelines were applied to the design of Heysham 2 and Torness, the UK's latest AGRs, due to raise power later this year, and they have also been applied to the proposed PWR at Sizewell 'B'. The criteria and guidelines require, inter alia, the use of PSA techniques in defining the design basis of the plant. The paper reviews the experience in applying the criteria and guidelines and discusses the areas where greatest difficulty was experienced in achieving compliance with the targets. The paper also discusses current developments in assessment techniques insofar as they are or will be applied in the UK. Such developments include common mode failure modelling, treatment of human factors and hazards. Finally the paper discusses possible developments to the criteria and guidelines, and the role of risk assessments in future safety cases.

1. INTRODUCTION

The extent and depth required of safety assessments for the UK's nuclear power programme have been steadily increasing over the past few years. There are various reasons. Some are industry inspired: the gradual accumulation of experience and development of safety thinking have inevitably led to changes. The UK's licensing authority, the Nuclear Installations Inspectorate, NII, has had a major influence on developing safety standards and levels of assesssment which have to be provided. The UK's utilities which operate nuclear stations (i.e. the Central Electricity Generating Board, CEGB, in England and Wales, and the South of Scotland Electricity Board,

SSEB, in Scotland) have to satisfy the NII before a license is issued. However, a number of developments have also been initiated by the CEGB and SSEB since, in the UK, the responsibility for ensuring safety also rests with the utilities.

These industry inspired developments stemmed partly from the decision in the early 1970s to lay down detailed design safety criteria and guidelines which inter alia embody numerical reliability targets [1]. These criteria and guidelines were applied to the design of Heysham 2 and Torness, the UK's latest AGRs, due to raise power later this year. They have also been used to develop the design and safety case for the PWR recently authorised at Sizewell B. The paper reviews the experience in applying the criteria and guidelines and discusses the areas where greatest difficulty was experienced in assessing compliance with the numerical targets.

However, there are other influences on safety assessment and future requirements which have come from outside the industry. The main influence has been the Sizewell 'B' Public Inquiry and its effects will be long felt. One outcome will be the need to prepare assessments which are more readily comprehended by the general public. The Chernobyl disaster has, of course, underlined this need, if the public is ever to balance correctly the risks from nuclear power against its benefits.

This paper finally summarises current developments in assessment techniques and gives some thoughts on the role of risk assessments in future safety cases.

2. Influences on UK Safety Assessments

2.1 UK's Design Safety Guidelines and Criteria

2.1.1 Background

As explained in ref [1] the CEGB decided in the early 1970s to set down safety requirements and criteria which future nuclear plants would have to meet in the UK. There had been a history of major delays to the completion of the early AGRs due to judgements on the acceptable level of safety hardening during the construction programme. The CEGB therefore decided that, before any further project could be authorised, it would first be necessary to agree the safety requirements in considerable detail with both the designers and with the licensing authority. In this way it was hoped that delays in future projects would be avoided.

In order to implement this objective, it was considered appropriate to collate all such requirements in stand alone documents. It was also considered opportune to review the existing CEGB requirements against those in other countries and to incorporate other concepts and standards where it was judged safety could reasonably be enhanced.

Safety requirements and criteria documents were first set down in 1974 for the proposed Steam Generating Heavy Water Reactor, SGHWR. When this project was abandoned the requirements and criteria were simply adopted for the next project which was the AGR at Heysham 2, now about to raise power.

The original intention in avoiding construction delays due to changes in safety requirements has been met. There have been no such delays.

These requirements were also subsequently adopted for the PWR at Sizewell 'B' and the design was developed to meet the requirements. These requirements are now embodied in two documents. The Design Safety Criteria [2] highlight the fundamental criteria which are then developed according to reactor type into Design Safety Guidelines, such as those for the PWR [3].

2.1.2 Summary of Requirements

The Criteria and Guidelines promote a disciplined, structured approach to safety. They include where appropriate previously developed safety practices and various deterministic requirements. However, one development of major importance was the use of numerical targets for determining the redundancy and reliability required of the reactor safeguards systems[1]. The fundamental criterion is that the overall frequency of an uncontrolled release of radioactivity must be less than 10^{-6} p.a.

This requirement was set recognising inter alia the need to limit any additional risk to a member of the public to a level low compared with everyday risks. The Guidelines use this fundamental criterion as the basis for deriving the reliability targets for the safeguards systems.

Numerical targets were introduced into the safety requirements because experience on the early AGRs had shown the

[1] In this context, the term "safeguards" system is used to denote any system required following a fault to ensure satisfactory shutdown and cooling.

value of probabilistic methods. For example, the greatest overall risk did not necessarily emanate from low probability, extreme accidents which apparently had the greatest potential for a release. More frequent accidents could dominate the risk unless suitably redundant and diverse safeguards were provided. It was in fact found necessary to consider the whole spectrum of possible accidents and the various sequences that could follow.

In deriving the reliability targets, however, a conservative approach was advocated for the early design stage. The designers were asked to identify conservative fuel and plant limits below which the fuel and plant behaviour can be stated with confidence. This defines a design basis boundary and the probability of exceeding the design basis also should be less than about 10^{-6} p.a, although exceedance can be tolerated provided the fundamental criterion for an uncontrolled release is still satisfied. In effect, the result is to ask the designers also to avoid having a core melt frequency appreciably greater than 10^{-6} p.a.

As well as numerical targets for defining permissible system failure rates, numerical targets also have to be met before system success can be claimed. For example, the total permissible frequency for sequences with releases giving rise to doses in the range 0.1 to 1 ERL[2] must be less than 10^{-4} p.a.

Other safety requirements include the need to consider all possible internal hazards (fire, disruptive failure of pressure parts or of rotating machinery etc) and external hazards (seismic events, flooding etc).

As well as the numerical targets, emphasis was also given to the need for good engineering, the use of a conservative design and the use of well proven components for which well validated data existed. The importance of Quality Assurance was highlighted including the need for independent design verification to be a part of the design process.

2.1.3 Application at Heysham 2/Torness

The Heysham 2/Torness design is based upon the earlier successful design at Hinkley Point B and Hunterston 'B'. However, application of the design safety requirements led to a redesign of the reactor safeguards systems with more formally claimed diversity, and greater redundancy and improved

[2] **The ERL level refers to the lower level (0.1 Sv) of dose specified in the UK at which evacuation should be considered.**

segregation against fire and other hazards. Inevitably, more supporting documentation had to be provided because a more comprehensive approach was called for than had been common in the past, and judgements had to be supported by evidence. In particular, documentary evidence for the validation of all computer codes used had to be provided.

The only real problem area lay in demonstrating that the 10^{-6} target limit for exceeding the design basis had been adequately met. To some extent this was due to constraints imposed by the safety guidelines themselves, for example on the maximum reliability that can be claimed from a non-diverse system. Such constraints were originally imposed to make the attainment of the targets more secure. The final assessment has produced a value around 5×10^{-6} p.a., the shortfall being deemed acceptable given the fact that a margin exists between exceedance of the design basis and the uncontrolled release of radioactivity. Areas where greatest difficulty was experienced are discussed in Section 3.

Although the numerical targets are not acceptance criteria as such, the numerical assessments performed are supportive of the safety case. As a consequence, there is a need to consider the extent to which the targets are met during operation, since there is a requirement, of course, to operate the station in conformity with its safety case. There is thus a need to ensure that, for example during operation, suitable constraints are applied to the permitted outages of essential equipment. In the past such constraints were specified on a deterministic basis, but reliability considerations potentially impose more onerous standards. In order to ensure that such constraints are not specified in a unnecessarily restrictive manner the CEGB has developed an interactive computerised device for assisting in maintenance planning [4]. Because of the complexity of the safeguards systems, and the many interactions, there could otherwise have to be a large number of operating instructions for operation under differing maintenance regimes. The device incorporates a fault tree model of the safeguards systems and can be continually updated to take account of plant outages and plant changes.

2.1.4 Application at Sizewell 'B'

The Sizewell 'B' PWR was also developed to meet the safety requirements as now embodied in the Design Safety Criteria and Design Safety Guidelines. The reactor is a standard Westinghouse 4 loop model with the balance of plant based upon the SNUPPS layout.

It was recognised that there would have to be a number of modifications to the SNUPPS design to accommodate UK practices and Sizewell site-specific requirements. The major influence, however, was the need to incorporate the reliability requirements of the guidelines. Again, there is a shortfall in meeting the 10^{-6} guideline target limit for exceeding the design basis, but the containment provides substantial additional benefit in limiting any radioactive release to the environment following core melt. The areas which caused difficulty in demonstrating compliance with the guideline target were those encountered at Heysham 2, and are those summarised in Section 3.

2.2. Impact of Sizewell 'B' Public Inquiry

The Sizewell 'B' Public Inquiry will have a major long term effect on safety assessment attitudes in the UK. This Inquiry included the first public examination in the UK of the safety of nuclear power stations. It was a very extensive inquiry lasting from 11 January 1983 to 7 March 1985. It was set up by the Government with very wideranging terms of reference including need, the economic case, site and local environmental consequences as well as safety.

The Board's evidence was voluminous and was considered by the Inspector to be "fully adequate to allow a thorough and searching examination of its safety case". There was in addition a vast amount of evidence from local interested parties and objectors as well as dispassionate bodies such as the NII. The range of examination was also wide, although inevitably the examination had to be selective.

The Inspector reported in January this year [5]. After Parliamentary debate and due consideration, the Secretary of State for Energy accepted the Inspector's recommendation that he gave his consent and deemed planning permision.

Although the Inspector's report contained a comprehensive review of both the CEGB's and the objectors' arguments, the Inspector did not find it necessary to add many caveats in his final recommendations relating to safety. He did, however, indicate a number of areas where, in his opinion, further action would be desirable. Such areas are reviewed below insofar as they have an impact on assessment requirements.

2.2.1 Risk Criteria

The Inspector noted that probabilistic analyses of risk only played a supporting role in the CEGB's safety case for licensing, and that the primary assurance of safety was to be sought in well-established engineering criteria and use of deterministic methods. The Inspector agreed with that opinion and also noted that accident probability analysis is a relatively recent development, and that its results are inevitably subject to substantial uncertainty.

The case, as set out in the Pre-Construction Safety Report, also concentrated on the probability and consequences of design basis accidents. However, the report also included an assessment of the probability of beyond design basis accidents with the potential to lead to an uncontrolled release of radioactivity. That is, a level I-type PRA analysis was performed.

As a separate exercise, an analysis of degraded core accidents and their consequences including health risks was mainly commissioned to test for "cliff-edge" effects[3] as the design basis was exceeded. However, the risk analysis was also of value in putting the risks from Sizewell 'B' into context with other risks, and to demonstrate the adequacy of the Board's Design Safety Guidelines and Criteria. It approximated to a full-scale level III PRA. The risk analysis for such beyond design basis events attracted much debate at the Public Inquiry.

In his report the Inspector noted that risk evaluation is highly subjective and that the public's views must be reflected in estimating a tolerable level of risk, in the full knowledge of the nature and size of the risks and benefits. The Inspector noted that it is important to consider social as well as individual risks. He did not consider that the CEGB had adequately justified the imposition of an additional risk of one in a million and thought that the CEGB had paid too little regard to social risk in setting its design targets. However, he also observed that there was no authoritative guidance from the UK Government or Parliament on what such a level of tolerable risk from a new nuclear power station might be, nor on how it should be determined. Nevertheless, he accepted the CEGB's uncontrolled release target as being "reasonable".

Notwithstanding the fact that he accepted that too much weight should not be placed on the results of probabilistic risk analyses, he recommended inter alia that

"The Health and Safety Executive (HSE) should formulate and publish guidance on the tolerable levels of individual and

[3] A ''cliff-edge'' effect refers to a sudden worsening of consequences or behaviour in response to a small change in initiating conditions.

social risk to workers and the public from nuclear power stations. Guidance should also be given on the application of the ALARP principle, and in particular on the relative priority of ALARP and fixed levels of tolerable and intolerable risk".

and "The HSE should, with other relevant organisations and individuals, develop means of applying cost-benefit analysis to criteria for those nuclear safety measures intended to reduce the probability of accidents or mitigate their consequences or both".

and "As a first step in formulating guidance on tolerable levels of risk and ALARP, the HSE should publish a consultative document to enable public, expert and Parliamentary opinions to be expressed; the preparation and publication of this document should receive high priority".

This supports the CEGB's view that, in future, probabilistic risk assessments will play an increasing role in supporting safety arguments. It is the authors' view, however, that it will be some time before cost benefit analysis is used in setting safety criteria.

2.2.2 Need for independent reviews

The need for independent assessments, of those parts of the design or argument crucial to safety, has long been recognised by the CEGB. Indeed, this is one reason why the Board has always had its own Health and Safety Department, independent of its construction and operating arms, to approve safety submissions and to ensure that adequate independent analyses have been performed. The approach is adopted in a very systematic manner for Sizewell 'B' where, for example, totally independent assessments are commissioned for the reactor protection system and the fault transient analysis. Even so, the Inspector thought it worth noting that the CEGB should consider making even wider use of peer review for major aspects of its safety analysis.

2.2.3 Public presentation of information

A very important area is the need for public acceptance of safety is recognised. It is not sufficient to be safe; the public also needs to be assured that this is the case. In this context the Inspector noted that the CEGB's evidence on the consequences of accidents was presented in a form which did not permit easy understanding. He noted that if successor PWRs to Sizewell 'B' were built, they would be more likely to win public acceptance if information on consequences were presented in a more readily comprehensible form. He also considered that reports should be published later in the licensing process which are suitable for lay public and Parliamentary scrutiny. Such documents should not be over-simplified, but should clearly explain the principal safety aspects thoughfully and clearly.

3. Areas of difficulty in assessing compliance with numerical targets.

3.1 Completeness

The main problem in demonstrating compliance with the numerical targets was that of "completeness". It is not possible, of course, to prove in any mathematical sense that there is no significant accident sequence which has been overlooked. This is one reason why the CEGB attaches great importance to the need for independent assessments and the continual need to appraise and reappraise the safety of its reactors. It is also why the CEGB ensures there is adequate feedback from operational information both within and outside the UK.

3.2 Human Factors

In its reliability assessments to date, the CEGB has chosen not to explicitly model errors of omission or commission, although the contribution of such errors to plant unreliability data was implicitly included. The CEGB's reasoning is that it is more valuable to use probabilistic assessments to identify the reliability with which the operator must act if human errors are not to dominate, and then to provide sufficient control and instrumentation to minimise the likelihood of errors. In his report on the Sizewell 'B' Inquiry, the Inspector concluded that the effect of human error on safety is potentially large but also noted that there are no reliable methods yet for quantifying human error.

3.3 Hazards

During the application of the guidelines to the design process, the CEGB's approach is initially to exclude the contribution from internal and external hazards. This is because it is considered that the redundancy and diversity to be provided is best determined from consideration of plant generated faults. Consideration of hazards then influences the qualification required of equipment (e.g. due to seismic considerations) or to the extent of barriers and segregation required (e.g. due to fire considerations). There is also the problem that certain hazards have a large associated numerical uncertainty. This is an area where the CEGB's Criteria and

Guidelines allow a deterministic approach. The Guidelines give the designer a level of hazard against which to design, using conservative design methods, and set a reliability target for safe shutdown. The judgement is that by designing to such standards a comparable level of safety is achieved as that for plant faults. However, some probabilistic assessments have been carried out to confirm the acceptability of the design approach. A commitment has also been made to the NII that hazards will be included in the PRAs when the methodology has reached an appropriate stage of development.

3.4 Common mode/common cause failures

The treatment of common mode/common cause failures is an area where difficulties are encountered since there is no universally agreed way of modelling such effects. By virtue of the guidelines requirements for diversity, however, the overall design of Heysham 2 and Sizewell 'B' are not overly sensitive to uncertainties in absolute values.

4. Developments in Assessment techniques

The areas identified in Sections 2 and 3 above relate to the need to improve techniques for carrying out probabilistic assessments, and the need for sound independent assessments. Examples of the approach being followed in the UK are given below:

4.1 Probabilistic assessments

The CEGB is developing an independent understanding of additional factors which may contribute to the assessment of overall risk. These factors include common cause failures, human factors and hazards.

Participation in the data benchmark tests on the EPRI classification procedure for common cause failures provided experience in root cause analysis of data. Collaboration with the Safety Reliability Directorate (SRD) of the UK Atomic Energy Authority in the European Reliability Benchwork Exercise provided a practical comparison of the various models for the analysis of Common Cause Failures (CCFs). A two year study programme on data and methods for CCF analysis with SRD is well advanced. This programme includes a data collection exercise specifically designed for the analysis of actual or potential CCF's. The programme also includes the development of a method of analysis of CCF's. This method allows benefit to be claimed for plant features which have been designed to minimise CCF's. CEGB has also carried out in-house research into a method of analysis which treats dependent and independent failure data in an integrated manner. A programme of applying this method to practical systems is currently being carried out.

For Human Factors (HF) CEGB has been developing an in-house understanding in two broad areas. These are in modelling and data. A theoretical model of the impact of HF's has been applied to a practical system for CEGB by the National Nuclear Corporation, NNC. This used the THERP modelling technique and gave a detailed understanding of the factors and assumptions required. CEGB is also monitoring the SRD contribution to the European Reliability Benchmark Exercise on HF, which will benchmark a number of different methods. For data, CEGB has produced guidelines for reporting data on human errors and these are being implemented into an existing CEGB event reporting scheme. Existing event report data are being assessed for data which may be suitable for HF error assessments. Theoretical aspects of human factors and man-machine interface are currently being investigated by ergonomics specialists in the CEGB. CEGB is a UK sponsor of experimental work at the Halden Project.

For the treatment of hazards effort has been concentrated on methods. A theoretical method has been developed and applied to a practical system for CEGB by NNC. More advanced methods than the current deterministic methods used by CEGB are being assessed. These more complex methods specify hazard distribution functions and obtain component fragilities and vulnerabilities. Difficulties in these methods are in the specification of input data.

4.2 Independent assessments

Specific independent assessments are confined to those areas crucial to the minimisation of a radioactive release, although the Board's Health and Safety Department ensures that the overall case has also been adequately assessed. One such area is the need to ensure that the functional capability of systems is correctly assessed. For Sizewell 'B', therefore, the CEGB has decided to assess the design basis transient analysis calculations in the formal safety submissions by performing additional calculations with alternative codes. The computer codes to be used have either been developed in the UK or obtained from the United States under licensing or information exchange agreements. To give confidence that these codes are capable of describing reactor behaviour in accident conditions with acceptable accuracy, they are compared with relevant experimental data from test facilities. The CEGB and the UK Atomic Energy Authority are participants in the code development and validation process such as the International Code Assessment Program sponsored by USNRC.

Another example of an independent assessment for Sizewell B is that of the ability of the Primary Protection System to meet its functional and integrity requirements. The system proposed for Sizewell 'B' makes use of programmable electronic devices. This is the first time a reactor protection system has used such devices in the UK. This has faced the assessors in the UK with a number of novel problems. The main one is the lack of any generally agreed methodology for assessing the integrity of programmable electronic systems against numerical target. Such assessments are not considered to be mathematically meaningful and so alternative assessment techniques including those for appraising the software integrity are being developed.

5. Possible Future Developments

It is expected that the move towards greater use of probabilistic assessments will probably continue in the UK. Such techniques are seen as valuable in supporting safety cases. However, such techinques cannot be regarded, at least for the foreseeable future, as the sole basis for constructing a safety case. There will remain many areas where numerical quantification is not possible in any rigorous mathematical sense. Thus, whilst attention to "bottom-line" risk numbers may well increase in future, it is important that such assessments are kept in context and are not regarded as the sole means of assessing the adequacy of a safety case.

It is still considered, therefore, that the CEGB's approach as set down in their criteria and guidelines remain acceptable, even after the experiences in designing to meet the targets and after the rigorous examinations at the Public Inquiry, although aspects are under review.

ACKNOWLEDGEMENT

This paper is published by permission of the Central Electricity Generating Board.

REFERENCES

[1] JENKINS, P.D., ''The use of probabilistic safety assessments methods for CEGB reactors'', Implications and Probabilistic Risk Assessment (Proc. Sem. Blackpool, England 1985), Elsevier Applied Science, London and New York (1985) 523-539.

[2] CENTRAL ELECTRICITY GENERATING BOARD, Design Safety Criteria for CEGB Nuclear Power Stations, Rev. Rep. HS/R-167/81, CEGB, Health and Safety Dept, London (1982).

[3] CENTRAL ELECTRICITY GENERATING BOARD, Pressurised Water Reactor Design Safety Guidelines, Rep. DSG-2, Issue A, CEGB, Generation Dev. and Construction Div., London (1982).

[4] HORNE, B.E., ''The introduction of probabilistic evaluations into the operation of a CEGB nuclear power station using the 'ESSM' facility''(Proc. Conf. PSA '87, Zurich, 1987), TÜV Rheinland, Cologne (1987).

[5] LAYFIELD, F., Sizewell B Public Inquiry Report, HM Stationery Office, London (1987).

PSA MODELS FOR SYSTEM AND HUMAN RELIABILITY: RESULTS OF BENCHMARK EXERCISES AND NEEDS

A. AMENDOLA, U. BERSINI, P.C. CACCIABUE,
G. MANCINI, A. POUCET
Systems Engineering and Reliability Division,
Joint Research Centre, Ispra Establishment,
Commission of the European Communities,
Ispra

Abstract

PSA MODELS FOR SYSTEM AND HUMAN RELIABILITY: RESULTS OF BENCHMARK EXERCISES AND NEEDS.

The results of international comparative exercises (benchmark exercises) are described. These exercises have been organized by the Commission of the European Communities-Joint Research Centre (CEC-JRC), Ispra, on selected probabilistic safety assessment (PSA) topics such as the analysis of system reliability, common cause failures and human reliability. The project involves major organizations dealing with nuclear power reactor development and operation, and is aimed at evaluating methods and uncertainties linked with PSA techniques and eventually achieving a consensus on procedures to be adopted, as a result of a common learning process. The results offer a quite comprehensive state of the art review of existing data, models and procedures. They also permit R & D efforts to be concentrated on the topics for which need for priority has been identified. In particular, the analyst should be able to deal more appropriately with the dynamic aspects of incident development, especially when studying man-machine interaction problems. Researches in progress at JRC-Ispra on this topic are described.

1. INTRODUCTION

Since 1982 the Commission of the European Communities-Joint Research Centre (CEC-JRC) has been co-ordinating a rather large project [1] aimed at:

– Assessing state of the art models, data and procedures for probabilistic safety assessment (PSA) of NPPs;
– Identifying the magnitude and nature of the associated uncertainties; and
– Eventually achieving a consensus on procedures to be adopted.

The project is based on a set of benchmark exercises (i.e. comparative studies independently performed by different teams on a same reference subject) [2–7]; in this way consensus procedures are achieved through the collective learning process, which is the natural consequence of work on the common study case, and subsequent comparison and discussion of the results obtained.

TABLE I. OUTLINE OF THE PROJECT AND PARTICIPATING TEAMS

Exercise	Systems reliability	Common cause failure	Human reliability
Timing	1982-1983	1984-1986	1986-1987
Reference systems	AFWS (EDF)	Auxiliary and emergency feed pumps (KWU)	– Test procedure – SG level control in case of multiple failures
Incident scenario	Main feed line break	Emergency power mode	Emergency power mode
EUROPEAN ECONOMIC COMMUNITY TEAMS:			
Belgium	KUL	Vinçotte-Tractebel-KUL-ULB	Vinçotte-Tractebel-ULB
Denmark	RISØ	RISØ	RISØ
Spain	–	–	CIEMAT
France	EDF; Framatome-Novatome	EDF-CEA-Framatome	EDF; CEA-Framatome
Germany, Fed. Rep. of	GRS; KWU-Interatom	GRS; KWU; KFA	KWU
Italy	ENEA-ENEL-Ansaldo	ENEA-ENEL-Ansaldo	ENEA; ENEL; Ansaldo
Netherlands	ECN; KEMA	–	ECN
United Kingdom	EWE (sponsored by NII)	UKAEA-CEGB	EWE (NII); UKAEA-CEGB
OTHER TEAMS:			
Sweden	RTI (sponsored by SKI)	RELCON (sponsored by SKI)	RELCON (sponsored by SKI)
Finland	–	–	Imatran-Voima
United States of America	–	PLG (sponsored by EPRI & NRC)	NUS (sponsored by EPRI & NRC)

The concrete significance of the project is grounded on the analysis of systems and/or incident scenarios of real NPPs, such as Elecricité de France-Paluel and Kraftwerk Union AG-Grohnde (see Table I), rather than on academic study cases, and on the involvement of a wide participation of experts belonging to all parties interested in construction, licensing and operation of NPPs, i.e. vendors, authorities, utilities, research establishments, certification and consultancy organizations.

As Table I shows, the participation has been extended to cover an increased number of teams and countries, even outside the European Economic Community (EEC). Recently, a co-ordinated research project, 'Reference Studies on PSA Accident Sequence Modelling', has been proposed by the International Atomic Energy Agency; this shows the lessons learnt from the EEC project, as well as on a smaller scale, experiences gained in the nordic area [8] and extends the experiences worldwide to other reactor types.

In addition to the review of the work performed and the results obtained, this paper describes the needs for further R & D activities. In particular, the largest uncertainties appear to be those connected with the modelling phase. This suggests that procedures for structured qualitative analysis prior to probabilistic modelling should be adopted and that knowledge based tools for supporting the analyst in developing a systems model are worth developing.

Furthermore, dynamic models and techniques to deal more appropriately with stochastic aspects of the incident development should be encouraged, especially for the analysis of man–machine interaction. Therefore, research in progress at JRC along these directions is briefly summarized.

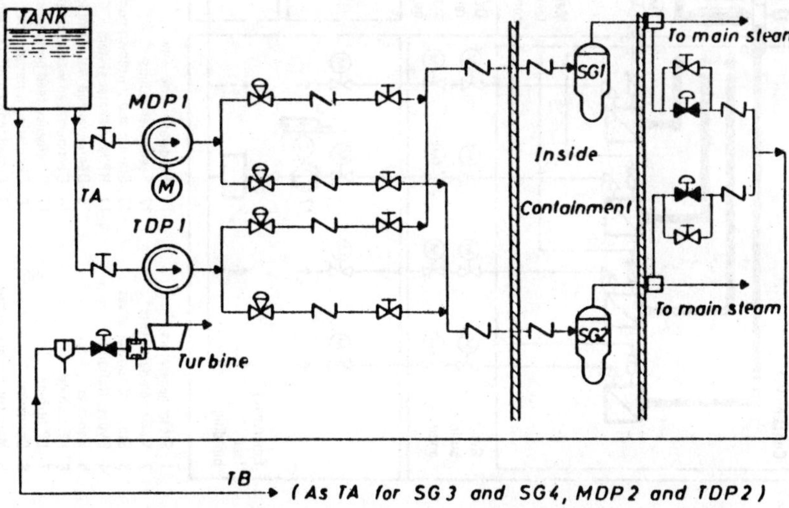

FIG. 1. Simplified scheme of the Electricité de France auxiliary feedwater reference system.

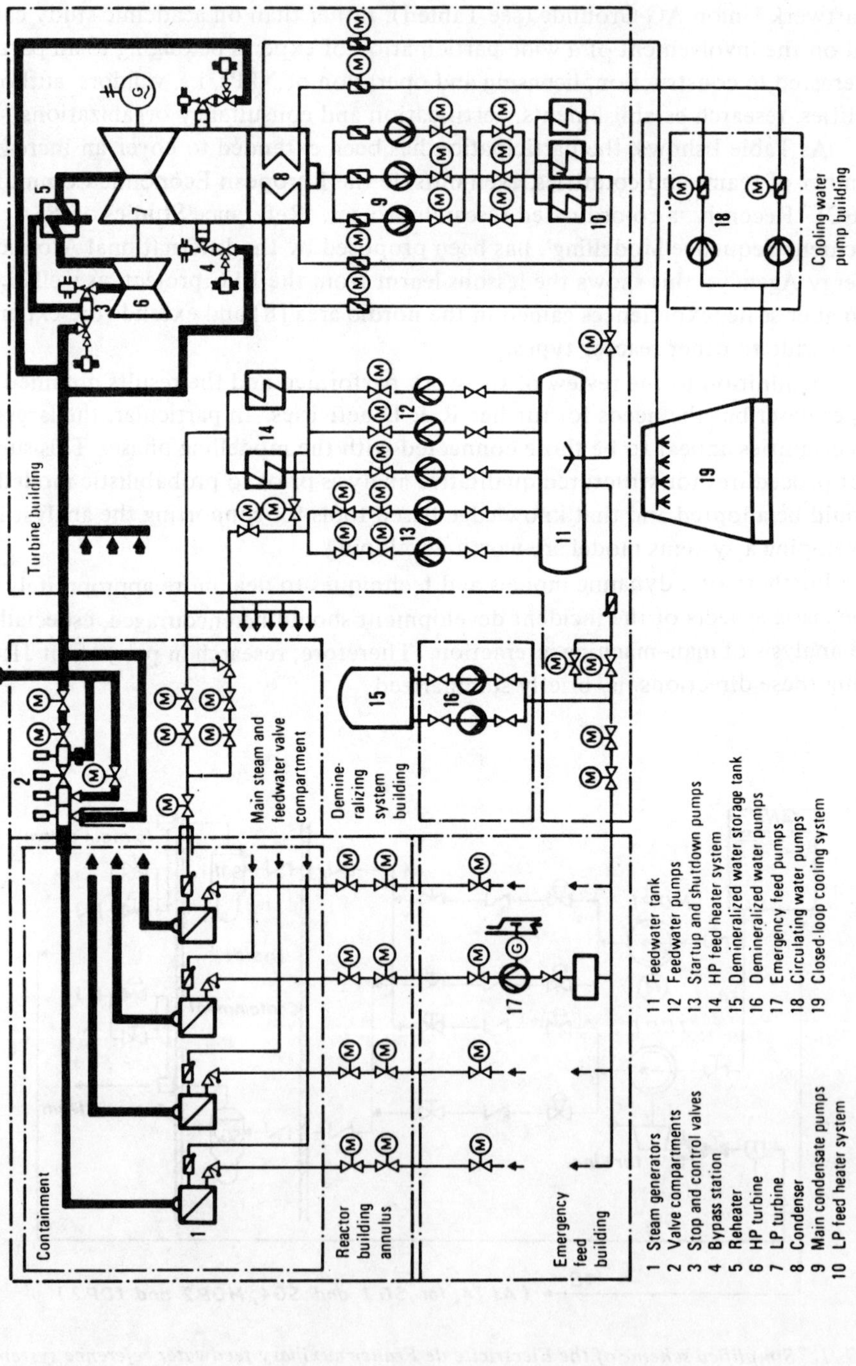

FIG. 2. Overview of Kraftwerk Union AG systems for secondary side heat removal.

2. CASES ANALYSED

The first study referred to the reliability assessment of the auxiliary feedwater system (AFWS) of the EDF 1300 MW(e) (Paluel unit), shown in Fig. 1. The system consists of two independent trains TA and TB. Each train allows auxiliay feedwater flow to feed two out of four steam generators by a motor-driven pump (MDP) and by a turbo-driven pump (TDP). Each driving turbine is fed by steam spilled from the two corresponding steam generators. Availability of only one steamline is sufficient to drive the turbine at the rated speed.

The relevant AFWS mission was the startup of the system on demand (because of a main feedline break in one out of the four steam generators) and the delivery of a prescribed flow for 7.5 h (i.e. the time required to put the reactor in a different state). Conservatively, success flow configuration was assumed to correspond either to the three safe steam generators, each fed at a rate of at least 50 t/h of water, or to only two steam generators fed at a rate of 100 t/h.

During this benchmark exercise it became clear that no consensus could be established on the analysis of dependent events and human reliability [2, 3]. As a consequence, a programme of further exercises was established [1] and namely on:

(a) Common cause failure analysis-reliability benchmark exercise (CCF-RBE);
(b) Human reliability (HF-RBE);
(c) Accident sequence modelling and assessment;
(d) Component data analysis.

The exercises (a) to (c) assume the Grohnde PWR designed by KWU and operated by KWG (principal owner: Preussen Elektra) as a reference plant. The three exercises were to be situated in a same overall accident scenario, namely the emergency power mode.

Two different and separate systems can deliver auxiliary feedwater in the case of a loss of preferred power. These systems are:

– A startup and shutdown system with two 100% redundant electrical motor driven pump trains (13 in Fig. 2); and
– An emergency feedwater system with four 100% redundant diesel pump trains (17 in Fig. 2).

The CCF- and HF-RBE had the objective of achieving a first consensus on these controversial subjects before approaching the final exercise of this first programme; that is the analysis of an accident sequence.

The CCF-RBE has already been completed and the general results made available to the technical community [5, 6], whereas the HF-RBE is at an advanced stage [7].

The concluding HF-RBE meeting is planned to be held by December 1987; at the same meeting the definition of the exercise on accident sequence will be finalized; a participation scheme similar to that of the HF-RBE is expected (Table I).

The benchmark on component data analysis is now starting, but in a different framework; it will indeed involve the members of EuReDatA (European Reliability Databank Association), an Association supported by the CEC.

The terms of reference have been drafted and reference raw event data will be supplied by the JRC-CEDB (Component Event Data Bank) [9].

The TOP event analysed in the CCF-RBE was the failure of supply of feed-water to steam generator in the emergency power mode. Availability of only one pump flow in one of the two systems mentioned above assures the success of the mission. Because of the high degree of redundancy built in, the system is very suitable for the study of the CCF problems. To avoid spread of modelling and data with respect to 'independent' events, the exercise assumed a common fault tree with common data; the subject of the study was how the unavailability figure of this fault tree is modified by dependent failures. Excluded from the CCF-RBE were external events or malfunctioning of external systems.

The HF-RBE has basically addressed the following issues [7]:

(1) The analysis of routine functional test and maintenance procedures; with the aim of assessing the probability of test induced failures, the probability of remaining undetected and the potential to initiate transients because of errors performed in the test; and

(2) The analysis of human actions during an operational transient, with the aim of assessing the probability that the operators will correctly diagnose the mal-functions and take proper corrective action.

Case (2) involves more challenging issues such as the analysis of human errors in diagnosis and in selection of response strategy, and a dynamic situation in which timing and time constraint play dominant roles.

Since the KWU plant is principally controlled by automated systems, it was necessary to frame the operator action in very degraded and improbable system conditions such as a CCF of the four isolation valves which should open on demand at the output of the emergency feed trains. The scenario has interesting features:

- A diagnosis of the incident taking place has to be made by the operators and a choice must be made between alternative strategies to prevent the steam generators from drying out;
- There is a time constraint aspect: the operators have about 1 hour to under-take and conclude the correct actions before the SG level drops below a dangerous level; and
- The actions to be undertaken involve different members from the shift staff at different physical locations, so that the aspects of communication need to be taken into account.

3. WORKING SCHEME

All exercises have been characterized by a similar working scheme, adequate for achieving the objective assumed. In a first documentation phase, system functional description, P and I diagrams, topology, procedures, etc. are distributed to the participants. Questions and needs for further documentation are discussed in a meeting following the visit of the plant under study; the visit allows the participants to familiarize themselves with the systems and to discuss with designers and plant operators. In the case of the HF-RBE, video tapes showing control room display and the sequence of operator actions for the case studies have been made available. Further time is left for clarification questions that are replied to by the plant experts and the answers are distributed to all teams, so that each team shares the same information.

Then a first working phase is used to achieve results not biased by any further communication among the teams. The results are summarized by the JRC and commented upon in a general meeting which is also used to define a second working phase that should ensure a convergence process, i.e. some common boundaries are assumed, selected topics are repeated in order to single out the most significant factors contributing to the overall spread, special exercises are designed to assess particular models and/or data.

A qualitative analysis is included into the terms of reference; the participants are asked to describe the assumptions assumed and the underlying reasons.

More details can be found in Refs [2–6].

4. SUMMARY OF THE RESULTS

The *systems reliability RBE* [2, 3] demonstrated how independent evaluations resulted in probability values being rather sparse; modelling uncertainties dominate data uncertainties. The uncertainty factors usually evaluated by propagating uncertainties on data throughout the fault trees predict probability ranges about one order of magnitude less than the range within which best estimates by different analysts fall. Use of the data available, difficulty in defining boundary conditions for a system when isolated from the rest of the plant, interpretation of the documentation and of the success conditions, completeness, assumptions performed by the analyst when information is lacking or for simplifying the model by introducing more or less conservative hypotheses, etc., are all factors influencing the modelling and the probability assessment.

As a consequence, better results (or better use of the results) can be achieved when:

– The decision process in which the analysis must be situated is clearly defined;

- Structured qualitative analysis procedures (such as FMEA) are used before any probabilistic modelling, especially if accurate results are needed such as for cost-benefit analyses;
- Computer-aided fault tree construction algorithms might be used to increase consistency and completeness of the models;
- Possibly fault tree techniques should be complemented by tools giving a better account of the process physics, especially when operator actions are to be taken into account.

The CCF-RBE [4, 5, 6] was most successful in understanding the different categories of dependent events and the different treatment they should deserve:

- Dependent events due to a clear deterministic effect even if triggered by a random event such as the unavailability of support functions, cascade failures, human errors, should, in principle, be modelled explicitly in the fault trees and event trees;
- Other potential multiple related failure events for which a numerous set of random causative factors (such as material defects, environment, etc.) can be imagined can be accounted for by parametric modelling; this is a cost-effective way of proceeding, with respect to the amount of data and the analysis effort that an explicit model of all possible factors could imply.

The CCF-RBE underlined the importance (already identified in the previous exercise and which will be again confirmed by the HF-RBE) of systematic and structured qualitative analysis for identifying potential mechanisms of dependence between failures, for screening and ranking potential dependent failure events and for matching the (possible) subsequent quantitative analysis with qualitative insights in system design and operation. A close link between qualitative analysis and quantification is absolutely necessary to obtain meaningful and consistent results. Proposed procedures in this direction can be found in Refs [3, 5].

In the first working phase, the quantitative methods used included parametric models such as the beta-factor model [10], the multiple greek letter (MGL) model [11], the Marshall-Olkin model [12], the binomial failure rate (BFR) model [13]; methods based more on judgement such as the cut-off technique and the partial beta-factor model [14, 15]; and the use of failure rate coupling as implicit treatment of dependence.

The numerical results obtained in Phase I show a significant spread as can be observed in Fig. 3. It was, however, difficult to obtain a clear picture of the causes of the variation between analyses from the Phase I results only.

The poor integration of the qualitative analysis with the modelling and quantitative analysis and the different degree of structure of the various qualitative analyses have, in our opinion, had a major effect on the large spread observed. This is confirmed by the fact that there was, in general, quite a good agreement between the qualitative conclusions reached by the different participants.

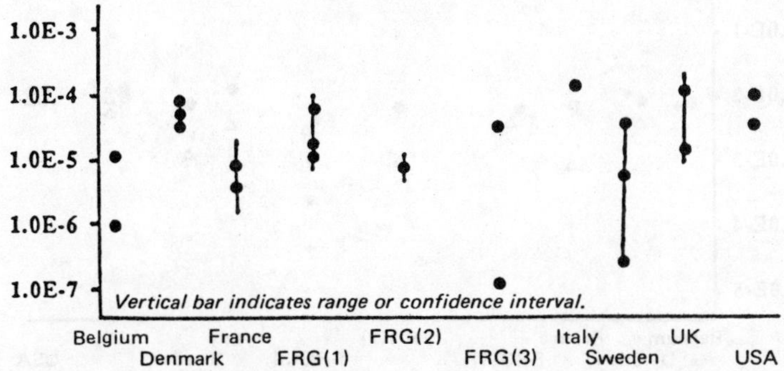

FIG. 3. Quantitative results obtained in phase I of common cause failure analysis-reliability benchmark exercise (CCF-RBE) for the emergency feed system only.

Working Phase II was organized to provide a deeper understanding of the sources and magnitude of uncertainties and spread in the quantitative results. The scope was limited to the four train emergency system only and, based on the qualitative analysis in Phase I, a common set of components to be included in the CCF analysis was agreed.

The following tasks were then performed:

- Repetition of the calculation according to the new scope but using the data assumed in Phase I (allowing to investigate the effect of the boundaries);
- Calculation of the unavailability using a common set of parameters estimated in a consistent way for different models (allowing the spread due to different models to be analysed);
- Assessment of CCF model parameters from a common set of 'raw' event data and calculation of the unavailability using this parameter (assessment of spread due to parameter estimation).

The calculation of the emergency system unavailability according to the restricted scope agreed for Phase II was first performed using the original data assumed in Phase I (see Fig. 4). A calculation was then performed using a common set of CCF parameters, assessed in a consistent way for both the MGL and the BFR models (see Fig. 4). The comparison of the results of these two calculations shows that, given a well defined scope and boundary for the system to be analysed, different analyses can achieve consistent results on the condition that the database is the same. Indeed, the results of the second calculation show a very good agreement even if different parametric models were applied (the MGL method seems to yield somewhat wider uncertainty bounds). Given this, it transpires that the spread observed in Fig. 4 for the first calculation is due to the spread in the original data (parameter values) used.

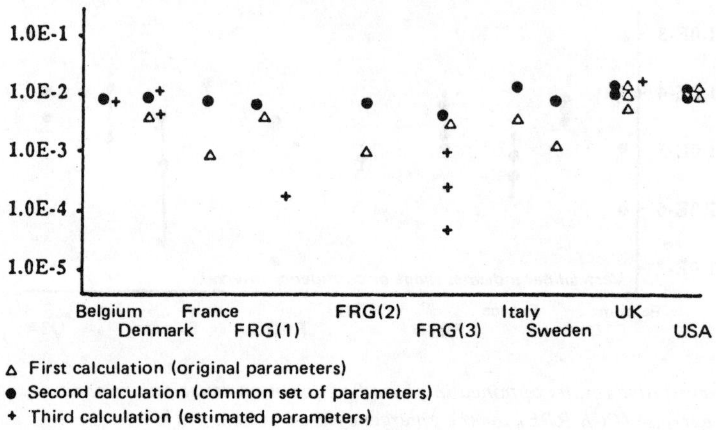

FIG. 4. Quantitative results obtained in phase II of CCF-RBE for the emergency feed system only (for a correct interpretation of the results see Ref. [5]).

Another example of the effect of the spread in the parameters is given by the results of the third calculation in Phase II (Fig. 4). For this calculation, a set of event reports was provided to the participants. This set was the same as the one used to assess the common parameters applied in the previous task. It can be observed that the parameter estimation is a key contributor to the variability observed in CCF quantification. The major source of the variability is the subjective judgement applied during the screening and assessment of the events, i.e. the decision whether some event is applicable to the plant under consideration or not, and to what extent redundancies could be affected. In order to reduce the spread introduced in parameter estimation, it was judged helpful to have some pre-established guidelines for performing the event screening process.

During the CCF-RBE, two important issues related to this transfer of experience from one plant to another were identified [16]:

(1) The extrapolation of experience in a system with a lower redundancy to a system with higher redundancy, e.g. how can an event concerning a CCF in a two-train system be used for assessing the probability of CCFs in a four-train system;

(2) The use of information on successes especially when extrapolating to lower redundancy systems; how to extrapolate the success information contained in an event regarding, for instance, a double failure in a four-train system to a two-train system.

Apart from the uncertainty arising from the subjective judgement used in processing the event data, other causes contributing to uncertainty of the data were also identified but were found to have lesser effect. These include the event versus component statistics issue (i.e. the artificial increase in the strength of the evidence if one uses the counting of component failures rather than the counting of CCF events) [17], the uncertainty in the number of independent events because of underreporting and the choice of prior distribution of a Bayesian estimation of the parameters is performed.

There was an agreement that the basic parameter model has, from a theoretical point of view, some advantages related to these issues.

The *HF-RBE* has not yet been completed and it would be premature to draw definite conclusions. Ref. [7] gives a rather complete survey of the methods used by the participants for assigning success/failure probabilities to the tasks the operators must accomplish. They include the systematic human action reliability procedure (SHARP) [18] frame, THERP [19], the success likelihood index methodology (SLIM) [20], the human cognitive reliability model (HCR) [21], the human error assessment and reduction technique (HEART) [22], the absolute probability judgement (APJ) [23] and the maintenance personnel performance simulation (MAPPS) model [24, 25]. The exercise will, therefore, result in a very comprehensive 'state of the art' review of all available models.

For the time being it can be anticipated that the factor which has the largest effect on the spread of the numerical results is the decomposition model of the overall task in a series of subtasks, with possible recoveries. Therefore, even by using the same database [19] the overall results might differ by several orders of magnitude, as found when analysing the test procedure study case. It appears that more repeatable results can be achieved when not attempting to breakdown a task performance procedure into too many elementary events, but when focusing on some principal aspect such as fault detection, diagnosis, action required.

Indeed, results obtained for the study case of operator intervention for SG level control appear to be much less sparse than the previous ones; this is due only partly to the learning process operating during the exercise, but principally because attention in the analyses was more concentrated on the judgement of the diagnosis/action process without too fine a breaking down of the procedure.

A further problem that deserves important consideration is the time dimension; models such as HCR give adequate credit to the possible dependence of success probability on time available for diagnosis and actions; such times can better be identified through stochastic dynamic simulation instead of using static techniques such as fault trees/event trees, incorporating cognitive models [26].

5. TOWARDS COGNITIVE MODELLING OF OPERATOR BEHAVIOUR

The difficulties already discussed in breaking down operator behaviour in a sequence of different tasks can be better understood when seen from a psychological viewpoint: the sequence of actions is the result of a previously developed planning process which, in turn, is the consequence of a diagnosis of the situation. This implies that the operator's primary intentions and different internal representations of the system, in terms of structural and functional schemes, have to be taken into consideration in order to evaluate the overall process of decision-action performed by an operator of any plant or system.

The analysis of operator errors during a man–machine interaction event must be concerned with the existence of these schemes, accounting not only for the executional aspect of errors (slips) but also and substantially for their activation (mistakes) at cognitive level. The tendency should be to evaluate cognitive processes which may introduce a 'dependence structure' among the single actions as well as quantifying the external manifestations, or slips, made in performing the intended plan. Consequently, human reliability analysis in man–machine systems can be improved by complementing classical reliability models with such detailed cognitive models, which have to precede the final construction and the numerical development of the tree; thus dependences and recoveries will be taken into account.

In the model, the main tendency will be towards an integrated simulation allowing to tackle all the activities of the operator in the same framework [27].

Given that intention is a fundamental aspect of cognitive processes, the formulation of sequences of intentions or goals and the ordering of such goals in a hierarchical goal-oriented structure is proposed as the general framework within which the strategies of operators are developed and carried over [28, 29]. Moreover, the dynamic allocation of goals in the working memory of the operator and the simulation of the attainment of a goal during the flow of events is based on fuzzy logic, which is a well suited theory for representing the operator's approximate knowledge and allows a semantic interface to be created between the system simulation and the operator cognitive model.

The hierarchical goal-oriented structure and the fuzzy logic are the foundations on which the proposed model is based [30, 31, 32].

6. CONCLUDING REMARKS

The paper discussed results obtained and lessons learnt by the JRC PSA benchmark exercises in terms of the 'state of the art' and needs; in particular, it shows how the project has succeeded in the establishment of a common attitude of the analysts towards a controversial topic such as CCFs.

As far as human reliability is concerned, since the exercise is still under way, the conclusions presented can only be of a preliminary nature. However, these encourage the development of models and techniques to deal more appropriately with the dynamic aspects of the incident development, according to research directions which have already been undertaken by the JRC and which have been briefly described in the paper.

REFERENCES

[1] AMENDOLA, A., "Results of the reliability benchmark exercise and the future CEC-JRC programme", ANS/ENS Int. Top. Mtg. on Probabilistic Safety Methods and Applications, San Francisco, CA, 1985.

[2] AMENDOLA, A., Uncertainties in systems reliability modelling: Insight gained through European Benchmark Exercises, Nucl. Eng. Des. **93** (1986) 215–225.

[3] AMENDOLA, A., Systems Reliability Benchmark Exercise, Final Rep. EUR-10696, CEC-JRC, Ispra (1985).

[4] AMENDOLA, A., "Common cause failure analysis: Insight gained from reliability benchmark exercises", Reliability Modelling and Applications, (COLOMBO, A.G., KELLER, A.Z., Eds), Reidel, Dordrecht (1986).

[5] POUCET, A., AMENDOLA, A., CACCIABUE, P.C., Common Cause Failure Reliability Benchmark Exercise, Final Rep. EUR-11054 EN, CEC-JRC, Ispra (1987).

[6] POUCET, A., AMENDOLA, A., CACCIABUE, P.C., "European benchmark exercise on common cause failure analysis", ANS/ENS Int. Top. Conf. on Probabilistic Safety Assessment and Risk Management, Zurich, 1987.

[7] POUCET, A., "Survey of methods used to assess human reliability in the Human Factors Reliability Benchmark Exercise", Int. Post SMIRT-9 Sem. on Accident Sequence Modelling: Human Actions, System Response, Intelligent Decision Support, Munich, 1987.

[8] MANKAMO, T., PETERSEN, K.E., PÖRN, K., ERICSSON, G., "Experiences from the Nordic Benchmark Analyses", ANS/ENS Int. Top. Mtg on Probabilistic Safety Methods and Applications, San Francisco, CA, 1985.

[9] BALESTRERI, S., "The component event databank and its uses", Reliability Data Bases, (AMENDOLA, A., KELLER, A.Z., Eds), Reidel, Dordrecht, Netherlands (1987).

[10] FLEMING, K.N., "A reliability model for common mode failures in redundant safety systems", 6th Ann. Conf. on Modelling and Simulation, Pittsburgh, PA, 1975.

[11] FLEMING, K., MOSLEH, A., DEREMER, R.K., A systematic procedure for the incorporation of common cause events into risk and reliability models, Nucl. Eng. Des. **93** (1986); presented at the Int. Post-Conf. Sem. of SMIRT-8, Brussels, 1985.

[12] MARSHALL, A.W., OLKIN, I., A multivariate exponential distribution, J. Am. Stat. Assoc. **62** (1967).

[13] ATWOOD, C.L., Estimators for the Binomial Failure Rate Common Cause Model, Rep. NUREG/CR-1401, USNRC, Washington, DC (1980).

[14] WATSON, J.A., Analysis of dependent events and multiple unavailabilities, SRS Q. Dig. (Feb. 1987).

[15] JOHNSTON, B., "Dependent failures", 16th Ann. Mtg. of the Systems Reliability Service Associates, [illegible] 198[illegible]

[16] DOERRE, P., "Possible pitfalls in the process of CCF event data evaluation", ANS/ENS Int. Top. Conf. on Probabilistic Safety Assessment and Risk Management, Zurich, 1987.

[17] APOSTOLAKIS, G., MOIENI, P., "On the correlation of failure rates", Reliability Data Collection and Use in Risk and Availability Assessment (WINGENDER, H.J., Ed.), Springer-Verlag, Heidelberg.

[18] HANNAMAN, G.W., et al., Systematic Human Action Reliability Procedure (SHARP), Rep. NP-3583 (Jun. 1984).

[19] SWAIN, A.D., GUTTMAN, H.E., Handbook of Human Reliability Analysis with Emphasis on Nuclear Power Plant Applications, Rep. NUREG/CR-1287, Sandia Nat. Lab. Albuquerque, NM (1983).

[20] EMBREY, D.E., HUMPHREYS, P., ROSA, E.A., KIRWAN, B., REA, K., SLIM-MAUD, an Approach to Assessing Human Error Probabilities Using Structured Expert Judgement, Rep. NUREG/CR-3518, Vol. 2, Brookhaven Nat. Lab. Upton, NY (1984).

[21] HANNAMAN, G.W., SPURGIN, A.J., LUKIC, J.C., Human Cognitive Reliability Model for PRA Analysis, Rep. NUS-4531, EPRI draft Rep. (1987).

[22] WILLIAMS, T.C., "HEART – a proposed method for assessing and reducing human error", Proc. 9th Advances on Reliability Technology Symp. Univ. of Bradford, UK.

[23] SEAVER, D.A., STILLWELL, W.G., Procedures for Using Expert Judgement to Estimate Human Error Probabilities in Nuclear Power Plant Operation, Rep. NUREG/CR-2743, Decision Science Consortium Inc. Falls Church, VA (1982).

[24] SIEGEL, A., et al., Maintenance Personnel Performance Simulation (MAPPS) Model: Description of Model Content, Structure and Sensitivity Testing: Summary description, Rep. NUREG/CR-3626, Vol. 1, Oak Ridge Nat. Lab., TN (1984).

[25] SIEGEL, A., et al., Maintenance Personnel Performance Simulation (MAPPS) Model: Description of Model Content, Structure and Sensitivity Testing, Rep. NUREG/CR-3626, Vol. 2, Oak Ridge Nat. Lab., TN (1984).

[26] AMENDOLA, A., "Accident sequence dynamic simulation versus event trees", Int. Post SMIRT-9 Sem. on Accident Sequence Modelling: Human Actions. System Response, Intelligent Decision Support, Munich, 1987.

[27] CACCIABUE, P.C., BERSINI, U., "Modelling human behaviour in the context of a simulation of man–machine systems", Human Decision Making and Control (PATRICK, J., DUCAN, K., Eds), Elsevier, Amsterdam and New York, 1987.

[28] BAINBRIDGE, L., "What should a good model of the NPP operator contain?", Int. Top. Mtg. on Advances in Human Factors in Nuclear Power Systems, American Nuclear Society, Knoxville, TN, 1986.

[29] RASMUSSEN, J., The role of hierarchical knowledge representation in decision making and system management, IEEE Trans. Syst., Man, Cybern, **15** 2 (1985).

[30] MANCINI, G., "Modelling humans and machines", Intelligent Decision Support in Process Environments (HOLLNAGEL, E., MANCINI, G., WOODS, D.D., Eds), NATO ASI Series, Springer-Verlag, Berlin (West), 1986.

[31] AMENDOLA, A., BERSINI, U., CACCIABUE, P.C., MANCINI, G., "Modelling operators in accident conditions: Advances and perspectives of a cognitive model", Post NATO-ASI Conf. on Intelligent Decision Support in Process Environments, Ispra, 1986.

[32] BERSINI, U., CACCIABUE, P.C., MANCINI, G., "Cognitive modelling: A basic complement of human reliability analysis", Int. Post SMIRT-9 Sem. on Accident Sequence Modelling: Human Actions, System Response, Intelligent Decision Support, Munich, 1987.

DEPENDENT FAILURE ANALYSIS IN PROBABILISTIC SAFETY ASSESSMENT

G.M. BALLARD,
UKAEA Safety and Reliability Directorate,
Culcheth, Warrington, Cheshire,
United Kingdom

Abstract

DEPENDENT FAILURE ANALYSIS IN PROBABILISTIC SAFETY ASSESSMENT.

Dependent failure events involving multiple failures of equipment and breaching designed redundancy are now recognized as significant to reactor safety. In order to more accurately reflect this significance in the probabilistic safety assessment it is necessary to develop improved methods of dependent failure analysis. Work is in progress to improve data availability but in parallel a framework for incorporating both data and engineering judgement is required. The paper outlines an anatomy of a dependent failure event which stresses the role of "coupling mechanisms". Defences against dependent failures are then related to various aspects of diversity. An initial quantitative indication of the significance, in current reactor operation, of some of these aspects is given together with a basis for modifying historic data for application to new plant designs.

1. INTRODUCTION

The basic design principle used in ensuring the safe operation of nuclear power plant is "defence in depth". This normally takes the form of redundant equipment and systems which provide protection even if a number of equipment failures occur. Such redundancy is particularly effective in ensuring that multiple, independent equipment failures with the potential for jeopardising reactor safety will be rare events. However the achievement of high reliability has served to highlight the potentially dominant role of multiple, <u>dependent</u> failures of equipment and systems. Analysis of reactor operating experience (ref 1,2) has shown that dependent failure events are the major contributors to safety system failures and reactor incidents and accidents. In parallel probabilistic safety assessment (PSA) studies have shown that the results of a safety analysis are sensitive to assumptions made about the dependent failure (CCF) probability for safety systems. Thus a Westinghouse analysis (ref 3) showed that increasing system dependent failure probabilities by a factor of 5 led to a factor 4 increase in core melt frequency.

1.1 In the past common cause/mode failure analysis (dependent failure is a preferable description) has not been widely recognised as being of particular concern or significance. Thus a number of PSAs have not included any modelling of dependent failures.

This lack of awareness or recognition was, in part, exacerbated by the great difficulty of modelling dependent failures in a satisfactory way. Generally methods were crude and undiscriminating and lacked any real capability to account for the design features of any specific plant. The models also lacked a clear engineering basis which could contribute to the understanding of dependent failure events.

Finally there was the problem of data. As one would expect from high reliability systems, data from operating experience is not abundant. However considerable data does exist but unfortunately it has not, until recently, been compiled into a useful form. This lack of good quality data has undoubtedly inhibited the use of dependent failure methods in PSA.

1.2 Progress on dependent failure analysis is however, being made. This is helped by the much greater awareness of the significance of dependent failure problems, resulting from analyses of operating incidents and accidents, and sensitivity studies on PSA results.

There is now a clear recognition of the need to improve our understanding of the mechanisms underlying dependent failure events, together with an estimation of both the impact of such events on any specific plant design and the means of mitigating or defending against such events. In co-operation with EPRI and the USNRC, SRD has been active in developing improved methods and data in pursuit of these objectives. Many aspects of this work has been reported elsewhere (ref 4) and this paper concentrates on the specific aspect of dependent failure modelling in PSA. This concentration does not imply that such modelling is the most important aspect of dependent failure analysis. Indeed it was stressed in reference 5 that a structured qualitative analysis of dependent failure potential is an essential and vital aspect of any overall analysis. Nevertheless a quantified assessment which could reflect accurately the qualitative analysis has been a major problem in the past and therefore deserves attention.

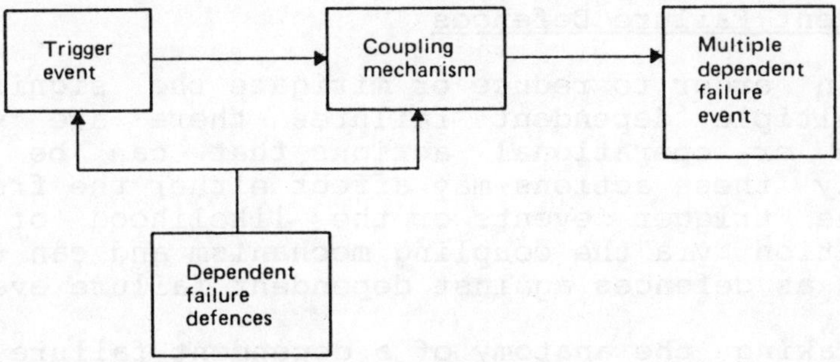

FIG. 1. Diagram of operational events which involve multiple dependent failures.

2. DEPENDENT FAILURE EVENTS

By definition a dependent failure event is one in which there are failures, of multiple equipment or systems, which are not independent. As a consequence the probability of occurrence of such an event is not accurately reflected by the normal multiplication of individual failure probabilities and may in practice be of significantly higher probability than would be indicated by such a calculation which assumes failure independence.

Analysis of operational events which have involved multiple dependent failures suggests that the anatomy of such events may be described as in Fig 1.

Taking these parts in turn:-

Trigger Event

Any multiple failure event involves at least one equipment failure. Such an equipment failure event is always a potential trigger event for a multiple failure event. No chronological order of failures is implied and the trigger event is really the cause of the equipment failure rather than the failure itself.

Coupling Mechanism

Given that a cause for one equipment failure exists (the trigger event) there has to be some reason why other equipment should fail in a dependent manner. Thus the coupling mechanism is the link between one equipment failure and other equipment failures within the multiple failure event.

Dependent Failure Defences

In order to reduce or mitigate the significance of multiple dependent failures there are various design or operational actions that can be taken. Clearly these actions may affect either the frequency of the trigger events or the likelihood of event escalation via the coupling mechanism and can thus be viewed as defences against dependent failure events.

2.1 Taking the anatomy of a dependent failure event described above various points arise:-

i) Any cause of equipment failure is a potential trigger event and trigger events are simply the normal causes of equipment failure.

ii) In order for a trigger event to lead to a multiple dependent failure event there has to be some coupling mechanism. If there is no coupling then a multiple failure event can only involve independent failures.

iii) A coupling mechanism may be a deterministic or probabilistic effect and may propagate different trigger events to different degrees. The coupling mechanism determines the conditional probability of multiple failures given the cause of one failure.

From this description it is clear that the driving force behind multiple dependent failure events is the existence of a coupling mechanism. So what is a coupling mechanism? Consider the following situation.

Suppose there are two exactly identical components being operated in exactly identical environments. Then if one component fails the other component must also fail because the two are, in every way, the same. However we know in practice that two components are never exactly identical and they are never operated in exactly identical environments. Thus if one component fails there will be some probability of the other component failing which depends on the cause of the failure and the extent to which the components and their operating environments really are identical. The cause of the failure can be identified with the trigger event and the degree to which the components and their operating environments are the same with the coupling mechanism. By simple extension the defences which affect the coupling mechanism are those features which reduce the extent to which the components are the same.

2.2 Consider some actual operational incidents as an illustration of the principles discussed above.

2.2.1

With the reactor at 20% power two reactor coolant pumps (RCP) were shut down for testing. The reactor protection system (RPS) failed to see either pump trip. The two RCPs were restarted and the other two RCPs were tripped. Again the RPS failed to see either pump trip.

Investigation of this incident, which involved the failure of 4 underpower monitoring circuits to detect pump trip, revealed that the relays had been adjusted two days earlier in accordance with the relay manufacturers specification. However, for this particular application, the relays had to be adjusted to operate at a specific pre-determined power level.

Analysis of this incident from a dependent failure point of view would indicate:-

. the design involved redundancy of relays which were of the same type and the same manufacture.

. the relays were reset at the same time

. the relays were probably reset by the same staff using the same procedure in each case.

Consequently most aspects of the design and operation of these relays were the same. The only elements of diversity were those achieved by using multiple redundant components whose failures would have a degree of randomness associated with the manufacturing and operational history. Since inappropriate setting/calibration of relays is a recognised cause of relay failure (trigger event) the high degree of sameness between the relays (coupling mechanism) indicates a high susceptibility to multiple dependent failure of these relays.

2.2.2

With the reactor at 84% power the high pressure coolant injection system (HPCI) was found to be inoperable. This situation required that the automatic depressurisation system (ADS) be checked. Due to improper use of the surveillance procedures all the ADS valves were inadvertently opened.

Analysis of this incident from a dependent failure perspective would indicate:-

. the valve surveillance was performed by the same staff member

. the valve surveillance was performed using the same procedure

. all the valves were surveilled at essentially the same time.

The multiple valve failures were due to an operator error (trigger event) propagated by a common timing, procedure and staffing of the valve surveillance process (coupling mechanism). In practice there was very limited diversity in this design but it is unlikely that diverse ADS valves would have materially influenced the incident because the sameness of the surveillance was the dominant mechanism for the incident.

2.2.3

During operation at 60% power surveillance testing was carried out on the Emergency Core Cooling System (ECCS). The suction valves in the two parallel lines from the water storage tank to the pumps failed to open on demand. Investigation revealed that this was due to a build up of ice on the valve stems as a result of a storm earlier in the day. Analysis of this incident from a dependent failure perspective would indicate

. the valves were subject to the same environment

. the valves were in the same location

. the valves were of the same design

The multiple valve failures were due to a storm (trigger event) which caused two failures because of the common design, environment and location (coupling mechanism) of the valves. The prime coupling appears to be the same environment/location of the valves but the lack of protection on the valves may have been due to a common design.

2.3 From these example incidents the concept of a multiple dependent failure event being driven by a coupling mechanism which is related to the "sameness" between redundant items can be seen. It is the elements of a redundant system which are the same which cause a trigger event, which could just lead to one component failure, to propagate to multiple component failures. Since generally the trigger events are the well-known causes of component failure and are not specifically related to multiple dependent events, the analysis of dependent failure events most usefully concentrates on the coupling mechanism. In addition it will be clear that dependent failure defences are actions to reduce the "sameness" of redundant trains and as such relate to the degree of diversity amongst such trains. Such diversity could extend from the randomness of the manufacturing and operational history which exists for components in simple redundant systems to the almost complete diversity existing in a system with designed equipment, functional, location and environment differences.

A short list of coupling mechanisms would comprise

. same engineering hardware

. same procedures - operation, test, maintenance

. same staff - operation, test, maintenance

. same environment

. same location

These categories could be sub-divided to add more discrimination if required. Thus for example difference in hardware may be present to various degrees as a result of equipment with different manufacturers, different parts/materials or different functional principles.

It should be noted that multiple failures which occur because of a designed connection in the system, e.g. a common valve, can be fitted into this framework but in practice are better left within normal reliability logic models. In addition there is the question of whether failures occur at essentially the same time (the relevant time interval will depend on the situation). While the actual coupling mechanisms will remain as in the previous list a defensive action will often entail ensuring that failures are separated in time. Thus for example component failures may be caused by a common test procedure but such failures may be separated in time (and thus not safety critical) if the application of the tests is staggered in time.

To provide initial information on the strength or existence of coupling mechanisms in current nuclear power plants a pilot study was made on 100 incidents from US nuclear plants. These were incidents, reported under the LERs, on systems/components such as motor operated valves, safety injection pumps, service water pumps, containment spray pumps and diesel generators. Thus the selection is largely from a mechanical orientation. All the incidents involved multiple dependent failures of equipment and an analysis of the primary coupling mechanism in each case indicates the following spectrum.

- Same hardware/design ~ 55%
- Same staff (operations/maintenance etc ~ 25%
- Same procedures ~ 10%
- Others (environment, location etc) ~ 10%

This pattern is reasonably consistent within the different systems/components represented in the data; it may be however that other types of system, e.g. electrical, instrumentation, could show a different pattern.

In parallel with assessment of primary coupling mechanism a judgement was made as to the most effective aspect of diversity which would provide a defensive action against the incidents analysed. As might be expected diversity of hardware, at various levels from manufacture to fundamental design, depending on the incident, was judged the most appropriate defence in over 60% of cases, with diversity of staff being the next most appropriate defence.

3. QUANTITATIVE MODEL DEVELOPMENT

To use the concepts discussed in the previous section it will be necessary to develop a quantitative model for application in a safety analysis. Such a model will need to address questions including the value in dependent failure terms of plant design and operational features such as redundancy, equipment/functional diversity and diversity in operational and maintenance staff. In this section the framework of a model is presented.

3.1 Prior to the quantification of a dependent failure frequency/probability the information from the qualitative reliability analysis should be formulated in a systematic way. The qualitative analysis will have generated minimum cut sets for system failure probabilities and event sequence frequencies. The cut sets will contain component failure probabilities. The task of the qualitative analysis is to identify those features of the components within individual cut sets that are the same. Thus each cut set is viewed as a mini-system in which we are concerned about potential dependent failures breaching the redundancy. In reference 5 a matrix formulation was used to display some of the features of the cut sets. More widely, however, it is necessary to be systematic in the identification of characteristics of the components that are the same including not only manufacture, type etc but also location, environment, test/maintenance procedures and timing, operational staff, etc.

3.2 In the past the available quantitative dependent failure (CCF) models have had great difficulty representing the output of the qualitative reliability analysis. This has partly been because the models have been simple but also because the models lacked an engineering basis in which to represent the design and operational features of plant components and systems. Detailed data on a wide variety of systems and components could be used to make existing simple models discriminating but generally such data does not exist. In the absence of such data it is important that any model must use available data but also have the capability of structuring, guiding and using engineering judgement.

A first order model following the discussion of Section 2 would be as in Fig 2.

The trigger event is some particular category of component failure causes which will depend on the data available. At the simplest level it may be the total failure rate of the component while, if more detailed data is available, it may be broken down into classes such as internal failure, human error in operations, test or maintenance, environment including all external influences.

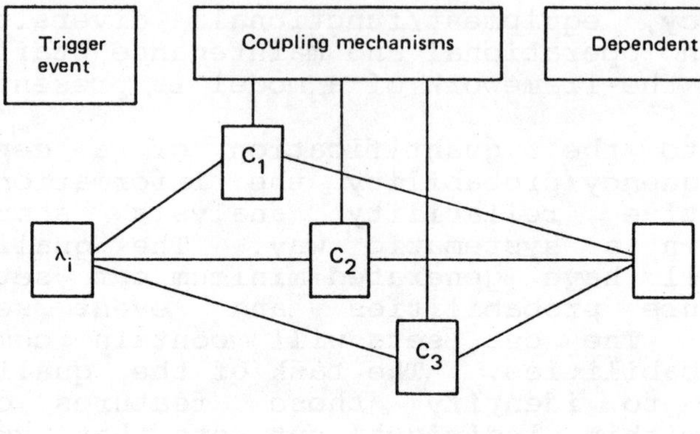

FIG. 2. Diagram of operational incidents.

TABLE 1. RELATIVE CONTRIBUTIONS TO THE COUPLING MECHANISMS

Same hardware (all aspects)		0.47
Same staff	— operation	0.08
	— test	0.05
	— maintenance	0.05
Same procedure	— operation	0.06
	— test	0.11
	— maintenance	0.04
Same environment		0.11
Same location		0.03
		1.00

The coupling mechanisms are those features of sameness which may be expected to couple component failures caused by the particular category of trigger events under consideration. Thus if λ_i is the total failure rate of the component all coupling mechanisms will need to be included. By contrast if λ_i represents component failures due to an internal fault then the coupling mechanism likely to be most effective is the same hardware.

TABLE 2. DATA ANALYSIS

Trigger event	Coupling mechanism
Internal component failure	Same hardware — manufacturer — type — design principle Same environment Same location
Externally caused component failure (including environment, energetic events, etc.) but *not* human error	Same hardware — manufacturer — type — design principle Same environment Same location
Human error caused component failure (operations, test, maintenance, etc.)	Same hardware — manufacturer — type — design principle Same staff — operations — test — maintenace Same procedures — operations — test — maintenance

Thus for example an analysis of 100 reactor events involving dependent failures indicated that the relative contributions of the coupling mechanisms, irrespective of trigger event or type of component (within a largely mechanical orientation) were as in Table 1.

For this analysis coupling mechanisms which made a significant contribution to the occurrence of the event were assessed (not just a primary mechanism) and the relative frequency of occurrence of each coupling mechanism calculated. If the spectrum of trigger events was assumed fixed this table gives the relative significance, in dependent failure terms, of each coupling mechanism and thus measures the relative potential improvements in reliability that can be achieved by the use of various aspects of diversity.

Further data analysis (using averaging across classes of components or engineering systems) should make significant progress towards the enumeration of a more detailed model (Table 2).

It should be possible to calculate

i) a representative value of λ_i for a few trigger event categories

ii) the conditional probability of additional component failures, given the trigger event, due to types of coupling mechanism. This conditional probability may be at various levels of detail, for example a single β factor or the various parameters of the Multiple Greek Letter (MGL) model, depending on the adequacy of the data.

3.3 If it is assumed that the nuclear power plants, covered by the event data utilised, are typical of an industry average then the parameters derived above can be viewed as a base case reflecting past design practice. The motivation for modelling is to be able to assess the effectiveness of design and operational actions taken to reduce the impact of dependent failure events on plant safety. The task is therefore to modify the base case in a structured and systematic way to reflect the specific defensive actions taken. Since diversity has not been a widely used aspect of plant design in the past there is little data available and thus the assessment process will be largely based on engineering judgement guided by the structure and parameters of the coupling mechanism model.

In accord with the set of coupling mechanisms listed in Table 1 the list of defensive actions will be seen in Table 3.

The potential effectiveness of these individual defensive actions is to eliminate entirely the contribution of a coupling mechanism whose industry average strength has been derived from available

TABLE 3. DEFENSIVE ACTIONS

Diversity of hardware	— manufacture — type — design principle
Diversity of staff	— operations/management — test — maintenance
Diversity of procedures	— operations — test — maintenance
Diversity of environment	
Diversity of location (segregation)	
Diversity of timing	

data. Within an event sequence cut set it should be possible to judge the extent of diversity, of the various types, using an engineering understanding ofthe potential failure modes of the components. On a scale of 1 (no diversity) to zero (total diversity) a multiplier on each coupling mechanism would thus be derived and would appropriately modify the strength of that coupling mechanism in establishing the conditional failure probability of a trigger event propagation. The application of engineering judgement is structured because the questions related to identified aspects of "sameness" between components in a cut set and not to an unspecific spectrum of potential dependent failure causes.

3.4 Using the previous discussion the framework for performing a qualitative and quantitative evaluation of dependent failure probabilities for a specific design involves the following steps:-

i) Identify the event sequence or system cut sets of interest.

ii) Use available data to quantify the failure rate for appropriate trigger event categories and the conditional probabilities for coupling mechanisms within the trigger event categories. This is then the base case

iii) Identify the aspects of hardware, staff, procedures etc which apply to the components within each cut set

iv) Within the coupling mechanisms (aspects of "sameness") assess the degree of coupling for the specific cut set and triggger event category. This assessment would be on a scale of 0 - 1 and reflect the extent of diversity achieved by specific intent.

v) Combine ii) and iv) to assess the dependent failure probabilities for each cut set.

3.5 Intentionally no specific quantification method has been discussed because the above framework can be used in association with many such methods including the β factor, Binomial Failure Rate, MGL etc. The intent has been to provide an engineering framework to assess the parameters of these methods and to make them more specific to any particular application. This is in accord with the European Benchmark Exercise (ref 6) which showed that the methods will give equivalent answers but major differences can be caused by parameter estimation. In that exercise SRD used a "partial β factor" approach which involved a combination of data and engineering judgement. The framework discussed here is an evolution from that approach to provide an improved basis for the incorporation of the more extensive data available and the essential engineering judgements.

4. CONCLUSIONS

There is no doubt that events involving multiple dependent failures are of major significance to reactor safety. Such events occur because redundant elements within the design are the same in a design or operational sense. (The strongest example would be the use of identical computer software in a redundant configuration). This underlying model of a dependent failure event provides the basis for combining actual data and engineering assessment in an analysis of the dependent failure probabilities for any specific plant. In this way the assessment of the plant in a qualitative sense will better reflect the defensive measures taken to guard against dependent failure events. A simple model, which should be capable of support from existing data while capturing the major aspects of dependent failure events, has been discussed and further work in the UK will develop and test this model and possible extensions.

REFERENCES

[1] BALLARD, G.M., "An analysis of dependent failures in the ORNL precursor study", ANS/ENS Int. Top. Mtg on Probabilistic Safety Methods and Applications, San Francisco, CA (1985).

[2] FLEMING, MOSLEH, KELLEY, On the analysis of dependent failures in risk assessment and reliability evaluation, Nucl. Saf. (Sep.–Oct. 1983).

[3] VAVREK, A., "Sensitivity Study of Common Mode Failure Rates for Sizewell B", ANS/ENS Int. Top. Mtg on Probabilistic Safety Methods and Applications, San Francisco, CA (1985).

[4] ELECTRIC POWER RESEARCH INSTITUTE, PRA Procedures for System Level Dependent Events Analysis, Draft Rep., EPRI, Palo Alto, CA (1987).

[5] UNITED KINGDOM ATOMIC ENERGY AUTHORITY, Common Cause Failure Reliability Benchmark Exercise, Rep. SRD-R-383, UKAEA, Warrington, UK (1987).

[6] POUCET, A., AMENDOLA, A., CACCIABUE, P.C., Common Cause Failure Reliability Benchmark Exercise, Final Rep. EUR-11054-EN, CEC-JRC, Ispra (1987).

PLANT-SPECIFIC RELIABILITY AND HUMAN DATA ANALYSIS FOR SAFETY ASSESSMENT

K.E. JÄNKÄLÄ, J.K. VAURIO, U.M. VUORIO
Imatran Voima Oy,
Helsinki, Finland

Abstract

PLANT-SPECIFIC RELIABILITY AND HUMAN DATA ANALYSIS FOR SAFETY ASSESSMENT.

A plant-specific reliability and human data collection and analysis system is under development at the Loviisa Power Plant. Historical failure data of the plant have been collected using a form developed to gather all relevant information about events. Equations for all possible combinations of answers to items on the form have been developed to obtain unavailability parameters needed in basic quantification and in time-dependent calculations. A set of computer programs has been designed to make possible storage of the failure data, their search in various ways, estimation of the reliability parameters and finally transfer of the unavailability parameters into the input file of the SETS fault tree program. A robust parametric empirical Bayes method has been developed to emphasize plant-specific data. Many of the problems encountered with maximum likelihood and earlier parametric empirical Bayes methods have been avoided. Learning models have also been developed for trend analyses. A survey of human errors has been carried out as part of the human reliability quantification. Plant records on human errors during 15 plant years have been collected using a special questionnaire focusing on causes, impacts, means of discovery, situation etc. Plant records on human errors are rather limited, therefore operating and maintenance staff have been interviewed, anonymously, using the same questionnaire. Some statistical summaries have been made of about 180 error events.

1. INTRODUCTION

The reliability data collection and analysis system under development for the Loviisa Power Plant has two main goals:

(1) Obtaining plant-specific data and parameters for the probabilistic safety assessment (PSA) project under way, and

(2) Establishing an on-line data collection and analysis system for the future.

The final system allows the updating of parameters for a living PSA, and facilitates trend studies and feedback from experience to optimize maintenance planning, spare parts inventory, technical specifications and limiting conditions for operation. This paper is focused on the details of a PSA data system which has been completed, and on human reliability data analysis.

Objectives set for the PSA data system have been described in Ref. 1. These objectives have led to the development of robust parametric empirical Bayes methods [1,2,3,4]. Many of the problems encountered with maximum likelihood and earlier parametric empirical Bayes methods have been avoided. The method is robust in the sense that it yields realistic estimates in all kinds of practical cases. Up-to-date formulation of the method is described in Ref. 4 for the estimation of failure rates and in Ref. 5 for the estimation of both failure rates and unavailabilities per demand. The reasoning of the method is discussed in Ref. 4 for the estimation of failure rates.

A special failure data collection form designed to gather all relevant information about events for basic quantification and time-dependent calculations will be described.

An automatic data processing system has been designed to facilitate the storage, search, sorting and parameter estimation of the failure data. The features of the system are illustrated in examples where historical data of the plant have been used.

A survey of human errors has been carried out as part of the human reliability analysis. Plant records on human errors during 15 plant years have been collected using a special questionnaire focusing on causes, impacts, means of discovery, situation etc. This questionnaire and some statistical summaries are presented.

2. FAILURE DATA COLLECTION

A special data collection form has been developed to gather all relevant information about events. A multiplicity of failure entering, detection and exit mechanisms is taken into account to enable both base case quantifications and time-dependent calculations. A computer program has been designed to enable storage of the failure data into an ADABAS data base of the VAX 11/785 computer. Thus, the valuable historical data will not be lost. Failures that entered during the operation of the plant from 1977 through 1985 have been collected using plant records like work orders and instructions, control room log-books and test records.

A translated version of this form is shown in Fig. 1. The work report and component identifications are given in the first two lines. The work report is identified by the items WORK NO and TK. Items LO and KZ specify the plant unit, system and location in the system, i.e. the process component. Each component belongs to one of the three categories LT: (A) Normally standby, (B) normally operating or (C) alternating standby and operation. Items with bold letters are described below.

The effect on the operating capability of the equipment is asked in item B. The answer is B1 also in a standby state if the failure has prevented operation.

Item H in the failure type row indicates the failure mode like "does not start", "does not operate given start", "fails to open", "fails to remain open" etc.

```
LO: _   WORK NO: ____   TK: __   FOREMAN: ____   FILLED BY: ____

KZ: ____________   EQUIPMENT NO: ____________   ROOM NO: ________   LT: _

OPERATING SITUATION AT DETECTION:
    A _ During operation B _ Before shutdown   C _ Turbine   shutdown
    D _ Hot shutdown     E _ Cold shutdown     F _ Ref.shutdown

WORK NAME: ________________________________________________

ACTION TAKEN: EK _ Repair, cleaning, lubrication, "exercising"
              ES _ Adjustment, calibration, repair of trip level
              EQ _ QA/QC
              EV _ Replacement with same kind of equipment or parts
              EM _ Alteration = replacement with diff.kind of equip.
              EL _ Repair of position, level etc.
              EH _ Maintenance (check, test, cleaning, lubrication, oil/
                   grease/resin replacement), no replacement of parts
              EE _ Undone (information)
              ET _ Other action (information)

INFORMATION: ______________________________________________
___________________________________________________________
___________________________________________________________

FAILURE TYPE: A _ Mechanical B _ Elect. C _ Instr. D _ Process.   H _

FAILURE DETECTION METHOD:
A1 _ Alarm                            A6 _ On demand
A2 _ Symptoms (immediately detected)  A7 _ QA/QC inspection
A3 _ Shift inspection                 A8 _ Repair or test after repair
A4 _ Periodic test                    A9 _ Random check
A5 _ Scheduled maintenance            A0 _ Work order without failure

EFFECT ON THE OPERATION OF THE EQUIPMENT (IF DEMAND HAD OCCURRED)
 B1 _ Prevented   B2 _ Did not prevent, repair prevented   B3 _ No effect

TRIP: A _ Reactor  B _ Turbine 1  C _ Turbine 2

EFFECT OF FAILURE ON PLANT OPERATION:
    GA _ Shutdown to cold standstill due to failure
    GB _ Shutdown to cold standstill according to tech.spec.
    GC _ Shutdown to hot standstill due to failure
    GD _ Shutdown to hot standstill according to tech.spec.
    GE _ Turbine 1 off the grid due to failure
    GF _ Turbine 2 off the grid due to failure
    GG _ Power limit due to failure
    GH _ Power limit according to technical specifications
    GJ _ Interruption of startup/shutdown
    GK _ Op.limit.according to tech.spec. plant operation
    GL _ No effect on plant operation

FAILURE/FAULT ENTRY (LIKELY)
    C0 _ Before previous 1)test 2)maintenance
    C1 _ During previous 1)test 2)maintenance 3)operation
    C2 _ After previous  1)test 2)maintenance 3)operation
    C3 _ 1)Immediately before detection 2)After previous shift inspection

HAS THE SAME FAILURE CAUSED OTHER FAILURES OR UNAVAILABILITY IN
REDUNDANT COMPONENTS
D _ No _ Yes  Work no:   ________ ________ ________ ________
            Component:   ________ ________ ________ ________
               Signal:   ________ ________ ________ ________

F1 _ DIDN'T FULFIL ORIGINAL DESIGN CRITERIA F2 _ DESIGN CRITERIA CHANGED

SHUTDOWN/POWER LIMITATION ____ MW   ____ h   ____ min

T __ WEEKS, CHECK OR TEST INTERVAL FOR THE FAILURE
     TEST GROUP ____________________________________

t0 DATE __.__.__ HOURS __.__ PREVIOUS 1)TEST 2)MAINTENANCE 3)OPERATION
t1 DATE __.__.__ HOURS __.__ FAILURE/FAULT DETECTION
t2 DATE __.__.__ HOURS __.__ WORK PERMISSION, SAFETY MEASURES TAKEN _
t3 DATE __.__.__ HOURS __.__ WORK STARTED
t4 DATE __.__.__ HOURS __.__ WORK COMPLETED
t5 DATE __.__.__ HOURS __.__ RETURN TO USE

WORKHOURS ____h  SHIFT SUPERVISOR ______  CHECKING: ____  DATE __.__.__
```

FIG. 1. Data collection form.

Any unavailability of a component in PSA models falls into one of the following classes determined by the answers to items B and H:

K: Fatal hardware failures; unavailability begins when the failure enters and continues through repair until the work is completed and component returned to use.

L: Incipient hardware faults that need to be fixed but the component is unavailable only during repair (unscheduled maintenance).

M: The unavailability is due to scheduled periodic maintenance or testing, taking a component out of service for some time.

H: The unavailability is due to human errors made in a periodic maintenance, test, calibration or restoration thereafter.

In terms of modeling, the events in classes K and L enter due to a failure rate that is usually different in standby and operating modes. The events of class H enter with a certain error probability per maintenance action or attempt. Class M events are fixed periods of unavailability at fixed times. So, the data of individual M events have not been collected but average times have been estimated.

The failure detection method A and the likely failure entry time C are extremely important in calculating the unavailability time of each event. For example, the unavailability time for the class L events is $t_4 - t_3$ for both standby and operating components, while $t_4 - t_1$ is valid for all class K failures that are detected immediately due to alarms or other symptoms. Periodic test interval T is needed for standby components for failures detected in tests or entering due to testing. If safety measures have been taken, times t_5 and t_2 are used instead of t_4 and t_3, respectively.

All the items LT, A, B, C and H are interconnected so that when you choose an answer to one item only certain answers are possible for the other items. The storing program makes these coherence checks of ABCH and LT combinations.

Item D with answer "yes" means that this event has caused failures of redundant components. This event should be modeled as one basic event that includes all the failed components listed in the item.

3. FAILURE DATA ANALYSIS

3.1. Failure rate and unavailability per demand

The failure rate λ and the unavailability per demand q are important characteristics needed for reliability and availability analyses. Test intervals are fixed constants and need no estimation. Average repair times are also relatively accurate. Repair rates can be estimated using the same formalism as for failure rates, just by replacing observed times to failure by observed repair times.

For plant-specific studies it is important to effectively utilize data collected at the plant. A family of estimation methods for that purpose, called empirical Bayes methods, has been largely developed by Efron and Morris since 1973.

In this study, a special robust empirical Bayes method [1,2,3,4,5] is used. In the method it is assumed that the failure rate or unavailability per demand of a component can vary within a population of similar components. This variability is modelled by a probability distribution. When this prior distribution, or its parameters, has been determined, posterior distributions can be determined for λ or q. For practical reasons we select the versatile gamma and beta distributions to describe the prior distributions, because they are the conjugate prior distributions for the Poisson (λ) and binomial (q) sampling models, respectively. In this method the prior parameters are estimated using a special moment matching method in which all data and only the data are used. Values to be used in risk calculations are based on individual posterior distributions of each component: posterior mean values are used in the base case calculations, and the whole posterior distributions are used in uncertainty calculations.

3.2 Unavailability

Two important component characteristics used in current risk assessments are

(1) Failure rates for components that cause initiating events i.e. their failures activate one or more safety functions, and

(2) Unavailabilities (fractional dead times) for components needed after initiating events to accomplish a safety function.

The unavailabilities used in base case calculations are time-averaged values. Time-dependent analysis will be performed for important failure combinations to verify that the scheduling of testing etc. does not change results excessively. An advanced version of the well-known FRANTIC code [6] will be used in such calculations. While failure rates are relatively easy to estimate from the observed numbers of failures, unavailabilities depend on several parameters due to many possibilities of failures entering and being detected and repaired.

Exact analytical expressions have been derived in Ref. 7 for time-dependent and asymptotic average unavailabilities of periodically tested simple standby systems. The average asymptotic unavailability of a periodically tested standby component was found to be

$$u = 1 - \frac{1 - e^{-\lambda(T-\Delta t)}}{(1 + e^{-\lambda(T-\tau)} - e^{-\lambda(T-\Delta t)})\lambda T} \qquad (1)$$

where T = the test interval, Δt = the time for which the component is unavailable during a test, λ = the failure rate and τ = the repair time during which the component is unavailable.

Because Δt is very small compared with T at the Loviisa Power Plant, Eq. 1 can be written in the form

$$u = 1 - \frac{1 - e^{-\lambda T}}{\lambda T} \cdot \frac{\mu}{\lambda + \mu} \tag{2}$$

where $\mu = 1/\tau$. This expression has the advantage that compared with its normally used approximation $u = \lambda\tau + 0{,}5\lambda T$, it always yields values between 0 and 1 in sensitivity and uncertainty calculations.

The unavailability of a normally operating or continuously monitored component is $u = \lambda/(\lambda + \mu)$ (as if $T \to 0$ in Eq. 2) and the unavailability of a component alternating between standby and operation is calculated using both this form and Eq. 2. Because failures in the standby mode may be immediately detected this form has to be used for those failures of standby components, too. Actually the unavailability class L failures should be calculated using this form for all kinds of components whereas the formula to be used for the unavailability class K of standby components depends on the failure detection method.

The probabilities of the class H events, i.e. human errors, are quantified with the unavailability per demand parameters q. The unavailability times of these failures have to be taken into account in order to obtain the mean unavailability values:

$$u = q\tau_u/T \tag{3}$$

where τ_u = the mean unavailability time.

Distributions of the unavailabilities are needed in uncertainty calculations. Their variances are estimated using Taylor expansions. For the unavailability of Eq. 2 the variance is

$$\mathrm{Var}(u) = (1+\mu T/2)^2 \frac{\mu^2}{(\lambda+\mu)^4} \mathrm{Var}\lambda + (1-\lambda T/2)^2 \frac{\lambda^2}{(\lambda+\mu)^4} \mathrm{Var}\mu \tag{4}$$

The variance of the unavailability of a normally operated or continuously monitored component is obtained from Eq. 4 by setting T = 0. The variance of the unavailability class H (Eq. 3) is assumed to be determined by parameter q because τ_u is usually more accurately known.

3.3 Data processing

A computer program has been designed to search failure data in various ways. In addition to any item or combination of items in the form, component type, manufacturer and room identification can be used as search criteria. Searches can also be made for a particular time interval, component, group of components or a subsystem. It is advisable to always have at least one component or subsystem as a search criterion in order to save computer running time.

Equations for all possible combinations of answers to items in the data collection form have been developed to obtain failure rates and unavailability parameters needed in basic quantification and in time-dependent calculations. A computer program has been written to estimate the parameters. The program reads the coded failure data from the file created by the search program. Then it calculates component- and group- specific failure rates and unavailabilities per demand, repair rates and finally unavailability values for selected failure modes. The unavailability values are transferred into the SETS fault tree program input file after comparison with generic data.

3.4 Examples

An example is given in Table I of the estimation of the reliability parameters of alternating redundant pumps. Two of these service water pumps are running, one month at a time. Results are shown only for the failure mode D, i.e. does not operate given start. Firstly, the program lists the detection date, component identification, work identification, coded items and repair time (EA) of the failure data found. Secondly, it lists the estimated parameters of the prior distribution, group parameters and component-specific posterior parameters of failure rates. Group parameters to be used in "coupling" sensitivity studies are estimated by pooling the failure data of different components together. Notice that the estimated failure rate of the second pump is high and the distributions of the pumps with no failures are wide and the upper fractiles are close to the group mean value. Thirdly, the corresponding estimates of repair rates are listed. In this case group parameters have to be used, which means now that the mean repair rate of the failed pump is used for all the other pumps. Finally, the component- and group-specific unavailability parameters are listed.

Table II gives another example with more usual data of the failure mode Z of the same pumps as in Table I. The listing has been cut to show only the failure rate and unavailability parameters. The second pump has failed also in this case more often than the others. However, the estimated failure rate of the second pump has been shrunk only a little towards the common mean value in this case because the other components have failed quite often, too. The distributions are now much narrower due to more data.

If no failures have taken place in those pumps, the sample can be increased to include other similar pumps with the same manufacturer and roughly the same characteristics and use. Expansion of the sample usually leads to the widening of the prior distribution, but it guarantees realistic non-zero failure

TABLE I. AN EXAMPLE OF THE PARAMETER ESTIMATION LISTING

```
FAILURE TYPE D:   DOES NOT OPERATE GIVEN START
EQUIPMENT TYPE C: ALTERNATING

FOLLOWING FAILURES WERE ACCEPTED:
 DATE   KZ         EQUIP.NO WORK NO. TK A  B C  DEFG H K JOS V T   SIT   EA   L
 830125 11VF12D001 LSM0795  48370    01 16 1 31  K K D -     B 001 A    15.67-
 850119 11VF12D001 -        7V       01 16 1 31  T L D -     B 001 A     0.37-
 850128 11VF12D001 LSM0795  62240    01 16 1 31  K K D -     B 001 A    18.52-
   22 RECORDS FOUND     3 ACCEPTED

FAILURE RATES (1/h) / PROBABILITIES (1/DEMAND)
METHOD = GAMMA, CONSERVATIVE

PRIOR PARAMETERS     x : -.675000E+00
                     y : 0.960000E+04
GROUP PARAMETERS     OBSERVATIONS : 0.300000E+01
                     TIME/TRIALS  : 0.128000E+06
                     PROBABILITY  : 0.312500E-04
                     VARIANCE     : 0.244141E-09

POSTERIOR PARAMETERS
COMPONENT   OBS.    TIME     PROBABILITY  VARIANCE     FRACTILE 5%  FRACTILE 95%
 11VF11D001  0.0    32000.00 0.781250E-05 0.187800E-09 0.168780E-08 0.347231E-04
 11VF12D001  3.0    32000.00 0.799279E-04 0.192134E-08 0.237481E-04 0.162010E-03
 12VF13D001  0.0    32000.00 0.781250E-05 0.187800E-09 0.168780E-08 0.347231E-04
 12VF14D001  0.0    32000.00 0.781250E-05 0.187800E-09 0.168780E-08 0.347231E-04

REPAIR RATES (1/h)
GROUP PARAMETERS ARE TO BE USED FOR ALL THE COMPONENTS

GROUP PARAMETERS     OBSERVATIONS : 0.300000E+01
                     TIME/TRIALS  : 0.345600E+02
                     PROBABILITY  : 0.868056E-01
                     VARIANCE     : 0.251173E-02

UNAVAILABILITIES
COMPONENT        UNAVAILABILITY  VARIANCE
 11VF11D001      8.99919E-05     2.76131E-08
 11VF12D001      9.19922E-04     5.35612E-07
 12VF13D001      8.99919E-05     2.76131E-08
 12VF14D001      8.99919E-05     2.76131E-08
      GROUP      3.59870E-04     7.54913E-08
```

TABLE II. AN EXAMPLE OF THE PARAMETER ESTIMATION WITH MORE DATA

FAILURE TYPE Z: FAILURE DID NOT PREVENT OPERATION, REPAIR PREVENTED
EQUIPMENT TYPE C: ALTERNATING

FAILURE RATES (1/h) / PROBABILITIES (1/DEMAND)
METHOD = GAMMA, CONSERVATIVE

POSTERIOR PARAMETERS

COMPONENT	OBS.	TIME	PROBABILITY	VARIANCE	FRACTILE 5%	FRACTILE 95%
11VF11D001	2.0	72888.00	0.313682E-04	0.219093E-09	0.115754E-04	0.587214E-04
11VF12D001	4.0	72888.00	0.453373E-04	0.316661E-09	0.205379E-04	0.779045E-04
12VF13D001	1.0	72888.00	0.243836E-04	0.170309E-09	0.753059E-05	0.478106E-04
12VF14D001	2.0	72888.00	0.313682E-04	0.219093E-09	0.115754E-04	0.587214E-04

UNAVAILABILITIES

COMPONENT	UNAVAILABILITY	VARIANCE
11VF11D001	3.97400E-04	6.83466E-08
11VF12D001	6.88767E-04	1.43152E-07
12VF13D001	2.72201E-04	4.09510E-08
12VF14D001	4.00698E-04	6.94853E-08
GROUP	4.27151E-04	3.64605E-08

rates, is consistent with plant-specific spectrum of manufacturers and use, with minimal dependence on generic (perhaps atypical) worldwide data. That is why we can get plant-specific estimates for almost all the components. Of course, running time failure rates cannot be estimated for standby components.

3.5 Other options of the search program

The search program includes a routine that sums up the number of occurrences of each item in the failure data search. For example, the search of the failures of certain closing valves of the low- and high-pressure injection and spray systems yielded 98 failures with the following

- failure modes: 54 failures to open, 22 failures to close, 1 failure to remain open, 3 trip level changes, 1 external leakage, 2 failures to operate automatically, 1 failure to operate (in general) and 14 failures of unavailability class L,
- detection methods: 82 periodic tests, 6 scheduled maintenances, 2 repairs or tests after repairs, 4 shift inspections,
- likely failure entry times: 79 after previous test, 3 after previous maintenance, 3 after previous shift inspection.

45 failures triggered the counting of the outage time according to the technical specifications of plant operation and the rest of the failures had no effect. Design criteria were changed once. Most of the failures detected in periodic tests were stickings because of boron crystallization and most of them could be repaired by "exercising".

The search program also includes a routine for the seeking of potential common cause failures. When a search using some key words has been made, the program checks for each failure if there are other failures with an unavailability time overlapping with the unavailability time of the comparison failure. No results have been obtained yet.

3.6 Trend analysis

Significant learning has taken place since the beginning of the operation of the plant. Thus, some trend analysis is needed to be able to use current estimate values. For a general trend analysis in a PSA one can assume that the expected number of failures, x, behaves like $x = At^B$ as a function of the total time on duty, t. For $B > 1$ the model describes increasing failure rates due to ageing. Several formulations have been derived in Ref. 8 to estimate parameters A and B of such a Weibull model. Studies of major accidents in the nuclear industry have indicated that models other than Weibull might better describe the learning process. A fractional learning model assumes that the failure rate is reduced by a certain fraction because of each accident or failure [9]. Such a model seems to fit many kinds of failure data, not just data on serious accidents. A smooth version of this model has been introduced in Ref. 10. This method has been applied to several failure data sets, and it seems to fit the experience with a variety of equipment at the component, system and plant levels. The error estimation of these methods and testing for verification of learning are being developed.

4. HUMAN RELIABILITY ANALYSIS

In the Human Reliability Analysis (HRA) of Loviisa 1, PSA errors in human actions are divided into three main groups: (A) errors made prior to an initiating event, (B) errors that initiate an accident or a transient and (C) errors that occur during an accident. The group A errors are modeled in system fault trees, the group B errors as initiating events and the group C errors at the event tree level. Up-to-date techniques are used for the three main categories of human errors. Advancements or plant-specific modifications have been made to some state-of-the-art techniques applied in earlier PSA studies [11].

A survey of human errors has been carried out as part of the HRA identification and quantification. Plant records on human errors during 15 operating years of Loviisa 1 and 2 units have been collected using a special questionnaire. Plant records on human errors are rather limited because special event reports are prepared only if safety systems are degraded beyond technical specifications, or when consequences are significant. To obtain a more complete picture of the types and frequencies of errors made, the operating and maintenance staff have been interviewed anonymously. About 180 questionnaires on error events have been gathered, of which about 80 events are based on interviews and 100 based on plant records. The human error data collection was performed by one of the shift supervisors released from shift duties for the PSA project.

The questionnaire focuses on root causes and impacts of human failures, means of detection, which working group was involved in the inappropriate action, what the situation was when an error was made and what was done to prevent the recurrence. The list of categories is shown in Table III. Multiple causes can be recognized for a single error, and many additional questions are included for cross-checking purposes to be able to draw proper conclusions from the statistics obtained.

Preliminary statistical analysis indicates that the situation for an error was mostly (60%) normal operation. Startups and shutdowns accounted for 22% of the total number of errors and 18% of errors were made during refuelling periods. The time fractions for plant stages are the following: power operation 90%, startup and shutdown 3% and refuelling period 7%. Out of the total number of inappropriate measures, 12% were made during periodic testing.

Table IV shows the means of detection and the measures taken to prevent inappropriate action in each activity situation. According to Table IV most of the errors (over 60%) were discovered immediately via alarms, equipment trips or the plant protection system or via symptoms. About 15% of the errors were revealed by casual inspection. This proportion was remarkably larger during the refuelling period, about the double compared with all other cases. Modifications of procedures were more common (20%) when the error was made in periodic testing, or startup/shutdown. Modifications to maintenance or operating staff training were rather rare in view of the fact that one of the most common root causes of an error was the lack of training or insufficient training as shown in Table V. Table V indicates that slips and carelessness were involved in about half of the total number of events. One interesting point can be noticed: the proportion of erroneous action or carelessness during startup was remarkably smaller, about 35% of the startup events involved this category. In 40% of all events this root cause category was the only reason for an error. The most common additional causes were the lack of time or stress (18%), the procedure was not followed (13%) and the training does not cover it (13%). The total number of error events with slip or carelessness involved was about 140.

Table VI compares the number of errors in action with a procedure available or unavailable. When a procedure was available, 35% of the cases with a procedure available occurred because of defects in the procedure. In over one half of the events one partial cause for malfunction was that the procedure was not followed. The proportion of slip or carelessness seems to be slightly larger when a procedure was available (70%) than not (50%). The lack of training was remarkably common when a procedure was not available, over 60%. The connection seems to be quite obvious: if there is not a procedure available, the importance of training is also undervalued. The most common action taken to prevent the error recurrence was none in both cases, about in 70% of the events. A large fraction of this category (no preventive action) reflects the fact that few, if any, of the consequences were severe. But it could also indicate that means of reducing human errors are not well known or the importance of the reduced number of errors is not fully realized.

The results of statistical analysis of plant human errors will be taken into account in PSA quantification and identification of human errors.

TABLE III. QUESTIONNAIRE FOR HUMAN RELIABILITY DATA

A. 1. OBJECT (KZ)________ 2. UNIT L01 ___ 3. UNIT L02 ___

4. YEAR: ________

B. ERRONEOUS ACTION: ________

C. SITUATION IN WHICH THE ERROR WAS MADE:

1. Periodic testing ___
2. Preventive maintenance ___
3. Failure search ___
4. Equipment removal/return ___
5. Failure repair ___
6. Normal operation ___
7. Startup ___
8. Shutdown ___
9. Annual revision ___
10. Response to a transient initiated by other reasons ___
11. Other ________

D. WORKING GROUP INVOLVED:

1. Operation ___ 2. Instrumentation ___
3. Electrical maintenance ___ 4. Mechanical maintenance ___
5. Outside contractor ___ 6. Other ___

E. IMPACT OF THE ERROR:

1. Reactor trip ___ 2. Turbine 1 ___ 3. Turbine 2 ___
4. Loss of power ___ % 5. Other ________

In response to a transient (C = 10):

6. Diagnostic error ___ 7. Action error ___

Impact on safety system: ________(KZ)

8. Red 1 ___ 9. Red 2 ___ 10. Unavailability time ________

F. MEANS OF DETECTION OF THE ERROR:

1. Alarms/trip ___ 2. Symptoms (immediately) ___
3. Shift inspection ___ 4. Periodic testing ___
5. Preventive maintenance ___ 6. On demand ___
7. Quality control ___ 8. Repair ___
9. Random casual inspection ___ 10. Inspection before startup ___
11. Other ________

G. PROBABLE CAUSES OF ERROR (AND RELATED QUESTIONS):

1. Written procedure exists ___ 2. Procedure defective ___
3. No written procedure ___ 4. Performed acc'g to procedure ___
5. Procedure misunderstood ___ 6. Procedure not followed ___
7. Wrong procedure selected ___ 8. Task too difficult ___
9. Situation misunderstood ___ 10. Acc'g to oral instructions ___
11. Oral instr. misunderstood ___ 12. Oral instr. not followed ___
13. Training doesn't cover ___ 14. Poor communication means ___
15. Slip, carelessness ___ 16. Lack of time, stress ___
17. Inadequate display ___ 18. Poor control means (in C.R.) ___
19. Correct action too late ___ 20. Other ________

H. PREVENTIVE ACTIONS TAKEN:

1. Modification of a procedure ___ 2. Modified training ___
3. System modification ___ 4. Addition of inspections ___
5. None ___ 6. Other ________

I. RECOMMENDATION: ________

J. SOURCE OF INFORMATION: ________

TABLE IV. MEANS OF DETECTION AND MEASURES TAKEN VERSUS TYPE OF ACTIVITY

		F1	F2	F3	F4	F5	F6	F7	F8	F9	F10	F11	H1	H2	H3	H4	H5	H6
C1	Periodic testing	8	6	2	3		2			1		1	5			1	16	1
C2	Preventive maintenance	8	1							2			1				10	
C3	Failure search	4	5										1	1			7	
C4	Equipment removal/return	2	1				1										4	
C5	Failure repair	1	2	1													4	
C6	Normal operation	25	11	5				2		12		8	6			2	48	5
C7	Startup	8	16							3	4	3	5	1			26	2
C8	Shutdown	3	1				1						1				4	
C9	Annual revision	2	10	3			2	2	2	11	1	2					32	3
C10	Transient response	3	3														6	
C11	Other	3	2											1			3	
	Total	67	58	11	3		6	4	2	29	5	14	19	3		3	160	11

MEANS OF DETECTION:
F1 Alarms/trips
F2 Symptoms, immediately
F3 Shift inspection
F4 Periodic testing
F5 Preventive maintenance
F6 In demand
F7 Quality assurance
F8 Repair
F9 Casual inspection
F10 Inspection before startup
F11 Other

PREVENTIVE MEASURES TAKEN:
H1 Modification of a procedure
H2 Modification of training
H3 Modification of a system
H4 Addition of inspections
H5 None
H6 Other

TABLE V. ROOT CAUSES VERSUS TYPE OF ACTIVITY

		G1	G2	G3	G4	G5	G6	G7	G8	G9	G10	G11	G12	G13	G14	G15	G16	G17
C1	Periodic testing	4			2				1		3		15					
C2	Preventive maintenance	1									2		8	2				
C3	Failure search		2					1			3		7	1				
C4	Equipment removal/return										1		3					
C5	Failure repair								1		1		3					
C6	Normal operation	6	5	3	6		1	1	3	1	6		44	5				1
C7	Startup		5		8			2			14		19	6				1
C8	Shutdown	1	1										3	1				
C9	Annual revision	1	1		9						1		31	8				1
C10	Transient response							1	1				4					
C11	Other							1					4					
	Total	13	4	3	23		1	6	6	1	31		141	23				3

G1	Defective procedure	G12	Slip, carelessness
G2	No procedure available	G13	Lack of time, stress
G3	Procedure misunderstood	G14	Defective display
G4	Procedure not followed	G15	Defective remote control in control room
G5	Wrong procedure	G16	Correct action done too late
G6	Task too diffucult	G17	Other
G7	Situation misunderstood		
G8	Oral instruction misunderstood		
G9	Oral instruction not followed		
G10	Lack of training		
G11	Defective connection		

TABLE VI. AVAILABILITY OF PROCEDURE ROOT CAUSES AND PREVENTIVE MEASURES

ROOT CAUSES		PROCEDURE AVAILABLE (43 events)	NO PROCEDURE (34 events)
G1	Defective procedure	15	
G3	Procedure misunderstood	4	
G4	Procedure not followed	24	
G6	Task too difficult		2
G7	Situation misunderstood	4	2
G8	Oral instruction misunderstood		1
G2	Oral instruction not followed		1
G10	Lack of training	7	22
G12	Slip/carelessness	32 (6)[a]	17 (3)[a]
G13	Lack of time/stress	8	8
G14	Defective display		2
	PREVENTIVE MEASURES TAKEN		
H1	Modification of a procedure	11	6
H2	Modification of training	1	1
H3	Modification of a system		
H4	Addition of inspections	2	
H5	None	28	25
H6	Other	1	2
		43	34

a) The number of events in which G12 was the only reason for an inappropriate action.

5. SUMMARY

A plant-specific reliability and human data collection and analysis system for the PSA of the Loviisa 1 power plant has been described. The PSA data system has been designed to be part of the future on-line data collection and analysis system.

Historical raw data have been collected using forms designed to gather all relevant information about events. Equations for all possible combinations of answers to items in the form have been developed to obtain failure rates and unavailability parameters in base case quantification and in time-dependent calculations. A multiplicity of failure entry, detection and exit mechanisms is taken into account. Parameters are estimated using a new robust parametric empirical Bayes method. A set of computer programs has been designed to facilitate the data storage, sorting, search and parameter estimation.

The parameter estimation for PSA and the comparison of the parameters with generic data is going to be completed in the near future. Advanced models are being developed and tested for verification of learning.

A survey of human errors has been carried out as part of the human reliability quantification using a special questionnaire. The results of the statistical analysis will be taken into account in PSA quantification and identification of human errors.

REFERENCES

[1] VAURIO, J.K., Objectives and Methods of Reliability Data Collection and Analysis at Loviisa Power Plant. Workshop on Probabilistic Safety Assessment as an Aid to Nuclear Power Plant Management, May 20-23, 1986, Brighton, U.K. OECD, NEA, CSNI, Task Force 7, Principal Working Group 5.

[2] VAURIO, J.K., LINDÉN, G., On robust methods for failure rate estimation. Reliability Engineering 14(1986)123-132.

[3] VAURIO, J.K., JÄNKÄLÄ, K., Comparison of Methods for Estimating Failure Probabilities. The 7th SRE-Symposium, Espoo, Finland, Oct. 14-16, 1986. Society of Reliability Engineers, Scandinavian Chapter.

[4] VAURIO, J.K., On analytic empirical Bayes estimation of failure rates. To be published in Risk Analysis, an International Journal.

[5] JÄNKÄLÄ, K.E., VAURIO, J.K., Empirical Bayes Data Analysis for Plant Specific Safety Assessment. Paper to be presented in Int. Topical Conf. on Probabilistic Safety Assessment and Risk Management, Zurich, Switzerland, August 31 to September 4, 1987.

[6] VAURIO, J.K., HERRMAN, N., Trans. Am. Nucl. Soc., Vol 35, (1980)264-266.

[7] VAURIO, J.K., On exact and asymptotic unavailability expressions for standby components and systems. Reliability Engineering 10(1985)151-173.

[8] VAURIO, J.K., Learning Curve Estimation Techniques for Nuclear Industry. Proc. Intl. Conf. Numerical Methods in Nuclear Engineering, CNS/ANS, 1983 September 6-9, Montreal, Canada.

[9] VAURIO, J.K., Learning from nuclear accident experience. Risk Analysis 4(1984)2, pp. 103-115.

[10] VAURIO, J.K., ISAKSSON, J., LINDÉN, U., Applications of a New Reliability Model That Fits Data. The 7th SRE-Symposium, Espoo, Finland, Oct. 14-16, 1986. Society of Reliability Engineers, Scandinavian Chapter.

[11] VUORIO, U.M., VAURIO, J.K., Advanced Human Reliability Analysis Methodology and Applications. Paper to be presented in Int. Topical Conf. on Probabilistic Safety Assessment and Risk Management, Zurich, Switzerland, August 31 to September 4, 1987.

L'UTILISATION DES ETUDES PROBABILISTES DANS L'ELABORATION DES SPECIFICATIONS TECHNIQUES D'EXPLOITATION D'UN REP

D.B. LÉVY
Service central de sûreté des installations nucléaires,
Ministère de l'industrie, des P. et T. et du tourisme,
Paris

A. DUBREUIL-CHAMBARDEL
Electricité de France,
Clamart

J.-M. MATTEI, G. BARS
Institut de protection et de sûreté nucléaire,
Commissariat à l'énergie atomique,
Fontenay-aux-Roses

France

Abstract-Résumé

THE USE OF PROBABILISTIC STUDIES IN PREPARING TECHNICAL OPERATING SPECIFICATIONS FOR A PWR.

The technical operating specifications define the rules which must be complied with in the normal operation of a unit in order to ensure the safety of the public and the operating personnel. It is obvious that any non-availability, programmed or accidental, of a piece of equipment important from the point of view of security, results in an increase in risk which must lead to a limitation of the non-availability and/or its duration. This is as true for a unit under load as for those in a shutdown state. Risk studies relating to the loss of redundant safety systems have made it possible to evaluate, for many materials, authorized maximum durations of non-availability, either accidental or programmed. However, an evaluation of the probability of a core meltdown during cold shutdown has made it possible to pass judgement on the technical specifications applied for limiting the non-availability of materials for maintenance during the entire period of cold shutdown.

L'UTILISATION DES ETUDES PROBABILISTES DANS L'ELABORATION DES SPECIFICATIONS TECHNIQUES D'EXPLOITATION D'UN REP.

Les spécifications techniques d'exploitation définissent les règles techniques qui doivent être respectées en fonctionnement normal de la tranche pour garantir la sécurité du public et du personnel d'exploitation. Il est clair que toute indisponibilité, programmée ou fortuite, d'un équipement important pour la sûreté entraîne une augmentation de risque qui doit conduire à limiter les indisponibilités et/ou leur durée. Cela est vrai aussi bien pour une tranche en puissance que dans les états d'arrêt. Les études de risque relatives à la perte de systèmes de sûreté redondants ont permis d'évaluer, pour de nombreux matériels, les durées maximales autorisées d'indisponibilité fortuite ou programmée. D'autre part, une évaluation de la probabilité de fusion du cœur durant les états d'arrêt à froid a permis de porter un jugement sur les spécifications techniques mises en place pour limiter les indisponibilités de matériels pour maintenance durant toute la période d'arrêt à froid.

1. INTERET DES SPECIFICATIONS TECHNIQUES

Les spécifications techniques d'exploitation définissent les règles techniques qui doivent être respectées en fonctionnement normal de la tranche pour garantir la sécurité du public et du personnel d'exploitation.

Les spécifications rappellent les limites de sécurité, fixent les conditions limites de fonctionnement, la liste des matériels dont la disponibilité est requise dans les différents états de fonctionnement et imposent la conduite à tenir en cas d'indisponibilité d'un équipement ou d'un ensemble d'équipements importants pour la sûreté.

Les spécifications techniques établies pour la centrale de Fessenheim, première tranche du palier 900 MWe à partir des spécifications techniques de la centrale de référence (Beaver Valley), ont été analysées, en 1977, par les organismes de sûreté lors de la mise en service de la centrale.

Cette analyse, effectuée avec une approche déterministe, a mis en évidence la nécessité d'examiner la validité des délais de maintien en fonctionnement de la tranche en cas d'indisponibilité de composants importants pour la sûreté; cet examen a été appuyé par des études probabilistes.

La prise en compte des premiers résultats de ces études, du retour d'expérience et des enseignements tirés de l'accident de TMI a conduit l'exploitant à proposer une nouvelle version des spécifications techniques en 1980.

Dans cette version, en outre améliorée dans la forme pour en faciliter l'utilisation, un objectif probabiliste d'accroissement du risque de conséquences inacceptables a été utilisé dans le cadre de la justification des délais d'indisponibilité des principaux matériels importants pour la sûreté.

Il est clair que toute indisponibilité, programmée ou fortuite, d'un équipement important pour la sûreté entraîne une augmentation de risque qui doit conduire à limiter les indisponibilités et/ou leur durée. Cela est vrai aussi bien pour une tranche en puissance que dans les états d'arrêt.

2. SPECIFICATIONS TECHNIQUES EN PUISSANCE

2.1. Critères retenus

Les études de risque relatives à la perte de systèmes de sûreté redondants ont permis d'évaluer, pour de nombreux matériels, les durées maximales autorisées d'indisponibilité fortuite ou programmée.

On appelle indisponibilité fortuite d'un matériel toute indisponibilité provoquée par sa défaillance. Une telle indisponibilité entraîne un surcroît de risque d'autant plus élevé que sa durée est longue: on détermine, pour un matériel donné, une durée maximale autorisée d'indisponibilité tranche en puissance, correspondant à un accroissement de la probabilité de fusion du cœur de 10^{-7} par indisponibilité.

L'indisponibilité programmée est l'indisponibilité résultant d'une maintenance sur un matériel. On vérifie que l'ensemble des maintenances faites annuellement sur un matériel donné n'entraîne pas une augmentation de risque supérieure à 10^{-7} par an.

2.2. Présentation de la démarche

De 1983 à 1985, Electricité de France (EDF) a réalisé des études de risque relatives à la perte totale de systèmes redondants. Il convient de rappeler que, pour faire face à ces situations complémentaires de fonctionnement, il est prévu de mettre en œuvre des parades et des procédures de conduite dites procédures H, précisant quand et comment les utiliser.

Les scénarios d'accidents consécutifs à ces pertes de systèmes redondants ont été modélisés à l'aide d'arbres d'événements. La probabilité de perte de ces systèmes a été estimée à l'aide de graphes d'états (markoviens ou semi-markoviens).

Quand l'état d'un système est jugé suffisamment dégradé, les spécifications techniques d'exploitation imposent de faire passer une tranche initialement en puissance dans un état de repli.

C'est pourquoi, à l'aide des graphes modélisant la perte de systèmes redondants, on a déterminé toutes les séquences d'événements (défaillance ou répartition de composants du système en question) aboutissant à la perte de ce dernier.

Chacune des séquences ainsi obtenues est «reportée» sur les arbres «en puissance» et «en état de repli».

Soit par exemple la séquence correspondant à la perte des quatre pompes RRI (système de refroidissement intermédiaire) en fonctionnement; elle peut se dérouler de la manière suivante:

- perte de la première pompe RRI voie A (maintien de la tranche en puissance): (t_0);
- perte de la deuxième pompe RRI voie A (la spécification technique d'exploitation impose de passer la tranche en état de repli, ce qui prendra 12 heures): (t_1);
- perte de la troisième pompe RRI (voie B): (t_2);
- atteinte des conditions de l'état de repli: ($t_1 + 12$);
- perte de la quatrième pompe RRI (voie B): (t_3).

Suivant que la quatrième pompe RRI tombe en panne avant l'atteinte des conditions de l'état de repli ($t_3 < t_1 + 12$) ou après ($t_3 > t_1 + 12$), la séquence est considérée comme une séquence initiatrice de l'arbre «tranche en puissance» ou de l'arbre «tranche en état de repli».

De nouvelles méthodes de calcul ont été développées pour calculer la probabilité de pouvoir ainsi atteindre ou non l'état de repli avant la perte totale du système, séquence par séquence.

Les dépendances temporelles entre systèmes ont été prises en compte dans le calcul de la probabilité d'occurrence des séquences accidentelles de chaque arbre (dépendances comme le fonctionnement en normal-secours de deux systèmes, possibilités de réparation de certains systèmes au cours d'une séquence et prise en compte du délai avant conséquences inacceptables, etc.).

Pour mettre en œuvre ces méthodes, EDF a développé l'ensemble de logiciels Elsa.

2.3. Application aux principaux matériels électriques

A partir de l'étude probabiliste du risque lié à la perte totale des alimentations électriques secourues, les délais maximaux d'indisponibilité des matériels suivants ont été calculés: transformateur de soutirage, transformateur auxiliaire, diesels, turbo-alternateur de secours LLS. Le LLS permet de garder une tension pour assurer l'injection aux joints des pompes primaires et de conserver à long terme le minimum de contrôle-commande nécessaire au pilotage.

Ces calculs ont été faits pour les tranches REP 900 MWe et 1300 MWe, avec turbine à gaz de site (disponible en trois heures maximum) ou régionale (disponible en 24 ou 72 heures) en faisant l'hypothèse que les joints de pompes primaires tiennent ou non en l'absence d'injection et de barrière thermique.

Les délais maximaux autorisés d'indisponibilité obtenus sont cohérents avec ceux déjà estimés auparavant pour le palier REP 900 MWe. On a de plus pu estimer le gain apporté par le LLS et la turbine à gaz. A titre d'exemple, avec l'hypothèse de non-tenue des joints de pompes primaires en l'absence d'injection, les résultats pour une tranche REP 1300 MWe sont donnés dans le tableau I.

A partir de ces résultats provisoires, après simplification et regroupement en valeurs arrondies, on détermine puis discute dans le cadre réglementaire les délais maximaux d'indisponibilité fortuite.

2.4. Développements prévus

EDF réalise actuellement une étude probabiliste de la sûreté de la centrale de Paluel (Projet EPS 1300).

Cette étude sera entièrement informatisée. Le logiciel Lesseps développé à cette fin permettra de faire automatiquement toute étude de sensibilité au taux de défaillance d'un composant ou à l'indisponibilité de ce même composant. L'aide à l'établissement des spécifications techniques de tous les matériels pourra donc être très fortement automatisée.

TABLEAU I. DELAIS MAXIMAUX AUTORISES D'INDISPONIBILITE

Composant(s) indispensable(s)	Turbine à gaz régionale (h)	Turbine à gaz (TAG) de site (h)
TA	84	443
TS	77	427
1 diesel	720	2095 (TAG 4 MWe) 7019 (TAG 7 MWe)
2 diesels	22	35 (TAG 4 MWe) 73 (TAG 7 MWe)
LLS	235	413
TA + 1 diesel	4,5	30,2 (TAG 4 MWe) 30,5 (TAG 7 MWe)
TA + LLS	2	40
LLS + diesel	26	212 (TAG 4 MWe) 237 (TAG 7 MWe)

Le logiciel Lesseps actuellement limité aux tranches REP pourrait être étendu pour estimer la sûreté de n'importe quelle installation complexe et être un outil d'aide à l'exploitation permettant de contrôler le surcroît de risque résultant des maintenances, indisponibilités fortuites ou tests de matériel.

3. SPECIFICATIONS TECHNIQUES A FROID

Durant les périodes d'arrêt à froid du réacteur, les opérations de maintenance préventive et la réalisation de modifications des circuits entraînent des indisponibilités programmées de composants ou systèmes. Les indisponibilités les plus contraignantes sont celles qui affectent les pompes des circuits importants pour la sûreté: ASG, RRI, SEC, RRV, RIS et EAS, ainsi que les alimentations des tableaux électriques secourus.

En raison du risque lié à ces situations pour lesquelles des procédures accidentelles ont été mises en place, EDF a entrepris de limiter par des spécifications techniques les indisponibilités de matériels pour maintenance, pour toute la période d'arrêt à froid.

TABLEAU II. PHASES CONSIDEREES POUR L'ETUDE PROBABILISTE DU RISQUE

	Phase 1	Phase 2	Phase 3	Phase 4	Phase 5	Phase 6
				Plan médian des turbulences		Piscine pleine
Temps après arrêt du réacteur (h)	4 → 20	20 → 68	68 → 548	68 → 308	Ouvert. Fermet. 68→128	128 → 488
Pression (bar)	24–28	P < 28 (Monophasique)	1–28	RCP ouvert	RCP ouvert	RCP ouvert
Température (°C)	177–90	90–60	90–10	70–10	60–10	60–10
Nombre de GV nécessaires	2	2	1			
Disponibilité du PTR en secours du RRA	Non	Non	Oui (1VDP ouverte)[a]	Oui	Oui	Oui
Arrêt à froid normal	4 biph 7 monoph 20					
Arrêt à froid pour rechargement						
Arrêt à froid pour intervention A						
— ou — B						
Nombre de passages par an / Durée annuelle (h)	6 / 96	3 / 144	1 / 480	1 / 240	1 / 60 X 2	1 / 360

[a] VDP: vanne de décharge pressuriseur

Ce projet de spécifications techniques définissant les matériels nécessaires pour assurer un certain nombre de fonctions (inventaire en eau, évacuation de la puissance résiduelle, confinement, contrôle de la pression, contrôle de la réactivité) a été transmis aux autorités de sûreté.

Une évaluation de la probabilité de fusion du cœur correspondante a été entreprise au Département d'analyse de sûreté de l'Institut de protection et de sûreté nucléaire (IPSN).

3.1. Méthode d'analyse

Lors du fonctionnement du réacteur en arrêt à froid, l'évacuation de la puissance résiduelle se fait par le circuit de refroidissement à l'arrêt (RRA). Ce circuit est intérieur à l'enceinte et permet la circulation du fluide primaire par l'intermédiaire de deux pompes. La chaleur de ce fluide est transférée dans les échangeurs RRA vers le circuit RRI puis vers le circuit SEC dans les échangeurs RRI–SEC. Le circuit RRA permet également la protection du circuit primaire contre les surpressions par l'intermédiaire de deux soupapes de sûreté.

L'étude probabiliste du risque lié aux états d'arrêt à froid a été réalisée en considérant six phases dont les durées ont été tirées du retour d'expérience, qui se distinguent par les configurations du circuit primaire (fermé ou ouvert), les paramètres thermohydrauliques (puissance résiduelle, pression, température, niveau d'eau) et les moyens de secours disponibles (tableau II).

Pour chaque phase considérée, une évaluation de la probabilité de fusion du cœur a été effectuée en considérant, d'une part, que tous les matériels sont disponibles, et, d'autre part, que seuls les matériels requis dans le projet de spécifications techniques le sont. Pour cette seconde évaluation, il a été tenu compte de la récupération des matériels non requis sur un temps moyen de 40 heures.

La double quantification a donc permis de déterminer la valeur de l'accroissement de risque obtenu par l'application de chaque combinaison de matériels requis par le projet, pour chaque initiateur, considéré sur la durée annuelle de chaque phase.

Il est ainsi possible de juger de l'acceptabilité du projet et de définir le cas échéant les matériels supplémentaires dont la disponibilité doit être requise pour diminuer la probabilité de fusion du cœur.

3.2. Evaluation probabiliste

3.2.1. Définition des initiateurs considérés

Les initiateurs pour lesquels l'étude probabiliste du risque a été effectuée se regroupent en deux familles:

- les pertes de la fonction évacuation de la puissance résiduelle;
- les pertes de réfrigérant primaire.

3.2.1.1. Pertes de la fonction évacuation de la puissance résiduelle

Les systèmes suivants sont utilisés pour évacuer la puissance résiduelle:

- RRA (circuit de refoidissement à l'arrêt);
- PTR (circuit de traitement des piscines, utilisable dans les phases 3, 4, 5, 6 en secours du RRA):
- RRI–SEC (systèmes de refroidissement intermédiaire et d'eau brute);
- CFI (prise d'eau);
- LHA et LHB (tableaux électriques secourus).

Pour chaque phase de fonctionnement, nous avons déterminé les combinaisons de défaillances entraînant la perte de la fonction évacuation de la puissance résiduelle prenant en compte:

- les défaillances des systèmes;
- les défaillances croisées entre une file d'un système et la file d'un autre système alimentée par une voie électrique différente.

Durant les phases 1, 2 et 3, les procédures accidentelles en place (H1-2 et H3-2) préconisent de passer sur les générateurs de vapeur sur perte de la fonction; pour les phases 4, 5 et 6 (circuit primaire ouvert), un appoint doit être réalisé de façon à compenser le débit vaporisé.

3.2.1.2. Pertes de réfrigérant primaire

Les accidents de perte de réfrigérant primaire dans les états d'arrêt à froid sont caractérisés par l'absence d'automatismes destinés à maintenir un inventaire en eau suffisant pour éviter le découvrement du cœur, à l'exception d'un signal de pression haute enceinte qui pourrait intervenir dans la première phase d'utilisation du RRA.

La procédure A10 a été prévue pour servir de guide à l'opérateur dans les phases où le réacteur est encore en pression.

Dans les phases où le circuit primaire n'est plus en pression, l'exploitant exclut les pertes de réfrigérant et en conséquence n'a pas prévu l'ensemble des moyens (procédures et matériels) nécessaires pour y faire face.

Néanmoins, dans l'étude probabiliste, l'hypothèse suivante a été utilisée: les probabilités d'occurrence de rupture de tuyauterie ont été prises égales à celles prises en compte en puissance. Cette hypothèse semble confirmée par les résultats de l'étude de P. Janzen[1].

---

[1] JANZEN, P., A Study of Piping Failures in US Nuclear Power Reactors, AECL-MISC 204.

3.2.2. Méthode de quantification

Le traitement de chaque initiateur a été effectué à partir d'arbres d'événements. Les événements ont été quantifiés par arbres de défaillances. Les arbres d'événements ont été élaborés en tenant compte des procédures de conduite existantes et des actions les plus adéquates qui pouvaient être entreprises en cas de défaillance de matériels.

Les données élémentaires de fiabilité utilisées résultent du retour d'expérience français; en ce qui concerne le facteur humain, les abaques de Swain ont été employés.

3.3. Résultats

L'étude fournit les principaux résultats suivants.

3.3.1. Dans la situation de disponibilité de tous les composants

La valeur de la probabilité de fusion du cœur est essentiellement due au blocage en position ouverte d'une soupape RRA et aux brèches, du fait de l'absence d'automatismes de mise en service du système d'injection de sécurité.

3.3.2. Dans les situations d'application des spécifications techniques proposées

En ce qui concerne les pertes de réfrigérant, des contributions importantes au risque apparaissent avec l'application du projet de spécifications techniques dans les états circuit primaire ouvert. Ces contributions sont dues au fait que l'exploitant postule l'absence de perte de refrigérant dans ces situations, donc ne prévoit pas de devoir y faire face, en particulier par des recirculations.

Pour ce qui est des pertes de systèmes, l'étude met en évidence des situations où le nombre de matériels requis n'est pas suffisant: par exemple, lors de la phase 2 pour couvrir l'initiateur: perte d'un tableau électrique et d'une file d'un système parmi RRA, RRI ou SEC.

4. CONCLUSION

L'analyse des principales séquences contribuant à accroître la probabilité de fusion du cœur sur les REP 900 a conduit les autorités de sûreté à proposer que des matériels supplémentaires soient requis dans les spécifications techniques à froid, et a confirmé que le champ d'application de certaines procédures devrait être revu.

EDF a satisfait à la première demande en requérant la disponibilité de matériels supplémentaires en arrêt à froid. Il faut d'autre part noter que la mise en place de tandems de soupapes assistées sur le RRA devrait diminuer significativement la probabilité de fusion du cœur.

Dans le cadre de l'analyse de sûreté des tranches REP françaises, des discussions sur le projet de spécifications techniques à froid ont été engagées entre les autorités de sûreté et EDF à la lumière de l'étude probabiliste présentée dans ce mémoire.

SYSTEM ANALYSIS AND EXPERIMENTAL STUDIES RELATED TO INDIAN PRESSURIZED WATER REACTORS

B.F. CHAMANY, S.K. MEHTA, S.K. GUPTA, V. VENKAT RAJ,
A. KAKODKAR, P.R. DASTIDAR
Bhabha Atomic Research Centre,
Trombay, Bombay, India

Abstract

SYSTEM ANALYSIS AND EXPERIMENTAL STUDIES RELATED TO INDIAN PRESSURIZED WATER REACTORS.

India has a nuclear power programme of building pressurized heavy water reactors (PHWRs) of 235 MW(e) and 500 MW(e) to a total installed capacity of 10 000 MW(e) by the year 2000. A standardized design has been evolved with design and engineering analyses capability meeting the international safety standards. The importance of transient analysis is well recognized. A number of the subsystem models have been developed and tested against experimental data as well as data specially recorded during plant transients. These models have been integrated into system codes for analyses of various plant transients. At the Madras Atomic Power Station a series of total load rejection tests were conducted and important parameters were recorded on fast recorders and analysed. RELAP4/MOD-6 was used to analyse International Atomic Energy Agency standard problems as well as loss of flow accidents of reactors at Rajasthan Atomic Power Station having horizontal channels. RELAP5/MOD-1 with its ability to represent two-fluid models is being commissioned to study PHWRs. Study of natural convection in a figure-of-eight loop with horizontal channels is in progress. Non-dimensional expressions have been developed for flow and heat transfer under natural circulation conditions. Experiments on blowdown from horizontal pressure tube geometries are planned. Experimental and analytical work on the rewetting of hot horizontal channels has been carried out. The design of the PHWR containment has been progressively improved to a present day design of double containment with suppression pool. Experiments to determine the pressure and temperature transients in a scaled down containment model are being conducted. The paper discusses the above aspects of transient and safety analysis and the supporting experimental studies.

1. INTRODUCTION

India has a nuclear power programme of building pressurized heavy water reactors (PHWRs) of 235 MW(e) and 500 MW(e) to a total installed capacity of 10 000 MW(e) by the year 2000. A standardized design has been evolved with design and engineering analyses capability meeting the international safety standards. The Bhabha Atomic Research Centre has developed a number of computer codes. These are being constantly reviewed against data obtained during transients experienced in the operating plants and special tests conducted during commissioning stages, in addition to specially engineered experiments.

2. REACTOR DESCRIPTION

The reactor core consists of an array of horizontal pressure tubes housed in a cylindrical vessel called a calandria. The pressure tubes contain fuel bundles which are cooled by pressurized heavy water. The primary heat transport (PHT) system is arranged with two sets of heat exchangers and pumps on either side. The reactor is provided with a shutdown cooling system for removal of decay heat under shutdown conditions and high pressure D_2O accumulator and low pressure light water accumulators and recirculation systems as part of the emergency core cooling system (ECCS) to provide cooling under loss of coolant accident (LOCA) conditions. The reactors are provided with four reactivity control systems as well as two independent fast acting and diverse shutdown systems for prompt shutdown of the reactor. In addition boron is added to the moderator system to compensate for xenon decay when the reactor is tripped after power operation.

A feed and bleed system is provided to maintain the PHT system pressure. A pressurizer will also be provided for future reactors of 500 MW(e) size. Basically, a PHWR is operated as a base load station, steam extracted by turbine matching the steam production in the steam generator. The signal for this is generated by a control system called a boiler pressure controller (BPC). This gives the excess pressure in the boiler as compared with the required boiler pressure for the measured reactor power. The excess steam is dumped into the condenser, through a series of steam discharge valves, in a cascaded fashion. If BPC error opens the large set of discharge valves it automatically introduces a set-back signal to reduce the reactor power at half a per cent per second till the set-back is cleared with power above the xenon poison-out limit and at the same time it does not cause reactor trip due to high boiler pressure, high PHT temperature, or PHT high or low pressure.

3. TRANSIENT ANALYSIS

All trip set points are arrived at after careful transient analyses of the power plant so that chances of any unnecessary trip are avoided except under extreme conditions. Due to reactivity limitations any reactor trip can cause xenon poison-out, which should be avoided. The importance of transient analysis was therefore well recognized from the beginning of our programme.

A simulation model has been developed to study the plant responses to various operational transients. It models various plant subsystems, their interactions and feedback effects. The model assumes subcooled liquid in the PHT system and an equilibrium model with homogeneous two phase mixture in the steam drum. The PHT pressure is calculated from the state equation of heavy water and the steam drum pressure is calculated from mass and energy conservation equations. A number of the subsystem models have been developed and tested against experimental data as well as data specially recorded during plant transients.

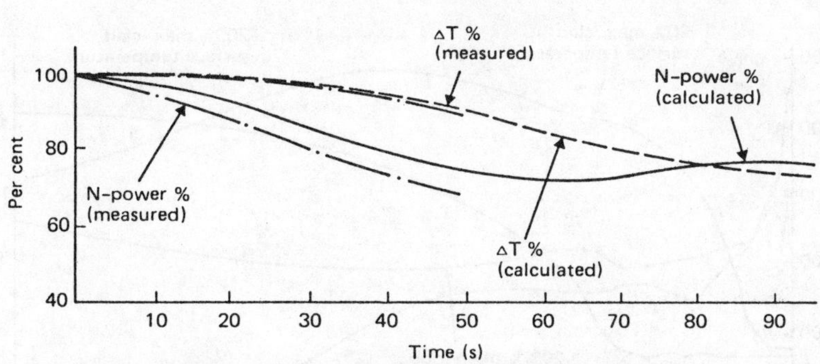

FIG. 1. Madras Atomic Power Station load rejection test.

The models developed include those for simulating the PHT system, boiler, turbine, condenser and the related controls for boiler level, boiler pressure, PHT system pressure, reactivity, reactor power set-back and turbine generator operation as well as reactor regulating systems. These models have been integrated into system codes for analysis of plant transients. The codes have been used to analyse various plant transients, including those from grid disturbances, recorded during plant commissioning and operation, and in understanding and predicting plant behaviour under various transients. Such analyses have also provided useful guidance in formulating the operator's action under various conditions and feedback for design improvements.

The large transients registered by disturbance recorders provide reliable data for identification and control without intentional disruption of the plant operation. During large power changes such as reactor trip, turbine trip, etc., data need to be recorded carefully with appropriate instrumentation. This provides valuable information to identify plant parameters which can help in effective tuning of the control systems so as to achieve the desired plant performances [1].

To illustrate the point it may be mentioned that at Madras Atomic Power Station a series of total load rejection tests were conducted and important parameters were recorded on fast recorders. These include PHT pressure, steam temperature, steam flow, feedwater flow, temperature differences across the core, etc. These were compared with the evaluations done by computer models and were found generally to agree within the limitations of data acquisition error bands for such tests. Figure 1 shows the measured and calculated parameters such as neutron power and core temperature rise during a transient. Controller simulation in the model was modified to make a better prediction of the above parameters. These tests have thus given us greater confidence in the use of the model to analyse postulated transients as well as any unusual incidents that might occur.

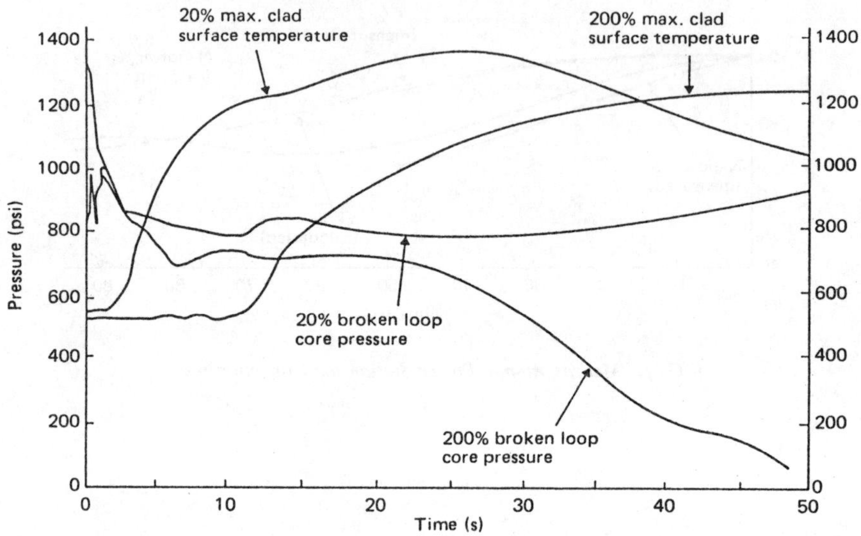

FIG. 2. Typical fuel temperature and core pressure transients for 20% and 200% break (1 psi = 6.895 × 10³ Pa).

The frequency and voltage changes of the electrical grid due to load changes or other faults in the electrical grid can affect the steady operation of nuclear power stations. This has been experienced in the initial operation at Rajasthan Atomic Power Station (RAPS) reactors mainly owing to their installation in a relatively weak grid. The operational transients experienced owing to a weak grid could lead to station isolation and even to a cascade effect resulting in power blackout for the whole grid.

The nuclear power station tripping due to underfrequency conditions, particularly if the station is of relatively large capacity, should be carried out only as a last resort, since it can only worsen the grid conditions. To avoid such a situation isolation of the region containing the nuclear power station can be done with advantage. For example RAPS, along with the adjacent hydro station and some of the bulk loads, have been considered for 'islanding' to avoid severe transients to the station. In ddition dedicated off-site power from a nearby hydro station ensures power to RAPS in case of total loss of power. Such a scheme is being studied for other nuclear power stations in India.

4. SAFETY ANALYSES

Several codes are presently available to do safety studies of LWRs, but very limited information is published analysing the safety of PHWRs. The major factors

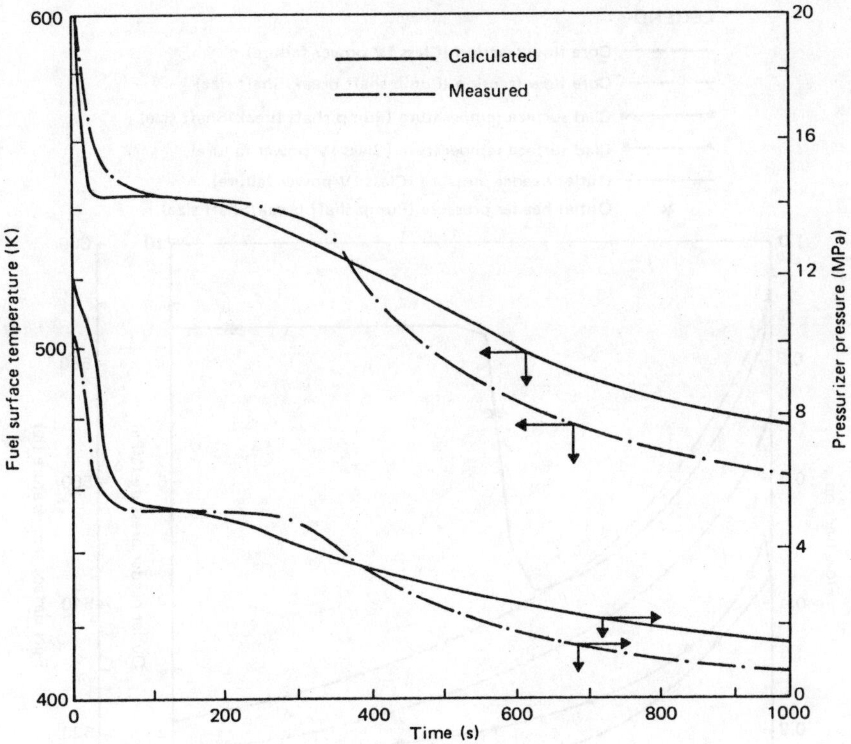

FIG. 3. Typical pressurizer pressure and fuel surface temperature transients.

that make the analysis of PHWRs different from that of LWRs are (a) the horizontal channel arrangements, (b) smaller amount of primary fluid in the PHT system, and (c) the presence of cold moderator as an alternative heat sink. To account for these an in-house computer code is being developed. At the present stage the code has models for homogeneous equilibrium flow, four quadrant two phase pump performance, enthalpy transport, and critical flow. It has a special model for estimating hydrodynamic transients in the reactor channel based on the distributed parameter approach.

Using this code, analyses were made for break sizes of 200% and 20% during the blowdown phase with ECCS being brought in. Subsystems simulated include fuel rods, fuel channels, steam generators, inlet and outlet headers and multiphase pump characteristics. The thermodynamic response of most significant parameters such as pressure, temperature and coolant flow in the hottest channels of reactors was obtained. Some of the results are presented in Fig. 2.

In parallel to the development of the above code RELAP5/MOD-6 a code has been adapted which is basically meant for safety analysis of LWRs. Using this code

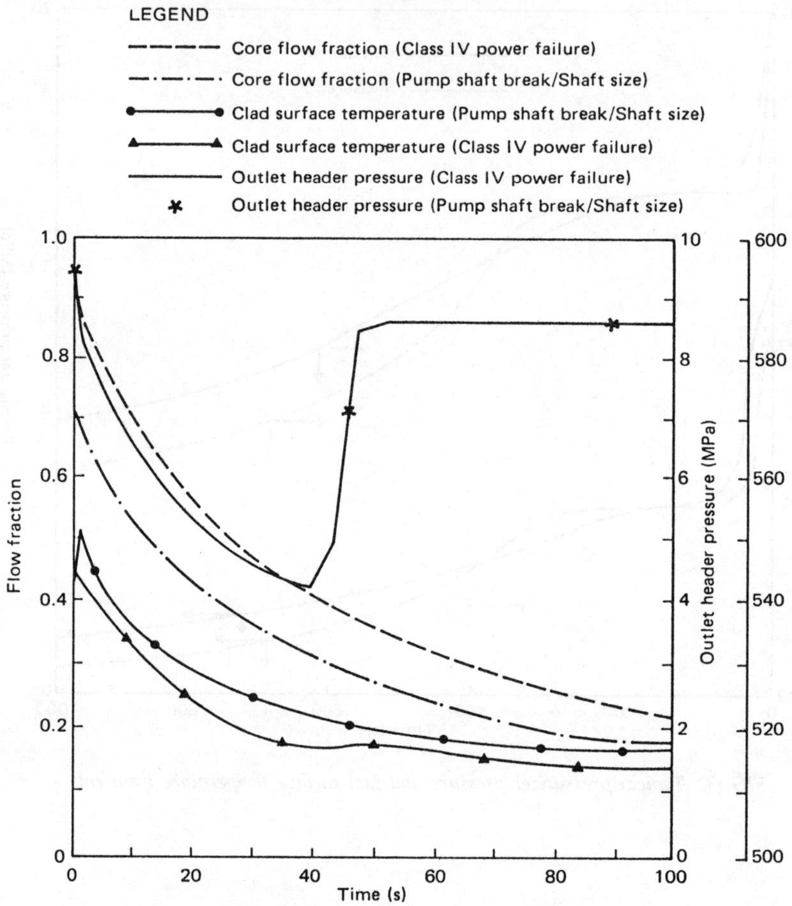

FIG. 4. Pump failure transients.

the International Atomic Energy Agency standard problem exercise for small break LOCA has been carried out and with changes in the modelling and options available in the code post-test analysis results were found to be very much in agreement with the experimental results. Some of the results are presented in Fig. 3.

The RELAP5/MOD-6 code was used to evaluate transients such as loss of flow accident or RAPS as the effect of horizontal channel for single phase flow conditions is not important for the case analysed. Different cases of LOFA considered are: (1) class IV power failure, (2) pump shaft break and (3) pump seizure. Some of the results are presented in Fig. 4. Efforts are being made to bring features of horizontal fuel channel geometry into the RELAP code thus making it suitable for PHWR analysis. In particular as RELAP5 have two-fluid models, it is more amenable for analysis of PHWR systems.

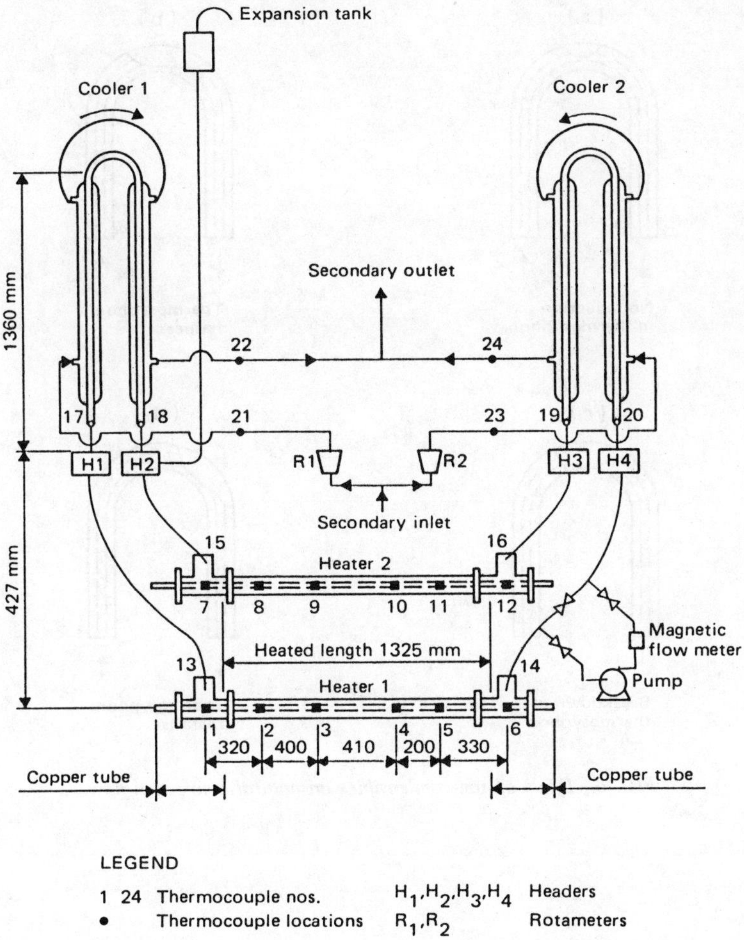

FIG. 5. Experimental loop.

5. THERMOSIPHON/NATURAL CONVECTION COOLING IN PHWR LOOPS

On complete loss of power to the coolant circulating pump, heat removal is effected by the thermosiphon cooling mechanism. A phased programme of studies on natural convection in a figure-of-eight loop with horizontal channels has been carried out. In the first phase, phenomenological experiments were carried out at atmospheric pressure in a transparent loop [2]. The loop (Fig. 5) consisted of two horizontal channels of annular geometry. The inner stainless steel tube was heated by electric currrent and the outer tube was of glass. The heaters were connected to transparent U-tube coolers through stainless steel headers.

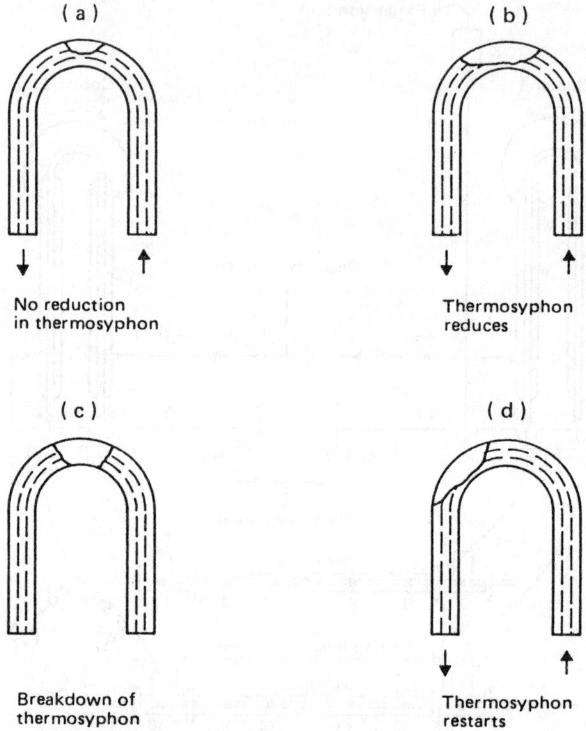

FIG. 6. Effect of non-condensibles on natural convection flow.

These studies indicate that from a stagnant initial condition, thermosiphon flow initiates in either direction. Below a certain power, flow is not initiated. However, after initiation at a higher power natural convection flow is maintained even if power is reduced to very low values. Non-condensibles hinder natural convection flow by collecting at the top of the inverted U-bend in the inner tube of the cooler (Fig. 6). When the volume of non-condensibles at the U-bend exceeds a critical value, thermosiphon flow breaks down temporarily and restarts after reorienting the air bubble. The experiments also indicated that the loop flow resistance under natural circulation conditions is greater than that under forced flow conditions (Fig. 7).

Non-dimensional expressions have been developed for flow and heat transfer under natural circulation conditions so as to extend and generalize the validity of experimental data generated. An analytical model has also been developed for the prediction of loop flow rates and temperatures. Comparison of measured and predicted flow rates was found to give reasonably good agreement. A computer code for transient thermosiphon analysis has been developed [3]. Typical calculated and

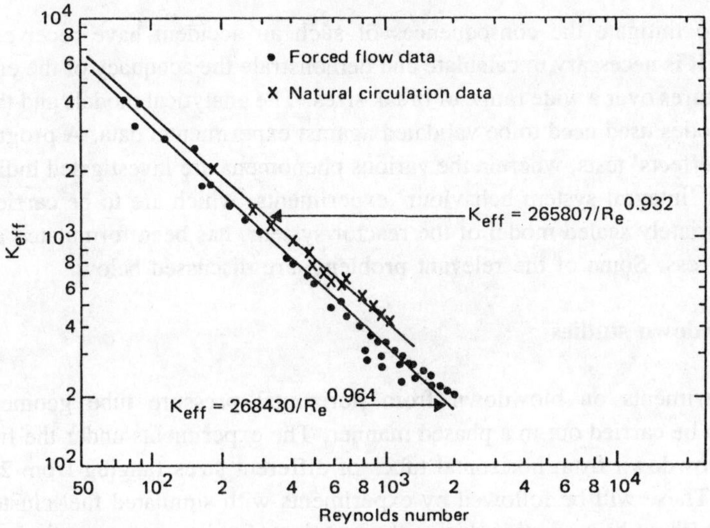

FIG. 7. Loop flow resistance.

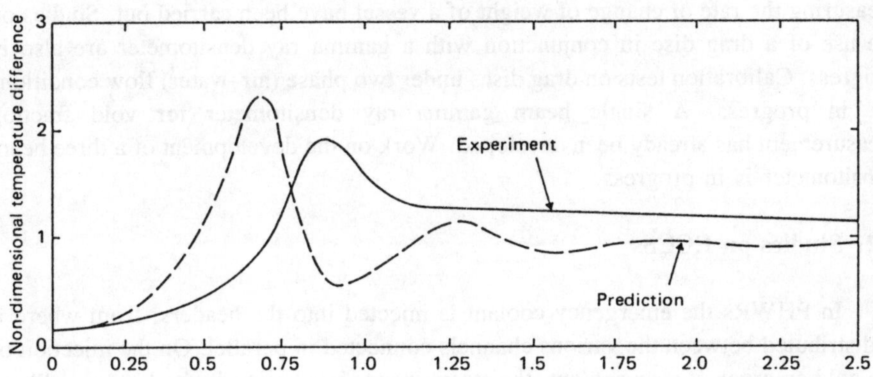

FIG. 8. Natural circulation transient.

measured transients for natural circulation flow initiation from stagnant conditions are shown in Fig. 8. A computer code for the analysis of natural circulation cooling of PHWRs has been developed and used to analyse the system behaviour.

6. SAFETY STUDIES RELATED TO PHWRs

While in the past attention has been centred mainly on the large break LOCA, the implications of a small break LOCA and the need to provide adequate safety features to mitigate the consequences of such an accident have received greater attention. It is necessary to calculate and demonstrate the adequacy of the engineered safety features over a wide range of break sizes. The analytical models and the system analysis codes used need to be validated against experimental data. A programme of 'separate effects' tests, wherein the various phenomena are investigated individually, as well as 'integral system behaviour' experiments, which are to be carried out on an appropriately scaled model of the reactor system, has been formulated and work is in progress. Some of the relevant problems are discussed below.

6.1. Blowdown studies

Experiments on blowdown from horizontal pressure tube geometries are planned to be carried out in a phased manner. The experiments under the first phase involve blowdown from horizontal tubes of different sizes ranging from 25 mm to 100 mm. These will be followed by experiments with simulated fuel clusters, both with and without heat addition during the transient. Studies are presently in progress to develop techniques for measuring transient two phase discharge during the extremely fast blowdown transient. Experiments on the use of load cells for measuring the rate of change of weight of a vessel have been carried out. Studies on the use of a drag disc in conjunction with a gamma ray densitometer are also in progress. Calibration tests on drag discs under two phase (air–water) flow conditions are in progress. A single beam gamma ray densitometer for void fraction measurement has already been developed. Work on the development of a three beam densitometer is in progress.

6.2. Studies on ECCSs

In PHWRs the emergency coolant is injected into the headers, from where it is distributed between the various channels connected in parallel. On the injection of the cold emergency core coolant, the steam–water interaction in the header is likely to cause condensation induced flow and pressure oscillations. This affects the emergency coolant delivery to the fuel channels. A phased programme of experiments in this area is in progress. In the first phase, experiments were conducted on a transparent one-tenth scale model of a portion of a typical PHWR header along with two tubes simulating feeder pipes [4]. The boiler branch was connected to the steam supply and cold water was injected through the ECC injection line. The experimental studies conducted gave valuable insight into the phenomenon of steam–water interaction. The steam water interface was found to oscillate as shown in Fig. 9. The frequency of oscillations varied from about 0.5 Hz to 1.3 Hz depending on steam and water flow rates. Frequency was found to increase with increase in steam flow rate. Pressure oscillations were also noted. Under certain conditions flow reversal in the feeder pipes was noted during the upward motion of the interface through the boiler branch. Further studies in this area are in progress. Scale model studies investigating

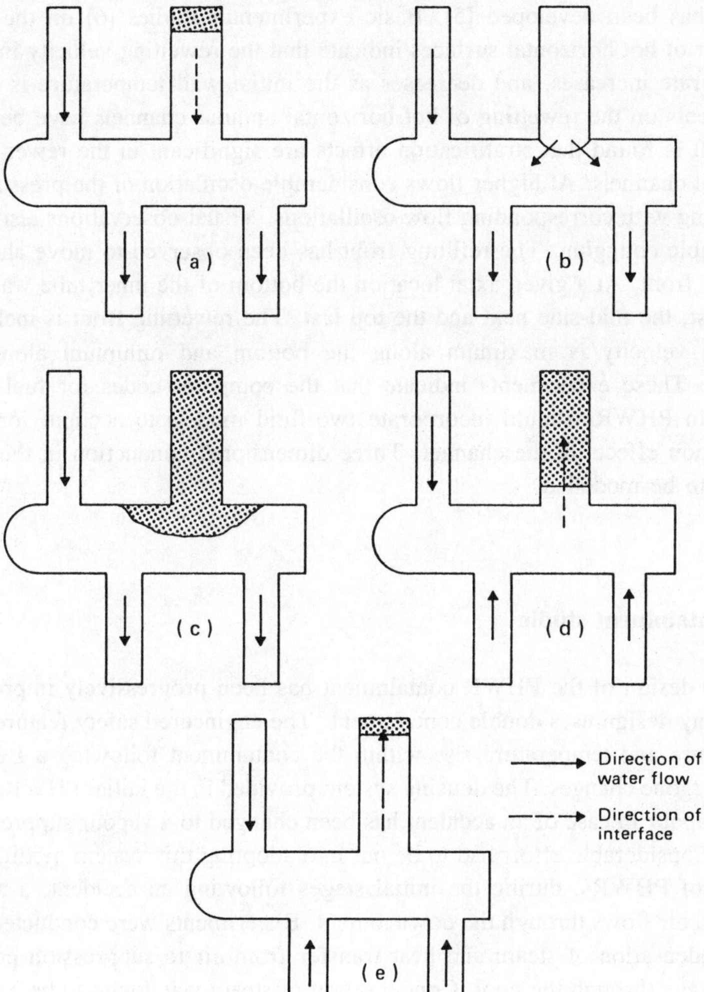

FIG. 9. Sequential position of steam–water interface.

the distribution of the injected coolant between the various fuel channels in parallel and containing fuel whose cladding is at high temperatures are also to be carried out.

By the time the injected emergency coolant enters the fuel channels, the cladding may reach very high temperatures owing to deterioration in the heat transfer. For effective cooling, rewetting of the fuel clad is essential. A phased programme of analytical and experimental work on the rewetting of hot horizontal channels has been carried out. A computer code for conduction controlled rewetting analysis has been developed [5]. Basic experimental studies [6] on the rewetting behaviour of hot horizontal surfaces indicate that the rewetting velocity increases as the flow rate increases, and decreases as the initial wall temperature is increased. Experiments on the rewetting of hot horizontal annular channels have been carried out [7]. It is found that stratification effects are significant in the rewetting of hot horizontal channels. At higher flows considerable oscillation of the pressure drop is noted along with corresponding flow oscillations. Visual observations also indicated considerable chugging. The refilling front has been observed to move ahead of the rewetting front. At a given axial location the bottom of the inner tube was found to rewet first, the mid-side next and the top last. The rewetting front is inclined. The rewetting velocity is maximum along the bottom and minimum along the top locations. These experiments indicate that the computer codes for fuel rewetting analysis in PHWRs should incorporate two-fluid models to account for the fluid stratification effects in the channel. Three dimensional conduction in the fuel rods also has to be modelled.

6.3. Containment studies

The design of the PHWR containment has been progressively improved. The present day design uses double containment. The engineered safety features limiting the pressure and temperature rise within the containment following a LOCA have also undergone changes. The dousing system provided in the initial PHWRs condensing the vapour in case of an accident has been changed to a vapour suppression pool system. Considerable effort had to be put into adopting this system modification. In the case of PHWRs, during the initial stages following an accident, a mixture of steam and air flows through the downcomers. Experiments were conducted to determine condensation of steam and heat transfer from air to suppression pool during their passage through the pool. Condensation of steam was found to be complete. It was observed that the air–steam mixture during its passage through different compartments of the containment is far from homogeneous. There is a need for greater understanding of the phenomenon with a view to correctly modelling it.

The phenomena of bubble growth, bubble detachment and bubble rise in the pool have been studied using a simplified analytical model [8]. The scenario postulated for the model is as follows: after the pool swell phase, a mixture of air and steam from the dry well comes to the suppression pool where the steam condenses. However, when the steam flow rate falls below the rate of condensation, chugging takes place. With condensation there is a reduction in steam mass and a sudden drop in steam pressure. The interface moves up the vent pipe when the steam volume decreases and the steam pressure starts increasing. When the steam pressure is sufficiently high the interface moves again, and the condensation and the chugging cycle repeat.

Experiments are being conducted to determine the pressure and temperature transients in a scaled down containment model. This mode is 1/1000 of the prototype by volume i.e. 1/10 on the linear scale. These experiments will yield data on the pressure–temperature history following a LOCA.

REFERENCES

[1] THANGASWAMY, S., CHAMANY, B.F., MURTY, L.G.K., RAY, R.N., "Sensitivity analysis: A tool for control system commissioning", Control System Tuning Commissioning and Dynamic Model Validation (Proc. Spec. Mtg Harrowgate, 1979), Central Electricity Generating Board, Gloucester (1979).

[2] VIJAYAN, P.K., VENKAT RAJ, V., MEHTA, S.K., DATE, A.W., "Phenomenological investigations on natural convection in a loop relevant to a PHWR", ASME Winter Annual Mtg, Florida, Nov. 1985, American Society of Mechanical Engineers, New York.

[3] VIJAYAN, P.K., DATE, A.W., VENKAT RAJ, V., MEHTA, S.K., "Transient behaviour of natural convection in a figure-of-eight loop case of initiation from rest", ASME/JSME Thermal Engineering Conf., Honolulu, Mar. 1987, American Society of Mechanical Engineers, New York.

[4] VENKAT RAJ, V., SAHA, D., DOLAS, P.K., MARKANDEYA, S.G., "Some separate effects tests pertaining to LOCA", Indo-German Workshop of Transient Analysis and Emergency Core Cooling Systems, Trombay, Jan.–Feb. 1985, Bhabha Atomic Research Centre, Bombay.

[5] VENKAT RAJ, V., DATE, A.W., "Analysis of conduction controlled rewetting of hot surfaces based on a two-region model", 8th Int. Conf. on Heat Transfer, San Francisco, Aug. 1986.

[6] VENKAT RAJ, V., SUKHATME, S.P., MEHTA, S.K., "Experimental investigations on the rewetting of hot horizontal surfaces", 7th Int. Conf. on Heat Transfer, Munich, Sep. 1982.

[7] VENKAT RAJ, V., Experimental investigations on the rewetting of hot horizontal annular channels", Int. Commun. Heat Mass Transfer **10** (1983) 299–311.

[8] GHOSH, A.K., "A simplified analytical model for pool swell", 9th Int. Conf. on Structural Mechanics in Reactor Technology, Lausanne, Aug. 1987.

PROTECTION AGAINST POSITIVE REACTIVITY EXCURSION EVENTS IN CANDU REACTORS

G.H. ARCHINOFF, J.C. LUXAT, G.M. FRESCURA
Ontario Hydro,
Toronto, Ontario, Canada

Abstract

PROTECTION AGAINST POSITIVE REACTIVITY EXCURSION EVENTS IN CANDU REACTORS.

Postulated accidents which involve addition of positive reactivity are included among the design basis events for CANDU (Canada deuterium-uranium) reactors. The paper describes the design features of CANDU which provide protection against such events, and presents results of analysis which demonstrate the adequacy of the design. Positive reactivity insertion can occur as a result of loss of reactivity control events, in which failures in the reactor regulating system lead to uncontrolled positive reactivity insertion from movement of reactivity control devices, and large break loss of coolant accidents or loss of electrical power events, which lead to addition of positive reactivity as a result of decreasing coolant density. For all accidents in these categories, protection is provided by two independent, diverse shutdown systems, each of which is capable of effectively shutting down the reactor prior to the occurrence of damage to the fuel. The maximum rate of reactivity insertion associated with these events is about 3 mk/s (reactivity (mk) = $(k_{eff} - 1)/k_{eff} \times 1000$). At this rapid rate of positive reactivity insertion, each of the shutdown systems is capable of adding approximately 6 mk worth of negative reactivity in the first second following initiation of the event, and more than 60 mk within 2 seconds. Because of their fast acting capability the shutdown systems provide effective protection over the full range of possible rates of positive reactivity insertion. It is also shown that the reactivity excursion experienced at the Chernobyl Unit 4 plant in April, 1986, is within the range for which the CANDU shutdown systems provide protection.

1.0. INTRODUCTION

Accidents which can cause rapid addition of positive reactivity to the reactor core have always been considered in the design of CANDU reactors. Indeed even prior to construction of the first large-scale station, Pickering NGS A, automatically initiated, fast acting shutdown systems which can cope with rapid reactivity transients had been an important feature of the CANDU design.

This paper describes the design features of CANDU which provide protection against positive reactivity excursions. Darlington Nuclear Generating Station, a 4 x 850 MWe station currently under construction on the north shore of Lake Ontario and scheduled to start up in 1988, is used as an example to describe the design. Results of the safety analysis performed in support of the application for the license to operate Darlington are also presented, to demonstrate the capability of the design.

2.0. CANDU DESIGN FEATURES

The core of a CANDU reactor consists of several hundred horizontal fuel channels - in a Darlington reactor there are 480. The pressure boundary of each fuel channel is a Zr-2.5 wt% Nb pressure tube, 10 cm in diameter and 0.4 cm thick. Each pressure tube contains 13 fuel bundles aligned end to end, as well as heavy water coolant pressurized to about 10 MPa and flowing at about 25 kg/s. Each pressure tube is contained concentrically within a Zircaloy-2 calandria tube, each of which is 13.1 cm in diameter and 0.13 cm thick. Spacing between each pressure tube and calandria tube is maintained by coiled spring annulus spacers, and the annulus is filled with gas and maintained at near atmospheric pressure.

The reactor core, consisting of the array of fuel channels, is supported in a horizontal cylindrical vessel called the calandria tank. Low pressure, heavy water moderator in this tank surrounds the fuel channels. The calandria tank is penetrated at each end by the fuel channels, which are connected to feeder pipes, then to headers and the remainder of the heat transport circuit. The calandria tank is also penetrated from the top to permit insertion of shutoff rods, reactivity control devices and in-core flux detectors, and at the sides to permit insertion of additional flux detectors and injection of liquid poison into the moderator.

The basic safety design objective relevant to positive reactivity excursion events is to assure the continued integrity of the pressure boundary of the heat transport system. With the heat transport system boundary intact, there can be no disruption of core geometry, and no release of fission products into containment, should fuel failures occur.

Continued integrity of the heat transport system is assured by preventing excessive overpressure and by preventing fuel overheating sufficient to cause rupture of the fuel channels. Several levels of protection are provided to ensure that these objectives are achieved in the event of inadvertent addition of positive reactivity. These include the reactor regulating system (RRS) and two fully independent, diverse shutdown systems (SDS1 and SDS2).

2.1. Reactor Regulating System

The Reactor Regulating System controls bulk reactor power and the spatial power distribution, and consists of neutron flux and reactor thermal power measuring devices, two digital control computers and reactivity control devices. The primary method of regulating both the bulk reactor power and the spatial power distribution is to vary the light water level in fourteen zone controller compartments. The information required to adjust the levels in the compartments is obtained from in-core detectors and thermal power measurements from selected fuel channels. If normal operating transients result in a decrease of reactivity in excess of that provided by the light water zone controllers, then groups of adjuster rods (stainless steel rods normally inserted in the core) are driven out of the core in a pre-specified sequence. Conversely, if an increase in reactivity occurs at a rate beyond the capability of the light water zone controllers, additional negative reactivity is provided by driving mechanical control absorbers (stainless steel - cadmium rods normally residing outside of the core) into the core.

The RRS also provides protection against deviations in important process and neutronic plant parameters. The Setback routine of the RRS automatically initiates a controlled reduction in reactor power if any of several monitored parameters deviates beyond a pre-specified limit. Monitored parameters which can initiate a Setback and which are relevant to positive reactivity excursions include high channel power, zone controller failure and spatial control off normal. The maximum Setback rate is 0.8% FP/s.

Further protection is provided by the Stepback program, which is a fast power reduction implemented in the dual control computers, independently of the RRS programs. The Stepback program monitors process and neutronic parameters which require rapid power reduction if they deviate sufficiently in the unsafe direction. Parameters monitored include zone flux, rate of increase of log power and reactor outlet header pressure. Stepback is achieved by either a controlled partial insertion, or by a full drop of the mechanical control absorber rods into the core, depending on the parameter initiating the stepback.

The Setback and Stepback functions provide plant equipment protection against mild positive reactivity excursions, and other anticipated plant transients, without requiring intervention by either the operator or the shutdown systems. For more severe transients, the shutdown systems will be initiated automatically to ensure that the safety design objectives are achieved.

2.2. Shutdown System Number One (SDS1)

SDS1 consists of thirty-two mechanical shutoff rods and associated sensing and signal processing instrumentation and initiation logic. The shutoff rods are inserted into the core vertically by spring assisted gravity action. About -60 mk is inserted within the first two seconds following initiation of a shutoff rod drop. (The unit of reactivity, milli-k, is defined as follows: reactivity (mk) = $(k_{eff} - 1)/k_{eff} \times 1000$. In CANDU reactors with fuel at equilibrium irradiation, 1 mk is equivalent to about 0.2$).

Parameters which can initiate a reactor shutdown by SDS1 include neutron overpower, high rate of logarithmic neutron power increase, heat transport high pressure, heat transport low pressure, heat transport coolant gross low flow, pressurizer low level, steam generator low level, steam generator feedline low pressure and moderator high level.

Several of the SDS1 trip parameters provide effective protection against reactivity excursions, as discussed later. However, the neutron overpower (NOP) trip, in particular, is designed specifically to protect against positive power excursions.

The SDS1 NOP system consists of fifty-four in-core flux detectors distributed at selected locations throughout the reactor core. The number and location of detectors is such that a reactor trip will be initiated prior to fuel sheath dryout in the event of an uncontrolled reactor power increase, for an extremely wide range of initial flux distributions in the core (546 flux shapes, each corresponding to a specific configuration of reactivity control devices, are considered in the design of the NOP system). Thus, the NOP trip parameter provides highly reliable, early detection of reactor power increases and initiates a reactor shutdown prior to dryout occurring on any fuel sheath.

2.3. Shutdown System Number Two (SDS2)

SDS2 causes reactor shutdown by injection of pressurized, concentrated gadolinium nitrate solution, a neutron poison, into the moderator through eight poison injection nozzles which penetrate the side of the calandria vessel. SDS2 has virtually the same trip parameters as SDS1. In particular, the NOP system of SDS2 provides equivalent protection to the

NOP system of SDS1. Reactivity insertion by SDS2 is slightly more rapid than SDS1, however the total reactivity depths during the first few seconds following initiation are similar.

SDS1 and SDS2 are physically separate shutdown systems. Each system has its own complete instrumentation, signal processing equipment and logic hardware located in a physically separated area of the generating unit. Several of the SDS2 trip setpoints are further from the operating points than the SDS1 setpoints in order that firing of SDS2 will not occur if shutdown by SDS1 is effective. Unnecessary firing of SDS2 is undesirable because the poison cannot be rapidly removed from the moderator. Consequently, the Xenon transient following the shutdown will prevent the reactor being made critical for approximately 36 hours following the reactor trip.

3.0. ASSESSMENT OF POSITIVE REACTIVITY EXCURSION EVENTS

Accident scenarios involving addition of positive reactivity include events in which unanticipated reactivity addition is the initiating event, and events in which reactivity addition is an indirect consequence of the initiating event. The former category includes loss of reactivity control (LORC) events, which involve postulated unanticipated movement of reactivity control devices leading to an undesired reactor power increase. The latter category includes loss of coolant accidents (LOCA) and loss of flow events which lead to positive reactivity insertion due to coolant voiding. Each of these categories of accidents is discussed in turn with emphasis on the design features which are effective in mitigating their consequences.

3.1. Loss of Reactivity Control

The primary safety design objective associated with LORC incidents is to maintain the integrity of the heat transport system. This is achieved by preventing:

(a) Excessive overpressure in the heat transport system, which could lead to a rupture in the coolant circuit, and

(b) Fuel centreline melting.

The concern associated with the second criterion is excessive overheating of the fuel leading to possible impairment of fuel channel integrity. However, there is no experimental evidence indicating that operation with fuel centreline melting conditions leads to pressure tube damage. Therefore, prevention of centreline melting is a sufficient rather than a necessary condition to maintain integrity of the pressure tubes, and is, therefore, a conservative safety objective for LORC analysis.

Fuel centreline melting can occur if the fuel element rating exceeds about 88 kW/m, even with normal heat transfer from the fuel to the coolant, or if the channel power is sufficiently high that the critical heat flux (CHF) is exceeded and fuel sheath dryout occurs. The first criterion sets a limit on bundle power, while the second places a limit on channel power.

For the Darlington reactors and more generally, for CANDU reactors using 37-element fuel, high channel power is the limiting condition. This is true even if the power distribution along the channel is distorted as in the case of LORC incidents postulated to occur from a reactor configuration involving reactivity devices in off-nominal positions.

Therefore, the derived safety target for LORC analysis is prevention of fuel centreline melting under dryout conditions. The channel power at which CHF occurs is referred to as the critical channel power. A typical value of the critical channel power for the Darlington units is 9.1 MW. The maximum channel power during normal operation is typically 6.7 MW, but can occasionally approach 7.5 MW.

LORC incidents are postulated as consequences of failures in the RRS that lead to an uncontrolled increase in reactivity due to movement of reactivity devices. The uncompensated reactivity excursion causes an exponential increase in neutron flux. This, in turn, increases the heat generation rate in the fuel and, hence, the rate of heat transfer to the coolant. Consequently a transient is initiated in the heat transport system in which the system pressure increases, driven by the increase in power transferred to the coolant. As the power to coolant increases, increased coolant boiling leads to increased coolant void reactivity. This tends to accelerate the positive reactivity insertion and, thus, the rate of increase of power to coolant. On the other hand, the increased fuel temperature introduces negative reactivity and acts to slow down the rate of power increase. The fuel temperature coefficient for equilibrium fuel is -0.42 mk per 100;C increase in fuel temperature.

Eventually, the shutdown systems are activated, reversing the reactivity excursion and shutting down the reactor.

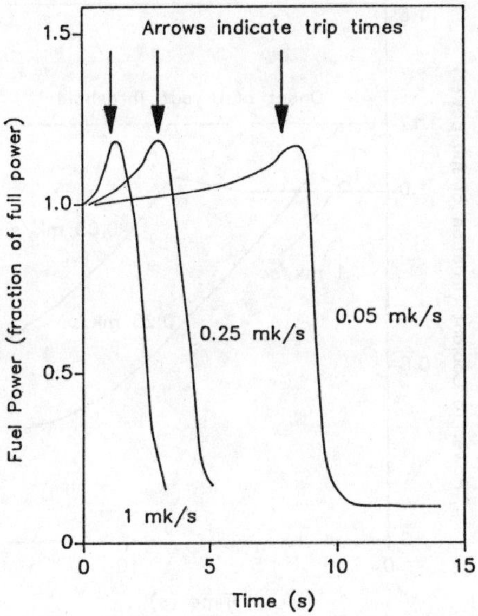

FIG. 1. Fuel power transients for loss of reactivity control from 103% power terminated by neutron overpower trip of SDS1.

The maximum reactivity insertion rate associated with a LORC, assuming a single failure of the RRS, is 0.26 mk/s. If multiple hardware interlock failures are assumed such that the failure of several devices is postulated, then the maximum rate of reactivity insertion can reach about 1 mk/s. LORC analysis is performed parametrically for reactivity rates up to 2 mk/s - a rate well in excess of the rates available even if multiple failures are considered.

Figure 1 shows fuel power transients for LORC events when the reactor is initially at full power. Transients for three different constant rates of reactivity insertion are shown, along with the time at which a reactor trip is initiated by the NOP system of SDS1. Figure 2 shows the corresponding power to coolant transients for the maximum rated fuel element in the core. It is clear from the figures that the NOP system provides early detection of the overpower transient, and effectively terminates the transient well before fuel dryout.

In order to determine the effectiveness of other trip parameters, analysis of LORC events is performed parametrically with reactor trip initiation times calculated

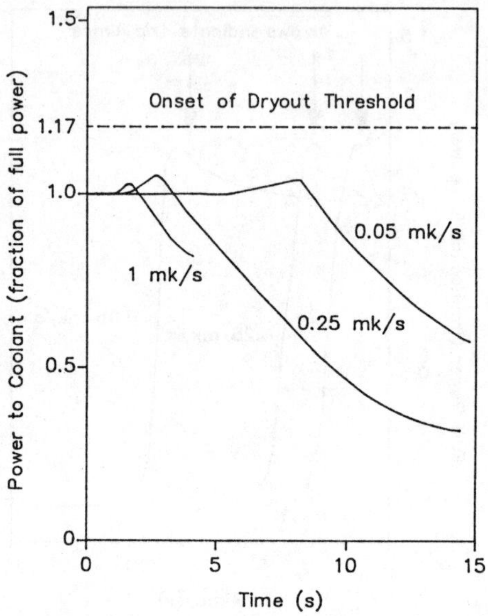

FIG. 2. Power to coolant transients for loss of reactivity control from 103% power terminated by neutron overpower trip of SDS1.

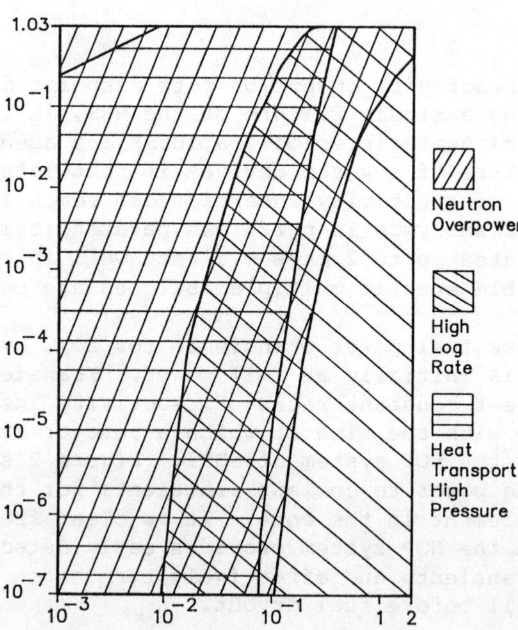

FIG. 3. SDS2 trip parameter effectiveness for loss of reactivity control as a function of reactivity rate and initial power level.

but with no reactor trip credited. Analysis is performed over the full range of initial reactor power levels from 10^{-7} to 1.03 times full power (the three percent margin above 100 percent full power is included to account for control uncertainty).

The range of initial reactor powers and net reactivity insertion rates for which various SDS2 trip parameters are effective is shown in Figure 3. Coverage by the same parameters on SDS1 is the same as or slightly broader compared to SDS2, due to the reduced margin to trip on the SDS1 high pressure and log rate parameters. Figure 3 shows that the NOP system provides effective coverage at high power over the full range of reactivity rates. The high log rate trip provides effective coverage for intermediate to high rates of reactivity insertion, while the high pressure trip provides protection from low to intermediate rates.

Overpressure protection is assessed through transient simulations in which the second (backup) trip parameter on each shutdown system is credited with shutting down the reactor. In these simulations, heat transport system pressure relief provisions are not credited in order to ensure a conservative over-prediction of peak pressure. The peak pressure reached in the reactor outlet header is compared to the design limit to determine the degree of overpressure, if any. The design limit is governed by the design pressure for the pressure tubes.

Analysis shows that the peak ROH pressure for any LORC is less than the acceptance criterion of 110 percent of the design limit of 10.55 MPa, when the SDS1 backup trip is credited, and less than the SDS2 limit of 120 percent when the SDS2 backup trip is credited. Thus, the maximum pressures are within Canadian regulatory limits on overpressure. These limits are based on, and are consistent with, the ASME Boiler and Pressure Vessel Code.

3.2. Loss of Coolant

Safety analysis of CANDU reactors addresses leaks and breaks in all piping components of the heat transport system. The full range of break sizes, from small leaks to guillotine ruptures, is considered.

A loss of coolant accident (LOCA) can lead to depressurization of the heat transport system due to coolant inventory loss, and a reactivity excursion due to coolant voiding. For all but large breaks in the largest diameter piping, however, the RRS is capable of compensating for the void induced reactivity, and reactor power is held constant. For example, constant power can be maintained even for a guillotine rupture of a fuel channel or the largest feeder pipe.

Thus, only LOCAs associated with large breaks in large diameter piping can induce overpower transients. The most rapid voiding, and hence the most rapid addition of positive reactivity, is associated with guillotine ruptures of reactor headers or pump suction piping. The associated reactivity insertion rate can be as high as 3 mk/s during the first second, reaching a maximum of about 8.5 mk when one of the two heat transport loops is fully voided.

The rapid increase in power generation associated with large break LOCAs causes sharp increases in fuel enthalpy as a result of the mismatch between power generation and heat removal. Transiently, heat removal is insufficient due to the thermal characteristics of UO_2 which prevent rapid transfer of heat through the cross section of the pellet and, for certain large breaks, due to poor heat transfer at the fuel sheath surface caused by early dryout.

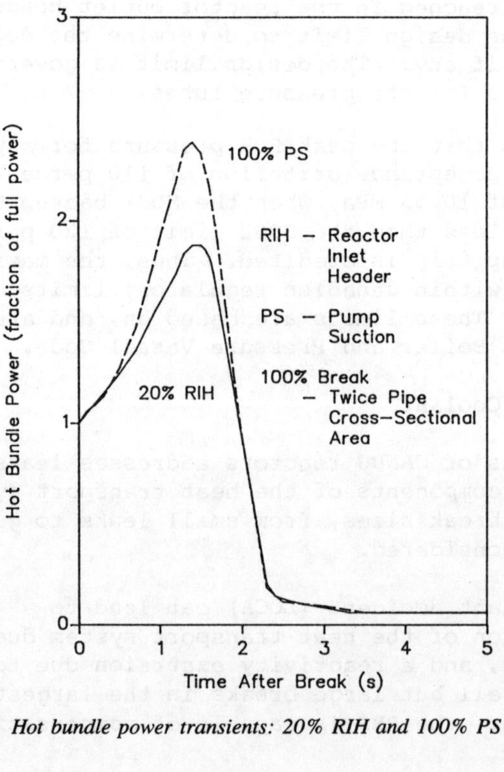

FIG. 4. Hot bundle power transients: 20% RIH and 100% PS breaks.

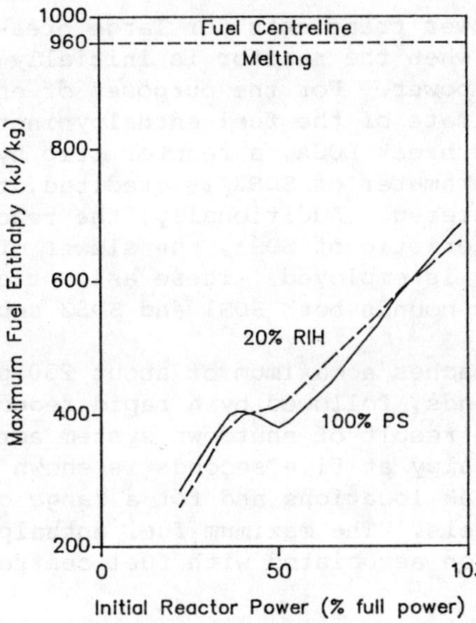

FIG. 5. Fuel enthalpy in the maximum powered element at 5 s following a large LOCA.

With the increasing fuel pellet enthalpy, the potential for fuel centreline melting exists, and possibly fuel breakup, if the power excursion is not terminated. The derived safety design objective for the shutdown systems following a postulated large break LOCA is prevention of fuel centreline melting. This is a conservative approach since fuel centreline melting occurs at significantly lower fuel enthalpy than does potential fuel breakup.

The specific criterion used in the assessment of shutdown system effectiveness for large break LOCAs is that the maximum fuel element enthalpy should not exceed 960 kJ/kg within five seconds after the break occurs, as fuel centreline melting is precluded at fuel enthalpies less than 960 kJ/kg. Trip parameter effectiveness is conservatively assessed by integrating the adiabatic fuel power generation over the first five seconds and comparing the result to the enthalpy criterion. Five seconds is selected as an appropriate interval because a reactor trip is usually completed within two seconds of the occurrence of the break. Hence, by five seconds the neutron flux is at low values with a stable spatial distribution and fuel power generation is close to decay power levels. It should be noted that avoidance of fuel centreline melting ensures that fuel breakup is avoided with a large margin, as the enthalpy required for fuel breakup is in excess of 1500 kJ/kg.

Figure 4 shows power transients for large break LOCAs assumed to occur when the reactor is initially at 103 percent full power. For the purposes of ensuring a conservative estimate of the fuel enthalpy increase following a large break LOCA, a reactor trip by the second neutronic trip parameter of SDS2 is credited, that is, the fourth trip registered. Additionally, the reactivity insertion characteristic of SDS1, the slower of the two shutdown systems, is employed. These assumptions ensure that the analysis bounds both SDS1 and SDS2 separately.

The peak power reaches a maximum of about 230 percent full power at 1-2 seconds, followed by a rapid reduction to decay power levels as a result of shutdown system action. The maximum fuel enthalpy at five seconds is shown in Figure 5 for different break locations and for a range of initial reactor power levels. The maximum fuel enthalpy is well below the criterion associated with fuel centreline melting in all cases.

The effective trip parameters for large break LOCAs are NOP and high log rate. At low initial powers, certain process trips such as heat transport low pressure and coolant gross low flow, are also effective. Regardless of the break size or location, the shutdown systems effectively mitigate the overpower transient induced by a large break LOCA. Fuel channel integrity is preserved, and fuel breakup is avoided with a large margin.

3.3. Loss of Flow

A positive reactivity excursion can be induced as a result of a total loss of electrical power to a unit. The most important consequence of such an event is a loss of power to the main heat transport pumps. As the pumps run down, coolant flow decreases to the point where boiling may occur the fuel channels. Void generation leads to an increase in power and to pressurization of the heat transport coolant. Reactor trips are initiated by the NOP, coolant gross low flow, or heat transport high pressure trips, terminating the power excursion. Subsequently, decay power is removed by thermosyphoning.

The safety design objectives for the shutdown systems for loss of flow events are similar to those for LORC events. The shutdown systems must ensure that fuel centreline melting is prevented in order to ensure that fuel channel integrity is preserved, and they must limit overpressurization of the heat transport system to acceptable levels.

Simulations of loss of electrical power from 103 percent full power show that the reactor is shut down within about five seconds after the power failure. The fuel centreline temperature of the maximum rated fuel element in the core increases by less than 100¡C, and remains well below the UO_2 melting temperature. The peak pressure is about 108 percent of the design limit when the SDS1 heat transport high pressure trip is credited, and only slightly higher when the SDS2 high pressure trip, which has a setpoint 100 kPa higher than SDS1, is credited. At the time of reactor trip, the coolant flow is sufficient to remove the fuel stored heat. Natural circulation is then established and is sufficient to remove decay heat.

The timing of reactor trips following loss of flow events is determined primarily by the heat transport pump rundown characteristic and the initial reactor power. The rundown characteristic is the dominant factor in the rate of flow reduction, while the initial power level governs the rate of voiding and, hence, the void-induced positive reactivity transient. Depending on the initial power level, a loss of flow results in a power excursion which can be relatively rapid or slow. Thus, if viewed in terms of transient power-cooling mismatch, the consequences of a loss of flow event are similar to those of LORC events, with the important exception that the low flow trip provides an effective additional trip parameter. The range of reactivity insertion rates considered in LORC analysis encompasses the rates associated with loss of flow events. Thus, the effective trip parameters for loss of flow events are coolant gross low flow, and the parameters which provide protection for LORC events.

3.4. Chernobyl Unit 4 Transient

In April, 1986, Unit 4 at the Chernobyl plant in the Soviet Union experienced a positive reactivity excursion which resulted in destruction of the reactor core. During the early portion of the transient, manual shutdown was initiated, but the shutdown system could not respond rapidly enough to prevent massive core damage.

Analysis has been performed of the Chernobyl transient, based on information supplied by the Soviets at the August 25-29, 1986 Chernobyl Post Accident Review Meeting in Vienna, for the purpose of assessing the response of a CANDU reactor to a transient similar to that experienced at Chernobyl. Analysis shows that the net reactivity insertion rate up to the time that manual trip was initiated, was less than 1 mk/s. This is well within the range of reactivity rates for which the high log rate trip (equivalent to a low reactor period trip) on either SDS1 or SDS2 of a CANDU provides effective protection. In fact, an equivalent reactivity ramp in a CANDU reactor would have been terminated via automatic shutdown system action initiated by the high log rate parameter. The peak reactor power would be about 15 percent full power (assuming an initial power level of 7 percent). Even if the timing of shutdown initiation were delayed until the time manual trip was initiated at Chernobyl, peak reactor power would be only 19 percent full power. In either case, the fast shutdown would prevent fuel damage and a stable shutdown state would be achieved.

4.0. DISCUSSION

The preceding section has shown the effectiveness of the shutdown systems in mitigating the consequences of positive reactivity excursion events. Results of analysis have been presented for specific scenarios characterized by specific initiating events and initial reactor operating conditions.

Space limitations preclude presentation of results for the full range of initial operating configurations. Analyses are, however, performed for all initial operating configurations to demonstrate the overall effectiveness of the shutdown systems. For example, the full range of initial power levels is considered, as is operation with a reduced number of heat transport pumps, operation in shutdown cooling mode, fresh and equilibrium fuelled reactor core states, and distorted flux shapes. In this way, assurance is provided that a positive reactivity excursion from any initial state within the operating range will be effectively terminated.

In addition to considering a wide range of initial operating states, the analysis also addresses the full range of possible severities of the initiating events. In some cases, this is achieved by performing the analysis parametrically, as in the analysis of LORC events where the rate of positive reactivity insertion considered ranges from very low rates to beyond the maximum rate achievable. In other cases, initiating events which pose the greatest challenge to the shutdown systems can be identified a priori. This is the case with large break LOCAs, where the largest break induces the most rapid coolant voiding and results in the fastest power excursion. Smaller breaks are also analyzed to ensure that any LOCA-induced overpower transient can be effectively mitigated. Loss of flow scenarios can be related to specific initiating events more easily. For example, complete loss of forced circulation is associated with a total loss of electrical power to the heat transport pumps. At the other extreme, partial flow reduction is associated with a single pump trip. Each of these events is analyzed specifically. As discussed earlier, the consequences of loss of flow events, with respect to reactivity changes, are similar to LORC events. Therefore, shutdown system effectiveness for events leading to loss of forced circulation is similar to that for LORC events, with additional protection provided by low flow trips.

Effective protection is provided by the shutdown systems as a direct consequence of the appropriate selection of trip parameters and setpoints, and the rapid negative reactivity insertion characteristics of the shutdown systems. The selection of trip parameters recognizes the governing phenomena associated with a power excursion that can adversely affect fuel cooling or piping integrity. These trip parameters include high power, high rate of logarithmic power increase (equivalent to a low reactor period), which is particularly effective at low initial power levels, high pressure, and low flow, which is effective for loss of flow events.

The trip setpoints are selected to provide coverage over as wide a range as possible, without adversely affecting production reliability by being too close to the nominal operating point. The installation of fast acting shutdown systems recognizes that rapid power excursions can potentially occur for certain postulated events. As shown earlier, the combination of appropriate trip parameters and fast shutdown effectively mitigates the consequences of even the most rapid power excursion.

5.0. CONCLUSIONS

All reactor designs must contend with accident sequences involving the addition of positive reactivity to the core. Furthermore, for the majority of designs there exist mechanisms, related either to inherent reactivity feedback or engineered reactivity devices, whereby rapid positive reactivity insertion can occur. A positive coolant void coefficient is only one of a number of such mechanisms. It was recognized from the outset of the CANDU program that the most effective means of coping with reactivity initiated transients is the provision of fast-acting shutdown systems capable of rapidly inserting a large negative reactivity. This paper has described accident scenarios in CANDU reactors which can lead to positive reactivity excursions, and discussed the design features in place to provide protection against such events. Results of analysis which demonstrate the effectiveness of the design have also been presented. The following conclusions have been reached:

(a) Safety analysis of CANDU reactors considers as design basis events positive reactivity excursions caused by postulated regulating system malfunctions or as an indirect consequence of other initiating events. The maximum reactivity insertion rate is about 3 mk/s (about 0.6 $/s) in the first second, and is associated with large break loss of coolant accidents.

(b) Protective design features are incorporated in the Reactor Regulating System. The RRS monitors important neutronic and process parameters and automatically initiates a reactor power reduction if a parameter deviates outside a pre-specified range. These protective features can adequately cope with common plant transients without requiring action by the shutdown systems.

(c) Appropriate diverse and redundant trip parameters and two independent, fast acting shutdown systems are incorporated into the design to mitigate the consequences of positive reactivity excursions. Each shutdown system inserts about 60 mk (about 10 $) worth of negative reactivity within two seconds of initiation. Each shutdown system, acting alone, is sufficient to preserve the integrity of the reactor core and heat transport system and prevent fuel breakup for the full range of possible reactivity excursions.

ANALYSIS AND REPORTING OF INCIDENTS AT A NUCLEAR FUEL REPROCESSING PLANT

H.S. RYCRAFT, P.W. BALL
British Nuclear Fuels plc,
Sellafield,
United Kingdom

Abstract

ANALYSIS AND REPORTING OF INCIDENTS AT A NUCLEAR FUEL REPROCESSING PLANT.

The paper briefly describes the reporting, recording and investigation arrangements for incidents, as used by British Nuclear Fuels plc, Sellafield. The main subject of the paper is a description of the methodology of the system for analysing incidents so that the underlying contributory causal factors are determined. The paper discusses the method, with reference to an illustrative example, and lists the factors considered by the analysts. The aim is to identify independently the primary causes of an incident and to classify these causes in terms of equipment, environmental, procedural, training, management, or operator deficiencies. From these analyses, information on accident prevention strategies may be derived and a more efficient allocation of resources determined to reduce the likelihood of similar events. Particular attention is given to incidents whose consequences could have been more severe, or that could have occurred in other plants. The paper goes on to discuss the differences in reporting criteria between radiological incidents and non-radiological incidents in terms of their impact on the public's perception of radiological accidents. A case is made that the public can over-react to minor radiological incidents. The roles of the media and the industry are explored in this context.

INTRODUCTION

Sellafield, in the north-west of England, covers a one square kilometre site, which is licensed under the British Nuclear Installations Act (1965). Thirteen thousand seven hundred people are employed on the Site, approximately six thousand on major building projects. One of them, the Thermal Oxide Reprocessing Plant (THORP), is said to be one of the largest construction sites in Europe. There are more than 30 installations on the site, for the storage of spent nuclear fuel, chemical reprocessing, fast reactor fuel manufacture, waste treatment and storage, development and analytical laboratories and four small Magnox reactors.

This paper describes the reporting, recording, investigation and analysis of radiological incidents occurring on the Sellafield Site; and the effect of these reporting requirements on the public's safety perceptions.

The requirements for such reporting, recording and investigation of incidents, include those imposed in the Nuclear Site Licence, and the rapid reporting of certain types of incidents and events to Government Departments. At Sellafield, British Nuclear Fuels plc have integrated these external requirements with a system that enables the Company to study the incidents occurring on its plants, so that action may be taken to reduce the likelihood of a recurrence.

THE RECORDING SYSTEM

BNF Sellafield has had a system for the recording of incidents and abnormal events (including non-radiological) since the 1950s and a comprehensive classification scheme for these incidents since 1976. Over the years the classification scheme and recording system have been developed and refined. The incidents are classified into type and severity (based on the consequences). A reference table is provided to operators with guideline descriptions of the severity of each type. Many of the descriptions follow the legislative requirements, such as the Ionising Radiations Regulations which specify dose limits to persons and levels at which they should be reported to legislative authorities and recorded in the 'Site Record'. Each type has up to four levels of severity which are, as far as possible, consistent between types. Incidents are sub-divided into Radiological and Non-Radiological for ease of classification and analysis.

Appendix 1 is a summary of the classification of type and an outline description of severity levels. The severity of the incident governs the minimum level of investigation required. The investigations can be conducted by formal teams of investigators that include trade and staff unions' representatives, or by asking one or two managers to investigate and report back in writing.

An up to date 'Record of Site Incidents' is maintained. This record is a list of the classified incidents described in detail with their investigation reports, a statement of the causes of the incident, and the recommended actions designed to prevent a recurrence. The last includes a status report of the actions taken.

The Record, therefore, contains much detailed information about the incidents. While this information is obviously useful, it does not give an insight into any common contributory factors for a plant or common trends for the site. Underlying contributory factors, once identified, may be eliminated or reduced, thus minimizing not only the same fault sequence but also other similar fault sequences. In addition the results of the analysis of contributory factors that have caused certain fault sequences in one plant can be used to improve the safety of other plants.

From October 1978 to September 1984 a simple weighted system of identifying causes of all recorded radiological incidents, using a set list, was undertaken by the manager responsible for investigating the incident. In 1984 a proposal by the Site's Central Safety Department to introduce a more comprehensive system of analysing causes of radiological incidents was accepted.

CONTRIBUTORY FACTORS ANALYSIS

The designers of the revised analysis of contributory factors started with a prior knowledge of the history of events both major and minor, and therefore knew what type of factors had been identified in the past. Nonetheless it was accepted that the new system could require modification during its first year to maximize its usefulness. A literature research failed to identify a tailor made system for use at Sellafield.

The Contributory Factors Analysis was developed using the then draft report "A Guide to Reducing Human Error In Process Operations", subsequently published by the UKAEA Safety and Reliability Directorate as SRD R 347 [Ref 1]. The causes of human error were supplemented by factors describing modes of equipment failures. It was decided that the analysis would include all the contributory factors identified, not just the primary cause of an incident.

Weighting points for the factors used in the earlier cause analysis were retained to give the managers an indication of the approximate importance of a factor in the individual incident, and the analysts an indication of the relative importance of that factor in the plant/management area. It must be stressed that the weighting factors are only indicators and cannot be treated with any statistical rigour.

The Contributory Factors Analysis was adopted and brought into operation on the 1 January 1985. Appendix 2 shows the factors considered by managers and analysts when assigning weighting points. It was also decided that the type of operation or situation needed to be noted. The factors given in the Appendix are those finally adopted from the end of 1985. The changes made during 1985 were not major but were introduced for clarification. For example one change was to specify more precisely the type and level of management which had "inadequately assessed an operation" so that training and communication could be improved where most required.

The people undertaking the Contributory Factors Analysis of incidents were auditors from the Central Safety Department. This had a number of advantages:

(1) They were independent of the operations management and therefore demonstrably unbiased,

(2) They had safety assessment/auditing experience and were therefore familiar with investigatory practices,

(3) Because of other duties they were sufficiently familiar with all buildings and processes on site to ask the appropriate questions and be aware of contributory factors that would be applicable in other areas and plants.

Using the same team of three auditors reduced subjectivity. One auditor would investigate and his findings would be agreed with a second auditor. Subjective judgement in any investigative procedure cannot be entirely removed, but the requirements for both another auditor to agree to the analysis results, and the justification of the analysis to the manager responsible, has improved objectivity.

During 1985 and 1986 a total of 227 radiological incidents were analysed, the majority (82%) were of the lowest severity recordable. Such incidents are minor and 65% of the total number of these events had not caused injury or release of radioactivity but were in the category "of management concern" i.e. an event had occurred which warranted investigation to ensure that safety had not been affected and would not be affected in the future. The majority in this group involved equipment failures or events that could have developed to give rise to more serious consequences.

From 1987 the analysis system has been extended to include the more serious non-radiological incidents and those which may have further implications than the immediate consequences.

ANALYSIS METHOD

The auditors required a detailed account of the incident. In most cases this was obtained from the investigation reports. If further information was required other people were consulted. The equipment and any procedures involved were also reviewed, and the auditor often observed the task procedure and equipment in use. In most analysis first hand experience was sought.

Attention is also given to the job organization and other environmental factors which may also have a significant effect both on human errors and equipment faults.

A case history serves to illustrate the procedure:

Example

An incident occurred whilst instrument mechanics were repairing a turbine flow meter. The mechanics restarted the pump before replacing a cover plate which held the assembly together, causing water to flow out of the connection between the flow meter and the process pipe, and into the operating area. On investigation it was found that the mechanics did not appreciate that the leak would occur at this stage if they tested the pump using the same testing procedures that had been used prior to the start of the job. The mechanics were unable to find correct drawings of the assembly and the hand sketches and procedures discussed with the plant engineer were inadequate.

The auditor's analysis concluded that:

4 weighting points to "Inadequate written instructions for the task" - this related to the out of date drawings and lack of approved instructions for the job.

4 weighting points to "Inadequate verbal instructions" - the instructions given by the plant engineer were inadequate and were considered to be misleading.

1 weighting point to "Inadequate assessment (plant "Managers)" - following the above allocation of the main contributory factors it was obvious that the plant engineer had not carefully assessed the task before allowing the job to proceed.

1 weighting point to "Inappropriate intended action" - the incident was directly caused by the incorrect dismantling and testing of the flow meter but this was not considered to be the primary contributory factor leading to the event.

Taking all the circumstances into account (they have been simplified for the purposes of this paper) the auditors decided that the instructional shortcomings were the most important factors leading to this event. The instrument mechanics proceeded with the job believing that they had adequate information. The obvious but salutary result from this incident was that no work on any equipment should begin unless properly researched and with written instructions available.

While for all incidents the auditors discuss their findings with the responsible managers, the auditors make the final decision on the contributory factors and weighting points.

The report is then issued to all relevant managers in the Company. This includes all Managers of plants at Sellafield and designers. Specific details of note are brought to their attention.

At the end of the year the auditors produce an analysis for the main Sellafield Safety Committee. This involves the analysis of the data base from the Contributory Factors weighting points. A rigorous statistical analysis using the weighting points is not undertaken as the number of incidents is small. Also, the results have to be interpreted bearing in mind the allocation of weighting points when considering individual plants. It was found that one incident, usually minor, could bias the results if only one factor had been identified i.e. 10 points allocated. This could overlay a true trend produced by a number of incidents of more importance where two or more factors were identified.

ANALYSIS RESULTS

One interesting trend was found in both 1985 and 1986. For incidents where there were no adverse consequences or whose consequences were minor, the Contributory Factors were found to be few i.e. only one or two, but in the case of incidents which had more severe or potentially more severe consequences, usually three or more factors were identified. This indicated a chain of errors or faults coupled with the failure of recovery procedures.

One encouraging aspect of the method was that the analysis gave a 'normal' profile when compared with other research in the field e.g. an approximation of the human error factors for the incidents gave a figure of 70%. Rasmussen [ref 2] has estimated that human error contributes from 50% to 80% for major accidents in industries with a high regard for safety of operations e.g. aircraft or airline industries. The contribution of human error appears as a high percentage because incidents caused primarily by equipment failures have already been reduced to a minimum in these industries. Therefore, the figure of 70% may not be reducible as a percentage but it does indicate that the greatest effect on reducing the number of incidents will be achieved by concentrating on reducing human errors.

ANALYSIS CONCLUSIONS

The Contributory Factors Analysis provides a basis from which to monitor trends of incident causes and as part of the analysis concentrates on the human contributory factors involved in incidents, it also gives an indication of the area in which resources may be concentrated to improve human performance to minimize incidents. This has become one of the industry's prime concerns since recent and past major accidents have identified human errors in both the cause of the accident and the failure to recover from a fault sequence [Ref 3 & 4].

The method is primarily a management tool for identifying factors that lead to incidents in plants, but also serves as a method for alerting other managers and designers to factors that should be considered during the operation or design of nuclear processing and storage plants.

Many organisations contacted during a review of accident databases by an international team [ref 4], expressed their interest in technical incident data. The majority was convinced that if collected and analysed properly the data could contribute substantially to incident prevention. Legal requirements for reporting do not give a complete basis for the collection of data as the requirements change from year to year and there is little guidance given as to the classification of incidents or preferred method of investigation. This has resulted in the collection of anecdotal data which cannot be used as a basis for preventing incidents and the development of in-house classification of incident schemes within all industries. The majority of these concentrate on details of the type and severity of the incidents with little information or data on the causes of the incidents. The method described above is the British Nuclear Fuels plc Sellafield attempt to collect and analyse data in order to use past experience in a structured way to minimize incidents on the site.

INCIDENT REPORTING

The Sellafield incident classification system, as described, separates the non-radiological incidents from the radiological incident mainly because the reporting requirements for radiological incidents are more onerous. In addition to the detailed and specific reporting requirements for radiological incidents, the UK Government has specified the incidents of which it requires rapid notification. Some of the specified incidents have minor health and environmental consequences but are considered to be of concern to the general public. For example, one type of incident that has to be reported to Government Departments is a release to the environment that could cause the most exposed member of the general public to receive a dose of 50 micro Sieverts. This is equivalent in terms of dose received to a flight from London to California or to eating 50 brazil nuts. None the less such an incident is an infrequent occurrence at Sellafield.

Direct comparison of severity of incidents between conventional and radiological incidents is therefore difficult but such a comparison may be attempted by members of the general public when incidents are reported to them. The understanding of probabilistic risk is central to the understanding of "severity" of consequence and is paramount in the case of radiological incidents. Risk and uncertainty are also inherently difficult to communicate to the public. A physical injury e.g. a broken leg as a result of a conventional accident is within the realm of most people's experience. However, understanding the effect on a person who has received a radiation dose which is small but has exceeded an annual dose limit, demands a deeper knowledge of radiological risk assessment and is therefore outside most people's experience.

THE MEDIA'S ROLE

A majority of the general public rely on the media to inform them of risks, so their perception of risk is significantly influenced by the media's treatment of information. Consequently if the media exaggerate, the public are likely to over-react to minor incidents. One example of this affecting Sellafield was seen in a 1986 editorial of a UK national newspaper. The article began "Yet another serious accident has happened at the Sellafield Nuclear Reprocessing plant .....". The event referred to in the article concerned the discovery, during routine analysis, that a batch of low active liquid waste in a sentencing tank would, if discharged, result in one limit in the discharge authorization, being marginally exceeded. The liquor was pumped to a plant for further treatment. At no time was there a release of radioactive material from plant containment, no dose was received by any person and the procedures and equipment operated as designed. How is the public able to make a judgement of the risk from this incident after reading the opening paragraph? In the media's defence, risk assessment, owing to its technical complexity, is not easy to report and the journalist also has to rely on "expert" sources whose opinions are too often divided on the consequence of radiation exposure. The debate on the effect of low level radiation is a prime example of the division of so-called expert opinion. Few journalists have the scientific background or the time to sort through such complexities and often use contradictory material to produce what we would call a "balanced report". An interesting study conducted by Cohen in the USA (ref. 5) supported the view that the public's fear of radiation is substantially greater than is realistic and that the media substantially exaggerates the dangers of radiation. The study also revealed that the media have often stated that the scientific community is "split down the middle" when in reality the radiation health science community in the USA clearly supported the "Establishment scientists" e.g. BEIR Committee, UNSCEAR, and ICRP Committee.

THE INDUSTRY'S ROLE

Not all the 'blame' can be laid at the media's door, in the past the industry has been reluctant to allow the media access to information. It has also been reticent in devising a different "language" to explain what it was really saying. The image of a 'secret industry', once established, is hard to dispel and can lead to "cover-up" claims. Recent major incidents, in both the chemical and nuclear industries, have affected the credibility of industrial "experts". The public require information presented in a way to aid their understanding of issues upon which to base their judgement of risk acceptability. The UK nuclear industry is attempting to break down the image of secrecy by increasing the freedom of information and by trying to promote an approachable image with the emphasis on being a safe industry. One part of this at Sellafield and other nuclear sites is making details of events that occur on the nuclear site freely available even though the most of these events are minor. The Company believes that in doing this it will eventually gain increased public acceptability.

THE WAY FORWARD

Through many projects such as the public information centre, guided tours around the site, and talks given to schools and other organizations, BNF has begun to improve the public's attitude to the safety at Sellafield. More generally the industry's strenuous endeavours to explain the Chernobyl accident and put it into proper perspective, appear to have been reasonably successful. However the authors believe that the major efforts to educate the public, and the media, on the true severity of reported radiological events must continue if the public are not to over-react to small consequence occurrences at nuclear establishments.

These actions together with the endeavours to reduce incidents to a minimum are considered to be the best ways to improve the public's perception of Sellafield's risk. The Contributory Factor Analysis improves the understanding of the causes of incidents and by applying the results should help to minimize the number of incidents.

APPENDIX 1

SUMMARY OF SELLAFIELD CLASSIFICATION SYSTEM FOR INCIDENTS

Incident Type

	Incident Type	Description
Radiological Incidents	1	Release or Spread of Radioactivity
	2	Radiation Dose from external or internal radiation
	3	Explosion or Fire affecting Radiological Safety
	4	Plant Abnormality
	5	Lost or Stolen Radioactive Material
	6	Incidents involving the Carriage of Nuclear Matter
Non-Radiological Incidents	7	Death or Injury to persons on or off the Site as a result of the Site's Activities
	8	Events of concern which do not involve injury
	9	Industrial Disease
	10	Release of Noxious or Offensive material to the environment

Incident Severity Description

Radiological Severity	Description
A	Major Incident e.g. involving death or serious injury to health; an uncontrolled Criticality excursion
B	Serious e.g. building evacuation, public advised to shelter
C	Incidents of concern e.g. statutory dose limit exceeded; breaching an Operating Rule
D	Minor events

Non-Radiological Severity	Description
1	Death or Major Injury
2	Dangerous Occurrence e.g. crane collapse
3	Injury that is reportable e.g. person off work for more than 3 days
4	Events of concern e.g. equipment failure and spillages

APPENDIX 2

ACCIDENT CONTRIBUTORY FACTORS ANALYSIS

Summary of factors considered and information recorded

1 All relevant details of the incident are recorded e.g. date, Description, and investigation undertaken.

2 The type of operation being undertaken at the time of the incident.

Normal
Abnormal e.g. recovery operations
Maintenance Operations
Infrequent operations (e.g. once a year but routine)
Stressful operations (e.g. Time constraints under abnormal conditions)

3 Factor List

EQUIPMENT FAILURES

Design Fault
Construction or installation fault
Age or Wear out
Random Failure
Inadequate or Inappropriate equipment provided

INSTRUCTIONAL ERRORS

Inadequate written instructions
Lack of/inadequate verbal instructions
Inadequate communications
Breach of instruction

MANAGEMENT ERROR

Inadequate assessment - Managers above plant manager
Inadequate assessment - Plant Managers and below
Inadequate assessment - Engineering Services
Inadequate assessment - Health Physics and Safety Services
Inadequate supervision
Job content - had the worker too much or too little to do? were manning levels adequate?
Administration Procedural error - incorrect paperwork

TRAINING

Lack of adequate training
Lack of experience

ENVIRONMENTAL PROBLEMS

Poor operating environment - temperature noise etc directly contributing to the incident

Poor housekeeping - untidyness, uncleared debris directly contributing to the incident

OPERATOR ERROR

Incorrect use of equipment

Inattention - distracted, over-familiar

Inappropriate Intended action - deliberate wrong act in error or poor judgement

Inappropriate automatic or Involuntary action

OUTSIDE OR NON-BNFL CAUSE

UNKNOWN CAUSE - where there is no explanation for an incident this factor may be noted

REFERENCES

[1] BAINBRIDGE, L., ''Guide to Reducing Human Error in Process Operation — Short Version'', Rep. R-347, UKAEA Safety and Reliability Directorate, Warrington, UK.

[2] RASMUSSEN, J., NATO Advanced Study Institute on Synthesis and Analysis Methods for Safety and Reliability Studies, Urbino, Italy, verbal pesentation, 1978.

[3] WOLF, C.P., SHELANSKI, V.B., Accident at Three Mile Island — The Human Dimensions, a Westview Special Study, Westview, Boulder, CO (1982).

[4] USSR STATE COMMITTEE ON THE UTILIZATION OF ATOMIC ENERGY, The Accident at the Chernobyl Nuclear Power Plant and its Consequences, Information compiled for the IAEA Experts' Meeting, Aug. 1986, Vienna.

[5] COHEN, B.L., A poll of radiation health scientists, Health Phys. **50** (1986) 639–644.

PROBABILISTIC METHOD TO CONSIDER FIRE ATTACK ON THE SEGREGATED SAFETY SYSTEMS IN A NUCLEAR PLANT

M. FINUCANE
UKAEA Safety and Reliability Directorate,
Culcheth, Warrington, Cheshire

D.D. DRYSDALE
Department of Fire Safety Engineering,
University of Edinburgh,
Edinburgh

United Kingdom

Abstract

PROBABILISTIC METHOD TO CONSIDER FIRE ATTACK ON THE SEGREGATED SAFETY SYSTEMS IN A NUCLEAR PLANT.

The paper describes a computer program called ARSSUN which models probabilistically the spread of fire through a nuclear plant. A representative matrix of the plant layout together with other graph theory techniques form the mathematical basis of the model. The paper describes the capabilities of the program, input data required, and some of the assumptions necessary to make the best use of the model. A simple worked example is presented to illustrate the principle of fire attack on the segregated safety systems of a plant.

1. INTRODUCTION

A fire is one of the more likely "external events" to place a nuclear plant at risk; up to the end of 1981 there were 100 reported fires in US nuclear plants. The fire at Browns Ferry (1) in 1975 dramatically demonstrated the potential a fire has for causing dependent (e.g. common-mode) failure of separate safety systems. Even though the lessons of Browns Ferry have been heeded, there still exists a need to fully quantify the improvement in safety provided by extra fire walls etc. This has proved difficult mainly because of lack of suitable data, e.g. fire frequencies and probabilities of fire spread. Existing probabilistic safety assessment (PSA) tools are also not readily adaptable for modelling fire spread through a plant; this is particularly so when considering fire spread simultaneously in more than one direction.

For the last few years fire safety engineers within the UKAEA have been developing, with the help of two UK

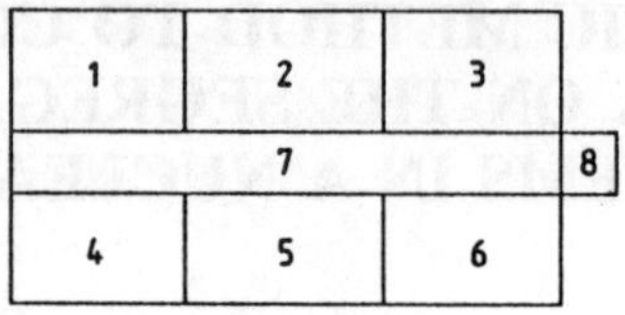

(a) Layout of plant

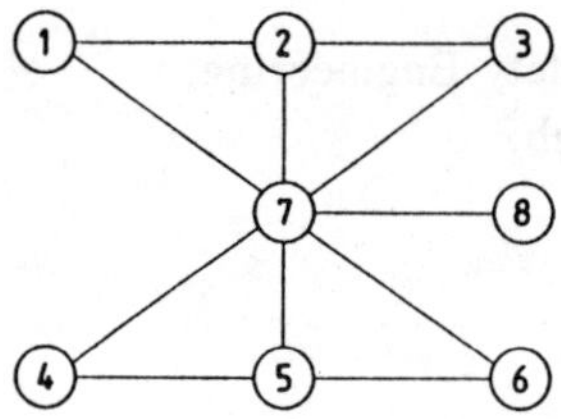

(b) Representative graph

FIG. 1. The transposition of a simple plant layout to a network graph.

universities, a probabilistic method to consider fire spread through a nuclear plant. The paper describes the computer code ARSSUN* which is the result of this effort.

2. PRINCIPLES OF THE MODEL

ARSSUN uses graph theory (2) to represent the layout of a nuclear plant in the form of a network or graph. A plant is divided into an appropriate number of spaces, each space is a node on a graph with all possible fire spread routes inserted as links between the nodes.

Figure 1 shows how the layout of a plant is represented by a graph. For each node of the graph the user needs to assign:

- Fire frequency;
- Probability of fire spread to adjacent nodes;
- Probability of fire spread throughout a node once fire has spread from an adjacent node.

---

*ARSSUN - Assessment of Risk to Segregated Systems Using Networks.

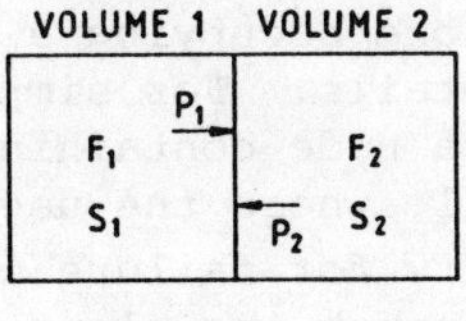

F_1 (or F_2) = Frequency of fire in volume 1 (or 2)

P_1 (or P_2) = Probability of fire spread from volume 1 to volume 2 (or volumes 2 to 1)

S_1 (or S_2) = Probability of fire spread throughout volume 1 (or 2) once fire has spread from volume 2 (or 1)

FIG. 2. Information required as input data.

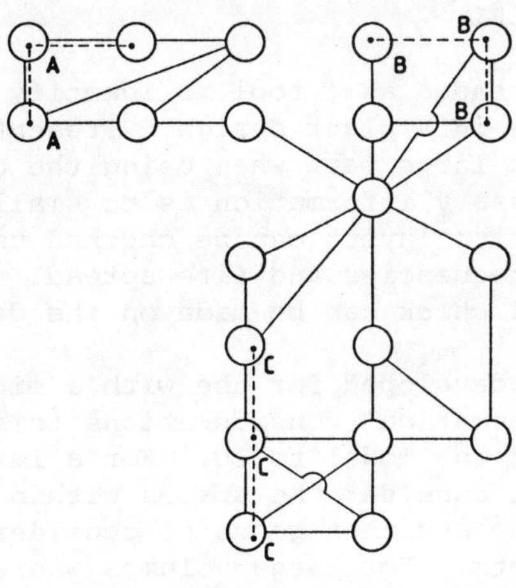

FIG. 3. Possible network graph of a plant with safety systems overlaid on graph.

Figure 2 illustrates these points. Section 4 of this paper discusses some of the difficulties when trying to predict fire frequencies and probability of fire spread.

On this network or graph of the plant, the user locates the independent safety systems. Typically there would be three or four such systems, each one occupying a number of nodes on the graph, see Fig. 3 for details. For simplicity in using the code once fire is present in a node containing a safety system the system is assumed to fail; where the user considers this unreasonable a probability for failure can be included. ARSSUN models spread of fire through the plant by generating all the paths (that have a reasonable probability of occurring) through the network, which fail two or more safety systems. The user of the code can specify any required combination of safety system failures as results, for example:

- Failure of any two systems;

- Failure of two specific systems;

- Failure of three from four systems;

- Failure of all four systems.

The code presents the results in the form of a short list of worst cases (i.e. potential fire paths of high probability), and an aggregate of all potential events.

3. USING THE MODEL

ARSSUN is intended as a tool to identify potential weaknesses to fire in a plant design. Presently, "engineering judgement" plays a large part when using the code because the data base of necessary information is so small. Early in the design of a plant its layout can be checked using "average" values for fire frequencies and fire spread. Once the layout is known then a final check can be made on the design.

The code was developed for use with a microcomputer and together with mathematical considerations this has limited the number of nodes in the model to 20. For a large plant it is necessary to first consider the spaces within a single fire compartment or zone and then go on to consider fire spread between compartments. For large volumes where there are a number of localised fire risks; then the space can be artificially divided to allow more suitable analysis with the code, see Fig. 4 for details. Similarly where there may be a high fire risk area at one end of a volume, with little chance of spread to the rest of the area, then the space can be divided as described above.

4. INPUT VARIABLES

A lot of historical data exists on fires, and from this information the frequency of a significant fire can be predicted

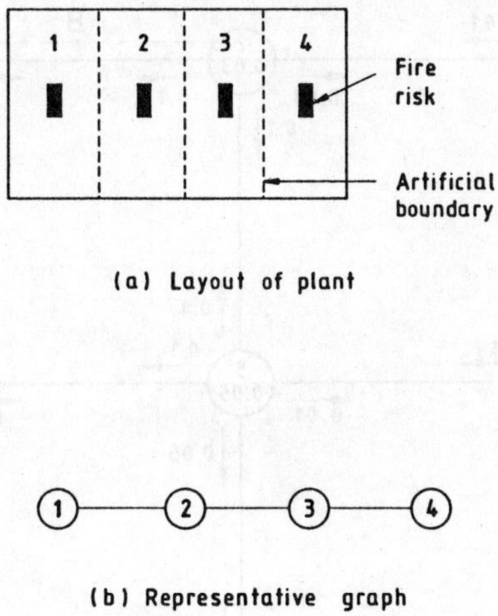

FIG. 4. Division of a large space into artificial volumes for the purpose of analysis.

for existing and new plant. An element of engineering judgement is necessary to divide a fire frequency for a whole plant into fire frequencies for small parts of the plant.

The probability that fire may spread between spaces is a function of many factors including the following:

- Fire severity;

- Construction of barrier

- Size of barrier;

- Penetrations (type, size, number, location).

Values of reliability currently used in PSA for adequately fire-rated and fully-protected barriers, (i.e. fire-stopping around cables and pipes etc) range between 10E - 02 to 10E - 04 probability of failure on demand. Edinburgh University are preparing, under contract to the UKAEA, a quantitative method to allow the reliability of a particular barrier to be calculated; the method will include the four factors mentioned above.

Computer fire growth models e.g. HARVARD (3) are necessary for this task to predict temperature and time histories which are then used to calculate fire severity. ARSSUN also includes a factor for fire spread throughout a volume. The data that exists for such an event is limited and often not applicable to

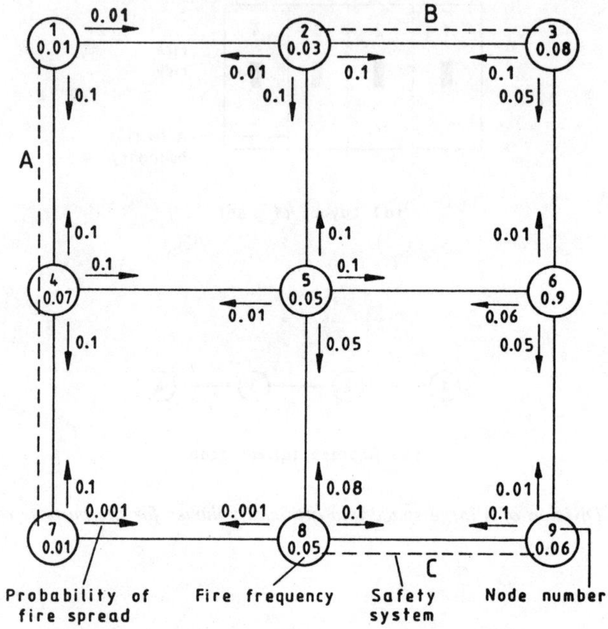

FIG. 5. Graph of a plant showing all input data necessary for ARSSUN.

nuclear installations. Therefore subjective judgement together with the results of fire growth models can be used to assign these probabilities.

5. WORKED EXAMPLE

Figure 5 shows a simple nine-node graph of a plant with all the necessary input data to run the program. Three safety systems are indicated (i.e. A, B and C) and the worst case results, for failing 2 from 3 systems, are as follows:

(1) To fail A and B

Route 452

Probability = 0.7×10^{-3}

(2) To fail A and C

Route 458

Probability = 0.35×10^{-3}

(3) To fail B and C

Route 63 and 69

Probability = 0.45×10^{-3}

(4) Aggregate probability (of all events) failing

any 2 from 3 systems = 0.44×10^{-2}

6. FUTURE DEVELOPMENTS

Research work co-sponsored by the UKAEA is soon to start at a U.K. university on including "time" as an input variable for the code. ARSSUN in its present form is solely probabilistic. In a real fire the time taken to spread from one space to the next may be long and so assumptions made by safety assessors about the response of the plant and personnel may be unduly pessimistic.

During the next six months, the code will be validated and a graphics package prepared to make the program user friendly.

REFERENCES

[1] COLLINS, H.E., et al., Recommendations related to Browns Ferry Fire, Rep. NUREG-0050, United States Nuclear Regulatory Commission, Washington, DC (1976).

[2] BOFFEY, T.B., Graph Theory in Operations Research, Macmillan, London and New York (1982).

[3] MITTLER, H.E., EMMONS, H.W., Documentation for DFC V, The Fifth Harvard Computer Fire Code, Home Fire Project Technical Rep. 45, Harvard University, Cambridge, MA (1981).

SAFETY AND SEVERE ACCIDENTS

Technical Sessions 3.2A and 3.2B

Chairman (Technical Session 3.2A)

F.P. MARTINEZ
Spain

Chairman (Session 3.2B)

W. LEPECKI
Brazil

THE INSAG SOURCE TERM REPORT

H. KOUTS*
Brookhaven National Laboratory,
Upton, N.Y.,
United States of America

Abstract

THE INSAG SOURCE TERM REPORT.

The International Nuclear Safety Advisory Group (INSAG), is a body appointed by the Director General of the International Atomic Energy Agency to advise him. INSAG has recently reviewed the state of research and understanding on the source term for hypothetical accidents for light water reactor nuclear plants for which probabilistic safety assessments have been performed. This has led to a report which comments on the subject and draws conclusions as to the probable outcome of the investigation under way. The principal conclusions, apart from those on the state of source term understanding, are that source term estimates must be viewed in terms of their effects on risk of the nuclear plants, and that the most effective way to reduce risk at this time is to take advantage of the opportunity for accident management. The paper reviews the INSAG report and its conclusions, with some interpretation and clarification being offered.

INSAG ACTIVITIES

The International Nuclear Safety Advisory Group (INSAG) came into existence in 1985, as the result of a decision by the Director General of the International Atomic Energy Agency which was first reviewed and discussed with the Board of Governors. Its objective is to advise the Director General on broad aspects of the safety of nuclear plants, and through issuing this advice, to provide what are believed to be useful safety comments and information to Member States of the IAEA. Its most important objective from the start has been the development of commonly shared safety concepts or principles which will ensure the safety of all nuclear plants.

INSAG consists of thirteen members, selected to provide a variety of backgrounds in national origins, technical interests, organizational affiliations, and responsibilities in the nuclear field. These members have been selected by the Director General and they do not represent their countries of origin. INSAG is in the third year of its three-year trial period. So far it has conducted seven meetings, at which a number of topics affecting safety were discussed.

It has so far issued two publications in the INSAG series (which is a part of the IAEA Safety Series) [1, 2]. INSAG-1 was an official report on the accident at the Chernobyl No. 4 nuclear plant, based on information provided by Soviet scientists at the post-accident review meeting in August, 1986. INSAG-2 is a report on the source term issue, which is the topic I shall discuss here.

* Member of INSAG.

In addition, INSAG has issued informal reports and has written letters to the Director General, covering other topics such as quality assurance, the importance of the Operational Safety Review Team missions, the IAEA reporting system on accidents and incidents, and the need for improved east–west dialogue on reactor safety.

THE SOURCE TERM DOCUMENT

The following is a summary of INSAG-2 [2] and a commentary and explanation of some of the points that it makes.

To understand the report it is of course necessary to know what the term 'source term' means. In the Reactor Safety Study, WASH-1400 [3], hypothetical severe accidents had been grouped into categories related to the amounts of fission products that might be freed to the environment by the accident, and the timing of this release. This combination of quantity of release and timing was loosely called the source term associated with the category of release. After the Three Mile Island accident, it was pointed out by several individuals that the quantity of fission products released to the environment as a result of that accident was substantially less than had been predicted by WASH-1400 for accidents causing severe core damage. The need to understand this observation led in many places to emphasis in reactor safety research programmes on those questions that could affect the fission product release from an accident leading to severe core damage.

The research has led to an impressive improvement in understanding of phenomena that would come into play during a severe accident. Some of these were not recognized or were not well understood at the time of WASH-1400. The improved state of knowledge has led to some revisions of source terms upward from WASH-1400 values, and some revisions downwards. However, there are still variations among safety analysts as to the full meaning of some research that has been completed. There are differences of views as to the completeness of some research, and the need for additional research.

The INSAG report provides the independent assessment by INSAG of the present status of source term research and analysis, and it presents the current INSAG views on the implications. On some points the report goes beyond the present to draw conclusions on the likely outcome of research or analysis under way, using engineering judgement for this purpose. In this way it draws some conclusions as to the relative importance of some phenomena.

The review was confined to light water reactors, because the accident phenomena would be decidedly different for other types of reactor. Moreover, it was confined to light water reactors of types that had been submitted to extensive probabilistic safety assessment (PSA) prior to the INSAG review. This latter restriction was needed because these PSAs serve in identifying and evaluating the importance of the accident phenomena covered by the review. The restriction also recognizes that the source term is very much plant dependent as well as accident dependent. In all that follows, remarks are to be understood as being directed only to those nuclear plants to which the INSAG report is addressed. That does not mean that statements are not valid for other plants, but that validity has not been looked at.

Nuclear plants are built and operated so as to prevent the escape of radioactive fission products. They are protected by a concept of defence in depth. Defence in depth is provided in two ways. The design, construction, and operation embody a set of measures that are a logical sequence of barriers to fission product release. These measures begin with proven design principles and methods, and continue through a succession of precautions to final mitigating features that would reduce the effect of any release that might occur if all other barriers failed. The second mode of defence in depth is a succession of true physical boundaries through which fission products would have to go to reach the environment. These are the fuel, the fuel cladding, the primary system boundary, and the containment structure. Of these, the most important is the containment. It is now recognized that the fuel and the fuel cladding are not independent barriers, because destruction of fuel would also lead to destruction of fuel cladding. Extensive damage to fuel and cladding would present a threat to the primary system boundary which might also fail if protective measures were not implemented.

The fundamental importance of containment is that as long as the containment is intact and functioning properly, the release of fission products from any severe accident would be very small. The study of the source term is therefore a study of phenomena that might threaten the integrity of the containment and its ability to serve as the effective ultimate barrier to fission product release.

The INSAG report draws attention at the start to the need to evaluate source term questions in terms of risk. Those accident sequences that would have a high probability of occurrence and also high consequences through fission product release would be important. Even accidents with high consequences would be unimportant if they had an extremely small chance of taking place. Thus the fact that some way to threaten the containment exists, with important results, does not necessarily make the threat important. There must be some real probability that the threat might be realized. WASH-1400 took care of this point by defining risk as the probability of an accident times the consequences of that accident.

Analysis has shown that if severe core damage were produced in an accident, leading to melt-through of the lower vessel head by the molten mass and liberation of fuel to the containment, and if the containment did subsequently fail, the consequences would depend dramatically on the length of time the containment had remained intact. Chemical reactions and deposition of aerosols that could have evolved would steadily reduce the amount of radioactive material in the containment atmosphere and available for dispersion through a containment breach. Only the accident sequences associated with early containment failure (say within one to two hours after melt-through of the lower vessel head) would contribute strongly to risk in the sense of the use of the term by WASH-1400.

Four modes of threatening the containment function of an LWR early in an accident have been identified. The first three are a result of phenomena that might occur as a consequence of formation of a molten core mass. This could lead to processes in the reactor vessel or processes outside the vessel if the largely liquid core mass melted its way through the bottom head of the vessel and into the containment, processes that could uniquely threaten integrity of the containment structure.

The first early containment threat is possible steam explosions in the reactor vessel that could be caused by sudden contact between the large mass of molten core material and water that might remain in the lower plenum region of the vessel. Industrial explosions of this general kind have occurred in non-nuclear industries such as iron foundries, steel mills, and paper mills. Some of these have been very destructive. The concern is that an explosion intense enough could blow off the head of the reactor vessel, converting it into a missile that could penetrate the containment building.

The second threat would be from hydrogen that could be generated in several ways. One way is chemical reaction between steam and hot zirconium in the fuel cladding. This reaction would occur if the Zircaloy cladding temperature rose above 1000°C, and it would increase in rate rapidly with temperatures above this value. Other hot core components could also be oxidized by any steam present during the core damage phase: silver–indium–cadmium control rods, iron and nickel in steel, etc. If the molten core melted through the lower vessel head and fell on the concrete floor below, this high temperature heat source resting on the concrete would decompose it and liberate water, which could oxidize residual zirconium and other metals, releasing additional hydrogen. Hydrogen generated by all these means could burn on being mixed with the air in the containment. It could even detonate if the concentration of hydrogen in air became too high. The containment threat that would follow would be possible overpressurization of the containment, causing it to fail.

The third concern as to possible early containment failure is attached to a phenomenon known as 'direct heating'. Some conceivable accident sequences would lead to melting of the reactor core while the pressure in the primary vessel remained high, near the settings of safety and relief valves. If the molten core then melted a passageway through the lower vessel head, it would be expelled with great force in a stream that might fragment into fine particles as it encountered the atmosphere below. During transport of these particles through the containment, they could transfer their heat directly to the atmosphere, causing its temperature and pressure to increase suddenly. If this process were extensive enough it could cause the pressure to be increased above the ultimate capability of the containment to withstand it.

The fourth mode of failure might not be the result of any severe core damage. Though some such accident sequences might cause bypass of the containment, others could simply involve failure of complete closure and isolation of the containment, so that radioactive products of an accident could pass to the outside through an unsealed opening.

These are the only important sources of possible early containment failure that current analysis indicates may have to be considered as possibly leading to an important contribution to risk.

The INSAG report noted that the analysis of steam explosion scenarios has led to a widely accepted view that this phenomenon would not be an important contributor to risk. Though steam explosions could take place in the circumstances considered, an explosion severe enough to convert the vessel head to a missile penetrating the containment wall would require a coherent drop of most of the core in a molten form into a large amount of residual water in the lower plenum, under restricted conditions of geometry. The probability of this combination of events is regarded as extremely low.

The INSAG report concluded that the threat of overpressure from hydrogen deflagration or detonation is being met in two ways. New analysis shows that containment buildings are very strong and could absorb more overpressure without failure than had previously been thought. In cases where the strength might not be great enough to withstand the maximum effects of hydrogen, special precautions have been taken. Hydrogen igniters are used in some containment buildings to maintain the concentration of unburned hydrogen at an acceptable level. In some plants, the containment atmosphere is inerted by replacing the air with nitrogen. The report concludes that the appropriate precautions have been taken where analysis indicates they are needed, and that early containment failure from hydrogen deflagration or detonation is now considered to be of low probability.

I have to add as an aside, however, that more recent experimental results indicate that detonation is possible at lower hydrogen concentrations than had previously been thought, when the gas temperature is raised. This may indicate that precautions are advisable in some nuclear plants not presently provided with them.

Concerning direct heating, the INSAG report says that further analysis is needed, but the opinion is expressed that the probability of containment failure from direct heating will in time be found to be too low for this phenomenon to contribute appreciably to risk.

The report addresses the question of possible remobilization of some of the fission products that may have settled out as aerosols, if the containment fails at a later time. These could have ended up in the primary system or in the containment. In either case, they could contribute to the source term for late containment failure. The extent of re-evolution is believed to be not large.

The report adds that suppression pool bypass may lead to a non-negligible contribution to risk for some boiling water reactors. The mode of bypass referred to is not passive, in that it would also be a result of interactions involving the molten core. Another similar possibility of containment bypass for pressurized water reactors is steam generator tube failure from superheating of the residual steam in the primary system by the overheated reactor core.

However, some modes of containment bypass would be the result of incomplete isolation of the containment during an accident. This possibility of human error or of isolation failure seems to set a lower limit to containment effectiveness for any nuclear plant requiring active isolation measures in the event of an accident.

Before drawing final conclusions, the report comments on some areas in which information is particularly needed from research programmes. These are: phenomena and effects associated with interaction between concrete and molten core materials, the radiation stability of chemical compounds involving fission products (especially CsI), surface chemistry involving fission products and their compounds, the extent to which zirconium might be oxidized in-vessel during an accident, and the extent of release of transuranic elements from molten fuel. These do not constitute all the research that is thought to be necessary, but they are areas in which uncertainty is unusually high.

One of the most important points made by the report is a comment on accident management. It is stated that the most valuable measures to reduce risk to the public from nuclear plants are those that prevent accidents from happening. Beyond the measures afforded by good design, good construction, informed and careful operation, proper maintenance, and effective regulation is the largely untapped field of actions that can be taken by an informed and alert operating crew to prevent off-normal conditions from developing into a severe accident. These are measures that would take advantage of normal plant systems to remedy the abnormal problems. For these possibilities to pay off, several preparatory actions are needed. First, the operating staff must have in the control room, easily available and interpretable, the information which tells what is wrong and what its effects are. Second, the operating staff must be so familiar with the plant that remedies in the form of beneficial use of alternate plant systems naturally come to mind. Third, ready ability must be present to manage these uses from the control room. And fourth, the training programmes for operational staff must contain numerous exercises that test and develop these capabilities.

The principal sequences of this kind could be part of routine operator training, and could be incorporated in abnormal operating instructions. If confidence is developed in implementation of these instructions, that fact could be used to modify event trees in PSAs, thus reducing estimates of risk for the plant involved.

The INSAG report draws a number of conclusions.

(1) No need is seen for new immediate generic action to improve safety of existing nuclear plants.

(2) The most valuable way to maintain risk and the source term at low values is to prevent severe core damage. The next most valuable measures would be those that protect the containment function. Mitigation measures are also useful, but they would not be needed if fission product release were averted.

(3) The most fruitful path to follow to reduce risk at this stage is through accident management.

(4) It is believed that there is no new undiscovered phenomenon or feature related to source terms that might affect risk substantially. However, research is still

needed to resolve questions of safety significance on which engineering judgement needs further reinforcing by sounder physical understanding.

(5) Training of operating staff should include more in the line of dealing with abnormal events, and the evaluation of severe accident sequences. I extend this conclusion to include instruction in the content of the PSA for the plant through inclusion in the training curriculum.

(6) It is noted that base-mat melt-through (the so called China syndrome) does not contribute appreciably to risk because the release from such a sequence would be small.

The report ends with several recommendations for IAEA action that might help to resolve remaining source term questions.

This is the content of the report INSAG-2, on the source term from hypothetical accidents to light water reactor nuclear plants. I have tried to present it fairly and directly, though I have added interpretation when this seemed helpful. The report goes beyond most comments on the source term in the extent of judgements made and estimates of the impact of future knowledge. No doubt the report is dated, and new information from future research and analysis will require much of the content to be revised. But I am confident that the conclusions of the report are about as good as can be achieved at this time.

REFERENCE

[1] INTERNATIONAL ATOMIC ENERGY AGENCY, Summary Report on the Post-Accident Review Meeting on the Chernobyl Accident, Safety Series No. 75-INSAG-1, IAEA, Vienna (1986).

[2] INTERNATIONAL ATOMIC ENERGY AGENCY, Radionuclide Source Terms from Severe Accidents to Nuclear Power Plants with Light Water Reactors, Safety Series No. 75-INSAG-2, IAEA, Vienna (1987).

[3] UNITED STATES NUCLEAR REGULATORY COMMISSION, Reactor Safety Study: an Assessment of Accident Risks in the US Commercial Nuclear Power Plants, Rep. WASH-1400, USNRC, Washington, DC (1975).

SEVERE ACCIDENT MITIGATING MEASURES IN THE RINGHALS AND FORSMARK PLANTS AND RESULTING REDUCTION OF THE SOURCE TERM

R. ESPEFÄLT, G. LÖWENHIELM, L. GUNSELL
Swedish State Power Board,
Vallingby, Sweden

Abstract

SEVERE ACCIDENT MITIGATING MEASURES IN THE RINGHALS AND FORSMARK PLANTS AND RESULTING REDUCTION OF THE SOURCE TERM.

In the paper the severe accident mitigating measures in the Ringhals and Forsmark plants are outlined. A strong emphasis is put on both hardware and procedural measures to ensure that the reactor is brought back to a stable cooling state and that the containment integrity will not be endangered. Specific measures to protect the BWRs in the early stage of an accident sequence will be implemented such as protection of penetrations from molten core debris and the possibility to vent the containment in the case of a loss of coolant accident combined with the loss of pressure suppression function. An important part of the envisaged strategy after vessel melt-through is to supply water to the containment as soon as possible to avoid pressure build-up, to wash out aerosols and to reach a final stable state by covering the remaining core with water. In the case of a late pressure build-up a filtered vent will limit the release of radioactive products to less than 0.1% of the core inventory of Cs and I in a 1800 MWt reactor according to the requirements by the Swedish government. The analysis performed so far has shown that this limit is met with a wide margin.

1. INTRODUCTION

In 1981 the Swedish government requested that the potential for severe accidents in Swedish nuclear power reactors should be investigated and that the need for filtered vents or other mitigating devices should be evaluated. For the Barsebäck plant a filtered vent was required by 1985. For the other plants the time limit was set to the end of 1988.

In the MITRA (Mitigation of Reactor Accidents) project [1] the Swedish State Power Board (SSPB) presented plant specific analyses of the SSPB's Ringhals and Forsmark plants. The results of the MITRA project were supplied to the Swedish Nuclear Power Inspectorate in April 1985. Based on this review and the recommendation given by the Inspectorate and the Swedish Radiation Protection Institute regulatory requirements were issued by the government in February 1986. One requirement is that in a severe accident situation where venting of the containment is needed the release should be limited to 0.1% of the core content of Cs-134 and Cs-137 in a 1 800 MWt reactor.

As a result of previous studies and the new requirements a programme is now being carried out with the aim of a new strategy and mitigating devices for handling severe accidents in the Forsmark and Ringhals plants before 1989.

In the accident management strategy a strong emphasis is put on both hardware and procedural measures to bring the reactor core back to stable cooling and maintain the containment integrity during an accident. The modifications include meaures to prevent temperature or pressure induced early containment failure for the BWRs, reliable back-up water sources for containment spray and means for filtered venting of all plants to prevent late containment failure by overpressure.

Design work is near completion and construction at site has started. Analyses for the Final Safety Analysis Report are under way, aimed at a verification that the mitigating measures meet the requirements set. Separately a verification program for the filtering device chosen is being carried out.

In the following a summary is first given of the adopted accident management strategy and the hardware measures being implemented and then the analysis work is outlined and some preliminary conclusions are given.

2. SEVERE ACCIDENT MITIGATING MEASURES

2.1 Measures to prevent early containment failure

The containment has to be protected against early overpressurization. In the BWRs a threat of early excessive pressure build-up would be a large loss of coolant accident (LOCA) combined with a reduced pressure suppression function. The latter would be caused by a leak between the wetwell and drywell parts of the containment.

The protection system will consist of a large capacity pressure relief line comprising a rupture disc and valves for reclosure after activation. In the event where this relief line is activated, core cooling is available and only primary steam would be released. Unfiltered release to the atmosphere is therefore considered acceptable (cf Figure 1). In cases where core damage is anticipated the valves of this line will be securely closed to avoid inadvertent opening through the rupture disc during a core melt sequence.

For the BWRs of the second generation in Sweden, in which the condensation pool is annular, special measures are needed to prevent early melt-through of the containment floor. In a core melt accident, where the reactor pressure vessel melts through the molten core debris would drop into the lower drywell. Electrical and mechanical penetrations in the lower part of the drywell would then be exposed to and probably damaged by the hot core debris, opening a release path out

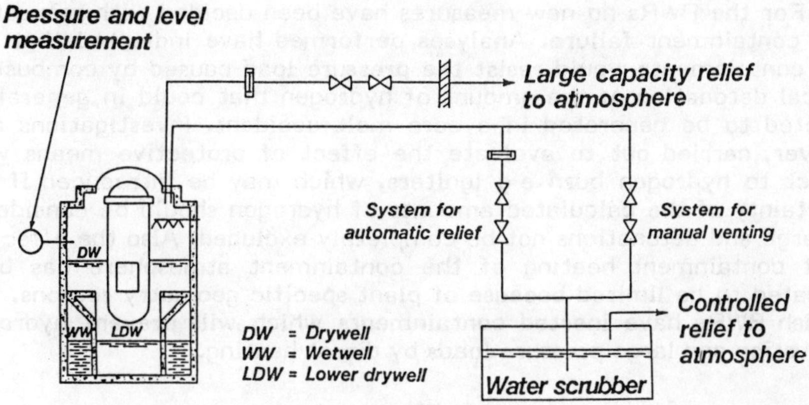

FIG. 1. BWR system for over-pressure protection and filtered containment vent.

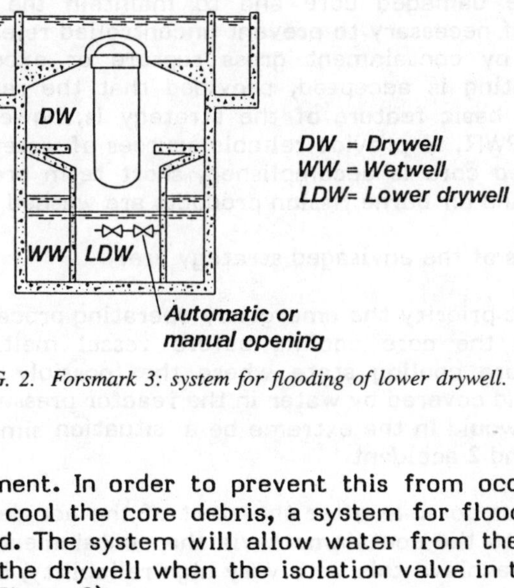

FIG. 2. Forsmark 3: system for flooding of lower drywell.

of the containment. In order to prevent this from occurring and to provide means to cool the core debris, a system for flooding the lower drywell is installed. The system will allow water from the condensation pool to flow into the drywell when the isolation valve in the connecting line is opened (cf Figure 2).

Flooding of the lower drywell will be initiated automatically or manually early in the accident sequence, so that the lower drywell is flooded before vessel melt-through occurs.

In addition devices for protection of the penetrations from direct contact with core debris and a protective lid on the drywell sump are being installed. By these means the debris cooling conditions will be equivalent to those of the first generation BWRs in Sweden, including Ringhals 1 where the condensation pool covers the whole bottom area of the containment.

For the PWRs no new measures have been decided with respect to early containment failure. Analyses performed have indicated that the PWR containments would resist the pressure load caused by combustion or local detonation of the amount of hydrogen that could in general be expected to be generated in a core melt accident. Investigations are, however, carried out to evaluate the effect of protective means with respect to hydrogen burn e g igniters, which may be introduced if the uncertainty of the calculated amounts of hydrogen should be considered too large and detonations not be completely excluded. Also the effect of direct containment heating of the containment atmosphere has been estimated to be limited because of plant specific geometry reasons. The Swedish BWRs have inerted containments which will prevent hydrogen combustion and large pressure loads by direct heating.

2.2 Accident management strategy

The goal of the accident management strategy is to restore stable cooling of the damaged core and to maintain the integrity of the containment. If necessary to prevent uncontrolled release of radioactive material e.g. by containment gross rupture or excessive leakage, a controlled venting is accepted, provided that the release limit is not exceeded. The basic feature of the strategy is, however, both for the BWR and the PWR, to provide reliable sources of water by which cooling of the damaged core is accomplished, short term pressure build-up is counteracted and air borne fission products are washed out.

The main items of the envisaged strategy are:

1. As a first priority the emergency operating procedures shall aim at restoring the core cooling before vessel melt-through. Thus a stable core cooling state where the (possibly damaged) core is cooled and covered by water in the reactor pressure vessel is aimed at. This would in the extreme be a situation similar to that in the Mile Island 2 accident.

2. As soon as possible after the start of the accident, water shall be supplied to the containment via the containment spray system. In order to achieve this with very high reliability, back-up sources of water to the spray system will be introduced. This will mainly consist of connections to diesel-driven pumps in the fire water system, but options for connecting mobile pumps will also be provided (cf Figure 3). By spraying water into the containment:

   - the pressure build-up by steam generation will be avoided and the pressure be kept below the design pressure of the containment for many hours

   - time is gained for actions to restore long-term cooling of the containment

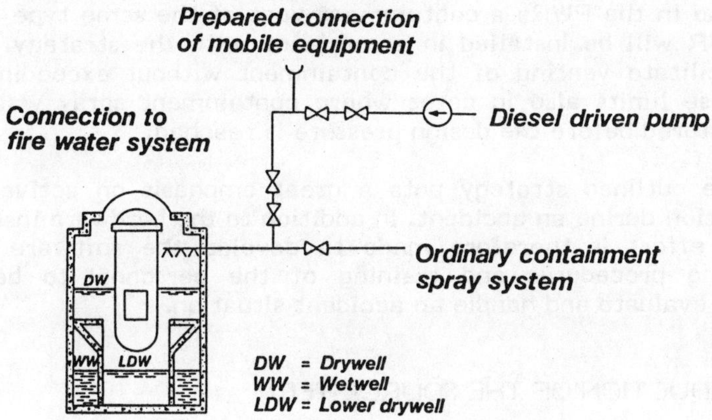

FIG. 3. System for redundant containment spray water source.

- radioactive aerosols are removed from the containment atmosphere and dissolved in the water

- by successively filling water into the containment up to core level all core material will be covered by water and cooled.

3. In BWRs, because of the relatively small containment volume, it will be necessary during the filling-up of the containment to relieve the compressed atmosphere to avoid exceeding the design pressure. For this purpose a venting system will be installed which will allow manual venting as well as automatic pressure relief (cf Figure 1). The vent line will contain a water scrubber (Multi-Venturi-Scrubber-System, MVSS), which will retain the fission products (excluding noble gases) so that in the design case the release limit is met with a wide margin. The venting is expected to be needed at a rather late time (around 20 h). The time depends on the hydrogen partial pressure and it can be significantly decreased if the accident progression is such that a very large hydrogen generation is obtained.

4. In the PWR there is no need to vent the containment for a long time if containment spray water can be supplied within 5-6 hours after the start of the accident. Without water injection to the containment the design pressure would be reached after this time and venting would be required. This is because in the Ringhals type PWR containment, flow paths exist between the containment and the reactor cavity, so that when pressure vessel melt-through occurs, there will be a considerable amount of water in the cavity. Steam will be generated by decay heat and a pressure build-up will occur in the containment.

Also in the PWRs a containment vent of the same type as for the BWR will be installed in accordance with the strategy. This will facilitate venting of the containment without exceeding the release limits also in cases where containment spray water is not restored before the design pressure is reached.

The outlined strategy puts a great emphasis on active operator intervention during an accident. In addition to the hardware installations, a great effort is therefore made to develop the software side e.g. developing procedures and training of the personnel to be able to properly evaluate and handle an accident situation.

3. REDUCTION OF THE SOURCE TERM

A Preliminary Safety Analysis Report (PSAR) for the new measures has been issued for review by the authorities.

In the PSAR a preliminary analysis was performed for total station black-out as design case. The black-out is not the dominant core melt sequence, but is assumed to envelop most other sequences which are less severe but more probable. The analytical tool for the analysis was MAAP 3.0.

It is found for the sequences which are covered by the black-out sequence, that the 0.1% release requirement is met with a very wide margin (noble gases are excluded). Also in the case in which the containment spray is not working a maximum release in the order of 0.1% is calculated.

The construction phase has started and all mitigating systems will be installed at the latest during the revision period of 1988. In parallel with the construction phase a Final Safety Analysis Report (FSAR) is prepared and will be supplied to the authorities in 1988.

In the FSAR work the following is being studied:

- Verification of the total black-out case as an envelope case.
- Studies of special issues, such as hydrogen production, core debris coolability and direct containment heating.

This work is under way and is planned to be completed during 1988. At present only preliminary analysis results have been obtained but it is expected that during the conference more definitive calculation results can be presented. Comparisons with the source term results in the MITRA project will be given.

As an example of how much the source term can be reduced by the proposed mitigating measures some preliminary analysis for the Ringhals PWRs is discussed below. The basis for the discussion is thorough PSA and source term study which was performed for Ringhals 2 [2] , and MAAP-calculations performed for the PSAR.

In a severe accident starting with LOCA or a transient the analysis shows that if containment spray can be activated before containment design pressure (equal to vent opening pressure) is reached, no venting is required and the release is very low and limited by normal leakage rates.

If, on the other hand, venting through the MVSS is necessary between 5 and 8 hours into the accident, when the spray is assumed to be activated the calculated source term is $6 \cdot 10^{-4}$% Cs and I of the core inventory. In the case that containment spray is not activated at all the calculated release is about 10^{-3}% Cs and I of the core inventory. In this sequence the release will occur twice, after 5.7 and 18.5 hours, each lasting about 1 hour.

Both these values are well below the government requirement.

The results of the Ringhals 2 Safety Study indicated that the by-pass sequences would give the largest source terms, but at a frequence $\leq 10^{-7}$ per year. It is expected, however, that the new mitigating systems will positively affect the source terms of the bypass sequences as discussed below

The by-pass sequences can be subdivided into four categories:

- Interfacing LOCA, event V.
- Loss of shutdown cooling sequences with failure to isolate containment.
- Steam generator tube rupture with stuck open secondary relief valve (a reseating valve gives considerably lower release).
- Containment isolation failure.

The Cs- and I-release is according to the Ringhals 2 study about 20 - 30% in the first two sequences and 6 - 7% Cs and I of the core inventory in the two last types of sequences.

It is judged that the highly reliable containment spray system will reduce the release for the environment considerably. There are two main reasons for this:

a) the transporting medium for the fission products, steam, disappears

b) the revolatilization of fission products due to the decay heat is prevented.

In this preliminary study it is found probable that the release in all four by-pass sequences will be in the order of 1 - 2% Cs and I of the core inventory, in the shut-down sequence even less. The condition is that the containment spray will start within 5 hours of the time when core cooling is threatened. This is not always the initiating event, e g in a steam generator tube rupture sequence this is considered to occur when switching from the Refueling Water Storage Tank phase to the recirculating phase has failed.

The containment isolation failure sequence also shows the flexibility of the filtered vent system. In this case the reactor operator must immediately open the manually operated vent path the moment he realizes that the containment is not isolated. This will be an effective measure to decrease the release.

To summarize the following preliminary conclusions can be drawn from the work with PSAR and above discussion:

- The government criteria are met with a wide margin for most accident sequences.
- A substantial reduction of the source term is also expected in bypass sequences not covered by the new mitigation measures, due to the injection of water, which limits the flows out of the containment to the environment.

Another important part of the current work is to analyse not only the environmental impact but also the radiological consequences at the site. Many systems in the reactor building will be highly contaminated and therefore access to important areas must be ensured. The impact of the noble gas release will also be evaluated.

REFERENCES

[1] ESPEFÄLT, R., MITRA — Final Report, Analysis of Severe Accident and Evaluation of Measures to Limit Releases of Radioactivity from Ringhals and Forsmark Nuclear Power Stations, Rep. KS-26/85, Vattenfall, Vällingby, Sweden (1985).

[2] NUS CORPORATION, Ringhals 2 — Safety Study Phase I and II, Reps NUS-4365 and NUS-4409, Gaithersburg, MD (1984).

BEYOND DESIGN BASIS ACCIDENTS FOR THE ITALIAN STANDARD NUCLEAR POWER PLANT: RESPONSE STUDIES AND PROVISIONS FOR PREVENTION AND MITIGATION

P. BARBUCCI
Research and Development Division,
Ente Nazionale per l'Energia Elettrica,
Pisa

B. CHIANTORE, L. LAZZERI, V. VALENTINI
Ansaldo SpA,
Genoa

A. GIURIATO
Construction Division,
Ente Nazionale per l'Energia Elettrica,
Milan

L. NOVIELLO, A. TURRICCHIA
Construction Division,
Ente Nazionale per l'Energia Elettrica,
Rome

G. SANDRELLI
Research and Development Division,
Ente Nazionale per l'Energia Elettrica,
Milan

Italy

Abstract

BEYOND DESIGN BASIS ACCIDENTS FOR THE ITALIAN STANDARD NUCLEAR POWER PLANT: RESPONSE STUDIES AND PROVISIONS FOR PREVENTION AND MITIGATION.

The paper presents the most significant and recent studies performed by Ente Nazionale per L'Energia Elettrica, and Ansaldo-Westinghouse in the field of severe accidents for the Italian Standard Nuclear Power Plant (PUN). In particular (a) the results of the severe accident analysis (AE, SE, TE) based on two different methodologies (United States Nuclear Regulatory Commission code suite made up of MARCH-3, TRAPMELT-2, VANESA, NAUA-MOD·5 and the MAAP 3.0 code) are presented and compared. In these analyses credit is given to a delayed recovery of containment safeguards (3 h to 12 h). Both methodologies show that the iodine and caesium release to the environment are, even in the worst case, four to five orders of magnitude below the core inventory; (b) a list of the main technical improvements introduced into the design to further prevent or mitigate severe accidents is given; and (c) the ratio between the failure pressure and the design pressure of the PUN primary containment (in reinforced concrete) has been calculated and found, with conservative assumptions, to be equal to 2.6.

1. INTRODUCTION

The Italian Standard Nuclear Power Plant (PUN) is a PWR that incorporates all the significant improvements dictated by the latest technological developments. Among other aspects:

– It meets all the design requirements laid down after Three Mile Island accident;
– It is designed to withstand external events, such as an airplane crash, with 10 h unattended operation;
– It has been the subject of a probabilistic safety study which has shown that it is a well balanced design without specific vulnerabilities and that the core damage probability is below 10^{-5}/year [1].

The first twin unit power station based on PUN is to be built at Trino Vercellese (Piemonte, Northern Italy). Here the external grid is well interconnected, also with foreign power supplies, and, in case of loss of off-site power, the plan of power recovery can rely, if necessary, also on hydro-electric and turbogas units.

The paper illustrates some of the most significant studies which have recently been performed and the improvements which have been introduced into the design to cope better with severe accidents. In fact, the widespread interest in beyond design basis accident scenarios, after Three Mile Island, has set the stage for similar evaluations also for PUN and for related considerations on the plant design.

The minimization of radioactive release, also in case of a severe accident, has been set as a design objective, in order to minimize land contamination and evacuation requirements even in such low probability scenarios.

2. ANALYSES

In order to evaluate the plant response to severe accidents and to verify the meeting of the design objective, thermal-hydraulic and fission product transport analyses have been performed by two independent methodologies:

– The methodology, used by Ente Nazionale per l'Energia Elettrica (ENEL), based on the United States Nuclear Regulatory Commission (USNRC) code suite (MARCH, MERGE, CORSOR, TRAPMELT, CORCON, VANESA, NAUA) and

– The methodology, used by Ansaldo and Westinghouse, based on the use of the integrated code MAAP 3.0 developed by the Industry Degraded Core Rulemaking Programme.

(Specifically, ENEL studies were performed using the thermal-hydraulic module MARCH-3 of the USNRC source term code package (STCP), the ENEL version of the TRAPMELT-2 code for the aerosol transport inside the reactor coolant system, the original VANESA and the ENEL version of NAUA-MOD·5 code for the aerosol production and transport in the containment). The results from the two sets of calculations have been compared and the differences explained in order to gain more confidence in the resulting plant responses.

Three sequences have been analysed: AE, i.e. large LOCA with loss of the emergency core cooling system (ECCS); SE, i.e. small LOCA with loss of the component cooling water system (CCWS); and TE, i.e. total station blackout (with unavailability also of the turbine driven emergency feedwater pumps).

In the AE sequence, the containment heat removal and filtration systems are assumed available from the beginning of the accident. In the other sequences, the containment heat removal and filtration functions are assumed to be recovered in time ranging from 3 to 12 h on the basis of conservative probability distribution functions of recovery time.

TABLE I. MAIN ASSUMPTIONS RELATED TO THE ANALYSED SEQUENCES

	AE	SE	TE
Availability of containment heat removal (one train made up of 1 spray pump and 2 fan coolers)	t = 0	t = 6 h (spray) t = 12 h (FC)	t = 6 h
Availability of secondary containment filtration system	t = 0	t = 3 h	t = 6 h
Primary containment leak rate	– 0.5% d^{-1}, constant, in USNRC codes; – variable, with 0.5% d^{-1} at design pressure, in MAAP		
Secondary containment filtration efficiency	DF = 2 with filtration system off DF = 50 with filtration system on		
Type of concrete in reactor cavity		Basaltic	

TABLE II. TIMING OF MAIN EVENTS (min)

EVENTS	AE USNRC Codes	AE MAAP	SE USNRC Codes	SE MAAP	TE USNRC Codes	TE MAAP
Core uncover	10	0.2	27	42	79	114
Start of core melt	35	12	50–182	60	115	150
Vessel failure	109	24	239	100	142	174
Cavity dry	–	–	–	276	206	426
Concrete attack	109	–	239	400	291	600

The results of the analyses, performed on the assumptions listed in Table I, are summarized in Tables II to VI. The timing of the key events is provided in Table II. The in-vessel fractional releases from the core are given in Table III. Table IV shows the retention factors in the reactor coolant system. The values reached by various containment parameters are given in Table V, while the fission product (FP) releases to the environment are shown in Table VI.

2.1. The AE Sequence

Owing to the assumed hot leg double ended break, the primary system very rapidly becomes empty, the reactor core uncovers and, after a while, starts melting. Differences, between the two models, in the start melt time (Table II) are due to the fact that in the MARCH analysis the water level inside the vessel at time zero was assumed to be that of the nozzles, while in the MAAP analysis it was set at the core top.

After melting, the molten core attacks the vessel bottom head. The reactor pressure vessel (RPV) fails more quickly in the MAAP model, because in this model the RPV attack is localized (in-core instrumentation tube failure).

The mass and energy release to the containment causes the actuation of the containment safeguards, so when the molten core drops into the cavity a large quantity of water is in the containment sumps and in the reactor cavity itself. This water mass, recirculated and cooled by the containment heat removal systems, completely quenches the molten core and avoids the concrete attack in the MAAP model, while in the MARCH model this quench is only partial and a limited core-concrete interaction takes place. As one can see

TABLE III. IN-VESSEL FRACTIONAL RELEASES FROM CORE (USNRC CODES)

Group[(a)]	Fractional Release, %		
	AE	SE	TE
Xe	100.0	99.6	99.8
I	100.0	99.6	99.8
Cs	100.0	99.6	99.8
Te	50.3	47.4	27.7
Sr	37.2	11.2	14.2
Ru	10.6	4.5	3.9
La	0.13	0.02	0.004
Core Materials[(b)]	2.4	1.3	1.5

(a) FPs are grouped according to WASH-1400.

(b) Fuel, control rods, claddings, core support plate. (In all cases the corresponding aerosols are composed mainly of control rod material.)

TABLE IV. PER CENT RETENTION IN THE REACTOR COOLANT SYSTEM (USNRC CODES)

Sequence	CsI		CsOH		Te		Aerosol *	
	(a)	(b)	(a)	(b)	(a)	(b)	(a)	(b)
AE	11.8	2.7	11.6	2.9	65.4	1.1	52.1	0.8
SE	42.6	42.5	44.7	44.6	32.6	32.4	52.2	40.0
TE	98.2	96.1	98.4	96.3	94.4	84.8	99.1	21.9

* All material different from CsI, CsOH and Te.

a Before RPV melt-through.

b After RPV melt-through.

in Table V, the hydrogen mass produced during the first 48 h, as calculated by MARCH-3, is much greater than the one calculated by MAAP, because of both a a larger extent of in-vessel zirconium oxidation and a remarkable contribution of ex-vessel reactions during the core-concrete interaction. The containment pressurization is limited (Table V) and well below the design pressure of 0.42 MPa. The decay power is removed by the spray and fan-cooler systems.

TABLE V. MAXIMUM VALUES REACHED BY VARIOUS CONTAINMENT PARAMETERS

PARAMETERS	AE		SE		TE	
	USNRC Codes	MAAP	USNRC Codes	MAAP	USNRC Codes	MAAP
Maximum pressure (MPa) absolute	0.29	0.32	0.28	0.43	0.39	0.43
Maximum temperature (K)	385	435(*)	398	425(*)	405	480(*)
H_2 produced in 48 h (kg)	795	55	890	237	685	210
Concrete erosion in 48 h (m)	0.55	0	0.65	0.1	0.60	0
Hydrogen combustion following recovery of containment heat removal	no	no	yes	no	no	no

(*) Lower compartment temperature.

TABLE VI. ACTIVITY RELEASE TO THE ENVIRONMENT (FRACTION OF CORE INVENTORY)

SEQUENCES		Xe(*)	I	Cs	Te	Sr	Ru	La
AE	USNRC Codes	4.3 E-3	4.6 E-7	4.6 E-7	3.4 E-7	1.0 E-7	9.6 E-8	1.0 E-8
	MAAP	3.8 E-3	7.0 E-8	7.0 E-8	6.5 E-8	2.8 E-11	–	–
SE	USNRC Codes	4.5 E-3	3.0 E-5	2.9 E-5	7.4 E-6	5.1 E-6	8.1 E-7	4.0 E-7
	MAAP	4.3 E-3	1.8 E-5	1.8 E-5	1.4 E-5	1.5 E-6	1.9 E-6	–
TE	USNRC Codes	4.5 E-3	1.2 E-5	1.1 E-5	2.1 E-5	2.0 E-5	9.7 E-7	6.1 E-8
	MAAP	4.1 E-3	7.6 E-6	6.5 E-6	2.2 E-6	2.4 E-7	1.6 E-9	–

(*) The noble gas release to the first 24 h.

According to both evaluations a hydrogen burn is predicted not to take place. It has to be noted, however, that by using the MARCH-3 code, the flammability correlation in Ref. [2], instead of the built-in 'default' value, a partial combustion could take place. On the other hand, even if this event took place it would not jeopardize the containment integrity since the peak pressure associated to a complete H_2 burn would be well below the containment design pressure. As regards the in-vessel emission from the core, CORSOR code calculates complete releases of volatile FPs, relatively high Te emission, and moderate releases of the less volatile elements (Table III).

The overall FP retention factor inside the reactor coolant system (RCS) turns out to be negligible if one considers the sequence through the RPV failure (Table IV).

Different retention mechanism modelling and plant nodalization in the MAAP analysis lead to RCS retention factors generally higher than those derived from the NRC codes.

The limited core-concrete interaction predicted by MARCH-3, coupled with the presence of a water pool on the melt, gives rise to a very small ex-vessel aerosol production.

The FPs and inactive aerosol are effectively removed by the spray system and the secondary containment filters. The releases to the environment are less than 10^{-6} of the iodine and caesium inventory according to both ENEL and Ansaldo-Westinghouse calculations (Table VI).

2.2. The SE sequence

In the present analysis a 2 in[1] diameter equivalent break on the primary pump seal was assumed. As a consequence of the small dimension of the break, the transient in the primary system is much slower than the one in the AE sequence. The primary pressure is controlled by the heat removal through the steam generator secondary side and the mass and energy release through the break.

As the primary pressure decreases, the accumulators begin to discharge their water into the circuit, which further delays the core melt. In the MARCH analysis such a delay is particularly long: in fact the overheated or molten nodes of the core are partially quenched, the steam generated from this quenching repressurizes the primary system and stops temporarily the accumulator injection. This sequence is predicted to occur twice before the accumulators are emptied and a complete core melt takes place. When the molten core drops into the reactor cavity the primary circuit is practically empty, so the core-concrete interaction immediately starts. In the MAAP-3 calculation the fuel melting and relocation, with the associated partial flow blockage, prevent extensive quenching of the core. Consequently, core slump and vessel failure occur

[1] 1 in = 2.54 cm.

earlier than in the MARCH-3 calculation, and less hydrogen is generated. According to the MAAP model a significant amount of water is still present in the primary system at the time of vessel failure; this mass floods and keeps the core debris quenched until about 4 h into the transient. After 6 h the containment spray is assumed to operate and that causes the containment pressure to decrease. At the same time the molten mass in the cavity is quenched also in the MARCH model and the core-concrete interaction is strongly reduced. After the refuelling water storage tank has emptied, the containment pressure increases until the recovery of the containment heat removal function by means of the fan coolers, assumed at 12 h.

The pressure transient calculated by MAAP-3 is steeper, due to steaming of the water mass present in the cavity in the first phase, and afterwards due to radiation and convection from the hot core layer to the containment atmosphere. Owing to the slower in-vessel transient and the ex-vessel reactions, the hydrogen production is higher than in the AE sequence. In the MARCH-3 analysis, after a couple of days the conditions for the H_2 burn could occur but the associated peak pressure is well below the containment failure pressure.

The concrete erosion in the basemat is limited to a maximum of 65 cm after two days; the progression is very slow after the recovery of the spray system and the flooding of the cavity.

Lower values for the ex-vessel hydrogen generation and the concrete attack are predicted for the MAAP-3 calculation because flooding by spray water occurs just after cavity dry-out.

Also in this sequence CORSOR code calculates a complete release of volatile FPs and a relatively high Te emission from the core. Because of the repeated in-vessel core quenchings, the release of less volatile elements is lower than that in the AE sequence. The overall FP retention inside the RCS is significant (about 40%), because in this sequence the SG acts as an efficient FP trap.

VANESA calculation shows that during the first hour of the core-concrete interaction all iodine and caesium still present in the melt are released; the aerosol is mainly composed of iron oxide, control rod and concrete materials. During the following hour the aerosol still contains a large fraction of iron oxide and concrete materials, while it is practically composed only of concrete materials later on.

The containment spray recovery quenches the melt and lowers the aerosol mass production by about two orders of magnitude, to which one more order of magnitude must be added, because of the decontamination due to the water pool. At the same time the spray system strongly removes the aerosol suspended inside the containment, while the subsequent fan coolers intervention has no significant effect on the remaining aerosol. The iodine and caesium releases to the environment, after secondary containment filtration, are about 2–3 10^{-5} of the core inventory, according to the ENEL as well as to the Ansaldo-Westinghouse evaluations.

2.3. The TE Sequence

This sequence differs from the other two mainly because the core heatup and melting take place at an elevated primary pressure. The reactor shutdown is followed by the boiloff of the water inventory of the steam generator secondary side. The loss of the steam generator heat sink leads to the heatup of the primary system and increases the primary pressure up to the pressurizer relief valve set point.

The primary coolant is then boiled off through the pressurizer relief valves, leading to the core uncovery and melting. As one can see in Table II, there are no big differences in the in-vessel timing of main events between the two models.

After the vessel failure, however, the MAAP code predicts a partial dispersion of the core outside the cavity. That explains the differences in the time of cavity dry-out and in the containment pressure and concrete erosion predicted by MARCH-3 and MAAP. The recovery of electric power and hence of one train of the containment safeguard system (1 spray pump + 2 fan coolers) at 6 h limits the containment maximum pressure to about 0.4 MPa in both analyses. No hydrogen burn is predicted by MAAP evaluation owing to the smaller calculated H_2 production. As far as MARCH-3 results are concerned, a hydrogen burn is not predicted using the built-in flammability limit, while it could happen using those of Ref. [2]. In this last case a pressure peak lower that 0.4 MPa would occur if the conservative assumption of a complete burn is made.

The core emissions calculated by CORSOR are similar to those of the SE sequence except for Te, for which a lower release is derived, because of the lower zirconium oxidation level. The FP retention factor in the RCS is very high for iodine, caesium and Te even after the RPV melt-through; 80% of the other aerosols is instead carried into the reactor cavity together with the molten core.

As regards the aerosol release during the core-concrete interaction and its transport inside the containment, the results of the VANESA and NAUA-MOD·5 calculations show trends similar to those derived for the SE sequence. The recovery of the containment safeguards, even if delayed by 6 h, allows the iodine and caesium releases to the environment to be limited to values of the order of 10^{-5} of the core inventory.

3. FEATURES FOR PREVENTION AND MITIGATION OF SEVERE ACCIDENTS

In order to reduce the probabilities and the consequences of severe accidents in PUN, several improvements have been introduced into the design, utilizing also the probabilistic safety study as a design tool.

The most significant prevention features are:

- Addition of a seal water injection system (ESIS), to reduce the probability of a small LOCAs upon loss of PCCWS;
- Addition of a steam generator overfill system (SGOS), to return water into the containment, in case of SGTRs and to prevent uncontrolled releases to the environment via safety and relief valves on the main steam lines;
- Installation of additional and independent on-site supplies of electric power;
- Installation of a turbogenerator, to feed an ESIS backup pump and to charge the batteries;
- Optimization of man-machine interface in the main control room, with special reference to plant supervision and diagnostics of accident situations;
- Installation of a third isolation valve on the reactors residual heat removal lines and leak monitoring;
- Installation of back-up protection system;
- Testing of safety injection system check vales;
- Redundancy and diversification of the recirculation valves from the sumps;
- Improvement on availability of service water and component cooling systems;
- Provisions for refuelling water storage tank refill;
- Provisions for scram frequency reduction;
- Systematic recycle of operating experience.

The most significant mitigating features are:

- Increased depth of the reactor cavity in the containment building (3 m), to be filled with concrete or other material, to ensure that the structural concrete of the containment basemat is not affected by molten core-concrete interaction;
- Provisions for continuous containment leak tightness monitoring;
- Provisions for alternative means of containment heat removal;
- Provisions for containment filtered venting, to avoid late containment over-pressurization;
- Hydrogen concentration measurements and provisions for its recombination.

All these technical improvements will be introduced together with a suitable set of emergency procedures to ensure that the important safety functions are recovered in case of initial loss.

4. MARGINS TO FAILURE PRESSURE FOR CONTAINMENT STRUCTURES

Structural analyses have been performed for the PUN containment and for all the other containments of the NPPs in Italy, in order to evaluate the margin to failure due to internal pressure.

It is very important to notice that, for the determination of the containment strength, all the penetrations placed on the shell surface are as significant as the main structure strength itself.

These penetrations can be classified as:
(a) Large openings (for equipment, personnel and emergency hatches);
(b) Mechanical penetrations (for piping);
(c) Electrical penetrations.

Consequently the ultimate pressure capacity for a containment is defined as the minimum value in the set of ultimate pressures calculated for the main structure and for each component listed above. In general the resulting safety coefficients are at the lower bound because of the assumptions made in the limit pressure calculations.

The following rules have been applied to the calculation of the ultimate pressure of the PWR containment structure in reinforced concrete:

(1) The yield strength of the materials is the nominal one;
(2) The liner is not considered to be pressure resistant with the structure;
(3) The influence of the temperature on the materials characteristics is considered;
(4) The calculations are performed using linear analyses; some non-linearities are taken into account by material characteristics modifications (i.e. concrete cracking);
(5) The ultimate pressure is reached when a general yielding of rebars appears in one section and the maximum strain is less than 1%.

In these conditions the ultimate pressure is 2.6 times the design pressure. Similar values are obtained for other plants which have the concrete containment; the coefficient can increase up to 5 for the steel structure containments. In this last case it has been shown that the leakage of some component appears at lower pressure values.

It is also possible to define an intermediate pressure (between the design and the ultimate) named 'no leakage' pressure, i.e. the pressure for which no substantial violation of the design leaktightness takes place. For the PUN containment this intermediate pressure is 1.4 times the design pressure and all the components forming a barrier against the leakage must remain in the elastic state under this load.

As a backup to the recovery of the containment heat removal function, which is capable of stopping the pressure increase, a controlled venting of the containment is probably the best way of preserving the global integrity at the containment and, at the same time, of fully controlling the associated leakage.

5. CONCLUSIONS

The analyses performed for PUN, with two independent methodologies, (USNRC code suite and MAAP code), have shown that it is possible to ensure containment integrity for the potential severe accident sequences, not only if the containment safeguards operate correctly but also if their function is recovered after several hours of unavailability. Even in the worst of the analysed cases, the activity release to the environment turns out to be four to five orders of magnitude below the core inventory for iodine and caesium, with lower fractions for the other isotopes.

Provisions have been adopted in PUN to achieve additional confidence that severe accidents are prevented or at least drastically mitigated. Substantial operational flexibility has been ensured for safety functions.

The ratio between failure pressure and design pressure of the containment is greater than 2.6.

REFERENCES

[1] NOBILE, M., PENNAZZI, M., SERRA, S., IAEA-CN-48/181, these Proceedings.

[2] MARSHALL, B.W., Jr., "Hydrogen: air: steam flammability limits and combustion characteristics at intermediate scale", Proc. ANS/ENS Top. Mtg on the Operability of Nuclear Power Systems in Normal and Adverse Environments, Albuquerque, NM (1986).

UNITED STATES DEPARTMENT OF ENERGY SEVERE ACCIDENT TECHNOLOGY AND ANALYSIS PROGRAM

G.D. McPHERSON
Office of Light Water Reactor Safety and Technology,
United States Department of Energy,
Washington, D.C.

R.R. HOBBINS, P. NORTH
Idaho National Engineering Laboratory,
Idaho Falls, Idaho

United States of America

Abstract

UNITED STATES DEPARTMENT OF ENERGY SEVERE ACCIDENT TECHNOLOGY AND ANALYSIS PROGRAM.

The United States Department of Energy program of severe accident technology has grown with the realization of the importance of severe accidents to the safety and survival of nuclear power. Today, it is comprised of the Three Mile Island (TMI) accident evaluation program, the Organisation for Economic Co-operation and Development Loss-of-Fluid Test (LOFT) program, and several experimental studies designed to reduce uncertainties to levels sufficient to support resolution of risk significant severe accident issues. In the TMI program, 10 core bores have recently been taken from the destroyed core, for the first time providing a look at what is contained beneath the cavity discovered in the upper portion of the original core region. A lens-shaped homogeneous volume of previously molten core material is generally enclosed in a 10–15 cm thick layer of metal–fuel agglomerate located above standing fuel rods in the lower regions of the core. An international program has been formed to share in various sample examinations and in an analytical benchmark study of the accident. The final LOFT experiment subjected a fuel bundle to the first stages of a V-sequence accident and achieved fuel temperatures above 2100 K for 4.5 minutes. The resulting release and transport of fission products and aerosols were measured. The damaged fuel is now being prepared for examination. The combined results of this controlled and instrumented experiment will provide a picture of events as they occurred in the early stages of the TMI-2 accident. In a supporting study on fission product revaporization, the physical, chemical, and transport properties of species that are likely to dominate the transport process are measured, new species, important to fission product transport such as the complex CsI·CsOH, are being discovered and studied and the resulting data is being correlated for use in severe accident codes.

1. INTRODUCTION

The U. S. Department of Energy (DOE) is conducting a severe accident technology and analysis program designed to assist, in cooperation with other funding agencies in the U.S. and throughout the world, in the development of an understanding of severe accidents. This understanding will be embodied in validated computer codes that will enhance the design and analysis of the safety features of Advanced Light Water Reactors (ALWRs), thereby supporting their certification in the licensing process. The program consists of a variety of elements the cornerstone of which is the Three Mile Island-Unit 2 (TMI-2) accident evaluation program. Other major elements of the DOE program include the Organisation for Economic Cooperation and Development (OECD) Loss of Fluid Test (LOFT) program, the Advanced Reactor Severe Accident Program (ARSAP), the Severe Accident Technology (SAT) program, and the fission product revaporization program. The TMI-2, LOFT, SAT, and revaporization programs are designed to provide data for the validation of computer codes to be used in ARSAP for the analysis of safety features of ALWRs and for support of their certification. Results from those program elements that provide insight into the understanding of severe accidents will be presented in this paper.

The examination of the TMI-2 accident represents the largest effort in the DOE program on severe accidents. Results from that work and results currently available from the LOFT LP-FP-2 test are summarized and results on revaporization are discussed.

2. TMI-2 ACCIDENT EVALUATION

The TMI-2 accident evaluation program is being conducted to (1) understand the physical and chemical end state of the TMI-2 core and related structures and the external influences which affected the accident, (2) understand what happened during the accident and to provide a qualified data base to permit an analysis exercise of the TMI-2 accident to benchmark severe accident analysis codes and methodologies, (3) understand the relationship between the phenomena and processes controlling the accident and the important severe accident/source term technical issues, and (4) assure that results of the program are effectively transferred to the nuclear community.

2.1 The End State Configuration

Knowledge of the end state of the damaged reactor is key to the development of a best estimate accident scenario. The known end state configuration within the TMI-2 reactor vessel is shown in Figure 1. The upper plenum appears to be remarkably undamaged but contains evidence of large temperature variations at different locations. Samples taken from control rod drive leadscrews indicate that temperatures during the accident were at a maximum of about 1250 K at the lowest part of the plenum and about 700 K at the top, based on the microstructure and microhardness of the steel [1]. An

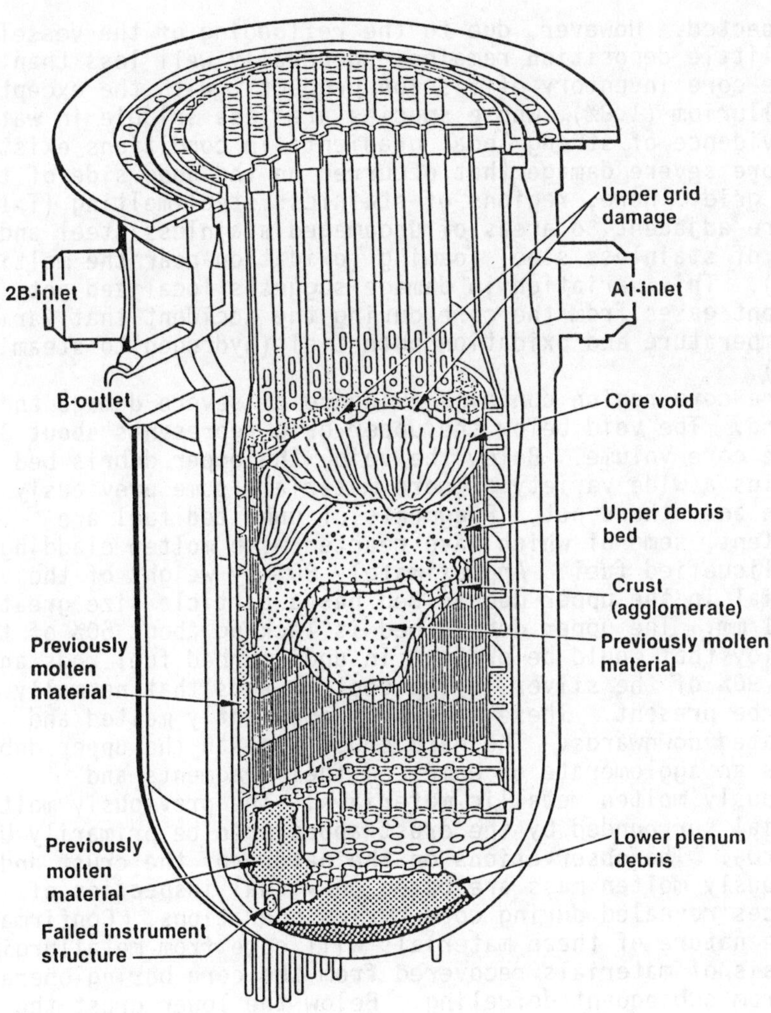

FIG. 1. End-state core configuration, 1987.

oxide scale exists on the leadscrew surfaces which is thicker at the lower, hotter end of the leadscrews. Oxide growth kinetics suggest that 67 minutes at 1250 K (virtually the entire heatup period) is required to produce the thickest oxide measured [2]. Additionally, the temperature at the bottom of the plenum is lower (970 K) at mid radius than at the radial center (1250 K), suggesting a convection current pattern in the gas exiting the core. In the temperature range exhibited by the upper plenum, deposition of a number of fission product species such as CsI, CsOH, Te, and SnTe would be expected. However, due to the reflooding of the vessel, very little deposition remains: generally well less than 1% of the core inventory of fission products, with the exception of tellurium (1.8%), whose species are less soluble in water.

Evidence of strong local gradients in conditions exists in the more severe damage that occurred on the underside of the upper grid. Here, regions of stainless steel melting (T>1700 K) were adjacent to areas of undamaged stainless steel and areas of stainless steel foaming (oxidation near the melting point). This variation in damage suggests localized jets of effluent gases from the core during the accident that varied in temperature and oxidation potential (hydrogen to steam ratio).

The core region contains evidence of severe damage and melting. The void below the upper grid represents about 30% of the core volume. Below the void, the upper debris bed contains a wide variety of core materials some previously molten and others not. Fragments of unmelted fuel are prevalent, some of which have reacted with molten cladding to form liquefied fuel. Approximately 80% by weight of the material in the upper debris bed has a particle size greater than 1 mm. The upper debris bed is lacking about 50% of the zircaloy that would be present in undisturbed fuel rods and about 90% of the silver in the control rods that normally would be present. These materials apparently melted and relocated downwards. The hard crust beneath the upper debris bed is an agglomerate of unmelted fuel fragments and previously molten metallic materials. The previously molten material surrounded by the crust appears to be primarily UO_2 and ZrO_2. The observations on the nature of the crust and previously molten mass are based on visual inspection of surfaces revealed during core boring operations. Confirmation of the nature of these materials will come from metallurgical analysis of materials recovered from the core boring operation and from subsequent defueling. Below the lower crust the rod stubs are essentially undamaged. The core support assembly appears to be basically undamaged although it contains solidified, previously molten core materials in some locations. The lower plenum contains debris composed primarily of previously molten ceramic material, mainly mixtures of UO_2 and ZrO_2 with small amounts of metallic structural and control materials. The amount of debris represents about 16% of the original mass of the core. Melt, through of the core baffle occurred in at least one location permitting molten core materials also to relocate within the core-former region inside the core barrel. Clearly, the accident scenario leading to these conditions contained significant core melting and a complex relocation process.

A best estimate accident scenario [3] has been developed based on these end state conditions, the response of reactor system instrumentation during the accident, the condition of instruments after the accident and engineering analyses of heat and mass balances.

2.2 Five Stages in the Accident Scenario

The accident is conceived to have five principal stages. The first is a period during which coolant was being lost with the reactor coolant pumps on (0 to 100 min., measured from the time of the turbine trip). Reactor coolant inventory was being lost through the pilot-operated relief valve (PORV) which, unknown to the reactor operators, had become stuck open after opening at 3 s into the transient. The reactor automatically scrammed on a high pressure trip at 8 s. As far as can be determined, the core was covered with a two-phase mixture and remained cool as long as the reactor coolant pumps were operating. The B-loop pumps were turned off at 73 min. and the A-loop pumps at 100 min. due to excessive vibration.

The second stage of the accident, the initial core heatup and degradation (100 to 174 min.) was initiated after the final pump shutdown. In this stage the reactor vessel liquid level decreased due to boiling in the core region, loss through the PORV and through the A-loop letdown line. The decreasing core liquid level led to core uncovery and subsequent core heatup sufficient to cause severe core degradation. The best estimate core liquid level at 174 min. is in the neighborhood of 0.5 to 1.0 m above the bottom of the core. Calculations with severe accident codes such as SCDAP indicate that fuel rod failure would be expected at about 140 min., in agreement with radiation measurements during the accident, and that fuel rod temperatures exceeded 2400 K where dissolution of UO_2 by molten zircaloy and relocation downward is expected. This liquefied fuel would tend to freeze near the level of the coolant in the core. Rapid oxidation of zircaloy cladding and generation of hydrogen is calculated in this same period.

The third stage of the accident was the pump transient at 174 min. One of the B-loop reactor coolant pumps pumped water for less than 15 s during which as much as 28 m^3 of water (enough to fill the reactor core) may have been introduced into the reactor vessel. However, due to the extensive core blockage, flow through the central regions of the degraded core was probably not appreciable. The coolant delivery to the reactor vessel rapidly pressurized the reactor system and caused thermal-mechanical forces that are believed to have fragmented the highly oxidized fuel rod remnants in the upper core region, forming a rubble bed on top of the relocated core materials in the central region of the core.

The period 174 to 224 min. is characterized by the heatup of the relocated and consolidated mass and the debris bed above it to form a molten mass of UO_2/ZrO_2 material surrounded by a crust. This heatup occurred despite the continued addition of coolant to the reactor vessel during this period because the degraded core geometry had become uncoolable.

Responses of reactor instrumentation indicate that molten core materials relocated, primarily through the core support assembly, into the lower plenum during the fifth stage of the accident from 224 to 230 min. The relocation process did not cause significant damage to the lower core support assembly, although there is evidence that instrument tubes were melted by contact with the debris in the lower plenum. The extent of thermal interaction with the lower head and the possible movement of molten core materials beyond the head through the instrument penetrations is currently unknown. There is evidence that some relocation also occurred by melting of the core baffle. The molten core materials that relocated to the lower plenum appear to be solid solutions of essentially stoichiometric dioxides of uranium and zirconium. The minimum melting temperature for such ceramic mixtures is 2800 K. It is most significant that the relocation of the melt into the water-filled lower plenum resulted in the formation of a debris bed with a coolable geometry. The interaction of the molten core materials with the water resulted in steam generation and an increase in the system pressure of about 2 MPa which lasted for approximately 15 min. Violent steam explosions do not appear to have occurred.

2.3 Results of Debris Examination

Analysis of data from the examination of materials removed from the damaged reactor has provided some insights into the chemistry of the environment within the reactor during the accident [2]. Based on the composition and thickness of the oxide on the surface of the control rod drive leadscrews and the metallographically-determined temperatures of the leadscrews, the H_2/H_2O ratio in the effluent gas from the core appears to have been less than unity for much of the time the core was uncovered. The presence of metallic cadmium on leadscrew surfaces suggests that the H_2/H_2O ratio was greater than about 0.01. Thermodynamic and kinetic calculations of gas phase chemistry and measurement of the ionic species of iodine in the water in the sump of the reactor basement suggest that radioiodine was transported from the reactor pressure vessel to the reactor building as an iodide rather than molecular iodine. There is evidence of the retention of cesium as a silicate within the oxide scale on the surface of leadscrews in the upper plenum, but the amount is small - less than 1% of the core inventory of cesium. Current information on fuel oxidation in TMI-2 suggests quite

TABLE 1. EX-CORE FISSION PRODUCT DISTRIBUTION (%)

	Noble Gases[a]	I	Cs	Te	Sr	Sb	Ru	Ce
Reactor building basement and tanks	47	21	42	3.5	2	0.3	0.5	.01
Reactor coolant system	NM	2	1	1.9	1	0.1	ND	NM
Auxiliary building	1	2	5	0.002	0.2	0.3	NM	NM
Decontamination resins	NM	NM	5	0.4	0.1	ND	ND	NM
Total outside core	48	25	53	5.8	3	0.7	0.5	NIL

NM = Not measured
ND = Not detected
(a) = In the reactor building atmosphere

TABLE 2. AVERGE FISSION PRODUCT RETENTION

	Percent of inventory retained*			
	Lower Plenum		Upper Debris	
Radionuclide	Average	Range	Average	Range
^{129}I**	2	0 - 10	22	10 - 28
^{137}Cs	16	9 - 22	21	6 - 32
^{125}Sb	5	3 - 10	28	18 - 38
^{106}Ru	7	4 - 9	55	35 - 86
^{90}Sr	100	40 - 160	93	79 - 102
^{154}Eu	85	75 - 94	90	60 - 108
^{144}Ce	114	106 - 124	114	90 - 130

* Compared with whole core average ORIGEN-2 analysis (μCi/gU) and average uranium content

** Average radionuclide content based on the average of a number of small (5-40 mg) particle analyses

wide ranges of oxygen to metal ratios in the upper debris bed and up to quite high values. This information suggests that pathways formed by oxidized structures may have transmitted essentially pure steam to local regions of the core at times during the accident.

The fission products outside the core reside primarily in the basement and tanks of the reactor building as shown in Table 1. It is interesting that, but not yet clear why, the iodine found outside the reactor coolant system is about half the fractional releases of the noble gases and the cesium. The release of tellurium is consistent with calculations using SCDAP with the Lorenz model [4]. The releases of strontium, antimony, ruthenium, and cerium are all very low. The release of antimony is lower than might be expected based on out of pile data, but the other releases are consistent.

The retentions of fission products measured in the upper debris bed and in the lower plenum debris are presented in Table 2. These data show considerable scatter due primarily to the use of a whole core average fission product inventory calculation, whereas individual samples may have quite different fission product inventories depending on their location (not well known) in the core before the accident (both uranium enrichment and neutron fluence vary with location in the core). In the upper debris bed which exhibited a wide range of temperatures from less than 1900 up to 3100 K but probably averaged around 2200 K, iodine and cesium were retained to about 20% of their inventories. These retentions appear high based on out-of-pile data (the CORSOR-M code calculates total release of iodine and cesium from fuel after 10 min. at 2200 K) but are in reasonable agreement with the results of in-pile severe fuel damage testing at the Idaho National Engineering Laboratory [5]. The retention of only 55% of ruthenium in the fuel in the upper debris bed would not be predicted by current fission product release models. The nearly 100% retention of strontium, europium, and cerium in the upper debris bed is consistent with current modeling. The retention of 2, 16, and 7% of iodine, cesium, and ruthenium, respectively, in the previously molten lower plenum debris is unexpected based on current models. The formation of the large mass of ceramic melt (with its low surface to volume ratio) in the core prior to relocation to the lower plenum may have had a significant influence on fission product release. However, the larger fractional retention of cesium compared with that of iodine suggests that the formation of some relatively stable form of cesium may also have played a role. Additionally, the retention of only 7% of the inventory of ruthenium is not predictable with current models. As was noted earlier, ruthenium and antimony were not found outside the core in large fractions of their inventories; therefore, it remains an important task to discover where in the reactor vessel these materials reside. Ruthenium was found to migrate to molten steel in out-of-pile meltdown experiments at Oak Ridge National Laboratory [6] and there are preliminary findings in the examination of debris from TMI-2 that ruthenium and antimony are associated with metallic components. Work on measuring the distribution of tellurium within the reactor vessel is proceeding based on a newly developed technique involving chemical separation and concentration of the tellurium to enhance its detectability [4].

2.4 Significance of Results to Severe Accident Analysis

Several of the findings have significance for severe accident analysis. The presence of water in the lower plenum and in the lower regions of the core had a significant influence on the course of the accident. This finding is particularly relevant for the small break and station blackout classes of accident sequences. Water in the TMI-2 core influenced the relocation of molten core materials to take place in three distinct stages: (1) the candling of liquefied fuel to a region in the core near the coolant level, (2) the reheating of the debris to form a large molten mass, and (3) the failure of the crust surrounding the large molten mass permitting the relocation of the molten material to the lower plenum. The relocation of the molten core materials took place primarily by flow through the holes in the lower core support assembly without melting the structures. The molten core materials were basically stoichiometric dioxides of uranium and zirconium containing small amounts of structural and control materials. With a ratio of hydrogen to steam in the core effluent between 0.01 and 1.0, it is clear that the accident was not steam starved. The molten debris interacted with the water in the lower plenum to produce steam and, most importantly, a debris bed of coolable geometry. The retention of the fission product cesium in the molten debris in the lower plenum is larger than would be predicted with current fission product release models. This means that significant fractions of the cesium inventory (16%) may be available for ex-vessel release in molten core-concrete interactions during unmitigated accident sequences. The analyses of unmitigated accident sequences need to model the observation that ruthenium largely migrates from the fuel to metallic components of the molten core debris.

Information regarding the processes of fuel pin damage and the release and transport of fission products, similar to the early processes in TMI-2, is expected from the OECD LOFT LP-FP-2 experiment [7,8], as discussed in the following section.

3. OECD LOFT FP2

The OECD LOFT LP-FP-2 experiment was a fission product release and transport experiment in which the center fuel module of the LOFT reactor was severely damaged by simulating a V sequence (interfacing systems loss of coolant accident). The 1.68-m long, shrouded 11x11 test bundle consisted of 100 fuel rods, 11 AgInCd control rods, and 10 instrument guide tubes.

Fuel temperatures above 2100 K were produced for about 4.5 minutes. Fission product and aerosol release from the core and transport through the upper plenum and the simulated low pressure injection system (LPIS) line were measured. The test bundle, deposition coupons, collection filters, and grab samples have been removed from the reactor. Data from various on-line instruments, including gamma spectrometers, and results from deposition coupons, collection filters and grab samples have been qualified and are being evaluated. The damaged bundle is now being prepared for examination. Results from that examination may modify the preliminary observations discussed below.

Visual examination of the shrouded fuel bundle revealed no indication of molten or solid debris at the bottom of the bundle although debris was observed on the upper end box. A gamma scan of the bundle gave indications consistent with some fuel rod relocation. Based on hydrogen generation, more than 14% of the zircaloy cladding was oxidized. Extensive damage to the bundle is also indicated by silver, indium, and cadmium control rod materials found on upper plenum surfaces. Maximum temperatures and the extent of melting and relocation of materials within the bundle will be determined by destructive examination.

Preliminary analysis of on-line noble gas release data indicates that only about 4% of the inventory of noble gas was released from the test bundle prior to reflood. These same data show the release rate of noble gases from the bundle decreasing by about 30% over a period of 150 s prior to reflood. It is not clear at this point whether the decrease in noble gas release at high temperature is due to a reduction in the release of fission products from the fuel or from a reduced flow of steam through the bundle. Such a flow reduction could be caused by the relocation and solidification of liquefied fuel and other molten bundle materials. Preliminary results indicate that only very small fractions of the volatile fission products iodine and cesium released from the fuel during the experiment were transported through the upper plenum and the LPIS line to the collection tank. Very small quantities of the fission products tellurium, barium and ruthenium were also measured in the collection tank. The majority of the iodine and cesium released from the fuel (about 16 and 20% of their inventories, respectively) ended up in the reflood water used to terminate the experiment.

These preliminary observations from the LP-FP-2 experiment will be evaluated in the context of the destructive examination of the test bundle, the TMI-2 accident evaluation and in light of revaporization experiments aimed at determining the details of how certain fission products are transported through the primary system.

4. REVAPORIZATION

Revaporization is difficult to measure in terminated accident sequences such as TMI-2 and the LOFT LP-FP-2 test because of the solubility of many fission product species in water, but it could be an important mechanism for the transport of fission products from the reactor coolant system to the containment in some unmitigated sequences. The DOE is sponsoring experimental work on revaporization at Argonne National Laboratory which is aimed primarily at the determination of gaseous species vaporizing from deposits of fission products and control materials. Using mass spectrometry in conjunction with a Knudsen cell, the vapor pressures of species above CsI, CsOH, and liquid solutions of these materials have been determined [9]. The vapor pressure measured for CsI is in good agreement with the literature; however, the vapor pressure measured for CsOH is an order of magnitude lower than the values in the JANAF tables [10]. A CsI·CsOH complex species has been found in the vapor above solutions of CsI and CsOH which increases the vapor pressure of I-bearing species above the solution. The vapor pressure of the CsI·CsOH complex has been determined over the temperature range 809 to 927 K for a solution of 0.135 mole fraction of CsOH in CsI. The revolatization of iodine from deposits containing CsI and CsOH is enhanced by the formation of the complex CsI·CsOH in the gas phase. New data indicates that the revaporization of CsOH itself is reduced compared with older, standard tabular data.

5. SUMMARY

The most significant insights from the DOE-sponsored research on severe accident technology are briefly summarized in this section. The presence of water in the lower plenum and in the lower regions of the core in the TMI-2 accident had a large influence on the progression and termination of the accident. Mechanisms, as yet undetermined, exist which can cause the retention of the volatile fission product cesium in molten fuel in amounts potentially significant for ex-vessel release in unmitigated accident sequences. Ruthenium tends to leave overheated fuel in favor of molten metal phases.

Geometry changes associated with the relocation and solidification of liquefied fuel may slow fission product release. The formation of the complex CsI·CsOH appears to enhance the revolatization of iodine.

REFERENCES

[1] VINJAMURI, K., AKERS, D.W., and HOBBINS, R.R., "Examination of H8 and B8 Leadscrews from Three Mile Island Unit 2 (TMI-2)", GEND-INF-052, September (1985).

[2] HOBBINS, R.R., et al., "Insights on Severe Accident Chemistry from TMI-2", Symposium on Chemical Phenomena Associated with Radioactivity Releases During Severe Nuclear Plant Accidents, Am. Chem. Soc., Anaheim, CA September (1986).

[3] TOLMAN, E.L., et al., "TMI-2 Accident Scenario Update", EGG-TMI-7489, December (1986).

[4] VINJAMURI, K., et al., "Tellurium Chemistry, Tellurium Release and Deposition During the TMI-2 Accident", EGG-TMI-6894, August (1985).

[5] MARTINSON, Z.R., PETTI, D.A., and COOK, B.A., "PBF Severe Fuel Damage Test 1-1 Test Results Report", NUREG/CR-4684, EGG-2463, 1 October (1986).

[6] PARKER, G. W., CREEK, G. E., and SUTTON, A. L., Jr., "Source Term Evaluations from Recent Core-Melt Experiments", Proc. Int. Symp. on Source Term Evaluation for Accident Conditions, Columbus, Ohio, October 28 - November 1 (1985), IAEA, Vienna (1986).

[7] MCPHERSON, G.D. and HICKS, D., "LOFT Goes out with a Severe Accident", Nuclear Engineering International, April (1986), 24.

[8] HICKEN, E.F., HICKS, D., and MCPHERSON, G.D., "The OECD LOFT Project, A Model for International Cooperation", (Proc. ENC 1986, Geneva).

[9] JOHNSON, C.E., "Chemical State of Volatile Fission Products in Nuclear Accidents", J. Nuclear Materials, to be published.

[10] JANAF Thermochemical Tables, 1947 Supplement, J. Physical and Chemical Reference Data, 3, 2 (1947).

CANDU FUEL BEHAVIOUR IN SEVERE FUEL DAMAGE CONDITIONS

J.H.K. LAU, C. BLAHNIK, O. AKALIN
Ontario Hydro,
Toronto, Ontario, Canada

Abstract

CANDU FUEL BEHAVIOUR IN SEVERE FUEL DAMAGE CONDITIONS.

During postulated severe fuel damage accidents in a CANDU reactor, several unique high temperature phenomena will occur which have the effect of limiting the fuel temperatures reached during the transient. These phenomena include the deformation of the fuel and fuel channels, the relocation of molten cladding material, and the rate of steam mass transfer onto the cladding surface for oxidation. In the paper, an integrated view of the roles of these processes and their inter-relationship during severe fuel damage accidents is presented. Owing to their impact on fuel temperatures, it is important that these phenomena are accounted for in the safety assessments of severe accidents, so that realistic consequences are predicted.

1.0 INTRODUCTION

In both the CANDU and LWR safety research programs, considerable attention is given to severe fuel damage accidents, because of their potential significant radiological consequences. In this type of accident (for example, a loss of coolant accident coincident with the postulated failure or impairment of the emergency coolant injection system), the fuel heats up at decay power under degraded steam cooling conditions and can remain at elevated temperatures for a prolonged period of time.

Under severe fuel damage conditions, the exothermic reaction between the Zircaloy cladding and steam can play a significant role in determining the severity of the fuel temperature excursion. With adequate steam supply, cladding oxidation obeys a parabolic rate law, with the resultant heat generation rate proportional to the reaction rate constant and the oxidizing surface area, and inversely proportional to the oxide layer thickness. Since the parabolic reaction rate constant increases exponentially with temperature, even at temperatures beyond the cladding's melting point [1], oxidation can provide a significant heat source during the accident. However, at severely degraded steam cooling conditions, the Zircaloy reaction rate can be limited by the mass transfer rate of steam onto the cladding surface. Furthermore, the results of heat-up experiments with CANDU and LWR fuel have consistently

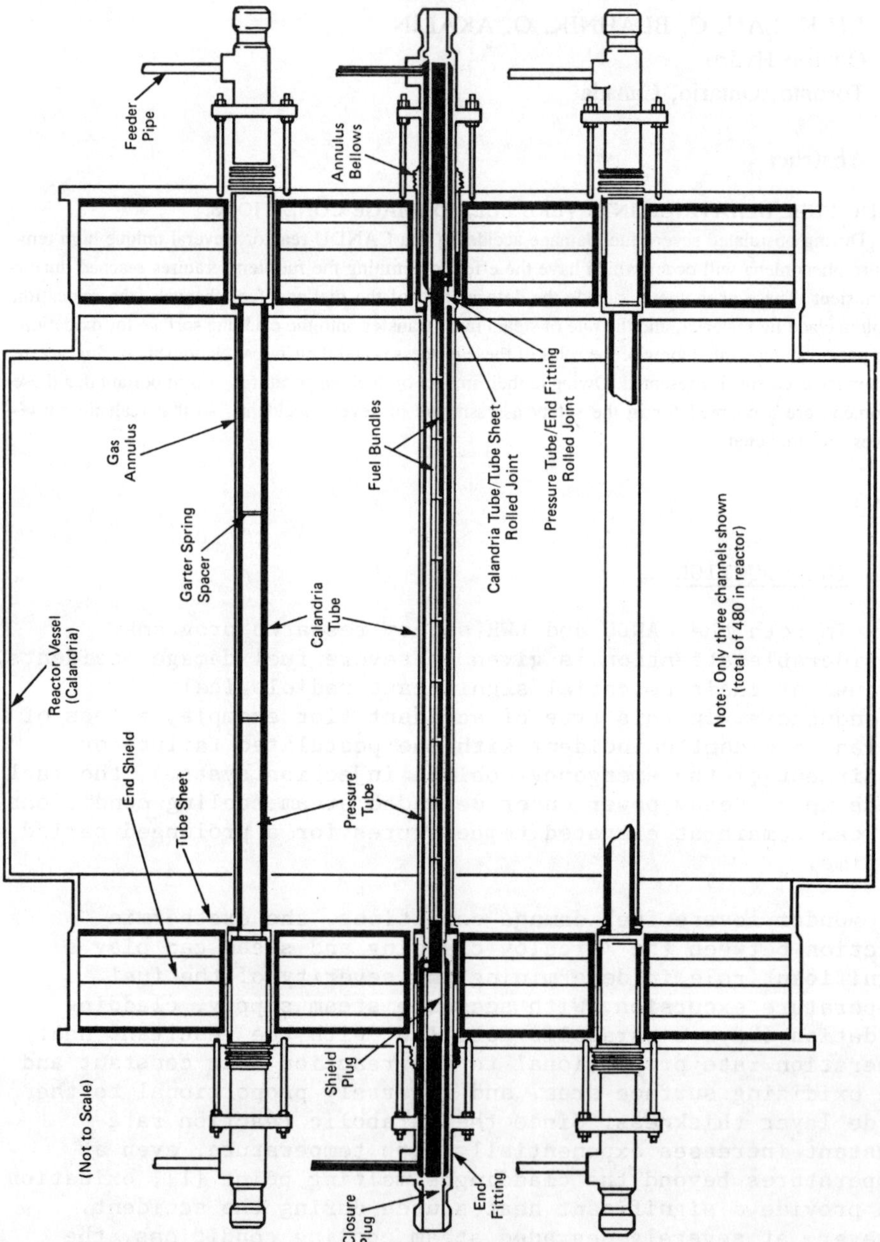

FIG. 1. Simplified diagram of CANDU fuel channel assembly.

shown that at rapid heat-up rates, extensive relocation of molten cladding material occurs, which leads to a surface area reduction of the Zircaloy metal relative to its nominal geometry [2-3]. Thus, even though the limitation in the rate of steam mass transfer to the cladding and melt relocation do not alter the total mass of Zircaloy available for oxidation, they reduce the overall reaction rate. There is an ongoing debate whether these processes (in particular, molten cladding relocation) can significantly limit fuel temperature excursions during a severe fuel damage accident.

At high temperatures, there are three processes which limit the severity of temperature excursions during a CANDU severe fuel damage accident. These processes include channel geometry deformation, melt relocation and the rate of steam mass transfer onto the cladding surface for oxidation. The purpose of this paper is to provide an integrated view of the roles of these processes and their inter-relationship during severe fuel damage accidents.

2.0 SOME DESIGN FEATURES OF THE CANDU REACTOR CORE

An 850 MW(e) CANDU reactor has 480 horizontal fuel channel assemblies, situated within a calandria vessel which contains cooled heavy water moderator. Figure 1 illustrates the channel assembly design. Each fuel channel consists of a horizontal Zircaloy-2.5% Nb pressure tube which contains thirteen fuel bundles and the high pressure, heavy water heat transport coolant. The pressure tube is rolled into the hub of an alloy steel end fitting at each end. Each assembly is supported by the end shield structure through a sliding bearing arrangement and partially by the calandria tube/pressure tube annular spacers. The Zircaloy-2 calandria tube is attached rigidly to the calandria vessel tube sheet, thereby separating the pressure tube from the heavy water moderator. The annulus between the pressure tube and the calandria tube is filled with low pressure CO_2.

The CANDU fuel bundle consists of short (0.5m), fuel elements welded at either end to Zircaloy-4 end-plates. The elements are separated from each other by brazed Zircaloy-4 spacers (refer to Figure 2). The Zircaloy-4 cladding is approximately 0.4 mm thick. The bundles rest on the pressure tube, separated from it by Zircaloy bearing pads brazed on the outer elements of the bundles.

3.0 CHANNEL GEOMETRY DEFORMATION AND HEAT REMOVAL

During a severe fuel damage accident, various heat sources and heat sinks interact to determine the thermal behaviour of

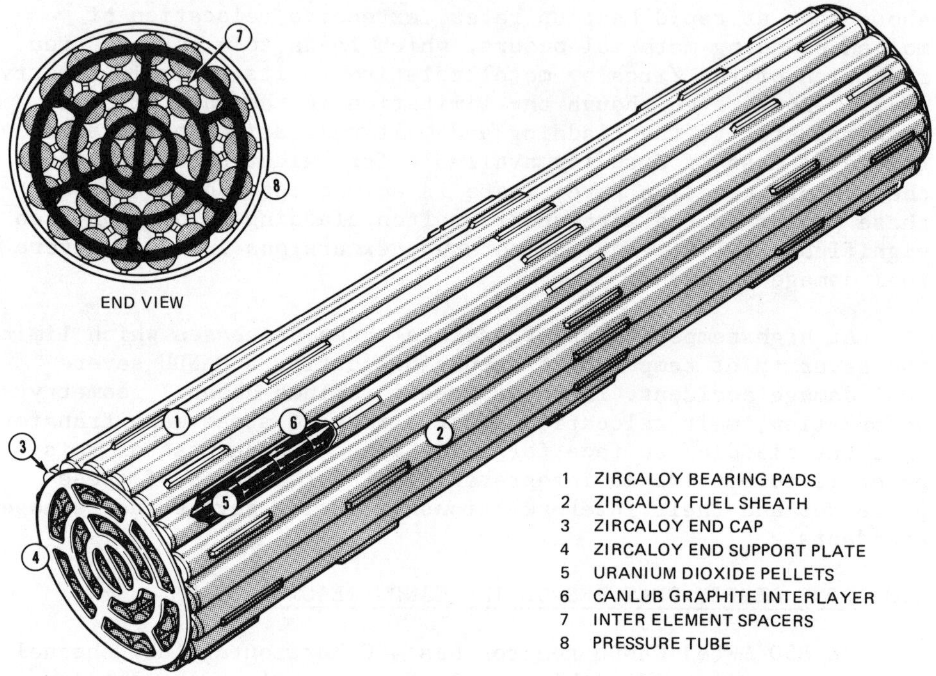

FIG. 2. CANDU fuel bundle.

the fuel and the fuel channel. The heat sources are the fuel's decay heat and the exothermic oxidation of the Zircaloy cladding. Heat removal is due to degraded coolant flow, as well as heat rejection to the moderator. During the transient, the fuel temperature increases to the level at which the heat removal rate balances the heat generation rate. Thus, the larger the thermal resistance for heat removal, such as the resistance through the gas annulus, the higher the fuel temperatures must be in order to reject the heat generated.

During normal reactor operation, the heat transfer out of a channel to the moderator is minimal, since the gas annulus provides an effective insulation. However, at elevated temperatures associated with severe fuel damage conditions, deformation of the channel components occurs and consequently the thermal resistance to heat transfer to the cool moderator is reduced. As shown in Figure 3, two modes of pressure tube deformation are possible under degraded steam cooling conditions. If the heat transport system pressure is high during heat-up, the pressure tube will strain radially (referred to as

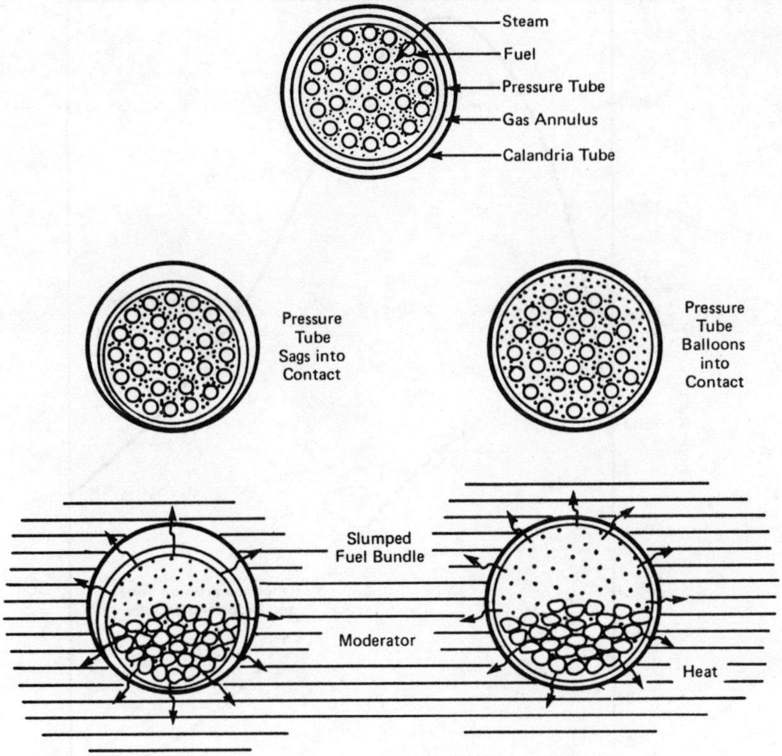

FIG. 3. Geometry distortions in a fuel channel (CANDU large LOCA with no ECI).

pressure tube ballooning) to uniformally contact the calandria tube. If the internal pressure is low, the pressure tube may sag under the weight of the fuel bundles to locally contact the calandria tube. At higher temperatures fuel bundles may also deform to establish extensive contact with the pressure tube, further promoting a more effective heat removal path from the fuel to the moderator.

Geometry distortions can also change the hydraulic resistances of the channel flow paths. Since the steam flow during a degraded steam cooling accident is driven by the pressure gradient across the fuel channel, any change in the channel's hydraulic characteristics will affect the steam flow rate. For example, with a given pressure gradient, the steam flow through a fuel channel increases when the pressure tube balloons, leading to an enhancement of heat removal from the fuel channel [4]. Furthermore, deformations can alter the local

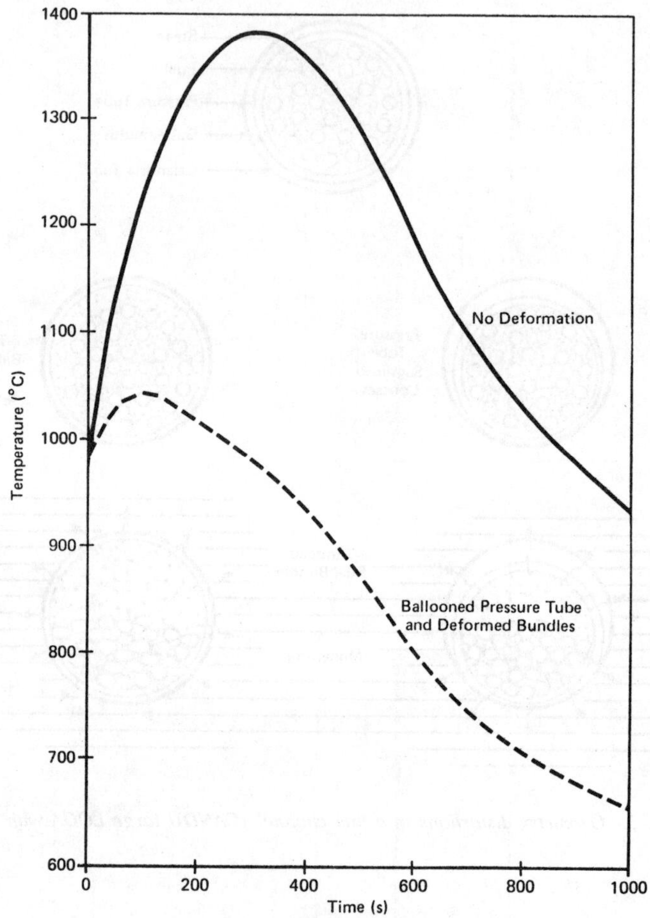

FIG. 4. Channel average fuel temperature transient — effect of deformation.

coolant flow pattern through a fuel bundle. In a deformed bundle, for example, the subchannel flow areas can be significantly reduced, limiting the access of steam to the bundle interior. Similar effects result also from pressure tube ballooning. The consequent increase in the flow area exterior of the bundle leads to a flow bypassing around the bundle, thereby limiting the access of steam to the bundle interior. Thus, bundle deformation and pressure tube ballooning can also result in steam starvation inside the bundle, which limits the potential cladding oxidation and its attendant heat generation.

A number of experimental and analytical assessments have been performed to quantify the effects of the deformation modes on the channel thermal behaviour during a transient [4-6]. The results of these assessments demonstrate that both pressure tube deformation modes provide significant heat removal from the channel to the cool moderator, with ballooning providing the more effective mode of heat removal. Furthermore, these assessments demonstrate the integrity of the channel, confirming that the channel is the end-point geometry for severe fuel damage accidents. Therefore, fuel response can be analyzed on an individual fuel channel basis.

Channel deformation has a significant limiting effect on the fuel temperature excursion. Figure 4 shows the results, calculated with the CHAN-II(MOD6) code [4], of the average fuel temperature transients in a 7.5 MW channel with and without channel deformation modelled. In these calculations, a large break in the heat transport piping is postulated, and the reactor is tripped immediately following the break. The channel experiences rapid coolant voiding, during which little stored heat is removed from the fuel. This is followed by a prolonged period of low steam flow which is driven by a small pressure gradient across the fuel channel of approximately 0.01 MPa. As shown in Figure 4, channel deformation results in an average fuel temperature reduction of about 300°C throughout the transient.

4.0 PROCESSES AFFECTING CLADDING OXIDATION

With adequate steam supply, the rate determining step for Zircaloy oxidation is the diffusion of oxygen through the surface oxide layer to react with the underlying metal. Hence cladding oxidation obeys a parabolic rate law. The rate is proportional to the oxidizing surface area and the reaction rate constant, but is inversely proportional to the oxide thickness. At high temperatures, there are two processes which can reduce the overall oxidation rate. The first process is molten cladding relocation, which reduces the surface area for oxidation. The second process involves the limitation of the rate of steam mass transfer onto the surface (referred to as the gas phase transport limitation), which limits the effective reaction rate constant.

4.1 Molten Cladding Relocation

Molten cladding relocation is observed in the heat-up tests of both LWR and CANDU fuel. With the vertical, thick-walled LWR fuel, melt relocation is governed by gravitational force. The melt runs down along the fuel rod, resembling the movement of wax on a candle [2]. However, with the horizontal, thin-walled

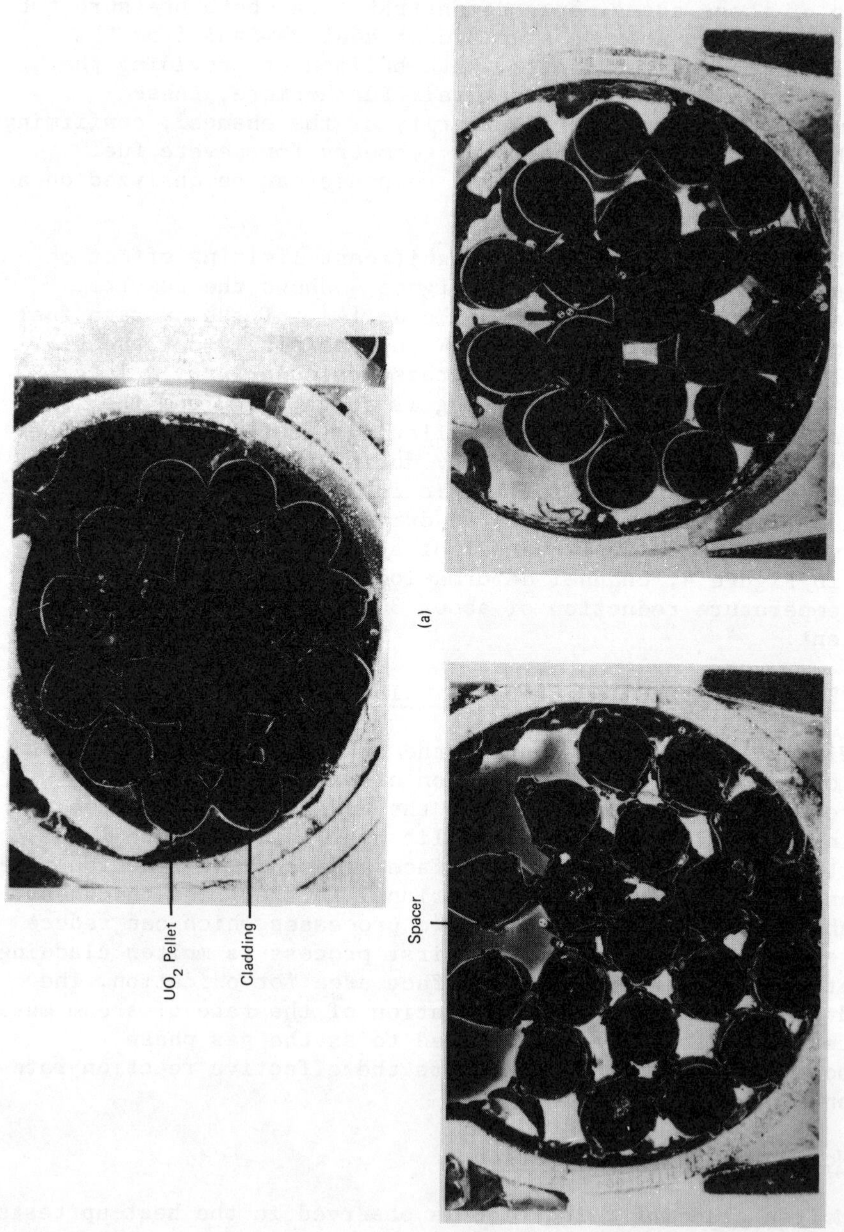

FIG. 5. Cross-sections of test bundle after high temperature transient: (a) upstream end, (b) pacer plant and (c) downstream end.

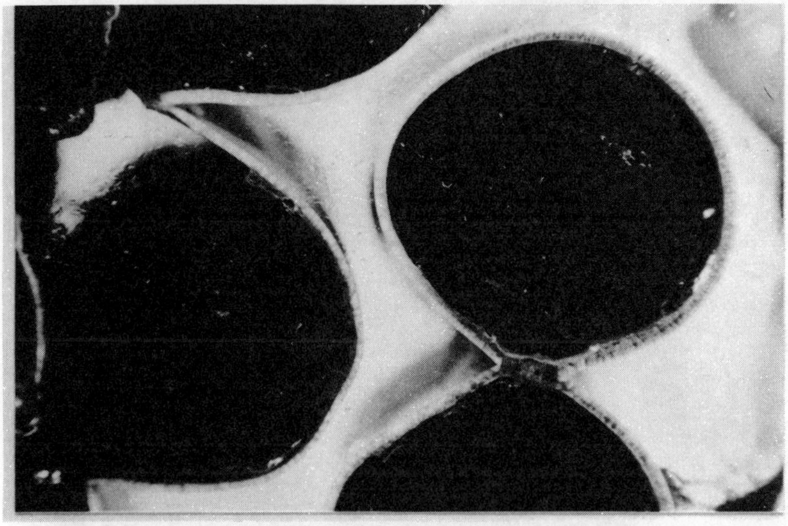

(a)

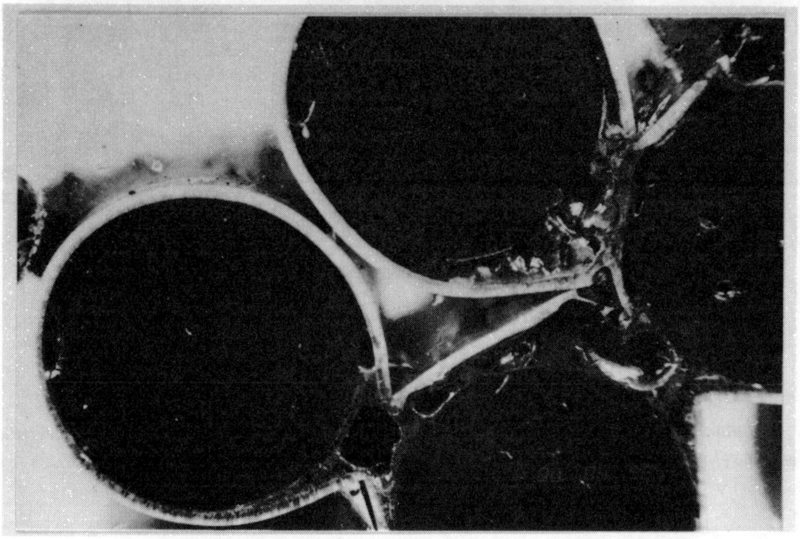

(b)

FIG. 6. Magnified view of inter-element contact locations: (a) rupture of contacting oxide shells and (b) melt relocation into inter-element gaps.

(a)

(b)

FIG. 7. Longitudinal sections of test bundle: (a) longitudinal section and (b) magnified view showing intra-element relocation.

CANDU fuel bundle, melt relocation is primarily governed by the melt's adhesive/cohesive properties relative to the adjacent solid boundaries. If the adhesion forces are greater than the cohesion forces, the melt will tend to spread on an adjacent solid surface. Otherwise, the melt will tend to run on the surface. The wetting properties of various

Zircaloy/Zircaloy(O)/UO_2 systems have been reported by Nikolopoulos and Dienst et al. [7-8]. They indicate that oxygen-free Zircaloy tends to wet ceramic UO_2 very poorly, while oxygen-saturated Zircaloy(O) wets UO_2 very well. The threshold oxygen concentration between the wetting and non-wetting regimes is approximately one weight percent of oxygen.

These properties are fully consistent with the experimental observations from heat-up tests of CANDU fuel bundles. Figures 5, 6 and 7 show the results of one of these bundle heat-up tests. In this test, a single unirradiated CANDU fuel bundle, with its outer ring of elements removed, was placed inside an insulating zirconium oxide tube [3]. An oxy-hydrogen torch was used to generate a superheated steam flow to heat up the bundle. The maximum bundle heat-up rate achieved in this test ranges from 25°C/s at the upstream end, to about 5°C/s at the downstream end. Peak temperatures of approximately 1900°C were attained along the bundle.

Figures 5a to 5c show the cross-sections of the test bundle at several axial locations after the high temperature transient. The fuel elements sag and contact each other extensively. At the points of contact the melt has moved into the available volume, causing obstructions in some flow subchannels. The presence of the element spacers did not prevent inter-element contact. As shown in the bundle cross-section at the spacer plane (Figure 5b), the spacers misaligned during the high temperature transient.

Figures 6a and 6b are the magnified views of the element-element contact areas. These figures show the pattern of relocation and illustrate that relocation into inter-element gaps would proceed as follows:

(a) By the time an inter-element contact occurs (at between 1200°C and 1400°C), an outer shell/layer of oxide has already developed on each contacting element. After contact, the oxide layers in the vicinity of the contact stop growing due to highly localized steam starvation. The contacting oxide layers, in fact, can become thinner as the heatup progresses, since the underlying metal will absorb oxygen from the oxide layers while oxygen is not replenished at the surface because of steam starvation. Depending on the magnitude of the external oxygen flux into the contact zone, the original oxide may be totally dissolved and the oxide shell breached by the time the metal melting temperature is reached.

(b) Following melting of the metal substrate, volumetric changes in the metal can produce large stresses in the oxide shell if it has not been previously ruptured. These stresses promote the rupture of the oxide shell at the thinnest cross-section, that is, in the vicinity of the inter-element contact. In addition, dissolution of the oxide layer will accelerate in the presence of the melt, further promoting rupture. Following rupture, the molten material is moved by capillary forces into inter-element spaces.

A different relocation pattern than inter-element relocation is also observed from the test results. Figures 7a and 7b show longitudinal sections of the test bundle. They show that molten material can also infiltrate into cracks in the UO_2, as well as the dish volumes at the ends of the pellets. This type of melt movement is referred to as intra-element relocation.

Whether melt movement is governed by inter- or intra-element relocation is dependent on the oxygen content in the melt. As explained earlier, the melt with a high oxygen content will tend to wet UO_2 surfaces well, and this behaviour tends to favour intra-element relocation. Conversely, a melt with a low oxygen content does not easily wet UO_2 surfaces and this behaviour favours inter-element relocation. Since the oxygen content in the melt will be higher if the cladding oxidizes for a longer time prior to melting, intra-element relocation is expected to be dominant under slow heatup conditions. In contrast, inter-element relocation would be dominant in rapid heat up scenarios.

Both types of relocation result in substantial surface area reduction of the Zircaloy relative to its nominal geometry. Measurements indicate that there is a reduction in the surface area exposed to the steam coolant by a factor of between one to three orders of magnitude [3]. This surface area reduction will, in turn, result in a reduction of the overall cladding oxidation rate in the fuel channel.

It should be noted that inter-element contact and relocation can cause large obstructions of the flow subchannels, expecially in the smaller triangular subchannels. Considering as well the flow by-passing effect due to pressure tube ballooning, it can be seen that the coolant flow through these subchannels will be small. The small coolant flow will also have a large hydrogen concentration due to Zircaloy oxidation. These are conditions which are conducive to limiting the rate of steam mass transfer to the cladding surface, which consequently reduces the oxidation rate.

4.2 Steam Mass Transfer Rate

In a flow of steam and hydrogen in which steam is continuously converted into hydrogen at the Zircaloy surface, there will be a flow boundary layer developed adjacent to the surface. This boundary layer has a concentration profile, with hydrogen having the highest concentration and steam having the lowest concentration at the cladding surface. Thus, steam in the bulk flow must first be transported through the boundary layer in order to access the Zircaloy surface. Under severely degraded cooling conditions, this steam transfer rate is very slow. Consequently, the steam supply to the Zircaloy surface may not be sufficient to sustain oxidation, especially at extremely elevated temperatures where rapid reaction rates prevail.

This mass transfer limitation is observed to some extent in the oxidation tests by Prater and Courtright of Pacific Northwest Laboratory [1], as well as in the tests by Chung of Argonne National Laboratory [9]. Molten Zircaloy oxidation tests at Whiteshell Nuclear Research Establishment also consistently show a strong effect of the rate of steam mass transfer on the oxidation rate at high temperatures.

Figure 8 shows the results of a parametric study which delineate the conditions at which mass transfer limitations may be dominant during CANDU severe fuel damage accidents. The data is obtained based on a mass transfer model utilizing the Chilton-Colburn analogy [10]. The steam flux that can be transported to the cladding surface is shown for a range of coolant temperatures. Coolant mass flow rates through the bundle of 20 g/s and 100 g/s, as well as hydrogen concentrations in the coolant of 25% and 75% are assumed. These are within the range of conditions predicted for the severe fuel damage accidents. It is observed that the larger the flow rate, (that is the more turbulent the flow), and the lower the hydrogen concentration, the larger the steam mass transfer rate.

Superimposed on Figure 8 are the steam fluxes required to sustain Zircaloy oxidation. Since the rate decreases with increasing oxide layer thickness, two curves are shown: that with ten percent of the cladding oxidized, and that with 50 percent of the cladding oxidized. These curves are intended to illustrate the behavioural trends, and will vary depending on the specific reaction rate correlations assumed. Comparison of the requirement and supply curves indicates that at elevated temperatures (in excess of about 2200 $^{\circ}C$), the cladding oxidation rate may be limited by the rate of steam mass transfer. Complementary to melt relocation, steam mass transfer acts in parallel to reduce the overall cladding oxidation rate in the fuel channel at elevated temperatures.

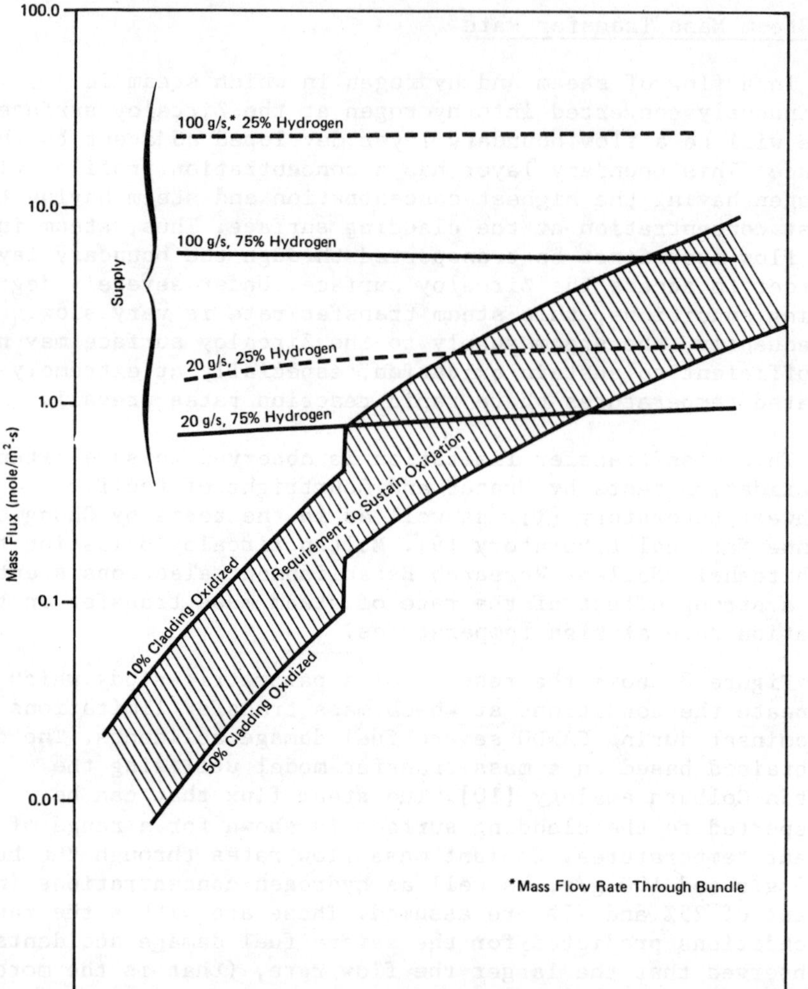

FIG. 8. Steam supply and requirement.

5.0 EFFECT OF CLADDING OXIDATION REDUCTION ON FUEL TEMPERATURE

Molten cladding relocation and the rate of steam transfer are capable of reducing the overall cladding oxidation rate in a fuel channel during a severe fuel damage accident. As a result, they have a moderating effect on the ultimate fuel temperatures. The analysis results shown in Figure 9 illustrate this moderating effect. The channel-averaged fuel temperature

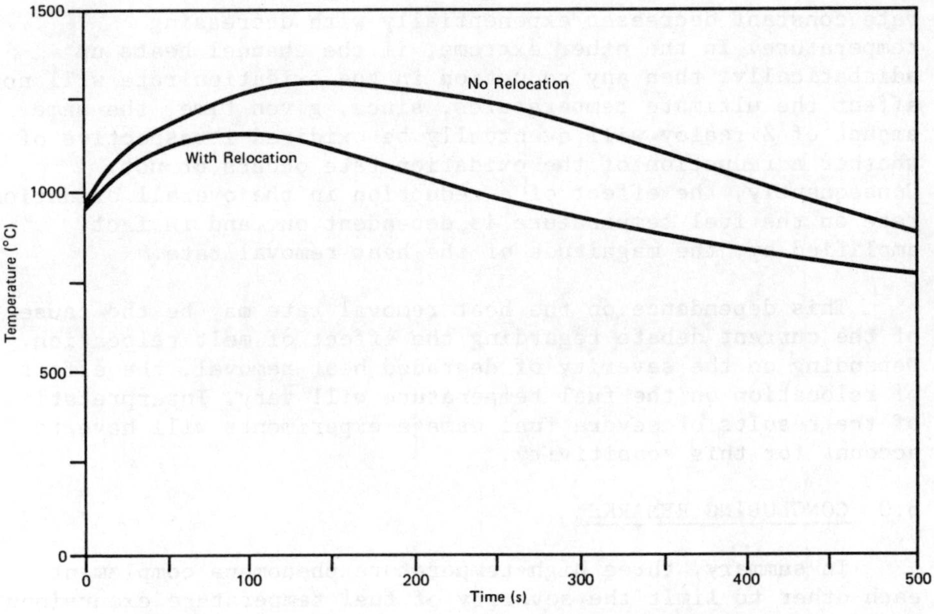

FIG. 9. Channel average fuel temperature transient — effect of relocation.

transient calculated for a 6.7 MW channel undergoing relatively rapid heatup at decay power is shown in Figure 9 for identical conditions with and without melt relocation. A constant steam flow of 100 g/s is assumed and the relocation model developed by Akalin et al. [11] is used. Pressure tube ballooning is modelled which provides a very efficient heat removal path to the moderator. The effect of relocation is significant, with a 200°C reduction in the predicted average fuel temperature in the channel. The effect on the peak fuel temperature is much greater.

The moderation of temperature arising from melt relocation and/or limited steam mass transfer is strongly coupled to the heat removal rate during the accident. Any relative reductions in the fuel temperatures due to Zircaloy relocation and/or steam mass transfer will become larger if the heat removal rate from the channel improves. Conversely, the effect will be smaller if the heat removal rate is degraded further (for example, pressure tube sagging rather than ballooning in the previous example). A reduction in the overall oxidation rate will allow the heat removal to become more effective in lowering the fuel temperature, since the oxidation heat source becomes smaller. The lowering of the temperature will feedback on a further reduction in the oxidation heat generation since the oxidation rate constant decreases exponentially with decreasing temperature. In the other extreme, if the channel heats up adiabatically, then any reduction in the oxidation rate will not affect the ultimate temperatures, since, given time, the same amount of Zircaloy will eventually be oxidized irrespective of whether a reduction of the oxidation rate occurs or not. Consequently, the effect of a reduction in the overall oxidation rate on the fuel temperature is dependent on, and in fact amplified by, the magnitude of the heat removal rate.

This dependence on the heat removal rate may be the cause of the current debate regarding the effect of melt relocation. Depending on the severity of degraded heat removal, the effect of relocation on the fuel temperature will vary. Interpretation of the results of severe fuel damage experiments will have to account for this sensitivity.

6.0 CONCLUDING REMARKS

In summary, three high-temperature phenomena complement each other to limit the severity of fuel temperature excursions during postulated CANDU severe fuel damage accidents. These include i) fuel channel deformation, ii) molten cladding relocation and iii) steam mass transfer. Due to their impact on fuel temperatures, it is important that these phenomena are properly accounted for in the safety assessments of severe accidents.

ACKNOWLEDGEMENT

The bundle heat-up tests described in this paper were conducted at Westinghouse Canada Ltd., Hamilton, Ontario, and were jointly funded by Ontario Hydro, Hydro Quebec, New Brunswick Electric Power and Atomic Energy of Canada Ltd. under the CANDU Owners Group program.

REFERENCES

[1] PRATER, J., COURTRIGHT, E.L., Oxidation of Zircaloy-4 in steam at 1300 to 2400 ^{o}C", Paper presented at the 7th Inter. Conf. on Zirconium in the Nuclear Industry, Strasbourg, France, June 1985.

[2] HAGEN, S., Out-of-pile experiments on the high temperature behaviour of Zry-4 clad fuel rods, Paper presented at the 6th Int. Conf. on Zirconium in the Nuclear Industry, Vancouver, Canada, June 1982.

[3] WADSWORTH, S.L., et al., Experimental investigation of CANDU fuel deformation during severely degraded cooling, Paper presented at the Int. ANS/ENS Topical Meeting on Thermal Reactor Safety, San Diego, USA, February 1985.

[4] LAU, J.H.K., et al., Feedback effects of deformation on fuel temperatures during degraded cooling accidents in CANDUs, Res Mechanica, 18, 4 (1986) 307.

[5] MUZUMDAR, A.P., et al., Fuel temperature excursions during accidents with degraded cooling in CANDU reactors, Proc. 2nd Int. Topical Meeting on Reactor Thermal-Hydraulics, Santa Barbara, USA, January 1983.

[6] GILLESPIE, G.E., Experimental investigation of heat transfer from a reactor fuel channel to surrounding water", Paper presented at the 2nd Annual Canadian Nuclear Society Conf., June 1981.

[7] NIKOLOPOULOS, P., et al., Determination of the interfacial energy and work of adhesion in the UO_2/Zircaloy-4 diffusion couple, Journal of Nuclear Materials, 124 (1984).

[8] DIENST, W., et al., Chemical interactions between UO_2 and Zircaloy-4 from 1000 to 2000°C, Nuclear Technology, 65, (1984).

[9] CHUNG, H.M., THOMAS, G.R., High-temperature oxidation of Zircaloy in hydrogen-steam mixtures, Paper presented at the 6th Inter. Conf. on Zirconium in the Nuclear Industry, Vancouver, Canada, June 1982.

[10] WELTY, J.R., et al., "Fundamentals of Momentum, Heat and Mass Transfer", Wiley, 1969.

[11] AKALIN, O., et al., Fuel temperature escalation in severe accidents, Paper presented at the 6th Annual Canadian Nuclear Society Conf., Ottawa, Canada, June 1985.

COMMISSION OF THE EUROPEAN COMMUNITIES JOINT RESEARCH CENTRE CONTRIBUTION TO LMFBR SAFETY ANALYSIS

A. BENUZZI, H. HOHMANN, A.V. JONES,
D. MAGALLON, H.U. WIDER
Joint Research Centre, Ispra Establishment,
Commission of the European Communities,
Ispra

Abstract

COMMISSION OF THE EUROPEAN COMMUNITIES JOINT RESEARCH CENTRE CONTRIBUTION TO LMFBR SAFETY ANALYSIS.

The current contributions from the Commission of the European Communities Joint Research Centre to the analysis of severe accidents in LMFBRs are reviewed. The principal tasks are the development of the European accident code 2 (EAC2), an advanced computer code for the analysis of unprotected whole-core accidents based on the existing EAC1, and the performance and analysis of a series of experiments to investigate post-accident heat removal problems exploiting the Faro uranium dioxide melting facility. Other Joint Research Centre contributions in related fields such as the analysis of in-pile subassembly accident and post-accident heat removal experiments are not reviewed. The status of EAC1 is discussed briefly with examples, and then the developments in course designed to lead to EAC2 are described. Models under development include a pin performance and failure model taking account of in-pin fuel motion, a post-failure material motion model and a fast-running three-dimensional transport neutronics package based on the nodal approach. In the Faro facility it is possible to melt up over 100 kg of uranium dioxide by Joule heating and deliver it to a desired test section. The various types of experiments planned are described and indications are given of the results of the first Blokker I tests in which a jet of uranium dioxide obliquely impacts a preheated steel plate. Computational tools have been developed for the analysis of the furnace itself and for the analysis of jet impact tests, fuel penetration experiments and the interaction of a stream of fuel with a sodium pool. The main assumptions of the models are presented together with some indicative results.

1. INTRODUCTION

The Commission of the European Communities (CEC) is responsible for a wide range of research actions in the field of nuclear reactor safety, a general review of which in the light of the Chernobyl accident may be found in the paper by De Bacci et al. [1] at this conference. Some of the research work is done by consultants and contractors, and some wholly by other organisations, with the Commission making a substantial contribution towards the costs, typically 50%. This paper is largely concerned with research performed by the Commission's own staff, mostly based at the Joint Research Centre (JRC) in Ispra in Northern Italy or under contracts supervised by them. Recent safety investigations there cover inter alia several aspects of LWR and LMFBR safety: examples of the kind of research done may be found in other papers at this conference.Here we indicate the present status and the trends in the area of severe accident analysis for liquid-metal cooled fast breeder reactors.

The types of accident considered are of low to very low probability, but are worthy of study within the context of a complete safety evaluation because of their potentially severe consequences both for the plant and for the general public. In practice this category of accident contains whole-core events such as unprotected loss of flow or reactivity transients, and also subassembly accidents with the potential for propagation to involve several rings of subassemblies. Over successive multiannual programmes the range of topics considered by the JRC has been progressively reduced in the interests of more thorough and complete investigations likely to have an impact on the European scale, and there are now just two research projects in the area of LMFBR severe accident analysis. One is the development of the European Accident Code (EAC), a highly modular computer code for the prediction of the initiating phase of whole-core accidents. The principal features of EAC, together with some sample results, are presented in Section 2 of this paper. Faro, the other research project, is predominantly experimental, and is designed to investigate certain post-accident heat removal problems out of pile. The Faro experimental apparatus and an account of the first successful experiments appears in Section 3. Also in the same domain are JRC contributions to the planning and analysis of a series of in-pile post-accident heat removal tests, and to the analysis of the Scarabee and Mol series of in-pile subassembly fault propagation experiments, but these topics will not be presented in this paper.

2. THE EUROPEAN ACCIDENT CODE

The European Accident Code (EAC) is being developed to improve predictions of low-probability unprotected loss of flow (LOF) and transient overpower (TOP) accidents. The first version, the EAC1 code system, has achieved reasonable agreement with similar codes in the recent Whole-core Accident Working Group LOF (WAC/LOF) comparative exercise [2]. However, some of the models of EAC1 are not state-of -the -art and/or are unnecessarily conservative. The present development version, EAC2, replaces earlier models by advanced models for pin mechanics, fuel motion and neutronics. Together with extensive validation efforts this development should lead to a benchmark type accident analysis code for LMFBRs.

The EAC1 code has been described in some detail in [3]. Its smoothly operating informatic structure as well as its two sodium boiling modules CFEM and BLOW will be retained in EAC2. CFEM is an efficient homogenous boiling model developed at JRC Ispra, while BLOW is a detailed heterogeneous boiling model developed and validated at Kernforschungszentrum Karlsruhe.In a recent extensive sensitivity study it was shown that EAC with the BLOW module runs very reliably and gives physically interesting results. As an example of this sensitivity study [4] two EAC/BLOW LOF calculations for different pump rundowns in the fictitious 700MWe "Europe" reactor are discussed. In the standard case of a pump rundown with a halving time of 6 s sodium boiling starts at 16.4 s, whereas with a 12 s halving time it only starts at 56.6 s. Fig.1 shows the comparison of the voiding and Doppler reactivities as well as the normalised powers in each case relative to the onset of boiling. This figure shows that the voiding reactivity, and thus the power, rise considerably slower in the case of the slower flow coastdown. The slight oscillations in the power in the latter case are due to the oscillations in the void reactivity. These in turn are caused by the chugging of the slowly voiding lower sodium interface in the lead

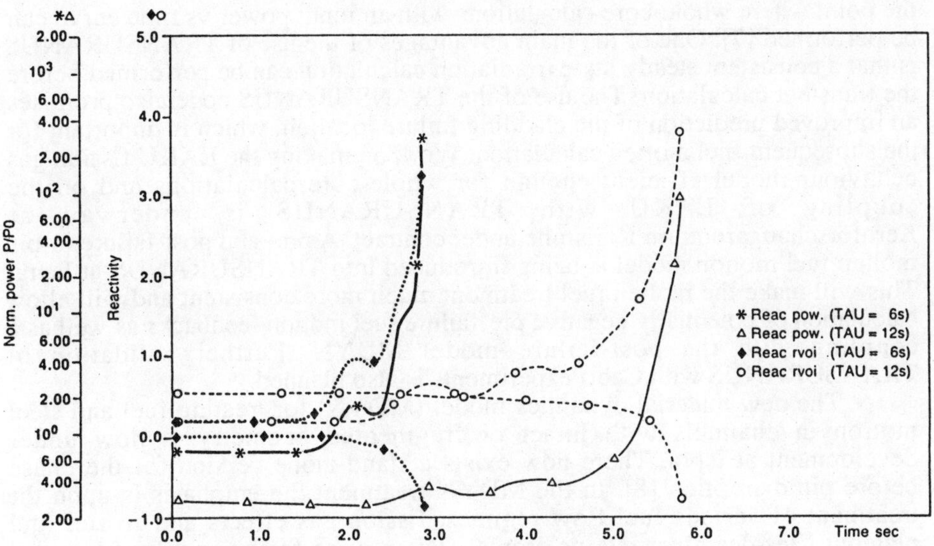

FIG. 1. Comparison of reactor powers and voiding and Doppler reactivities for different pump coastdowns.

channel of this whole-core calculation. For the case of the slower pump rundown the power rises later (which would give the reactor operator more time to react) and the eventual power rise is less rapid. Moreover the predicted cladding temperature profiles are more conducive to upward pin failure propagation and subsequent fuel dispersal. Therefore this calculation supports the traditional view that a slower pump coastdown is beneficial for reactor safety.

In the earlier WAC/LOF comparative exercise [2] the EAC1/CFEM and EAC1/BLOW whole-core calculations showed good agreement with the other participating codes as regards fuel pin temperatures, one- and two-phase hydraulics and the neutronics feedback calculations. The fuel pin mechanics predictions with the finite element model PINM were not fully satisfactory and the EPIC post-failure fuel motion predictions could not be considered very reliable because of the lack of fuel pin breakup and fuel freezout models in EPIC.

The development of the EAC2 code system includes the implementation of the detailed and partially validated TRANSURANUS pin mechanics and heat transfer models [5,6], and the development and inclusion of a detailed model (MDYN) for calculating the fuel and steel motion. Moreover an efficient 3-d hex-z nodal neutronics model (HEXNOD) for static diffusion or transport calculation as well as a dynamic treatment (HEXNODYN) are being developed under contract for EAC2.

The coupling of the TRANSURANUS fuel pin behaviour code with the EAC informatic structure and the CFEM boiling model has already advanced to the point where whole-core calculations with an input power vs time curve can be performed [7]. One of the main advantages of the use of TRANSURANUS is that a consistent steady-state irradiation calculation can be performed before the transient calculation. The use of the TRANSURANUS code also promises an improved prediction of the cladding failure location, which is important for the subsequent fuel motion calculation. Work on making the LAKU fission gas behaviour model efficient enough for whole-core calculations and on the coupling of LAKU with TRANSURANUS is underway at Kernforschungszentrum Karlsruhe under contract. A pre- and post-failure in-pin molten fuel motion model is being introduced into TRANSURANUS at Ispra. This will make the molten fuel treatment much more consistent and will allow calculation of potentially negative pre-failure fuel motion feedbacks as well as a coupling with the post-failure model MDYN. Further validation of TRANSURANUS with Cabri experiments is also planned.

The new material dynamics model (MDYN) for treating fuel and steel motion in channels with intact or fragmented geometry is now under development at Ispra. There now exists a stand-alone version for the phase before pin disruption [8]. In the MDYN treatment the emphasis is upon the treatment of various fuel flow regimes, fission gas effects and frozen fuel plateout. Besides a particle or droplet flow regime frothy annular films and bubbly slug flow regimes are considered. These latter are mostly relevant to voided channel situations in which fuel freeze-out may occur. Recently a more implicit numerical treatment has been developed which has halved the running time [8]. Coupling of MDYN with TRANSURANUS has just begun, and it is planned to validate the coupled MDYN/TRANSURANUS combination with Cabri experiments at a later stage.

The efficient 3-d hex-z nodal diffusion and transport neutronics code HEXNOD developed by Kraftwerk Union in the Federal Republic of Germany under a JRC contract will replace the 2-d diffusion neutronics code MARC of EAC, but the cross-section generation models from MARC will be retained in EAC2. The new hexagonal geometry of HEXNOD will allow the consideration of azimuthal assymmetries due to control rods and differences in burnup. Such inhomogeneities may make accident predictions milder through incoherence effects. The present version of HEXNOD has already been benchmarked for 2-d diffusion and transport problems, as well as for 3-d diffusion problems. 3-d transport benchmarking of this hex-z code will be done through comparisons with Monte Carlo codes. Of particular interest is how well the transport approximation of HEXNOD will perform for strongly disrupted core configurations.

For the time integration the point kinetics approach is only appropriate so long as no significant fuel motion occurs. Since we are interested in tracking extensive fuel motion a dynamic neutron calculation is needed. The Free University of Brussels is in the process of coupling the quasistatic model of the CASSANDRE code with HEXNOD under a JRC contract. The final outcome will be an advanced neutronics package for EAC2.

3. THE FARO PROJECT

This section is divided into two subsections. The first describes the Faro installation and the results of the most recent experiments, while the second

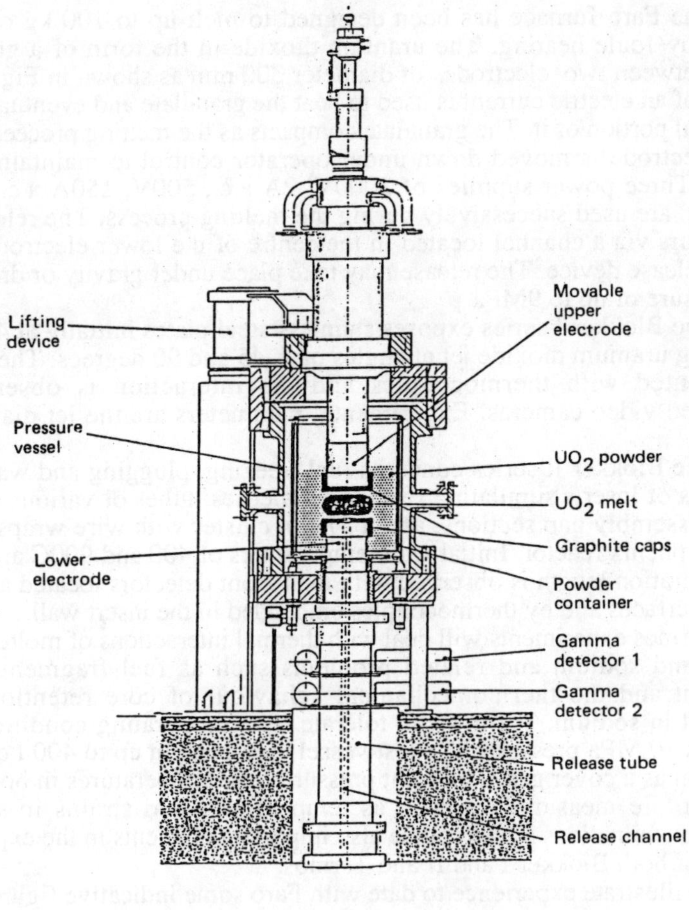

FIG. 2. Faro large-scale melt facility.

gives some account of the tools developed for the analysis of the Faro furnace and of the tests themselves.

3.1 The Faro Installation

Faro is a large multi-purpose test facility in which various phenomena involving molten uranium dioxide can be studied out of pile. The main objectives include studies on molten fuel pools within the Faro furnace, fuel impact and erosion against stainless steel structures in the Blokker I test section, fuel penetration and freezing in channels in the Blokker II test section, and fuel jet stability and the fragmentation of large masses of fuel in sodium in the Termos test section. Each topic and experimental series is supported by theoretical analyses, as discussed in subsection 3.2.

The Faro furnace has been designed to melt up to 100 kg of uranium dioxide by Joule heating. The uranium dioxide in the form of a granulate is placed between two electrodes of diameter 300 mm as shown in Fig.2, and the passage of an electric current is used to heat the granulate and eventually to melt the central portion of it. The granulate compacts as the melting proceeds, and the upper electrode is moved down under operator control to maintain electrical contact. Three power supplies of 2000V, 2A a.c., 500V, 150A a.c. and 60V, 15kA d.c. are used successively during the melting process. The release of the melt occurs via a channel located in the centre of the lower electrode, using a special release device. The release may take place under gravity or driven by an overpressure of up to 9MPa.

The Blokker I series exposes stainless steel plates initially at 400C to an impinging uranium dioxide jet at angles of 5, 45 and 90 degrees. The plates are instrumented with thermocouples and the interaction is observed with high-speed video cameras. Experimental parameters are the jet diameter and velocity.

The Blokker II series concerns fuel freezing, plugging and wall melting in a series of inserts simulating reactor geometries: tubes of various diameters, intersubassembly gap sections, and a 37-pin cluster with wire wraps based on the Superphenix reactor. Initial wall temperatures of 400 and 900C are planned. The penetration length is observed with melt front detectors located at the inner channel surfaces and by thermocouples embedded in the insert wall.

Termos experiments will deal with thermal interactions of molten uranium dioxide and sodium and related problems such as fuel fragmentation and settlement and the thermomechanical behaviour of core retention devices immersed in sodium. Termos can tolerate severe operating conditions, up to 900C and 10 MPa pressure. The test vessel may contain up to 400 l of sodium, with argon as a cover gas. Transient pressures and temperatures in both sodium and argon are measured, as well as temperatures and strains in structures. Post-test examinations are of course also important elements in the experimental analysis of both Blokker I and II and Termos.

To illustrate experience to date with Faro some indicative figures may be given. It has not proved possible on a regular basis to measure temperatures within the furnace, and the quantity of melt present is deduced from the descent of the electrode and from the stored energy (electrical input minus thermal losses). Experience indicates the 220 MJ of stored energy correspond to 100 kg of melt. Only in test 26 was it possible to measure the temperatures at various stations in the radial direction, using a tungsten ultrasonic temperature sensor (UTS) positioned horizontally 80mm above the lower electrode. Fig. 3 shows the measured temperature histories, indicating temperatures up to 3000C. In later tests early UTS failure has occurred due to mechanical actions and chemical attack. Various protective materials are being tried to prolong the life of the UTS.

3.1.1 Blokker I tests

As shown in Fig.4, in the Blokker I series actuation of the release device directs a stream of molten uranium dioxide onto an inclined stainless steel plate. Two tests have been performed so far, with angles of inclination of 5 and 90 degrees. 100 kg of melt were discharged in the 5 degree case (test 36) and 70 kg in the 90 degree case (test 37). Table I summarises the main experimental

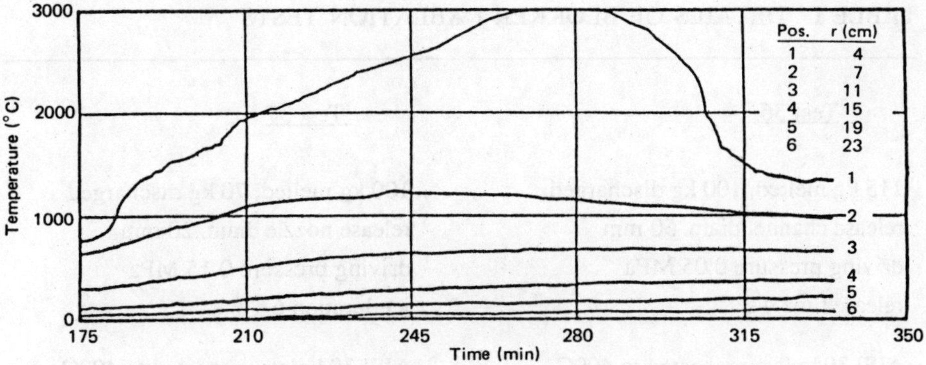

FIG. 3. Ultrasonic temperature sensor temperature histories at various radial stations.

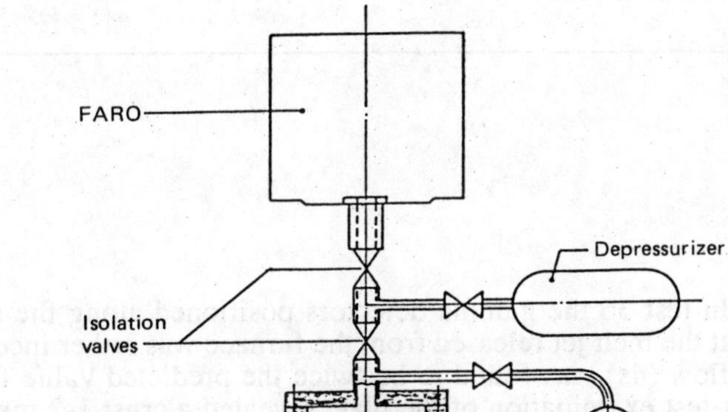

FIG. 4. Blokker I ablation experiments.

TABLE I. DETAILS OF BLOKKER I ABLATION TESTS

Test 36	Test 37
115 kg melted, 100 kg discharged	100 kg melted, 70 kg discharged
release channel diam. 50 mm	release nozzle diam. 20 mm
driving pressure 0.05 MPa	driving pressure 0.15 MPa
release time 4 s	jet duration 6 s
AISI 304 plate preheated to 400C	AISI 304 plate preheated to 400C
Plate inclination 5°	Plate inclination 90°
plate dimensions 840x200x40 mm	plate dimensions 280 diam.x40 mm
distance outlet to plate 100 mm min. 400 mm centre to centre	distance outlet to plate 95 mm

conditions. In test 36 the gamma detectors positioned along the release tube indicated that the melt jet released from the furnace was rather incoherent. The duration of flow (4s) was found to be twice the predicted value for a gravity release. Post-test examination of the plate revealed a crust 1-2 mm thick. The border was clearly detached from the plate. The shape of the crust gave an indication of the flow. At the upper end of the plate the jet was 80 mm broad, with steel ablation varying from 3 mm on the centre-line to zero at the edges. In the centre of the impact zone a small depression was created in the steel, 6mm deep and of diameter 10mm.These results appear similar to those from experiments performed in Kernforschungszentrum Karlsruhe in which no ablation was observed when a jet of molten alumina played upon a steel plate. The melt temperature was much lower in this case however (2300C) and the plate was initially at room temperature.

In test 37 a coherent jet of at least 55 kg of molten uranium dioxide was produced, of which 15 kg remained on the horizontal plate at the end of the test, in solid form. This probably came from the trailing edge of the jet, which was not expected to be coherent. No ablation of the plate was observed, suggesting that a certain crust formation occurred which protected the plate. Analysis of the internal structure of the crust will provide information about its formation and destruction. The central part of the plate appears very shiny over about 100 mm, suggesting that some melting took place. Metallographic analysis of the plate will test this hypothesis. The first Blokker series will conclude with a test using a 45° plate, and is expected shortly.

3.2 Computational Tools for Faro Analysis

Concerning the Faro furnace the main physical processes to be modelled are the electrical heating of the uranium dioxide powder, its sintering and melting, and the formation of a molten pool. Several computer codes have been developed for the simulation of these phenomena. PHEAT is a 2-d finite difference conduction model which calculates the preheating of the powder. The PHEAT temperature fields form one input to the codes JOULE and CONDIF. JOULE is a 2-d finite difference code developed for the analysis of formation of the molten pool and the subsequent convective motions. CONDIF [9] is an explicit 2-d finite element code with the same basic assumptions i.e. incompressible laminar flow, stationary electrodes and negligible electromagnetic effects. CONDIF also takes account of heat conduction in surrounding structures. An original finite element scheme, the Taylor-Galerkin method, has been implemented to calculate the convection-dominated motions of the melt [9]. A 1-d simulation of the Faro furnace, from the preheating phase through to melting and superheating of the uranium dioxide, is being developed, named SMURF. The code includes electrode motion, thermal and electrical contact resistances and also convective heat exchange between the molten and solid uranium dioxide. The main reason for this development was to supply a fast-running tool for sensitivity studies relating to furnace and material parameters, and to give guidance to the experimenter during the performance of a test. Fig.5 shows a calculated temperature profile within the furnace in a space-time plot. All the codes mentioned have been used for intercode comparisons, scoping calculations and experiment predictions.

To analyse the Blokker I ablation tests a suite of codes has been developed by MATEC Ltd. of Milan under a JRC contract, JET 3D, BOUNDY and MELT. JET 3D calculates the jet hydrodynamics by the FEM under the assumption of irrotational and steady flow. The (laminar) boundary layer near the stagnation point of impact on the plate and the heat transfer from it are calculated by the 2-d FD code BOUNDY, assuming an infinitely thick structure, constant material properties and a quadratic distribution of the velocities in the upper boundary of the layer. To complete the sequence the heat fluxes are used by the 2-d finite difference transient heat conduction code MELT to calculate the propagation of the melt front within the structure, using a boundary-fitted coordinate scheme. When geometry changes become significant the cycle of calculations is repeated to take account of the new surface shape. A satisfactory comparison with Kernforschungszentrum Karlsruhe simulant ablation experiments has been achieved. Fig.6 shows the calculated ablation after 2s resulting from a vertical jet of liquid iron at 2300C impinging on a steel plate at 45°, initially at room temperature.

Concerning the melt penetration experiments Blokker II the main physical phenomena to be modelled are the transient flow of a fluid in an initially empty channel with freezing of the fluid, possible wall melting and channel plugging. Two codes have been applied, TRACRUF, which assumes that transient freezing is primarily by conduction to the wall through the growing crust, and PLUG which makes similar assumptions but ignores inertia forces and assumes constant material properties. Both codes have been extensively applied to test precalculations, comparisons with available experimental data (Kernforschungszentrum Karlsruhe, Sandia National Laboratory) and comparisons with other codes. Generally agreement is satisfactory.

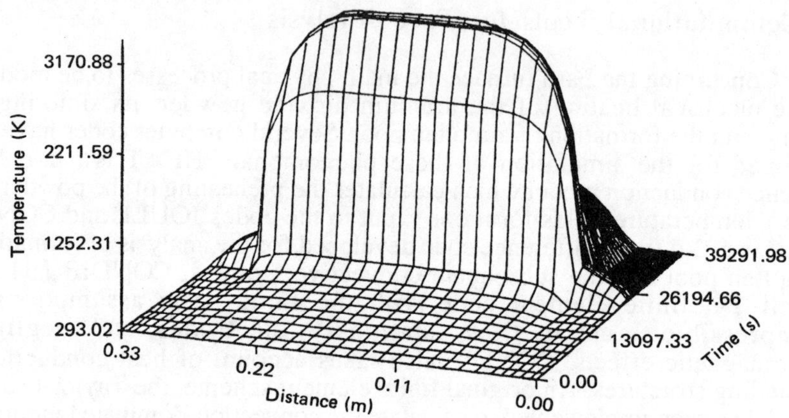

FIG. 5. Example of SMURF predictions for Faro furnace temperature profiles.

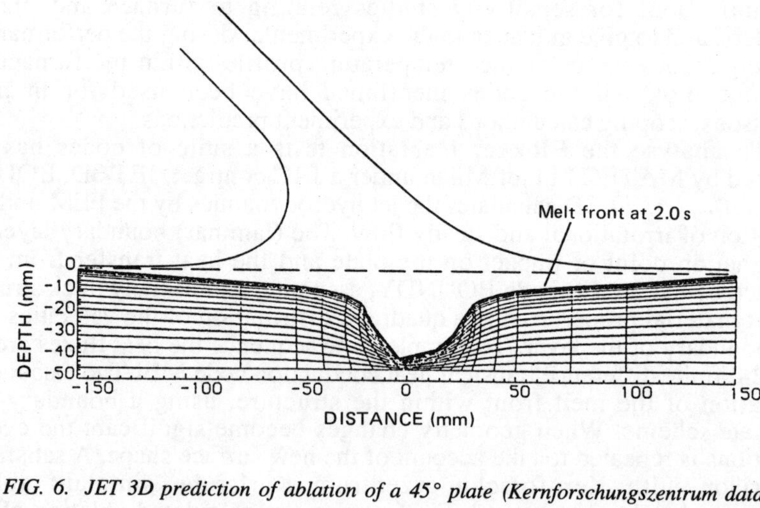

FIG. 6. JET 3D prediction of ablation of a 45° plate (Kernforschungszentrum data).

In the Termos tests the physical phenomena involved are fuel jet penetration and breakup in a sodium pool, boiling of the sodium and fuel freezing and fragmentation. The code JENA has been developed to treat this situation. It assumes a transition boiling regime leading to the formation of small vapour bubbles near the jet surface, and thermodynamic equilibrium. Fig.7 illustrates the jet profile when a 20mm wide plane jet of uranium dioxide falls into sodium. One case is without heat transfer, whereas in the case with heat transfer a stochastic distribution of local forces has been applied to the jet

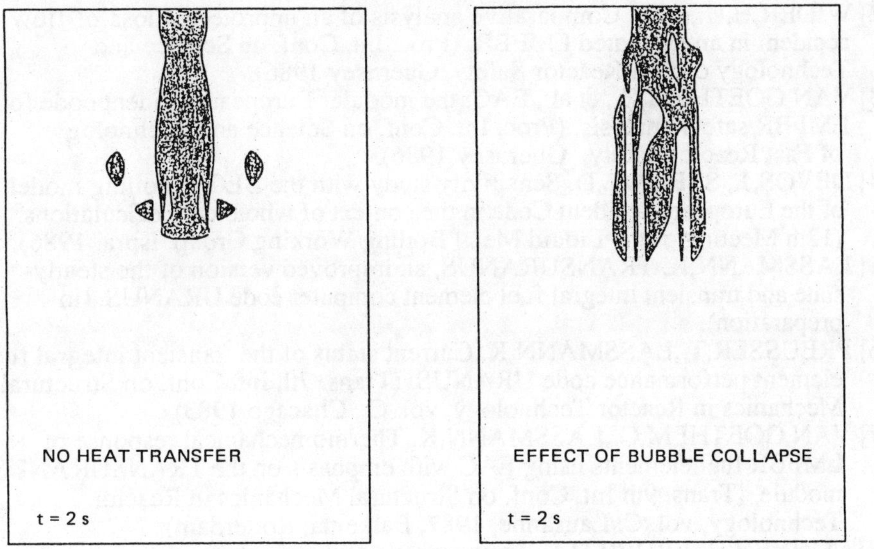

FIG. 7. JENA predictions of uranium dioxide jet breakup in a sodium pool.

surface to simulate the effects of bubble collapse. Apart from JENA it is also planned to apply the more sophisticated Advanced Fluid Dynamics Module [10] to the Termos tests when this code becomes available.

4. CONCLUDING REMARKS

The strategy of the Joint Research Centre of the CEC in the field of LMFBR safety analysis is to select a small number of topics and then to do work on them of sufficient scope and depth to have a perceptible impact on the European scale. Given continued commitment on the part of the member states the European accident code will when complete be the most ambitious and detailed of its type. After sufficient validation it is intended to serve as a benchmark for other initiation phase codes in addition to functioning as a safety analysis tool. The experimental results from Faro, when thoroughly analysed, will also constitute a unique body of knowledge. There is no other facility in Europe with the capacity to melt such large quantities of pure uranium dioxide. Comparisons with simulant or thermite tests will be one application of this corpus of data, but model development and code validation are the principal applications planned.

REFERENCES

[1] DE BACCI,M., et al., The views of the Commission of the European Communities on nuclear safety after Chernobyl, and its research activities.(Proc. International Conference on Nuclear Power Performance and Safety, paper 25, IAEA, Vienna, 1987).

[2] WIDER,H.U.,et al., Comparative analysis of an unprotected loss-of-flow accident in an irradiated LMFBR. (Proc. Int. Conf. on Science and Technology of Fast Reactor Safety, Guernsey 1986).
[3] VAN GOETHEM,G., et al., EAC: the modular European accident code for LMFBR safety analysis. (Proc. Int. Conf. on Science and Technology of Fast Reactor Safety, Guernsey 1986).
[4] DEVOS,J., STRUWE,D. Sensitivity study with the BLOW boiling model of the European Accident Code in the context of whole-core calculations. (12th Meeting of the Liquid Metal Boiling Working Group, Ispra, 1986).
[5] LASSMANN,K. TRANSURANUS, an improved version of the steady-state and transient integral fuel element computer code URANUS. (in preparation).
[6] PREUSSER,T.,LASSMANN,K. Current status of the transient integral fuel element performance code URANUS. (Trans. 7th Int. Conf. on Structural Mechanics in Reactor Technology, vol. C , Chacago 1983).
[7] VAN GOETHEM,G.,LASSMANN,K. Thermomechanical response of LMFBR fuel elements using EAC with emphasis on the TRANSURANUS module. (Trans. 9th Int. Conf. on Structural Mechanics in Reactor Technology, vol. C. Lausanne, 1987. Balkema, Rotterdam)
[8] NGUYEN,H.,WIDER,H.U. Improved algorithm for transient multicomponent multiphase flows. (5th Int. Conf. on Numerical Methods in Laminar and Turbulent Flow, Montreal, 1987).
[9] HOHMANN,H., et al., Contribution to FBR analysis: the Faro programme at JRC, Ispra. (Proc. Int. Conf. on Science and Technology of Fast Reactor Safety, vol. II, pp 139-144. Guernsey 1986).
[10] CASTELLANO,L., et al., Erosion effects of a hot liquid jet impacting obliquely on a solid plate. (Trans. 9th Int. Conf. on Structural Mechanics in Reactor Technology, vol. E pp 107-113. Lausanne, 1987. Balkema, Rotterdam).

BIBLIOGRAPHY

DONEA,J.,GIULIANI,S., The Computer Code CONDIF for Transient Convective Heat Transfer. (CEC Report EUR 6822/IEN, 1980).
BOHL,W.R., et al., Computational methods of the Advanced Fluid Dynamics Model. (Trans. Intl. Topical Meeting on Advances in Reactor Physics, Mathematics and Computation, vol. III pp 1625-1640. Paris, 1987).

SEVERE ACCIDENTS IN THE FRAMEWORK OF ITALIAN SAFETY PRACTICE

C. SENNIS, G. PETRANGELI
Direzione Sicurezza Nucleare e Protezione Sanitaria (ENEA-DISP),
Comitato Nazionale per la Ricerca e per lo Sviluppo
dell'Energia Nucleare e delle Energie Alternative,
Rome, Italy

Abstract

SEVERE ACCIDENTS IN THE FRAMEWORK OF ITALIAN SAFETY PRACTICE.

Safety practices adopted in Italy are based on two general principles: integration of safety requirements with health protection objectives; and extensive 'defence in depth', which includes redundancy and diversity of physical barriers, of types of safety measures and of types of safety criteria (deterministic and probabilistic). In this framework severe accidents are considered for introduction into safety analyses and external emergency planning in order to implement an additional barrier, founded on management of severe accidents, against uncontrolled releases of radioactivity. The criterion dealing with severe accidents, formulated in June 1985 and widely discussed, can be stated thus: "It is considered realistic to assure, by additional accident management provisions, with a confidence level of the order of 95%, that external releases of iodine and caesium, in situations otherwise leading to uncontrolled severe accidents, be limited within 0.1% of core inventory." This criterion is now in a final phase of trial-implementation in Italian plants. The combination of the Italian design criteria for design basis accidents and core melt prevention and of the above stated criterion concerning severe accidents can be viewed as a more realistic alternative to an unrestricted 'safety goal' approach to safety. This alternative consists of establishing limited quantitative goals based on sound knowledge, to be extended as knowledge grows, and in critically and responsibly submitting to engineering judgement, as a final verification, what is left beyond such limited goals, in the unlimited field of conceivable possibilities. The present status of the implementation in Italy of practices concerning severe accidents is discussed in the paper with reference both to accident management within the plants and to external emergency planning.

1. SAFETY PHILOSOPHY AND LICENSING REQUIREMENTS IN ITALY

Safety practices in Italy are broadly based on two general principles:

— Integration of safety requirements with health protection objectives;
— Extensive 'defence in depth'.

Concerning the first, the final purpose of any safety provision is the protection of workers, the public and the environment. The original ICRP statement concerning the principles of dose limitation and optimization was restricted to situations under control. However, in view of the contribution to the overall risk to the public from a nuclear power plant due to potential from accidental releases, the implementation of such principles has been extended to the latter conditions, up to and including design basis accidents (DBAs).

The second principle, 'defence in depth', implies the adoption of the maximum reasonable degree of redundancy and diversity both in safety provisions and in safety assessment methods. The kind of defence in depth adopted includes (besides external emergency plans):

- Physical *barriers*: fuel, primary pressure boundary, double containment;
- Types of *safety measures*:
  - Measures aimed at preventing the occurrence of malfunctions and of accidents (e.g. design margins, quality assurance);
  - Measures to avoid the escalation of malfunctions into accidents (e.g. plant status diagnostic systems, reactor protection system);
  - Measures to mitigate the release of radioactive substances under DBA conditions (e.g. ECCS, containment);
  - Severe accident management provisions (procedures, equipment, training).
- Types of *safety criteria*:
  - Deterministic (engineering judgement) criteria;
  - Probabilistic criteria.

A more extensive description is given in Section 3 of this paper concerning the safety measures issuing from the above mentioned principles with reference to DBAs and core melt prevention as applied in the Italian design criteria and standards for PWRs [1].

2. SEVERE ACCIDENTS

After extensive study, in June 1985 a tentative position was proposed as regards severe accidents in safety analyses and emergency planning. This position was also presented and discussed in international meetings [2].

We were looking for an answer to the following question: what are the maximum realistic releases, even under severe accident conditions, which are consistent with the best exploitation of the capabilities of plants of present interest, given the establishment of an accident management structure (procedures, equipment, training)?

The position we developed can be expressed as follows:

"It is considered realistic to assure, by additional accident management provisions, with a confidence level of the order of 95%, that external releases of iodine or caesium, in situations otherwise leading to uncontrolled severe accidents, be limited within 0.1% of core inventory". Both accident management actions and pre-existing openings in the containment have to be taken into account. This position is intended primarily to serve as a criterion to decide upon modifications to plant and procedures. It is now also being considered for amplified external emergency plans.

As a general technical choice for the control of a molten core by accident management actions, the 'wet' strategy was adopted. It consists of trying to supply the damaged core, either in-vessel or outside it, with cooling water at the earliest possible time. The only precaution which is under consideration is to carry out this operation gradually, in order to avoid sudden detrimental effects.

This strategy implies, of course, that in some sequences the hydrogen and pressurization problems may be aggravated and therefore they need additional attention. There is, however, a general consensus on the possibility of coping with these side-effects.

The Italian Electricity Generating Body (ENEL), is now completing the study of two NPPs, Caorso and PUN (Progetto Unificato Nucleare), in order to identify those plant and procedure provisions which are necessary to comply with the above criterion. A proposal is due at the end of 1987.

Some plant provisions presently under consideration are:

- for the BWR (Caorso) plant (inerted containment): improvements in the drywell spray system; large wetwell venting; long term filtered wetwell venting; modification of drywell sump;
- for the PWR (PUN) plants: alternative ways to assure component cooling in extreme emergencies; filtered venting; refractory lining of lower reactor cavity; improved control of containment leakage (pre-existing openings).

Figure 1 gives a picture of probabilities and phenomena for risk dominant sequences and for an ideal plant with a core melt probability of 10^{-6}/year. The implementation of the above stated "95% (or 5%) probability — 0.1% release" criterion is also illustrated.

Some additional comments are in order concerning the use of probabilistic techniques in the regulatory treatment of severe accidents and the

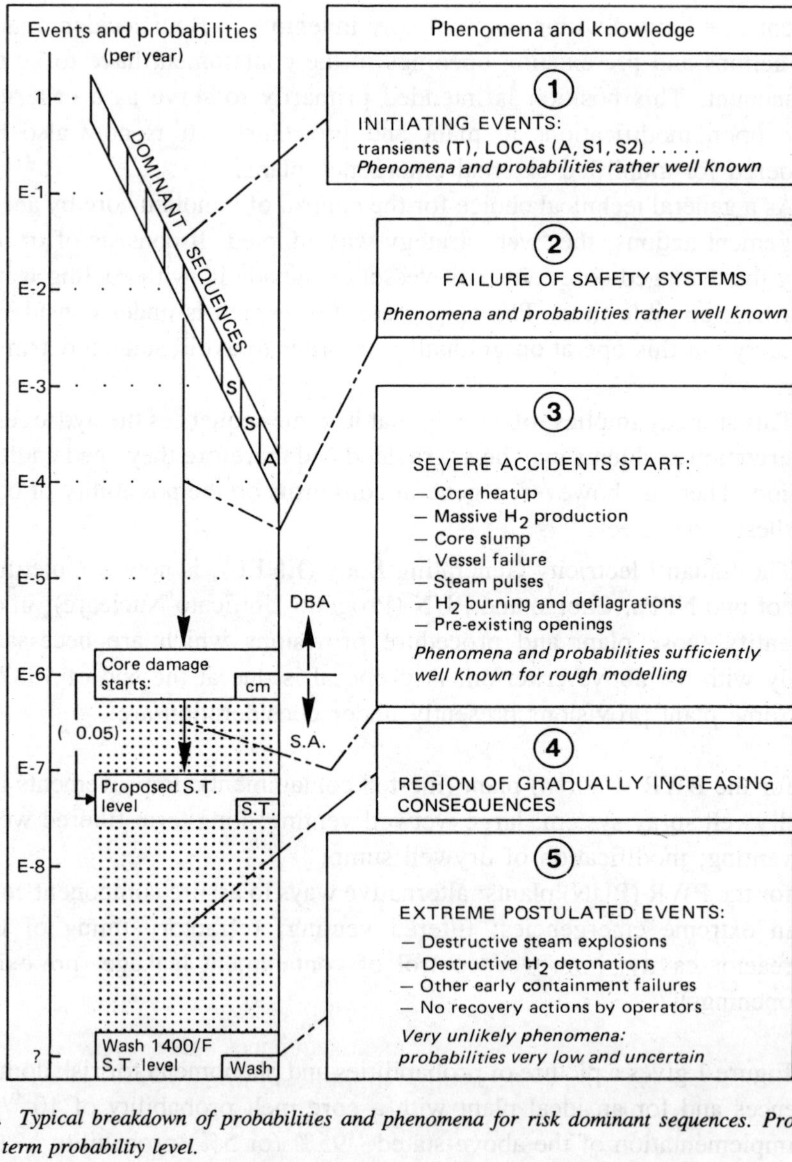

FIG. 1. Typical breakdown of probabilities and phenomena for risk dominant sequences. Proposed source term probability level.

present trends in Italy as regards severe accidents in relation to external emergency planning.

2.1. Probabilistic analyses

Probabilistic methods for the assessment of NPPs have been a subject of interest and of application in Italy since 1966, well before severe accidents became an issue of interest for regulatory purposes. Complete analyses of whole plants have, however, been completed only in recent years and are still under development as a tool.

More precisely, analyses for four NPPs are now available:

— ALSRA (Alto Lazio safety and reliability analysis), for the Alto Lazio, a twin 1000 MW(e) BWR6 station under construction;
— PUN-PSS (probabilistic safety study);
— Latina PSS, for the Latina gas-graphite plant;
— Caorso PSS, for the 850 MW(e) BWR4 Mark II Caorso plant.

Objectives of these analyses include:

— Balancing of plant design
— Assessment of safety level
— Assessment of residual life

and more recently:

— Availability of a rational basis for severe accident management provisions.

Some considerations emerging from the experience so far gathered in this area are the following:

— One of the most fruitful outcomes of probabilistic analyses is a complete picture of all the conceivable accident sequences and of their relationship with plant features. This picture supplies an in-depth knowledge of the plant which was not offered by the previous (binary) practice of a priori subdividing accidents into 'credible' and 'incredible' ones. Optimization of plant design and operation is, therefore, better guaranteed.
— As a measure of the safety level, 'bottom line' overall probabilistic figures are, naturally, more reliable for lower levels of accident severity. With a good control of analysis procedures it seems that the 'start of core

degradation' severity level is included in the field where numerical estimates are meaningful. The same holds for the more likely paths for the subsequent loss of integrity of the containment. Comparative evaluations, moreover, are, of course, more reliable than absolute ones.

— Areas exist where, for lack of a database or of adequate models, probabilistic estimates are not reliable enough and their operative use is not warranted: deterministic criteria should be used here. In this sense, too, probabilistic methods are to be viewed as an additional tool for safety assurance.

Although it is a powerful instrument, probabilistic analysis has, nevertheless, a limited range of reliable use. Sequences of events falling within extremely low probability levels, natural phenomena of extreme intensity and hypothetical single plant events are examples where the meaningfulness of a probabilistic treatment is doubtful.

The range of applicability of probabilistic analyses has to be defined by careful judgement on the basis of all the available evidence. The range itself depends on the purpose of the specific analysis. In particular, if the objective is solely to compare different plants as sources of risk or different design alternatives, some inherent uncertainty in the analysis is tolerable. The opposite is true for 'absolute' evaluations of risks, reliability, etc.

This is one of the reasons why there is some reluctance to use safety objectives in the form of probability of ultimate health effects (primary safety goals).

Some 'cutoffs' or limits to the applicability of probabilistic techniques in previous Italian practice are the following:

— failure probability on demand for each non-diversified safety function in a plant: higher than 10^{-5}/demand;

— failure probability on demand for the complex of non-internally diversified safety functions necessary to terminate a sequence of plant events: higher than 10^{-4}/demand;

— core melt probability: range from 10^{-5} to 10^{-6}/reactor-year;

— extreme natural events (like the most intense potential earthquake at a site) are considered more suitable for a deterministic treatment.

More recently, relative to the consideration of severe accidents in safety analyses, the merits of probabilistic methods as a valuable additional tool for decision making were reaffirmed. They provide elements for decision on issues such as the following:

— ranking severe accident sequences in an order of priority for additional safety (accident management) requirements;
— rational combination of intensities of independent detrimental parameters (such as delay to repair plant systems and size of pre-existing openings).

Conversely, new limits to the reliable application of probabilistic methods had to be established in the severe accident field. The limit finally chosen as a reference and for trial use is, as already mentioned, a 95% probability of not exceeding a 0.1% release, conditional upon core melt.

Various considerations entered into the decision concerning this 95% probability limit. First, as a number of probability evaluations for phenomena relating to severe accidents are rather subjective in nature (probability of repair within a given time, probability of destructive steam explosions and so on), the 95% probability level, as the commonly accepted expression of 'certitude' in subjective evaluations, was the natural candidate. Secondly, we sensed that any figure equal to or higher than the 99% level would have been embarrassing for experts requested to give probability evaluations in the field of theoretically hypothesized, yet never observed, severe accident phenomena. Thirdly, a reduction of a factor of 20 in the probability of large releases, as implied by the 95% limit, was considered significant enough for the establishment of a new barrier to releases based mainly on existing retention capability margins of the plants and on accident management.

It cannot, however, be demonstrated as yet that this additional barrier introduces a reduction factor higher than about 2 in the total accident related risk according to the current point estimates.

As previously noted, this fact is neither surprising nor discouraging per se, in view of the remarkable lack of justification inherent in some estimates of probability and intensity of the postulated most unlikely events (e.g. destructive steam explosions). In this field the danger of departing from common sense is evident.

The above illustrated practice concerning the use of the probabilistic method recognizes this danger and tries to pursue a realistic alternative to an unrestricted 'safety goal' approach.

This alternative consists of establishing limited, meaningful goals, possibly to be extended as knowledge grows, and in critically and responsibly submitting to engineering judgement, as a final verification, what is left beyond such limited goals, in the unlimited field of conceivable possibilities.

One obvious advantage of such an approach is that it offers a firmer conceptual basis for decisions on improvements of the plants' safety. Compliance with primary safety goals of the type so far proposed could very likely be demonstrated without any need to establish an additional safety barrier based on severe accident management. This demonstration could discourage any such effort, despite the clear perception that it certainly adds much to real plant safety. The reason for this apparent contradiction is the fact that current global accident risk evaluations tend to be dominated by the rather speculative low-probability, high-consequence type of events. The importance of the better known region of non-extreme sequences is thus obscured. Paradoxically, here the commendable attempt to include in risk analyses even less likely, rather speculative, events works, in the end, against the implementation of feasible improvements of plant safety. The concurrent negative effects on public perception of danger are to be noted, too.

2.2. Severe accidents and external emergency planning

Up to now external emergency planning around NPPs in Italy has been made with reference to a class of 'degraded' DBAs, leading to releases lower than 30 TBq (roughly 1000 Ci) of ^{131}I. For several years the idea of augmenting the severity of reference accidents for emergency planning has, however, been debated. This process has, understandably, been accelerated by the Chernobyl accident and, at present, the related studies and activities are proceeding at a quick pace.

A possible outcome of the present discussions is that DBA releases will continue to be the basis for local emergency plans, while the nuclear section of a new nationwide plan will be based on severe accident releases of some magnitude. Co-ordination of local and national plans has to be assured.

Regarding the size of emergency planning zones, it seems quite natural to refer primarily to the above mentioned 0.1% of core inventory for iodine and caesium coupled with the lower bound of IAEA emergency dose reference levels.

In this connection, the issue of the feasibility of such augmented emergency plans becomes important. It is directly tied to the adopted siting criteria and, in this respect, the Italian situation has been homogenized (as far as minimum requirements are concerned) by the preparation and adoption of a 'site chart' in recent years [3]. At the present stage of the related studies it seems that the 0.1% release is also a good compromise between the feasibility of rather exacting external emergency plans and realistic limits of internal accident management provisions.

In this framework the 0.1% release should be considered as an 'upper bound' for any plant, leaving space for improvements in specific plants to be considered in the future on an individual basis.

The concept of an 'upper bound release' for nationwide emergency planning seems, however, to fit in rather naturally with a homogeneous siting policy and with the adoption of reactors with similar technology.

3. ITALIAN DESIGN CRITERIA AND STANDARDS FOR PWRs

A brief description follows of design criteria in the field of DBAs and core melt prevention.

Quantitative targets in terms of effective dose equivalent (for individuals of the population) for a two unit site have been established for the different plant conditions, grouped in probability ranges, as follows:

— 0.1 mSv per year for normal operation and anticipated operational transients.
— 5 mSv for accidents whose annual probability is higher than 10^{-3}.
— 100 mSv for DBAs whose annual probability is lower than 10^{-3}.

In addition, it has been recommended to the designer to limit the consequences of this last category of DBAs to 5 mSv (and this is easily achievable in many sequences).

Similarly, an upper limit of 4 man·Sv has been established, with reference to each unit, for the annual collective dose equivalent for exposed workers in all normal and scheduled operations. This target has been fixed on the basis of pertinent international experience, taking into account that present nuclear plant technology is too complex and still not sufficiently robotized. However, simply reaching this objective is not considered satisfactory: by careful use of optimized design options (with regard to items such as layout, pressure boundary materials, primary coolant chemistry, component reliability, location and working life), significant further reduction of collective occupational doses will be presumably achieved.

Regarding the aspects directly related to the plant systems and components a few significant examples of the requirements established are as follows.

Pressure boundary. For the reactor pressure vessel, requirements cover material, design and construction techniques. No longitudinal welds are accepted; with a view to forged 'integral' vessel construction as a future objective, circumferential welds are not allowed in the high flux area.

To avoid embrittlement problems towards the end of the economic plant life, a target of 10^{19} n/cm^2 for maximum fast neutron fluence on the vessel wall has been specified. Compliance with this requirement has determined a few changes in the original design of the vessel, the adoption of thicker neutron pads and provision for low leakage fuel management.

For pressure piping, high quality material (low content of inclusions and segregations and high degree of toughness) is required, to prevent crack formation and growth; additional requirements have also been established for technological control on materials, pre-service and in-service inspection.

Containment. Following Italian practice, the plants are provided with double containment, both for accident release control and physical protection against external impacts (specified missiles and pressure waves).

Steam generator tubes. Requirements related to design, materials and monitoring of the integrity of steam generator tubes have been set up, in consideration of the steam generator tubes being part of the primary pressure boundary and containment. The steam generator overfilling protection system, with liquid discharge inside containment, is included. Reduction of challenge from anticipated transients without scram (ATWS) is achieved by increased capacity of pressurizer relief valves.

ATWS. Combined protective measures against ATWS have been required. They include, as a preventive measure, a backup trip function, and mitigating measures such as automatic turbine trip and feedwater actuation, increased capacity of the pressurizer valves and an alternate automatic shutdown system capable of plant shutdown in a few minutes after ATWS.

External man-made events. Besides the passive protection against external missiles and pressure waves (i.e. the double, strengthened containment), special emergency systems are provided to cope with possible transients caused by man-made events: plant control is automatically assured without operator intervention and without access to the control room for a period of ten hours.

In order to verify the results achieved in the design and the general balancing of the plant defences, a probabilistic safety analysis is required, to be developed along with the design from its preliminary stages, and guiding and controlling the choice of coherent design alternatives. This requirement is international: although the approach to the use of probabilistic methodologies is not homogeneous among the different countries, much experience has already been gained. In Italy, the probabilistic approach to safety has been progressively introduced, as an item of licensing, starting from the early 1970s, when quantitative reliability objectives were established for specific systems and components.

The verification of plant defence balancing is particularly relevant in connection with sequences beyond DBA that entail core meltdown. In this respect a quantitative safety design objective has been established in terms of maximum overall probability for core degradation per reactor-year.

Review of recent designs has shown that, with reasonable and careful use of existing technology, values of the order of 10^{-6} to 10^{-5} can be achieved. This objective has been incorporated in the general design criteria. It is commonly found that the results of probabilistic studies are strongly dependent on the way they are performed. In order to reduce this dependency, guidelines have been established.

Specific attention, through the use of suitable calculation methods, must be paid to the contribution of common mode failures to system unavailability. In order to stimulate the adoption of diverse redundant systems, a minimum failure probability of 10^{-5} per demand has been established for any single safety function performed by an individual non-diversified system. Also, as a direct consequence of this cutoff, a backup protection system, technologically diverse from the existing reactor protection system, is required for those safety functions whose availability is critical to the fulfilment of the probabilistic safety target. Reactor trip is included among these functions.

REFERENCES

[1] NASCHI, G., ''General safety criteria for the next Italian PWRs and their rational basis'', Thermal Reactor Safety (Proc. Int. ANS/ENS Top. Mtg San Diego, CA, 1986).

[2] PETRANGELI, G., ZAFFIRO, C., ''Regulatory implications of source term studies'', Source Term Evaluation for Accident Conditions (Proc. Int. Symp. Columbus, OH, 1985), IAEA, Vienna (1986) 747–757.

[3] COMITATO NAZIONALE PER LA RICERCA E PER LO SVILUPPO DELL'ENERGIA NUCLEARE E DELLE ENERGIE ALTERNATIVE, Nuclear Power Plant Site Requirements and Selection Methods, Rep. DISP (77) 2, Direzione Sicurezza Nucleare e Protezione Sanitaria, ENEA-DISP, Rome (1977).

IAEA-CN-48/237

РАДИОАКТИВНОЕ ЗАГРЯЗНЕНИЕ ОКРУЖАЮЩЕЙ СРЕДЫ БАССЕЙНА БАЛТИЙСКОГО МОРЯ С 1970 ПО 1986 гг.

Л.Н. ЛАЗАРЕВ, Ю.В. КУЗНЕЦОВ, Л.И. ГЕДЕОНОВ,
З.Г. ГРИТЧЕНКО, Л.М. ИВАНОВА, В.М. ГАВРИЛОВ
Радиевый институт имени В.Г. Хлопина,
Ленинград

С.М. ВАКУЛОВСКИЙ
Институт экспериментальной метеорологии,
Обнинск

Союз Советских Социалистических Республик

Abstract–Аннотация

RADIOACTIVE CONTAMINATION OF THE ENVIRONMENT OF THE BALTIC SEA BASIN IN 1970–1986.

The concept of nuclear power development adopted in the USSR provides for the reprocessing of spent fuel. The programme of power development in the USSR up to the year 2000 is concentrated primarily on the construction of nuclear power plants with WWER-1000 reactors but also includes the planned construction of some power plants with fast neutron reactors. Accordingly, means of transporting WWER-1000 fuel are now being devised and a plant is being designed to reprocess it. It is planned that the plant will be built in constructional stages, each with a capacity of 1500 t of uranium per year. These stages will be introduced as the need for plutonium arises in order to produce mixed fuel. The paper describes the technology adopted for this plant which is based on the water extraction system. The radioactive waste has to be solidified. It is planned to recover ^{85}Kr from the gaseous effluent. Extraction technology is being adapted in the USSR to the reprocessing of fuel from fast reactors with a view to establishing a highly reliable and safe process. This is guaranteed by the choice of regimes for implementing the individual stages of the technology, the use of radiation protection equipment and compliance with all radiation protection requirements.

РАДИОАКТИВНОЕ ЗАГРЯЗНЕНИЕ ОКРУЖАЮЩЕЙ СРЕДЫ БАССЕЙНА БАЛТИЙСКОГО МОРЯ С 1970 ПО 1986 гг.

Принятая в СССР концепция развития ядерной энергетики предусматривает переработку отработавшего топлива. Программа развития энергетики в СССР до 2000 года ориентирована на преимущественное строительство АЭС с реакторами ВВЭР-1000, предусматривается также строительство нескольких АЭС с реакторами на быстрых нейтронах. Поэтому в СССР в настоящее время создаются транспортные средства для топлива ВВЭР-1000 и проектируется завод для его переработки. Завод намечено строить очередями производительностью до 1500 т урана в год, ввод которых будет осуществляться по мере появления потребности в плутонии для изготовления смешанного топлива. В докладе дано описание технологии, принятой для этого завода, которая основана на водно-экстракционной схеме. Радиоактивные отходы подлежат отверждению.

Предусмотрено улавливание криптона-85 из газовых выбросов. Экстракционная технология адаптируется в СССР для переработки топлива реакторов с ориентацией на создание процесса с высокой надежностью и безопасностью. Это обеспечивается выбором режимов осуществления отдельных стадий технологии, применением ядернобезопасного оборудования и выполнением всех требований радиационной безопасности.

Относительно мелководный и полузамкнутый водоем, имеющий водообмен через узкие Датские проливы с Северным морем, Балтийское море интенсивно используется рядом государств для решения многочисленных хозяйственных задач, в том числе и для охлаждения своих ядерно-энергетических установок. В течение последних десятилетий этот водоем находится под тщательным экологическим контролем, поскольку по ряду показателей (например, по загрязнению морской воды углеводородами) относится к наиболее загрязненным в мире. Контролируются в Балтийском море и радиоактивные загрязнения.

Исследования охватывают также приземный слой атмосферы и атмосферные выпадения, поскольку последние являются одним из источников радиоактивного загрязнения водоемов.

Наблюдения ведутся непрерывно на протяжении многих лет. При этом ставились вопросы: какие радионуклиды определяют радиационную обстановку в бассейне Балтийского моря; какова связь между радиоактивной загрязненностью воздушных масс над Балтикой и самого бассейна; что управляет динамикой распределения радионуклидов в водной системе этого водоема; как быстро может произойти самоочищение водной массы Балтийского моря от дозообразующих радионуклидов.

ТЕХНОГЕННЫЕ РАДИОНУКЛИДЫ В АТМОСФЕРЕ

Радиационная обстановка в регионе до апреля 1986 г. определялась глобальными выпадениями от испытаний ядерного оружия. После подписания Московского договора о запрещении испытаний ядерного оружия в атмосфере (1963 г.), содержание техногенных радионуклидов в атмосферном воздухе стало уменьшаться со временем в результате очистки атмосферы и радиоактивного распада. Практически значимыми техногенными радионуклидами, дающими вклад в дозовые нагрузки, к концу 1985 г. остались лишь цезий-137 и стронций-90. На рис. 1 приведен ход изменения среднемесячной обменной активности цезия-137 в приземном воздухе вблизи Ленинграда с 1959 по 1986 гг.

На рис. 2 приведены кривые кумулятивного накопления стронция-90 и цезия-137 в период с 1957 по 1987 гг., и для примера – рутения-106 и церия-144, поскольку в атмосферных выпадениях в конце апреля–мая 1986 г. их вклад в общую загрязненность земной поверхности был существенным. Из данных, представленных графически на рис. 1, видно, что в годы, предшествующие 1986 г., величины среднемесячных

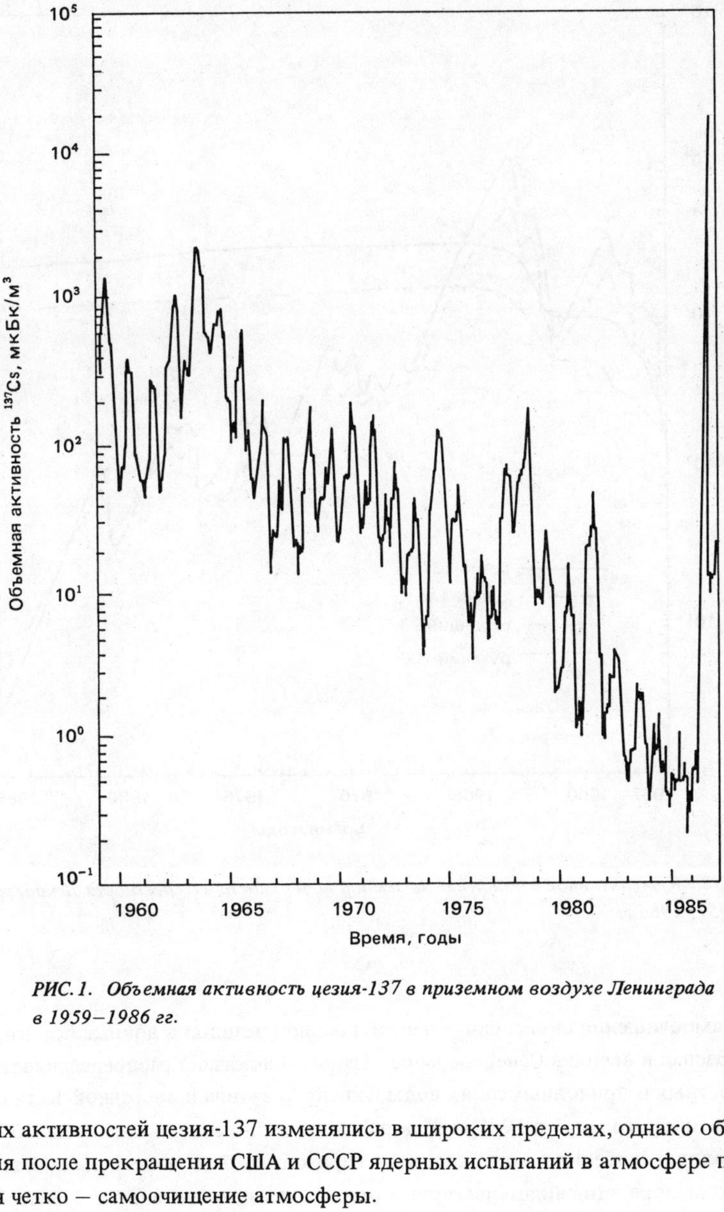

РИС. 1. Объемная активность цезия-137 в приземном воздухе Ленинграда в 1959–1986 гг.

объемных активностей цезия-137 изменялись в широких пределах, однако общая тенденция после прекращения США и СССР ядерных испытаний в атмосфере прослеживается четко – самоочищение атмосферы.

ТЕХНОГЕННЫЕ РАДИОНУКЛИДЫ В БАЛТИЙСКОМ МОРЕ

Как и для атмосферного воздуха, в водной массе Балтийского моря основными дозообразующими техногенными радионуклидами в период с 1970 по 1985 годы также были стронций-90 и цезий-137 (рис. 3.). В этот период времени имели место

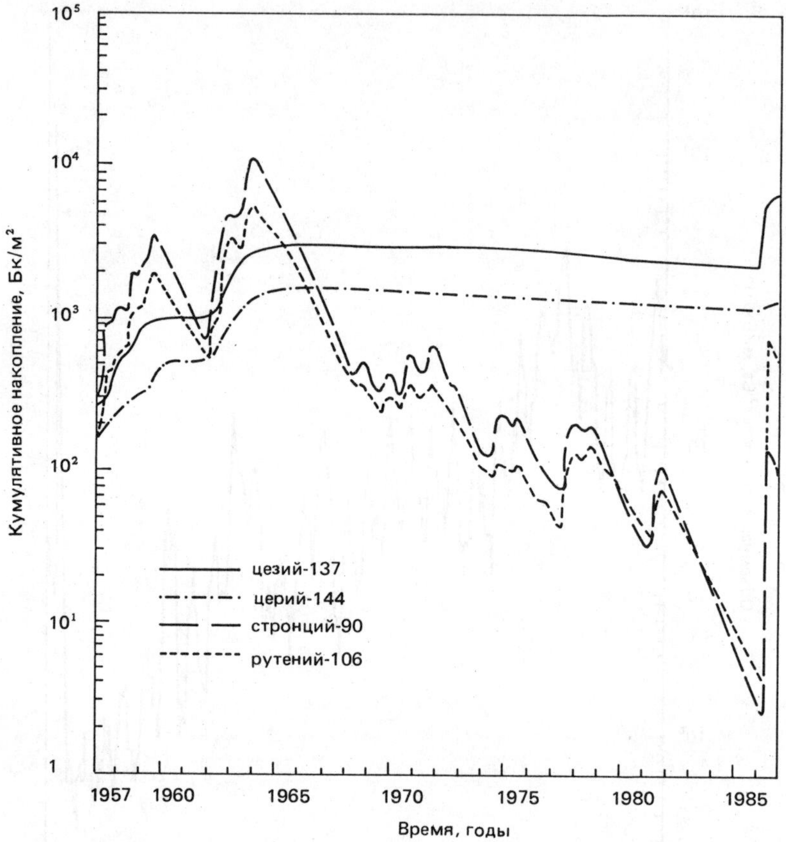

РИС. 2. Кумулятивное накопление на земной поверхности в окрестности Ленинграда в 1957–1986 гг.

процессы самоочищения акватории – переход радионуклидов в донные осадки, радиоактивный распад и вынос в Северное море. На рис.4 показано распределение цезия-137 в поверхностных и придонных слоях воды Финского залива и восточной части собственно Балтийского моря в 1985 г. Как видно из рис.4, содержание цезия-137 в Финском заливе, особенно в поверхностных горизонтах, было ниже, чем собственно в Балтийском море, что свидетельствует о том, что реки, впадающие в Финский залив, такие как Нева и Нарва, несут воды с меньшим содержанием цезия-137, ввиду сорбционного извлечения цезия речными взвесями и осаждения последних в приустьевых районах моря.

Характер распределения цезия-137 в собственно Балтийском море находился под влиянием двух противоположных по действию и по расположению источников. С одной стороны, менее соленых вод Финского залива, распространяющихся на запад в поверхностных горизонтах моря, с другой стороны, относительно тяжелых вод с

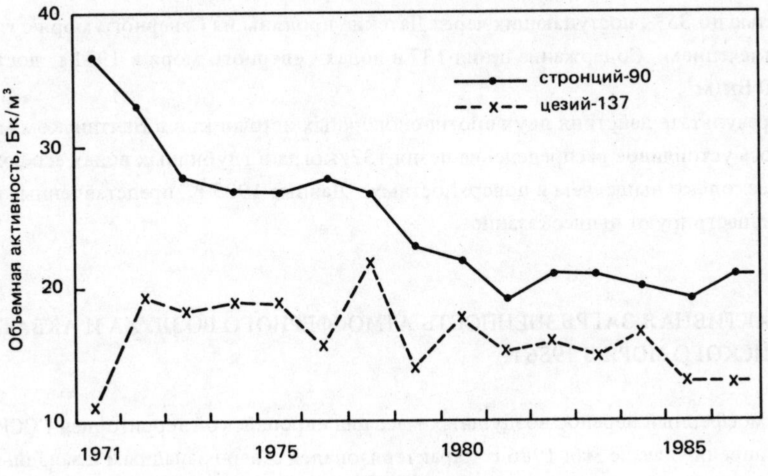

РИС. 3. Объемная активность поверхностных вод Балтийского моря в 1971–1985 гг.

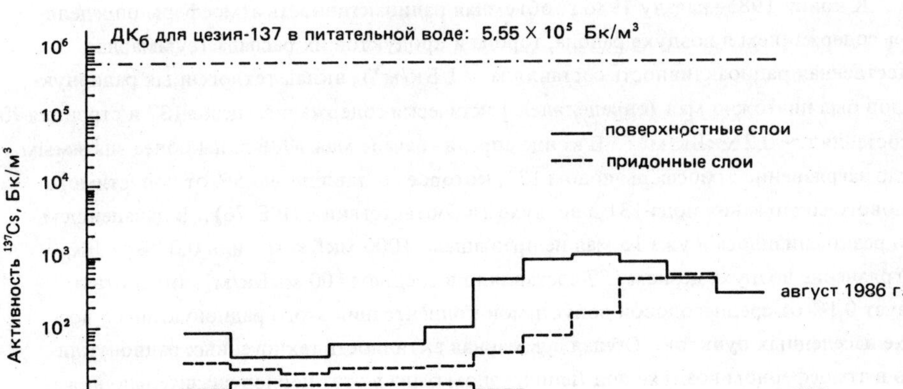

РИС. 4. Распределение цезия-137 (Бк/м³) в поверхностных и придонных слоях вод Финского залива и восточной части собственно Балтийского моря в 1985 и 1986 гг.

соленостью до 35%, поступающих через Датские проливы из Северного моря с глубинным течением. Содержание цезия-137 в водах Северного моря в 1981 г. достигало 380 Бк/м^3.

В результате действия двух противоположных источников в Балтийском море сложилось устойчивое распределение цезия-137, когда в глубинных водах его содержание несколько выше, чем в поверхностных. Данные 1985 г., представленные на рис. 4, иллюстрируют вышесказанное.

РАДИОАКТИВНАЯ ЗАГРЯЗНЕННОСТЬ АТМОСФЕРНОГО ВОЗДУХА И АКВАТОРИИ БАЛТИЙСКОГО МОРЯ В 1986 г.

Атмосферный перенос воздушных масс над европейской территорией СССР в конце апреля–начале мая 1986 г. характеризовался северо-западным и западным направлениями, что способствовало распространению радиоактивных примесей в направлении региона Балтийского моря. В этот период наблюдался подъем объемной активности отдельных радионуклидов в приземном воздухе скандинавских стран и северо-западной части СССР.

Выпадение и кумулятивное накопление на земной поверхности отдельных радионуклидов с конца апреля 1986 г. проходило с различной интенсивностью. При сравнении их с результатами, полученными в 1961–1963 гг., они были максимальными для цезия-137, рутения-103, рутения-106. Для стронция-90, церия-141, церия-144, циркония-95 они не превышали значений, определенных в выпадениях в 1970–1980 гг.

К концу 1985–началу 1986 г. объемная радиоактивность атмосферы определялась содержанием в воздухе радона, торона и продуктов их распада (суммарная естественная радиоактивность составляла ~ 1 Бк/м^3), вклад техногенных радионуклидов был ничтожно мал (определялся фактически содержанием цезия-137 и стронция-90 и составлял ~ 0,2 мкБк/м^3). В конце апреля–начале мая 1986 г. наиболее значимым было загрязнение атмосферы иодом-131, которое составляло до 5% от допустимого годового содержания иода-131 в воздухе (в соответствии с НРБ-76). В дальнейшем оно резко снизилось и уже 15 мая не превышало 1000 мкБк/м^3, или 0,07% от ДК$_б$. Загрязнение воздуха цезием-137 составляло в среднем 100 мкБк/м^3, что соответствует 0,1% от среднегодовой допустимой концентрации этого радионуклида в воздухе населенных пунктов. Общая суммарная активность техногенных радионуклидов в атмосферном воздухе под Ленинградом была в этот период значительно ниже его естественной радиоактивности, обусловленной радоном, тороном и продуктами их распада (~ 1 Бк/м^3).

Повышение плотности радиоактивных атмосферных выпадений привело к увеличению радиоактивного загрязнения части акватории Балтийского моря.

Первые определения, проведенные в конце апреля–начале мая 1986 года в диаметрально противоположных районах Балтийского моря – восточной части Финского залива и вблизи Датских проливов, показали, что в обоих случаях имело место увеличение радиоактивного загрязнения. В ряде водоемов прибалтийского побережья также было отмечено увеличение уровня радиоактивного загрязнения. Эти наблюдения побудили к более обширным исследованиям.

В июне 1986 г. нами было предпринято обследование вод Финского и Рижского заливов. Полученные результаты в их предварительном варианте были представлены на 13-е заседание НТС ХЕЛКОМ. Эти результаты обнаружили значительную неоднородность, "пятнистость" радиоактивного загрязнения, отражающую неоднородность атмосферных радиоактивных выпадений. В Финском и Рижском заливах было выявлено 2 зоны загрязнения, различающихся по величинам объемной активности изотопов цезия примерно в 10 раз: зона более загрязненная, расположенная в восточной и центральной частях Финского залива, и зона менее загрязненная, охватывающая западную часть Финского залива и Рижский залив. Полученные данные показали, что среди присутствующих радионуклидов доминирующая роль принадлежит изотопам цезия. При этом отношение цезий-134/цезий-137 на всей обследованной акватории составляло в среднем 0,5, как и в атмосферных радиоактивных выпадениях в этот период. Одновременно было установлено, что содержание стронция-90 не увеличивалось, оно осталось практически на глобальном уровне даже в самых загрязненных районах моря. В связи с этим в дальнейшем основное внимание было уделено исследованию распределения изотопов цезия.

В июле–августе 1986 г. было выполнено обследование вод восточной части собственно Балтийского моря, включая Куршский и Калининградский залив, а также повторное обследование Финского залива.

Данные обследования, проведенные в июле–августе, подтвердили июньские наблюдения. Была отмечена существенная разница в уровнях загрязнения различных районов моря. В августе район повышенного загрязнения в Финском заливе продолжал отчетливо фиксироваться, но его резкие границы размылись, более загрязненная область продвинулась на запад по направлению течения поверхностных вод.

Восточная часть собственно Балтийского моря оказалась наименее загрязненной среди обследованной нами акватории. Здесь содержание цезия-137 в поверхностных водах составляло около 40 Бк/м^3. В юго-восточной части моря и в Калининградском заливе, так же как и в Гданьской бухте, содержание цезия-137 было несколько повышенным, достигая 80–100 Бк/м^3.

Анализ водных проб из глубинных водных горизонтов однозначно свидетельствовал о проникновении изотопов цезия в глубинные слои, но интенсивность проникновения была не одинаковая для различных районов моря. На рис. 4 показано распределение цезия-137 в поверхностных и придонных слоях моря по всем районам обследования в июле–августе 1986 г. В мелководных районах Финского залива (25–30 м) к августу 1986 г. концентрация цезия-137 сравнялась в природных и поверхностных слоях, в то время как в июне существенно различалась. В районах с большими глубинами и в июле–августе существовали заметные различия в содержании изотопов цезия в поверхностных и природных слоях. Общая картина распределения цезия-137 в поверхностных и придонных слоях обследованной части Балтийского моря представлена на рис. 4.

В 1986 г., как и в предыдущие годы, прослежено поступление радиоизотопов цезия с водами рек Балтийского бассейна. Сопоставление результатов показывает, что реки, относящиеся к западной части прибалтийского побережья, оказались значительно менее загрязненными по сравнению с реками восточной части побережья – Невой и Нарвой. При этом воды Невы и ее источника – Ладожского озера имели значительно меньшие концентрации цезия-137 по сравнению с центральной частью Финского залива. Все это подтверждает неравномерность радиоактивного загрязнения водных поверхностей, обусловленную неравномерностью атмосферных радиоактивных выпадений, в результате чего в Финском заливе возникло обширное радиоактивное пятно, включающее районы различной интенсивности загрязнения. Полученные нами данные, относящиеся к ноябрю–декабрю 1986 г., свидетельствуют, что к этому периоду времени содержание цезия-134 и цезия-137 в водах Невы и Ладожского озера уменьшилось примерно в 20 раз по сравнению с началом мая.

В период июня–августа 1986 г. наряду с определением радиоизотопов цезия нами были выполнены определения плутония-239, -240 в водах Балтийского моря. Увеличение содержания плутония-239, -240 в июне 1986 г. по сравнению с 1985 г. было невелико и не превышало последнего более чем в 3–5 раз.

При исследовании радиоактивного загрязнения вод Финского залива нами всегда уделялось большое внимание району расположения Ленинградской АЭС – Копорской губе. В 1986 г. здесь проводились ежемесячные наблюдения. Копорская губа, как и весь Финский залив, подверглась дополнительному атмосферному радиоактивному загрязнению, хотя и в несколько меньшей степени, чем центральная часть Финского залива. Содержание цезия-137 в поверхностных водах Копорской губы с ходом времени снижалось от 400 Бк/м^3 в июне до 190 Бк/м^3 в ноябре 1986 г. В этот период были выполнены исследования фазового распределения цезия-137 в водах Копорской губы. Полученные результаты показали, что основное количество цезия-137 находится в водной фазе и около 2% на взвеси. Аналогичные опыты, проведенные в 1982 г., также свидетельствовали, что на взвеси находится 2,5–10% цезия-137 от его общего количества в воде. В целом ряде районов, например в восточной части собственно Балтийского моря, загрязнение радиоактивным цезием остается ниже уровня содержания цезия-137 в Северном море, загрязнение вод которого поддерживается на высоком уровне за счет систематических сбросов жидких радиоактивных отходов ядерными перерабатывающими заводами Западной Европы. Следовательно, и в дальнейшем воды Северного моря будут увеличивать запас цезия-137 в Балтийском море.

Таким образом, на основании исследований, проведенных в июне–ноябре 1986 г., можно полагать, что в результате атмосферного переноса радионуклидов в этот период увеличилось кумулятивное накопление только цезия-137 и цезия-134. Для стронция-90 и плутония-238, -239, -240 кумулятивный запас не увеличится. Повышение кумулятивного накопления цезия-137 может привести к увеличению ожидаемой эффективной эквивалентной дозы облучения населения региона (за 50 лет), где фиксировались максимальные радиоактивные выпадения, со 190 мЗв (среднее значение для Северного полушария) до 191 мЗв, т.е. всего лишь на 0,85%.

В большей части региона Балтийского моря вклад от цезия-137, выпавшего в 1986 г., будет ниже этой величины. Вклад остальных радионуклидов, учитывая их период полураспада и низкое содержание, пренебрежимо мал.

Следует ожидать также постепенное самоочищение водоема от цезия-137 вследствие сорбционного извлечения этого радионуклида из воды ее взвешенной фазой. Накопление цезия-137 в донных отложениях будет определяться скоростью осаждения взвесей и типом донных отложений.

NUREG-1150: THE REACTOR RISK REFERENCE DOCUMENT RESULTS AND REGULATORY APPLICATIONS

J.A. MURPHY, M.A. CUNNINGHAM
Office of Nuclear Regulatory Research,
United States Nuclear Regulatory Commission,
Washington, D.C.

M.T. LEONARD
Battelle Columbus Division,
Columbus, Ohio

United States of America

Abstract

NUREG-1150: THE REACTOR RISK REFERENCE DOCUMENT RESULTS AND REGULATORY APPLICATIONS.

The paper summarizes the recent comprehensive reassessment of reactor risk performed by the United States Nuclear Regulatory Commission (NRC), as presented in Draft NUREG-1150: The Reactor Risk Reference Document. Draft NUREG-1150 provides results of risk analyses for five different light-water reactors in the United States of America. For the plants analyzed, this document describes (1) the major factors related to internally initiated events (excluding fires, floods, and earthquakes) that contribute to severe core damage, (2) the frequencies and related uncertainty ranges of severe core damage events, (3) the major factors and severe accident phenomena that could lead to containment failure (and particularly early containment failure), (4) the conditional probabilities and uncertainty ranges of early containment failure, (5) the sequences and risks of severe accidents, including the sensitivity of these risks to factors such as evacuation or sheltering measures, (6) comparisons of the risks with NRC safety goals, and (7) cost and risk-reduction analysis of plant-specific measures that could reduce risk from severe accidents.

INTRODUCTION

Soon after the Three Mile Island accident in 1979, the NRC began to review existing U.S. nuclear plant designs and identify potential risks to the public from reactor accidents outside the envelop of the licensing design basis. Consequently, the TMI Action modifications to plant designs and operating procedures were made. In addition, major programs were begun to reassess the role that severe accident issues should have in NRC's regulatory process. These programs were related to the development of an appropriate regulatory policy, a set of decision criteria, and an expanded data base on severe accident likelihoods, phenomena, and risks [1].

The Commission developed and issued its Severe Accident Policy Statement in 1985 and followed with its policy implementation plan (Implementation Plan for Severe Accidents and Regulatory Use of New Source Term Information) in 1986. This plan provided for the resolution of severe accident issues through (1) a systematic examination by industry of plants for risk contributors and (2) regulatory use of improved source term information arising from studies such as those described in NUREG-0956, "Reassessment of the Technical Bases for Estimating Source Terms,"[2] and in Draft NUREG-1150: The Reactor Risk Reference Document[3].

Thus, NUREG-1150, with its detailed treatment of plant core damage frequencies containment performance, fission product source terms,and uncertainties, now provides a key technical data base to support implementation of the NRC severe accident and safety goal policies. This paper focuses on Draft NUREG-1150 results and their intended regulatory application.

Results of Plant Analyses

Draft NUREG-1150 describes the results of detailed risk analyses performed for the following five plants: Surry, a Westinghouse-designed three-loop PWR with a subatmospheric containment; Peach Bottom, a General Electric-designed BWR with a Mark I containment; Sequoyah, a Westinghouse-designed four-loop PWR with an ice condenser containment; Grand Gulf, a General Electric-designed BWR with a Mark III containment; Zion, a Westinghouse-designed four-loop PWR with a large, dry containment. The distinctly different containment designs for these plants were important in their selection since it was known that containment design differences have a major effect on plant risk.

A principal value of probabilistic risk analysis is that it represents the most complete compilation of information available to develop an integrated regulatory perspective on relationships among initiating events, human and equipment performance, mitigating features, accident phenomena, and health consequences.

Brief discussions of some of these factors and related perspectives that can be obtained from Draft NUREG-1150 are summarized as follows:

a. Plant Core Damage Likelihood. The term "severe core damage" is intended to cover all reactor fuel element states in which the cladding integrity has been destroyed either through melting or chemical reaction. In Figure 1,

TABLE I. PRINCIPAL CONTRIBUTORS TO MEAN CORE DAMAGE FREQUENCY

Percent contribution to core damage frequency	Surry	Zion	Sequoyah	Peach Bottom	Grand Gulf
Station blackout	38	2	5	86	99
Loss of offsite power	4	1	—	—	—
Loss of bus	20	—	3	—	—
LOCA	28	18	59	1	—
Interfacing LOCA	4	—	—	—	—
ATWS	6	—	1	12	1
Loss of component cooling water	—	79	31	—	—
Severe core damage frequency (a^{-1})	2.6×10^{-5}	1.5×10^{-4}	1.0×10^{-4}	8.2×10^{-6}	2.9×10^{-5}

Consequently, a determination of the significant contributors to severe core damage frequency for any plant other than those analyzed would require additional plant-specific examination. In addition, it was found that the principal accident sequences contributing to core melt frequency were not necessarily the dominant contributors to public risk.

b. Plant Accident Risk. In Figures 2 and 3, the wide variability in risk of early and latent cancer fatality for all the studied plants is displayed. The method of display used in these figures reflects the quantitative uncertainty analyses performed for Draft NUREG-1150. For a particular plant's risk, each horizontal line represents one outcome of the statistical sampling process. The density of lines is a measure of the distribution of samples within the range.

The primary sources of uncertainty that contribute to the range for each plant vary from plant to plant. For Surry and Zion, these uncertainties include the magnitude of direct containment heating with concurrent hydrogen burning. For Sequoyah, hydrogen generation and burning, the potential for delayed core-concrete interactions, and

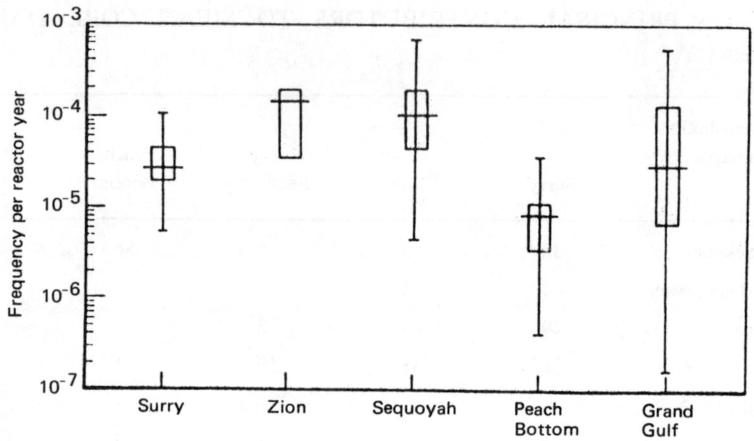

FIG. 1. Comparison of severe core damage frequencies.

the frequencies of such events for the five studied plants are displayed in a "box and whisker" format. The box's height represents the range of the plant's mean value of severe core damage frequency based on individual sensitivity studies, showing the potential effects of variations in the analytical modeling of plant design features and procedures. The mean can vary by as much as an order of magnitude within this range. The whiskers represent the extremes of the 5 to 95 percent confidence bands depicting variability in equipment failure data for the ensemble of sensitivity issues studied. A comparison of the Draft NUREG-1150 results with the Reactor Safety Study indicates a decrease in core damage frequencies for Surry and Peach Bottom. This is believed to have been caused by improvements in hardware, training, and testing procedures at these plants, as well as improvements in PRA modeling techniques.

The relative contributions of various accident sequences to the estimated total core damage frequency for each of the analyzed plants are identified in Table 1. Station blackout and small loss-of-coolant accidents (LOCAs) involving reactor coolant pump seal are often the major contributors. Anticipated transients without scram (ATWS) and large LOCAs appear to be relatively unimportant. The most significant factors driving these accident frequencies were the particular plant-specific design of support systems and plant-specific common cause failures.

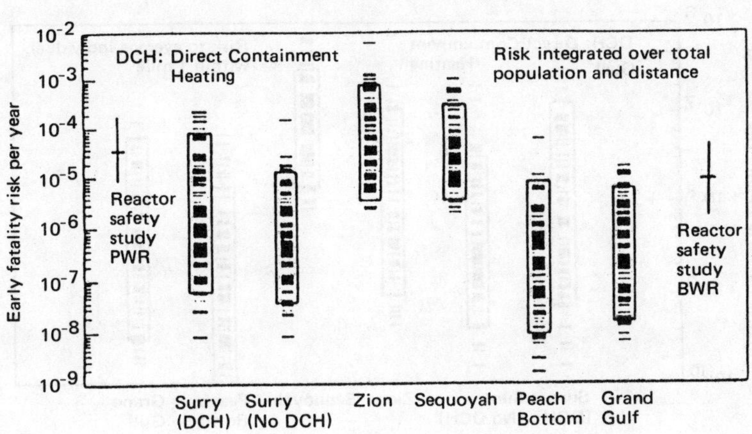

FIG. 2. Comparison of early fatality risks.

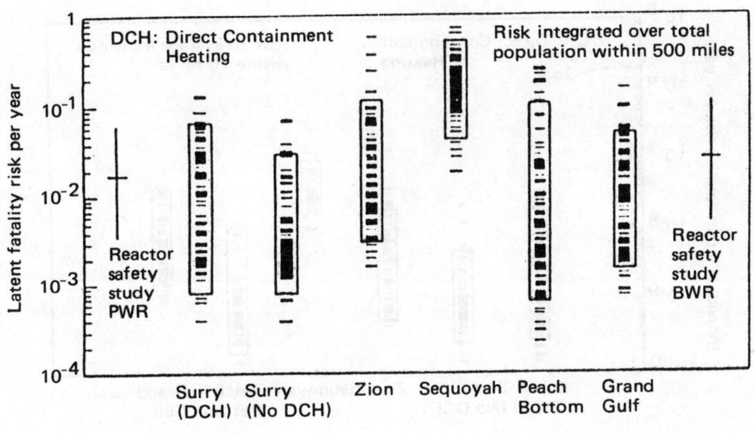

FIG. 3. Comparison of latent cancer fatality risks. 1 mile: 1.609 km.

the modeling of common cause failures in the component cooling water system are important. For Peach Bottom, the important factors are the potential for core debris attack on the drywell wall, the amount of radionuclide deposition in the reactor building, and the modeling of common cause failures to the station batteries. For Grand Gulf, the potential for late iodine reevolution from the suppression pool, revaporization of radionuclides originally deposited on the reactor pressure vessel internals, and the modeling

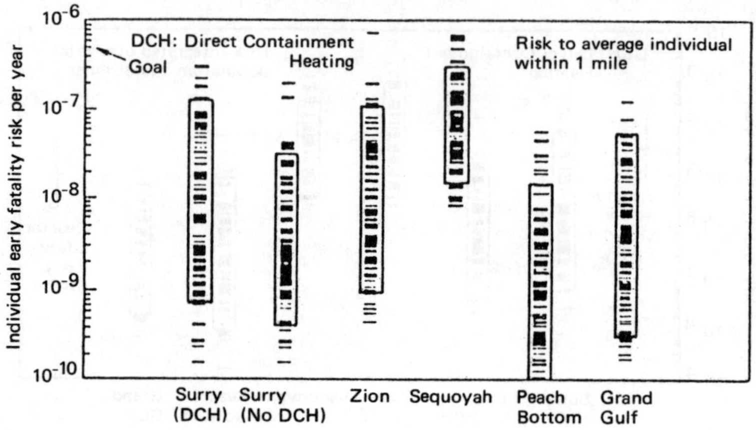

FIG. 4. Comparison with NRC safety goals (early fatalities). 1 mile: 1.609 km.

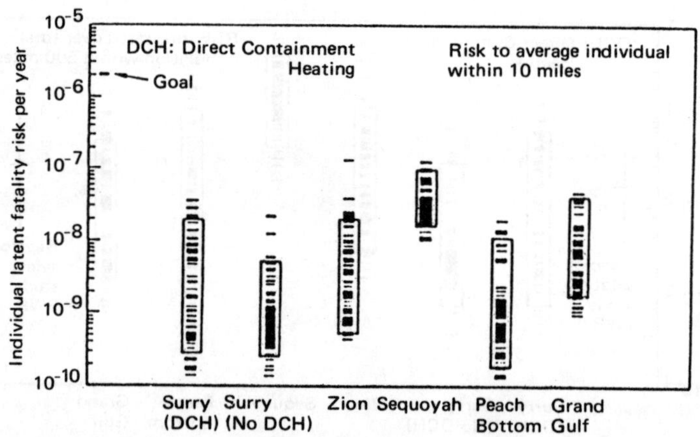

FIG. 5. Comparison with NRC safety goals (latent cancer fatalities). 1 mile: 1.609 km.

of diesel generator reliability are significant. In addition to these factors, the variability of risk results from plant to plant is influenced by other considerations such as population density (Zion has the highest, Grand Gulf, the lowest) and core melt frequency (Peach Bottom's and Grand Gulf's are lower than evaluated in previous risk assessments). For latent cancer fatality risks, variability among plants is less than that for early fatalities since latent risks are affected primarily by

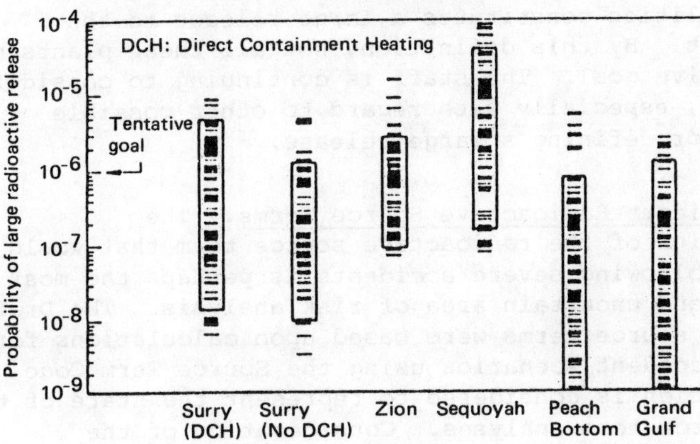

FIG. 6. Comparison with NRC safety goals (latent cancer fatalities) using a definition that a release sufficient to cause one or more early fatalities constitutes a large release to the environment.

source term magnitudes and not significantly by the timing and energy of releases or emergency response assumptions.

In general, these risk estimates show no particular trend when compared with risk data from previous risk analyses except to state that the Reactor Safety Study values often appear to be near the upper end of the risk range. This is attributable to the fact that the Reactor Safety Study generally ascribed relatively high likelihoods of early containment failure and large source terms. Further, plant changes since the publication of the Reactor Safety Study have resulted in a lowering of the severe core damage frequency, e.g., test and maintenance procedures for the Surry interfacing-system check valves and containment venting and ATWS procedures and equipment for Peach Bottom.

Figures 4 and 5 display the risk from internal initiators in terms of the NRC safety goals. These figures indicate that all the plants that were studied meet the explicit risk criteria stated in the NRC's Safety Goal Policy [4].

The Commission has also suggested a tentative criterion that "the overall mean frequency of a large release of radioactive material to the environment from a reactor accident should be less than 1 in 1,000,000 per year." In Draft NUREG-1150, the staff has evaluated the five plants against the suggested guideline (Figure 6) using a definition that a release sufficient to cause one or more early fatalities constitutes a large release to the environment. By this definition, not all these plants meet the tentative goal. The staff is continuing to consider this issue, especially with regard to other possible criteria for defining a large release.

c. Severe Accident Radioactive Source Terms. The determination of the radioactive source term that would be released following severe accidents is perhaps the most difficult and uncertain area of risk analysis. The Draft NUREG-1150 source terms were based upon calculations for selected accident scenarios using the Source Term Code Package, which is considered to represent the state of the art in source term analyses. Consideration of the phenomena not modeled in the Source Term Code Package and related uncertainty analyses relied heavily on expert judgment. Typical source term information displayed in Figure 7 indicates that the magnitude of the upper portion of the uncertainty band of the source terms that could be released to the environment is not significantly different from previous estimates, but the lower portion of the band is well below earlier estimates. In accidents where the containment does not fail, environmental releases are substantially lower than earlier estimates. It is anticipated that the source term information for the studied plants can be useful in evaluating the adequacy of the source term assumptions presently used in NRC licensing considerations, i.e., siting and environmental qualification matters.

d. Effect of Emergency Preparedness on Consequence Estimates. An important generalization supported by Draft NUREG-1150 analyses is that the offsite early fatality consequences and risk can be affected by the measures assumed in emergency evacuation scenarios. The relative effectiveness of alternative offsite emergency actions in reducing the average dose to individuals in the vicinity of a plant (Surry) following a severe accident is illustrated in Figure 8.

Surry analyses show that the variation in early fatalities based on reasonable assumptions for early evacuation and sheltering can be large depending on the nature of the accident. Latent cancer fatalities, total population dose, and offsite cost risks determined in Draft NUREG-1150 are less strongly affected by immediate actions such as evacuation near the accident site, as the long-term effects are driven primarily by the magnitude of the releases and the postaccident land and crop interdiction criteria.

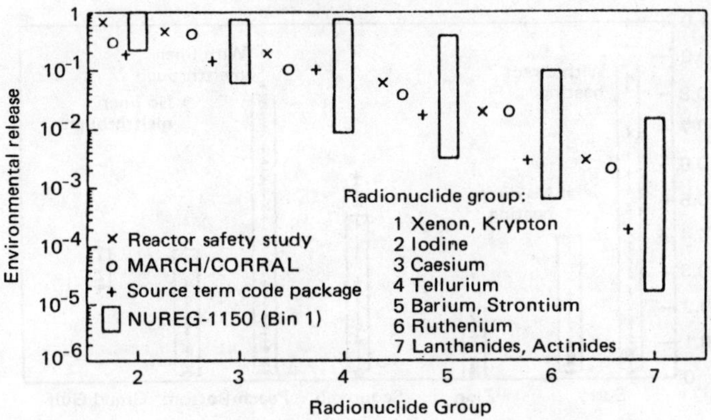

FIG. 7. Comparison of source term results: station blackout, early failure scenario — Surry plant.

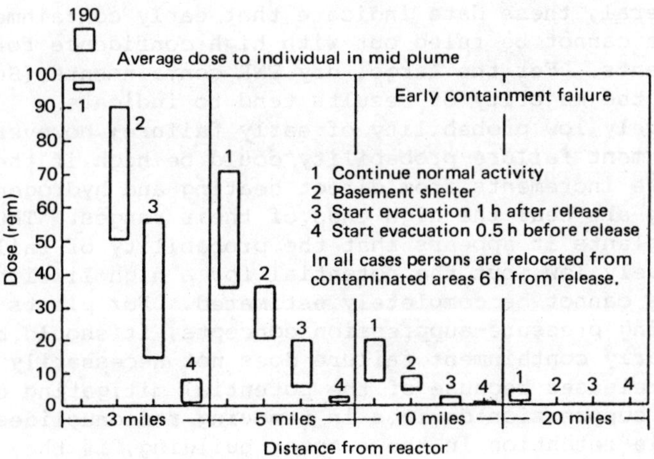

FIG. 8. Effectiveness of emergency response actions — Surry plant. 1 mile: 1.609 km.

e. Containment Performance. In Figure 9 the ranges of the estimated conditional probabilities of early containment failure, weighted by core damage frequency, are displayed for the five plants analyzed. The horizontal lines within the vertical bars represent the individual sample results from the uncertainty analysis and provide a qualitative indication of the concentration trends within the range, based on the judgment of experts.

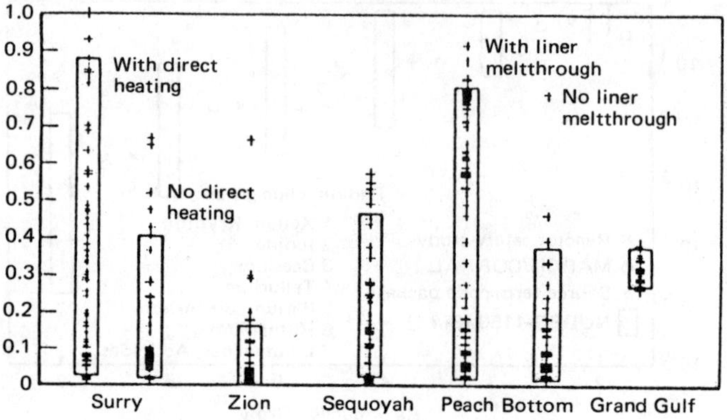

FIG. 9. Comparison of early containment failure probabilities.

In general, these data indicate that early containment failure cannot be ruled out with high confidence for any of the plants. For the large, dry PWR containments (Surry and Zion), the majority of results tend to indicate a relatively low probability of early failure; however, the containment failure probability could be high if the pressure increments from direct heating and hydrogen burning are near the high ends of their ranges. Thus, for these plants it appears that the probability of failure is relatively low, but the potential for a high likelihood of failure cannot be completely estimated. For plants employing pressure-suppression concepts, it should be noted that early containment failure does not necessarily lead to large releases because of the potential mitigating effects of the suppression devices in removing radionuclides and possible retention in the reactor building, if they are available.

f. Risk-Reduction Potential. One of the major objectives of the NRC's Severe Accident Research Plan was to provide cost-benefit information for a series of design and procedural options for potential preventative and mitigative modifications to the plants. These cost-benefit studies addressed such severe accident safety features as independent core cooling or decay heat removal systems, improvements to ac or dc power or component cooling reliability, improved testing and maintenance procedures, enhanced reactor coolant system depressurization capability, improved hydrogen control capability, containment venting systems, additional containment spray or cooling capability, provision for containment flooding, curbs to prevent core debris attack on containment structures, and missile shields. Costs were calculated using the MACCS code, or by assuming $1,000/person-rem averted, with and without onsite costs. The results of these analyses have shown that the benefits, in terms of risk reduction, do not clearly outweigh the costs of any of the proposed modifications. Of particular note, however, is that even given the large uncertainties associated with any estimated averted risk, the maximum amount of money that could be justified for backfit considerations on current NRC cost-benefit theory alone is limited to a few million dollars, considering only offsite costs, and about $20 million, considering both onsite and offsite costs. Safety options that are preventive in nature and relatively inexpensive were found to have the highest potential for being cost effective. For Sequoyah, these included procedural changes to increase the duration of core coolant injection, dc power improvements, and improvements in the component cooling water system; for Peach Bottom, improvements in the testing and maintenance of the station batteries; for Grand Gulf, modifications to the procedure for affecting long-term core cooling through the use of standby fire pumps, improvements of seal cooling for the high pressure core spray pump, and use of an existing small diesel to enable core cooling through the control rod drive system. Many of these address station blackouts, which are the dominant contributors to risk at several of the studied plants. Some could have negative benefits (competing risks) that are not fully evaluated in the present work.

Regulatory Applications

The Commission's Severe Accident Policy Statement acknowledges that accidents beyond the design basis constitute the major risk to the public associated with the radioactive releases from nuclear power plant accidents. The Commission expressed no need for immediate action, or generic rulemaking, or other regulatory changes for these plants because of severe accident risk. However, the Commission plans to order a systematic safety examination each nuclear plant to determine whether particular accident vulnerabilities are present and what cost-effective changes are desirable to ensure that there is no undue risk to public health and safety. Information and insights obtained from NUREG-1150 will provide the regulatory staff with needed benchmarks for their evaluation of these individual plant examinations. It provides plant-specific models that can be used to evaluate the risk significance of possible plant modifications, improved and optimized technical specifications, and possible improvements and expansions in emergency operating procedures. It also helps focus NRC's limited resources on the major contributors to plant risk and to the uncertainties associated with these assessments.

In addition, NUREG-1150 information can be used in conjunction with NRC safety goals as an aid to regulatory decision making. For example, comparison of the risk estimates for the five plants studied show that the risk range for each of these plants falls beneath the quantitative design objectives (Figures 4 and 5). However, the Commission is also considering a supplemental goal that the probability of "large release" of radioactive material would be less than 10^{-6} per year of reactor operation appears to be more difficult to achieve using the Draft NUREG-1150 definition of "large release". The Draft NUREG-1150 analyses provide a useful tool to explore other meaningful definitions and comparing them to the quantitative design objectives in the Safety Goals. Such work is now under way by the NRC staff.

Plans for Revised Analysis

Between the publication of Draft NUREG-1150 in February 1987 and the final publication in 1988, several reviews are being conducted and improvements are being implemented. Contacts are being made with plant and utility personnel to verify the representation of the plant configuration and procedures. The reports have been distributed to several hundred persons representing various organizations and interests domestically and abroad, and many of them will be providing their comments. An external peer review committee has been formed under the chairmanship of Prof. William Kastenburg of the University of California at Los Angeles. Other committees have been formed to evaluate the procedures for performing uncertainty analysis utilizing expert opinion. Comments and recommendations from these various sources will be assessed and the analyses will be modified as appropriate.

In addition, new technical information pertinent to the NUREG-1150 uncertainty issues, is being generated through on-going analyses and experiments sponsored by the NRC. This information and information from other sources, including foreign organizations, is being incorporated in the data base for reconsideration by the experts.

Analyses have been initiated to extend the assessment of risk for two of the five NUREG-1150 plants to include external initiating events. Work has also been started to augment NUREG-1150 with a sixth plant, a BWR with a Mark II containment.

Finally, a concerted effort is under way to provide a meaningful quality assurance review of the plant analyses and to improve the overall scrutability of the results.

REFERENCES

[1] UNITED STATES NUCLEAR REGULATORY COMMISSION, NRC Action Plan Developed as a Result of the TMI-2 Accident, Vols 1 and 2, Rep. NUREG-0660, USNRC, Washington, DC (1980).

[2] UNITED STATES NUCLEAR REGULATORY COMMISSION, Reassessment of Technical Bases for Estimating Source Terms, Rep. NUREG-0956, USNRC, Washington, DC (1986).

[3] UNITED STATES NUCLEAR REGULATORY COMMISSION, Reactor Risk Reference Document — Draft for Comment, Vols 1–3, Rep. NUREG-1150, USNRC, Washington, DC (1987).

[4] UNITED STATES NUCLEAR REGULATORY COMMISSION, Safety Goals for the Operation of Nuclear Power Plants; Policy Statement, Vol. 51, Federal Register, USNRC, Washington, DC (1986) 30028 pp.

THE DEVELOPMENT OF A PROTOTYPE EXPERT SYSTEM AS AN AID TO EMERGENCY OPERATING PROCEDURES

Young-Whan LEE, Hyun-Koon KIM
Korea Advanced Energy Research Institute (KAERI),
Tae Jon

Soon-Heung CHANG, Joon-On YANG
Korea Advanced Institute of Science and Technology (KAIST),
Seoul

Republic of Korea

Abstract

THE DEVELOPMENT OF A PROTOTYPE EXPERT SYSTEM AS AN AID TO EMERGENCY OPERATING PROCEDURES.

It has recently been recognized that the use of computer aided operation based on knowledge-engineering techniques for operators in responding to adverse plant conditions is becoming increasingly important. A prototype expert emergency reactor operations support (EROS) system, has been developed which will provide aid for emergency operating procedures (EOPs) of Westinghouse pressurized water reactors. Now that EOPs consist of a series of logic steps covering a broad spectrum of event sequences, these logic steps can form knowledge bases and can be represented by rules in an expert system. Most of the logic steps are programmed with a KAIST official production system tool, and performance of EROS has been verified using a plant simulator.

1. INTRODUCTION

One of the important lessons learned from Three Mile Island incident in 1979 was to ensure the operational safety of an NPP. In the event of an accident, the operators are facing conditions under stress owing to information overload that might lead to an unexpected human error of commission or omission. In order to minimize operator errors in coping with the accident, introduction of computer aided diagnosis is becoming essential.

When plant abnormal conditions occur, emergency operating procedures (EOPs) are utilized to diagnose the plant status and to mitigate the consequences of the event. The EOPs currently available for Westinghouse plants are the result of plant specific application of emergency response guidelines (ERGs), which were generically developed by Westinghouse Owner's Group. EOPs provide diagnostic treatment based on written logic steps which are, however, voluminous and somewhat complicated for effective reference in an accident situation. This necessitates the use of a computerized procedure monitoring system that could help the operators reach correct decisions and perform proper actions as required by EOPs. For this purpose an expert system is being developed using expertise in the domain of plant emergency operations. Many efforts have recently been made in the expert system application for nuclear industry in the Republic of Korea.

The emergency reactor operations support (EROS) is a prototype expert system currently under development for monitoring the status of EOPs and to provide simplified and better use of the procedures. The system will monitor plant off-normal signals and advise appropriate actions with minimum operator inputs. It will also display transparent explanation of the reasoned action steps.

Major impetus behind the development of this system is the existence of similarities between EOPs and an expert system in that logic steps in EOPs can easily be represented by rules to be used in an expert system. In this regard the EROS utilizes surface knowledge about plant emergency operations that facilitates implementation in a prototype phase of development.

2. FRAMEWORK FOR EMERGENCY OPERATIONS

Prior to the Three Mile Island (TMI) incident, the operator's emergency procedures primarily addressed design basis accidents. Since they were event oriented procedures, they might be inadequate for diagnosis when there are multiple failures or unidentified symptoms.

In response to post-TMI regulatory requirements, EOPs have been revised to overcome shortcomings. That is, the contents of ERGs are symptom-oriented guides providing the operator with emergency treatment capabilities facing two distinct types of situations; these are either problems that can be diagnosed properly, or problems that are not amenable to diagnosis.

The events pertaining to the first type can be recovered with event related procedures called optimal recovery guidelines (ORGs). The optimal recovery concept is the event related recovery necessary to achieve optimal end state which permits repair and return of the plant to normal operation. ORGs are organized into four categories: non-accident, loss of reactor coolant, loss of secondary coolant, and steam generator tube rupture. For each basic category of events, three classes of guidelines are provided based on the degree of difficulty in accident diagnosis. These are main guidelines (E-guidelines), sub-guidelines (ES-guidelines) and emergency contingency action guidelines (ECA-guidelines). For events pertaining to the second type, the function restoration concept is utilized to support the optimal recovery concept and is the function

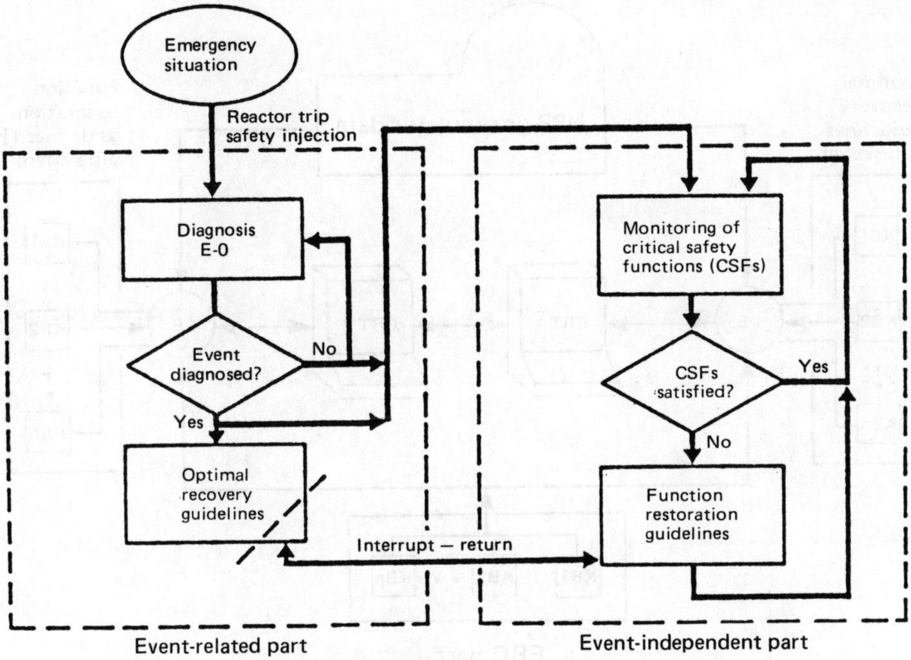

FIG. 1. Emergency operations concept.

related restoration strategy to be used when the optimal recovery is inadequate. For this type of accident, the plant can be recovered with function related procedures called function restoration guidelines (FRGs). FRGs are implemented only if critical safety functions (CSFs) are monitored as 'challenging'. CSFs should be maintained at a safe state in order to achieve the fundamental goal in nuclear safety. CSFs consist of six items: subcriticality, core cooling, heat sink, pressure boundary integrity, containment integrity, and reactor coolant system inventory. This means the prevention of uncontrolled release of radioactive materials into the environment. In order to improve the operator's job performance, a CSF status tree (CSFST) is provided for speedy and correct identification of necessary entry points to the function restoration or steps in-effect. For each of the six CSFs, a status tree exists, which has a single entry point and multiple exit points in order to support a branching structure for operator decision.

Figure 1 shows the optimal recovery concept and the CSF restoration concept incorporated in the operator's emergency response model.

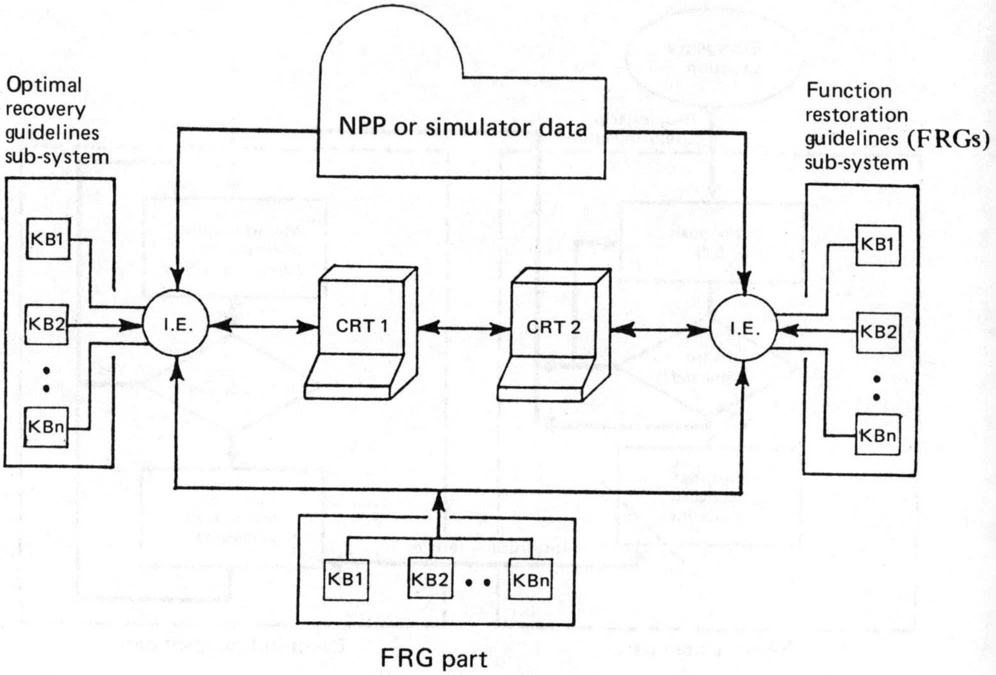

FIG. 2. Schematic of the emergency reactor operations support.

3. DESCRIPTION OF EROS

EROS is designed to deal specifically with EOPs for Westinghouse type PWRs. The proposed organization of EROS is illustrated in Fig. 2, where EROS consists of two sub-systems executing tasks of each sub-system independently. The one sub-system performs the ORG or FRG procedures, while the other performs CSF checking procedures. The two sub-systems are interconnected and can be mutually interrupted so as to allow the execution of function restoration according to the monitored status of CSFs.

The knowledge base of EROS is divided into three modules; each containing the knowledge of the ORGs, CSFSTs and FRGs. Each module is composed of several sub-knowledge bases which account for internally linked sub-guidelines. These knowledge bases are efficiently controlled by introducing a context limiting method. The context limiting method activates or de-activates the sub-knowledge bases according to the requirements of the system.

4. VALIDATION EFFORTS

The main purpose of a validation test of EROS is to verify the adequacy of the system by demonstrating that the system provides proper action guidances as per written procedures as well as helps the operators mitigate the consequences of the incidents.

In real situations, guideline E-0 is entered when a reactor is tripped or safety injection is initiated. At the initial stage up to step fourteen, operators have only to verify automatic actuation of reactor trip, safety injection, main feedwater pump trip, auxiliary feedwater initiation, main steam isolation, and so on. If any of these systems fails to work properly, the operators are required to take corrective actions immediately.

In EROS, this situation is carried out by use of working memory elements named 'autostart' and its corresponding fourteen attribute-value pairs which represent the status of auto-actuation of these safety related systems. For verification purposes, the attribute-value pairs are initiated as working memory data in the system.

A full scale plant simulator is utilized for validation tests against two design basis accidents; large loss of coolant and steam line break accidents. In addition, capability for checking CSFs and advising steps of function restoration has successfully been verified within the sub-programme. As an example, a series of large loss of collant accident (LOCA) procedures is simulated that requires optimal recovery in the following sequential procedures:

E-0	(Reactor trip or safety injection)
E-1	(Loss of reactor coolant)
ES-1.3	(Transfer to cold leg recirculation)
ES-1.4	(Transfer to hot leg recirculation).

The minimum number of operator decision steps necessary to achieve optimal plant end-states from the beginning of a large LOCA is about forty-four steps, while it is about thirty-seven steps for steam line break accident. For the first fourteen steps from the entry of E-0, the information provided by the EROS gives earlier responses than operators do in a simulated environment. The real time capability is, therefore, achievable by augmenting on-line monitoring of plant variables and component status. Following step fourteen of E-0 throughout the event sequences, EROS questions the operators about the information needed. This information is screened in order to avoid the operator's misunderstanding of a situation. Transfer to long-term recirculation phase (ES-1.4) is designed to be effective when the explicit entry criteria are satisfied. EROS responds to these accidents in appropriate fashion. In addition, the system exhibits capability of providing timely advisory information corresponding to the status of simulated transient. The tests reveal that the prototype expert system can offer proper guidelines to the operators in a clear and concise way under even a situation of stress. However, real time capability is not yet verified when the system will have thousands of production rules, as will be the case of enlarged, full scale ERGs expert system.

Future work will include knowledge extension and verification that the EROS performs generally with desired accuracy in respect of the real environment.

5. CONCLUSIONS

EROS is a prototype system currently under development. The following conclusions may be drawn from this study:

(a) Computerized ERGs, as an expert system, are good tools for enhancing the operational safety of an NPP. This can effectively reduce human errors in accident diagnosis.

(b) It is desired to refine and simplify the logic steps of ERGs that facilitates implementation of surface knowledge of emergency operations to rule-based expert systems.

(c) Capability of real time performance is important in designing expert systems for plant diagnostics. This can be achieved by utilizing more efficient tools than the Korea Advanced Institute of Science and Technology official production production system and, to a great extent, by directly monitoring plant variables through signal-to-symbol transformation techniques.

(d) In addition, automated interaction between EROS and NPPs are also desired in order that the operators avoid problems caused by misunderstanding of the plant status.

(e) Extension and generalization of knowledge bases of EROS should be performed to validate its adequacy.

(f) Transparency capability will be added so that the operator can confirm the conclusion drawn by EROS.

BIBLIOGRAPHY

FORGY, C.L., in OPS5 User's Manual, Carnegie-Mellon University, Pittsburgh, PA (1981).

HAYES-ROTHS, F., et al., Building an Expert System, Addison-Wesley, Reading, MA (1983).

MONTY, S.R., et al., "Symptom-based guidelines for response to emergency transients – the developmental process and necessary elements", Proc. 2nd Int. Top. Mtg on Nuclear Power Plant Thermal Hydraulics, Tokyo, 1986.

NELSON, W.R., "Reactor: an expert system for diagnosis and treatment of nuclear reactor accidents", Proc. Natl Conf. on Artificial Intelligence, CONF-820874-1, Idaho Natl Eng. Lab., Idaho, ID (1982).

SHARMA, D.W., et al., Design of an artificial intelligence system for safety function maintenance, Trans. Am. Nucl. Soc. **50** (1985).

STELLA, M.E., et al., "The Westinghouse emergency response guidelines considered as an expert system", Proc. 2nd Int. Mtg on Nuclear Power Plant Thermal Hydraulics, Tokyo, 1986.

WESTINGHOUSE OWNER'S GROUP, Emergency Responsive Guidelines, Rev. 1, Pittsburgh, PA (1983).

ROLE DE L'IPSN DANS LA PREPARATION AUX SITUATIONS D'URGENCE ET DANS LEUR GESTION

C. CHAULIAC, P. FACHE, J.-M. EVRARD
Institut de protection et de sûreté nucléaire (IPSN),
Commissariat à l'énergie atomique,
Fontenay-aux-Roses, France

Abstract–Résumé

THE ROLE OF IPSN IN PREPARATIONS FOR EMERGENCY SITUATIONS AND IN THEIR MANAGEMENT.

Within the French organization for safety, the Institute for Protection and Nuclear Safety (IPSN) is responsible for providing technical assistance to the public authorities. With respect to the state of preparation of the means for coping with emergency situations, this technical support is centred, in particular, around the three following axes: (1) safety analysis of emergency plans; (2) incorporation in the nationwide crisis organization; (3) improving the means of evaluating and forecasting accident situations. These three subjects are discussed in greater detail, particular emphasis being placed on the first and third.

ROLE DE L'IPSN DANS LA PREPARATION AUX SITUATIONS D'URGENCE ET DANS LEUR GESTION.

Au sein de l'organisation française de la sûreté, l'Institut de protection et de sûreté nucléaire (IPSN) est chargé d'apporter un soutien technique aux pouvoirs publics. En ce qui concerne l'état de préparation des moyens permettant de faire face aux situations d'urgence, ce soutien technique s'articule en particulier autour des trois axes suivants: 1) l'analyse de sûreté des plans d'urgence; 2) l'insertion dans l'organisation nationale de crise; 3) l'amélioration des moyens d'évaluation et de prévision des situations accidentelles. Ces trois sujets sont développés, un accent particulier étant mis sur le premier et le troisième.

1. INTRODUCTION

L'accident du réacteur américain de Three Mile Island, en 1979, a bien mis en évidence la difficulté de gérer une situation de crise lorsqu'une installation de technologie complexe, comme l'est un réacteur nucléaire, est en jeu: difficulté de l'établissement du diagnostic, difficulté de prévoir l'évolution de l'accident, nécessité d'intervenir le plus tôt possible mais à bon escient, nécessité d'informer les pouvoirs publics pour qu'ils puissent mobiliser leurs moyens d'intervention afin d'assurer la protection des personnes et des biens, et, enfin, difficulté de compréhension des intervenants nécessitant de «traduire» les événements depuis le langage technique vers celui des décideurs.

Plus récemment, l'accident de Tchernobyl a montré les difficultés de tous ordres qui se posent sur un site gravement contaminé, et en particulier les problèmes médicaux de traitement et de tri des personnes irradiées.

Ces deux exemples montrent à l'évidence que les accidents, qu'ils soient à cinétique relativement lente ou au contraire rapide, nécessitent une mobilisation immédiate et coordonnée d'un grand nombre de personnes dans des domaines très divers. Seuls des plans d'intervention définissant à l'avance les rôles et missions de chacun en cas de crise peuvent permettre, dans la première phase de l'accident, la mise en place la plus efficace possible des moyens de secours.

En général, les moyens locaux s'avèrent très rapidement insuffisants et il est alors nécessaire de les relayer par des moyens en hommes et en matériel rassemblés au niveau national.

Ainsi est mise en évidence la nécessité de disposer au niveau local de plans d'intervention tandis qu'une organisation nationale est mise en place pour accroître l'efficacité des intervenants.

L'organisation française en cas de situation d'urgence dans une centrale nucléaire est présentée ci-après. Le rôle tenu dans ce cadre par l'Institut de protection et de sûreté nucléaire (IPSN) du Commissariat à l'énergie atomique (CEA), qui sert de soutien technique aux pouvoirs publics, est décrit dans trois domaines où il développe son action:

- l'analyse de sûreté des plans d'urgence:
- la participation à l'organisation nationale de crise;
- l'amélioration des moyens d'évaluation des situations accidentelles et de prévision de leur évolution, destinés à être utilisés par les spécialistes de son Centre technique de crise (CTC).

2. ORGANISATION DE CRISE AU NIVEAU LOCAL: LES PLANS D'URGENCE INTERNE ET LEUR ANALYSE DE SURETE

Lorsqu'un accident grave risquant de conduire à terme à une nuisance radiologique pour la population se produit sur un réacteur, c'est d'abord au niveau local qu'une organisation de crise doit se mettre en place pour faire face aux événements.

2.1. Distribution des rôles à l'échelon local

Il faut distinguer ce qui est du ressort de l'exploitant et ce qui relève des pouvoirs publics.

L'exploitant met en place un Plan d'urgence interne (PUI) dont le but est de lui permettre de substituer à l'organisation habituelle une organisation de crise. Dans cette nouvelle organisation chacun connaît à l'avance les tâches et les responsabilités qui lui incombent.

La plaque tournante de cette organisation est le PC (Poste de commandement) Direction. A ce PC sont rattachées des unités opérationnelles: le PC local, qui regroupe l'équipe de conduite dans la salle de commande, le PC Mouvement, chargé d'assurer la logistique du site (gardiennage, balisage, matériels) et le PC Contrôle, qui centralise les mesures radiologiques et en assure la synthèse.

A cette organisation de l'exploitant répond celle des pouvoirs publics qui mettent en place un Plan particulier d'intervention (PPI) avec un PC fixe établi par le Préfet à la préfecture. Il rassemble les différents responsables des représentations départementales des ministères ainsi que des délégués de l'exploitant et des services centraux des ministères (Industrie, Santé). Un PC opérationnel est établi au plus près du site.

2.2. Objectif et contenu des PUI

Le PUI permet à l'exploitant de faire face aux conditions particulières d'exploitation en cas de situation d'urgence. Ses objectifs sont les suivants:
– replacer l'installation dans un état sûr;
– limiter les conséquences de l'accident sur les biens et les personnes;
– collecter toutes les informations permettant de juger de l'ampleur de l'accident et d'estimer son évolution en s'appuyant sur l'organisation de crise;
– informer les pouvoirs publics.

Chaque PUI est rédigé selon une maquette élaborée au niveau national par l'exploitant à partir d'un plan type conçu par le service concerné du Ministère de l'industrie (Service central de sûreté des installations nucléaires) avec l'appui technique de l'IPSN.

Il comprend cinq parties consacrées aux sujets suivants:

– description des dispositions pré-accidentelles (situations retenues, moyens logistiques, etc.);

– présentation des missions dévolues aux agents ayant à tenir un rôle actif dans l'application du plan, des documents d'application spécifiques à chaque mission élémentaire, des abaques d'évaluation du risque radiologique, etc.; c'est ici le corps opérationnel du document;

– méthode à suivre pour tirer les enseignements de l'accident (rapports à élaborer par l'exploitant, diffusion qui doit en être faite, etc.);

– justification du bien-fondé des dispositions prises, tant sur la forme que sur le fond;

– informations et consignes générales au personnel impliqué par la mise en place du PUI mais ne tenant pas un rôle actif dans ce cadre.

Ce plan type peut être adapté pour tenir compte des conditions propres à chaque site: sites REP (réacteurs à eau sous pression), sites UNGG (graphite-gaz), sites RNR (réacteurs à neutrons rapides), sites multi-exploitants nucléaires (Tricastin) ou dont l'environnement n'est pas uniquement nucléaire (Gravelines), sites à situation frontalière, etc.

Mais, malgré cette apparente diversité, la forme et le contenu des PUI sont simplifiés par deux caractères spécifiques de la situation nucléaire française: l'unicité de l'exploitant (qui implique une certaine cohérence entre sites) et la prédominance de la filière REP (à l'intérieur de laquelle les réacteurs d'un même palier sont homogènes).

2.3. Analyse de sûreté des PUI

La description précédente montre que la place et le rôle du PUI au sein de l'organisation de crise sont primordiaux et c'est ce qui lui vaut l'attention particulière des autorités de sûreté.

Dans ce cadre, l'ensemble des PUI fait l'objet d'une analyse détaillée de la part de l'IPSN. Le but de cette analyse est de s'assurer que la mise en place du PUI qui, «dans le feu de l'action», conduit à une modification de l'organisation habituelle du site, va permettre de gérer la situation accidentelle dans les meilleures conditions possibles. Parmi les thèmes clés de réflexion qui se sont dégagés au cours d'une récente analyse globale de ces plans d'urgence, on relève en particulier les points suivants:

a) *Détermination des seuils de déclenchement des PUI:* on souhaite actuellement privilégier des critères basés sur l'entrée dans certaines procédures accidentelles car ces critères sont moins subjectifs que ceux basés sur des niveaux de rejet qui étaient utilisés auparavant. Le choix de cette méthode suppose qu'une liste exhaustive des procédures entraînant le déclenchement des PUI soit établie pour chaque filière.

b) *Domaine des situations accidentelles couvertes par le PUI:* il doit inclure les accidents hors dimensionnement, notamment les accidents conduisant au rejet de type S3 [1]. Les conséquences de ce type d'accidents sur le site doivent être prises en compte (circulation sur le site, habitabilité des locaux, etc.).

c) *Adéquation entre situations accidentelles et logistique* (locaux, équipes, matériels, moyens de mesure, télécommunications, transmissions de données): le contenu du PUI est comparé à une grille d'évaluation construite à partir de la maquette présentée sous 2.2.

d) *Caractère opérationnel du PUI:* il s'agit de s'assurer que l'organisation prévue permet de faire face correctement aux situations accidentelles. Ceci doit être vérifié notamment au cours d'exercices.

e) *Cohérence entre PUI et PPI:* les informations transmises dans le cadre du PUI doivent permettre une correspondance directe avec les niveaux d'intervention du PPI et les actions de chacun des deux plans doivent être cohérentes.

3. ORGANISATION DE CRISE AU NIVEAU NATIONAL ET MOYENS CORRESPONDANTS

3.1. Organisation nationale

Il est clair que l'échelon local est seul à pouvoir faire face, dans un premier temps, à un accident conduisant, à plus ou moins long terme, à des rejets radioactifs notables dans l'environnement. Si l'accident présente des risques de rejets radioactifs notables, il est intéressant que la structure locale puisse être épaulée par une structure au niveau national avec un découpage similaire à celui de l'échelon local [2]. Des PC Direction sont établis au siège de l'exploitant, au centre opérationnel de la sécurité civile, au Service central de protection contre les rayonnements ionisants, au Ministère de l'industrie.

A des centres de décision, s'ajoute le «cercle de réflexion» représenté par les différentes équipes de crise qui sont constituées aussi bien au niveau local, sur le site, qu'au niveau national, chez l'exploitant, avec l'appui du constructeur de la chaudière nucléaire, ainsi qu'à l'IPSN agissant en tant qu'appui technique du Ministère de l'industrie avec la mission suivante:

- évaluer le risque de rejet de produits radioactifs dans l'environnement;
- prévoir l'évolution probable de la situation à court et moyen terme;
- évaluer les conséquences radiologiques potentielles notamment pour la population;
- formuler éventuellement des propositions pour l'application de contre-mesures par les autorités locales à partir de la prévision des rejets.

Enfin, bien que l'IPSN n'ait pas pour rôle d'intervenir dans la conduite de l'installation, il serait amené à donner plus formellement son avis sur certaines actions de l'exploitant, en particulier en cas de rejet concerté.

Pour assurer sa tâche, l'IPSN a mis en place, au Centre d'études nucléaires de Fontenay-aux-Roses, un centre technique de crise [3] dont la mise en fonctionnement est prévue avec un système d'astreinte. Ce centre est équipé de moyens de télécommunications et disposera prochainement de la retransmission télématique d'un nombre important de paramètres logiques et analogiques de fonctionnement en provenance du calculateur du réacteur accidenté. En outre, ce centre dispose de moyens pour évaluer l'état de l'installation, prévoir l'évolution de cet état et évaluer les conséquences radiologiques potentielles, notamment pour la population, en particulier grâce à des abaques en cours d'informatisation.

3.2. Amélioration des moyens d'évaluation et de prévision des situations accidentelles en cas de crise

Pour que les équipes nationales de crise puissent agir avec efficacité, elles doivent comporter des spécialistes compétents, expérimentés et familiers des installations et formés aux situations de crise. Compte tenu de la complexité des installations nucléaires et des situations auxquelles il faut faire face, ces spécialistes doivent être secondés par des outils d'aide à l'évaluation.

Ces outils doivent être adaptés aux situations et en particulier répondre aux exigences suivantes:
- rapidité de mise en œuvre,
- facilité et souplesse d'utilisation,
- exploitation optimale des informations disponibles,
- prise en compte des différentes incertitudes (mesure, modélisation physique, paramètres non accessibles, etc.).

A l'IPSN, c'est le but du projet SESAME (Schéma d'évolution des situations accidentelles et méthodes d'évaluation) d'améliorer ce type d'outils.

La démarche générale de ce projet comporte deux niveaux:

1) Au premier niveau, les grandeurs physiques mesurées sur la centrale et retransmises vers le CTC sont évaluées à l'aide de corrélations. Cette étape permet une interprétation des mesures et une évaluation des grandeurs caractéristiques concernant aussi bien le cœur (et les autres sources potentielles d'activité) que le confinement. La comparaison de ces grandeurs à des seuils significatifs permet de situer l'état de la centrale au sein d'une grille des états physiques et de répondre au premier objectif qui est le diagnostic de l'ampleur de l'accident.

2) Le deuxième niveau concerne le pronostic sur l'évolution de l'accident. Il intègre la prise en compte des actions des opérateurs et de la disponibilité des systèmes de sauvegarde et, éventuellement, des hypothèses sur l'origine de l'accident. Des corrélations permettent alors une prévision quantifiée de l'évolution à moyen terme de l'installation, en particulier des rejets possibles de radioactivité.

Les grandes lignes de cette démarche font apparaître trois rubriques pour lesquelles le projet SESAME est en charge d'améliorer le degré opérationnel des connaissances:

1) L'évaluation et la prévision de l'état du circuit primaire et du cœur. Cette étude utilise comme bases l'approche par états [4], des études paramétriques à l'aide de codes existants et diverses corrélations. Elle est étendue au combustible de la piscine de stockage du bâtiment du combustible.

2) L'évaluation et la prévision de l'état du confinement qui comportent une estimation de la radioactivité dans l'installation, une détermination des chemins de fuite et une quantification des rejets.

3) L'évaluation et la prévision de la disponibilité des systèmes de sauvegarde. Cette étude nécessite entre autres la connaissance de la disponibilité des sources électriques. Pour ce faire, un système expert gère les données en provenance de la centrale et signale les défaillances présentes ou prévisibles à terme.

Au fur et à mesure de la fourniture par SESAME des outils d'aide à la réflexion pour ces différents thèmes, ils seront testés dans le cadre d'exercices de crise. Leur mise à l'épreuve régulière permettra de les compléter, de les assouplir et de les améliorer.

4. CONCLUSION

L'IPSN agit comme soutien technique des pouvoirs publics en situation de crise dans une centrale nucléaire.

Avant la crise l'IPSN intervient de deux manières:

1) *En faisant l'analyse du PUI:* en particulier en veillant aux règles qui président à son déclenchement, au domaine de situations qu'il couvre, aux moyens dont la disponibilité est prévue, à son degré opérationnel et à sa compatibilité avec le PPI.

2) *En améliorant les outils de réflexion* nécessaires à l'évaluation des situations de crise. Cette démarche s'appuie sur une synthèse opérationnelle des connaissances relatives à l'évaluation et la prévision de l'état du circuit primaire et du confinement et de la disponibilité des systèmes de sauvegarde.

Pendant la crise, l'IPSN fournit son appui technique au Ministère de l'industrie au travers de son Centre technique de crise, grâce aux outils dont celui-ci dispose, et est en relation permanente avec les équipes de réflexion de l'exploitant. Le fonctionnement de l'organisation est régulièrement testé au cours d'exercices de crise.

REFERENCES

[1] L'HOMME, A., DUCO, J., «French current practice for source term», ANS Topical Meeting on Fission Product Behavior and Source Term, Snowbird (1984).

[2] CARTIGNY, J., MAJOREL, Y., «Organisation des pouvoirs publics en France en cas d'incident ou d'accident concernant la sécurité nucléaire», Emergency Planning and Preparedness for Nuclear Facilities (C.R. Coll., Rome, 1985), AIEA, Vienne (1986) 461.

[3] SCHERRER, J., EVRARD, J.-M., NEY, J., «Organisation technique des autorités de sûreté en cas d'accident dans une installation nucléaire», ibidem, 471.

[4] SUREAU, H., et al., «Application de l'approche par états à EDF: les procédures SPI, SPU, UI», Operating Procedures for Abnormal Conditions in Nuclear Power Plants (Journées d'étude AIEA, Munich, 1986).

IAEA-CN-48/286

USE OF THE GRADED RESPONSE IN EMERGENCY PLANNING

H. SPECTER
New York Power Authority,
New York, N.Y., United States of America

Abstract

USE OF THE GRADED RESPONSE IN EMERGENCY PLANNING.

In order to minimize the offsite health consequences of a severe nuclear accident it is important to have an efficient emergency response. One strategy to respond to large releases of radioactive material is to have a massive evacuation of citiziens within a circular area (360°) surrounding the plant, with a radius up to 16 km (10 miles). Studies presented in the paper show that an alternative to massive evacuation, i.e. the graded response, is simpler, highly conservative, and much more efficient than massive evacuation plans. The graded response almost totally uses sheltering, with selected relocation of a small percentage of the population near the plant. It is expected that no offsite early fatalities would occur if a graded response was fully implemented, yet no credit is taken for reduced source terms or the low probability of a release occurring. This type of emergency response is applicable to all water reactor types and to all sites. The paper compares sheltering to evacuation, displays the early fatality risk versus distance from the point of release, explains why overly large emergency planning zones can actually increase public risks, and provides the results of an analysis of a specific high population site which compares a massive evacuation response to the graded response. In this latter analysis use of the graded response was calculated to lower the early fatality risk by a factor of 25 000 relative to a massive evacuation response.

Emergency planning at some nuclear power plants has become a focal point for confusion and conflict.

In the United States a large number of new laws on emergency planning have been proposed at state and federal levels, certain operating plants have had their licenses challenged by this issue, and two recently completed power plants have yet to begin power operation because of emergency planning concerns.

Part of this conflict on emergency planning can be traced to a misconception about what actions are essential for survival in the event of a major nuclear accident. Many believe that a prompt massive evacuation - i.e., rapid evacuation out of 16 or more kilometers in every direction (360°), must be done without hesitation to avoid becoming a fatality. One can speculate that some people incorrectly equate nuclear power plant accidents to a rapidly expanding fireball of an atomic bomb explosion, and fear that they would be quickly engulfed. The widespread dispersal of radioactive material from the Chernobyl accident has been misinterpreted as a need to promptly evacuate in all directions out to considerable distances to avoid loss of life.

In some cases where nuclear power plants have been sited in areas where the neighboring populations are large, some citizens have opposed the operation of these plants. Believing that massive evacuation is necessary, yet concerned that such a response cannot be quickly and effectively implemented in congested areas, they understandably resist plant operation. It is therefore important to examine whether or not such a massive evacuation is the optimum response to a severe nuclear accident.

Studies presented in this paper show that an alternative to massive evacuation, i.e., the graded response, is simpler, conservative, and much more efficient. The graded response almost totally uses sheltering, with selected evacuation of a small percentage of the population near the plant. Sheltered people in the plume deposition area would be relocated within a few hours after plume passage. It is expected that few or no offsite early fatalities would occur if a graded response was fully implemented, yet no credit is taken for reduced source terms or the low probability of a release occurring. This type of emergency response is applicable to all water reactors and all sites. Thus it can be shown that there is room for agreement between the nuclear technical community and those who resist nuclear power because of emergency planning concerns. Both groups, for different reasons, reject the use of prompt, massive evacuations.

Nuclear power plant accidents can cause onsite and offsite economic and health consequences. Emergency planning's greatest potential benefit is in reducing offsite health effects. Offsite health effects from nuclear accidents can be divided into two broad categories, latent health effects and early health effects. Societal latent health effects principally arise from extended exposure to very low levels of highly dispersed radioactive material, assuming interdiction actions have been taken to minimize food chain effects. Even the most severe release of radioactivity from a nuclear power plant accident would only increase the societal latent fatality rate by a small increment, probably less than 1%. This small increment would likely be less than the year-to-year statistical variation in the normal societal latent fatality rate and would therefore be too small to be detected. Most important, keeping this societal latent fatality increment to such a small value <u>does not</u> depend upon taking emergency protective actions, such as sheltering or evacuation, during the time a radioactive plume is being released from a damaged plant.

Certain people near the point of release may have individual increases in their latent fatality risk which would be higher than the societal value. Their individual latent fatality risks might increase by approximately 1% per 100 rads of whole body exposure[1]. Emergency responses taken to reduce the early fatality risk, described later,

[1] 1 rad = 1.00×10^{-2} Gy.

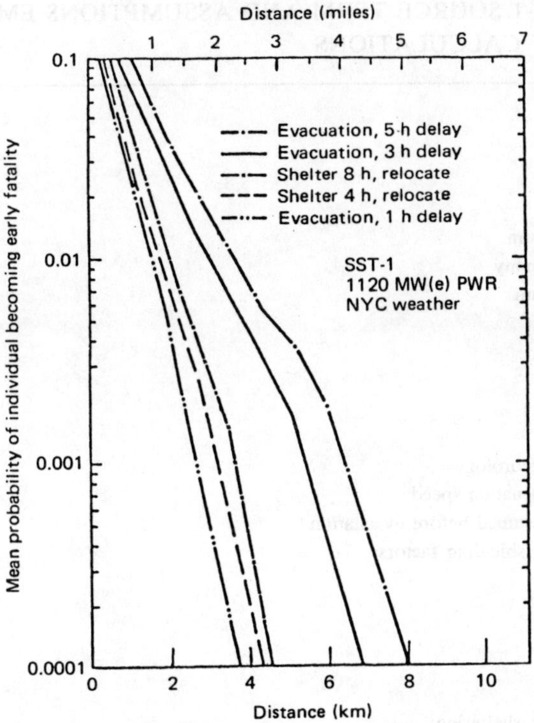

FIG. 1. Mean probability of individual becoming early fatality versus distance (SST-1). Source: Issues, Information Needs, and Programs for Improved Emergency Preparedness (Nov. 1984).

would also be useful in limiting this individual latent risk. In any case, an individual's latent fatality risk cannot increase by more than several percent from a nuclear accident.

Since latent fatality risks are inherently small, then it is the early health effects that could be most profoundly affected by emergency planning. Early fatality and societal latent fatality risks have opposite geographic and temporal distributions. Whereas the societal latent fatality risk is highly dispersed and radiation exposure would accumulate slowly over many years or even decades, the early fatality risk is geographically very short ranged and exposure time periods of even a few hours can be important. Therefore in order to develop an emergency response strategy to minimize early health effects, particularly the early fatality risk, it is important to understand how this risk is distributed as a function of distance, the importance of time of exposure, and the relative merits of evacuation versus sheltering. Much of this information can be found in figure one which plots the mean individual early fatality risk versus distance from the point of release for a very large source term,

TABLE I. SST-1 SOURCE TERM AND ASSUMPTIONS EMPLOYED FOR CONSEQUENT CALCULATIONS

A. SST-1

Group element	*Release fraction*
Xenon and krypton	1.0
Iodine	0.45
Caesium and rubidium	0.67
Tellurium and antimony	0.64
Barium and strontium	0.07
Ruthenium	0.05
Lanthanum	0.009

B. Assumptions

1120 MW(e) PWR
New York City meteorology
16 km per hour evacuation speeds
Normal activities assumed before evacuation with the following shielding factors:

	Factors
Cloudshine	0.75
Groundshine	0.33
Inhalation	1.00
Shielding factors for sheltering:	
Cloudshine	0.50
Groundshine	0.08
Inhalation	0.50

SST-1, [See Table 1]. A uniform population distribution was assumed in this analysis. Three of the lines in figure one are for evacuation responses where the time to initiate evacuation differs from case to case. The other two cases examine sheltering, one for four hours of sheltering and the other for eight. The four and eight hour sheltering periods were assumed to begin at the time of plume passage and end when people were relocated out of the plume deposition area.

Several conclusions can be drawn from these analyses:

A. The individual mean early fatality risk decreases very rapidly with distance. An evacuee (with a 1 hour delay time[2]) initially at two miles[3] from the point of

[2] Delay time is defined as the time between initiation of a public warning and the time when people begin to evacuate. In these analyses the warning time was assumed to occur 0.5 hour prior to the release of radioactive material. Thus in a 3 hour delay time, people would begin evacuation 2.5 hour after the release of radioactivity began.

[3] 1 mile = 1.6 km.

release has a risk several thousand times smaller than an evacuee at the point of release (on site). The Chernobyl experience is consistent with this analysis. All early fatalities and early injuries were due to onsite exposure.

B. Timing is important. There is about a 50 fold increase in the individual mean early fatality risk when the evacuation delay time is increased from 1 to 5 hours (for evacuees initially at two miles).

C. Sheltering, followed by relocation out of the contaminated areas within a reasonable time period, is an effective response to large releases. Sheltering is more effective than evacuation, except when evacuation has a short delay time or occurs prior to release of radioactive material.

D. Increasing the size of the Emergency Planning Zone (EPZ) could actually increase early fatality risks. There is virtually no contribution to the mean early fatality risk beyond 16 kilometers (10 miles). If even more massive evacuations (larger EPZ's) were utilized, the evacuation process could be slowed down, possibly to the detriment of those nearest the plant.

The next step is to apply the insights gained from figure one to an actual site where the population distribution is non-uniform. A U.S. Nuclear Regulatory Commission (NRC) study of the effectiveness of different emergency response strategies at a high population site is instructive. Table 2 reports NRC results which compare a massive evacuation strategy to a graded response strategy. Massive evacuation is modeled as 30%, 40% and 30% of the population evacuating with 1, 3, and 5 hour delay times, respectively. It is assumed that prior to evacuation the population did not respond to the accident and acted as they normally do, i.e, about 85% of the time indoors and 15% outdoors. In the graded response strategy, greatest emphasis is given to those closest to the plant to have them take an emergency response as soon as possible. The NRC's graded response analysis assumed a timely evacuation of an area, 3.2km in radius, immediately next to the plant. This inner zone was surrounded by an outer zone where sheltering was assumed. Those sheltered people not in the plume deposition area would be radiologically unaffected by the accident. Sheltered people in the plume deposition area were assumed to stay sheltered for four hours after the plume passed and then relocated out of the contaminated area. Except for the differences in these emergency response strategies, all other aspects of these NRC analyses - meteorology, demography, the computer codes used, etc. were identical.

Assuming an SST-1 source term, these NRC analyses calculate a 25000 fold reduction in the mean early fatality risk [32.8 vs. 0.0012] when the graded response is utilized instead of massive evacuation. There are also large

TABLE II. UNITED STATES NUCLEAR REGULATORY COMMISSION ANALYSIS

		Mean	SST-1 99%	Peak		Mean	SST-1 99%	Peak
Massive evacuation[a]	Fatalities	32.8	1100	2160	Fatilities	8.64	200	2100
	Injuries	191	2700	8090	Injuries	55.8	1700	3280
Graded response[b]	Fatilities	0.0012	1.0	7.8	Fatilities	0	0	0
	Injuries	15.5	570	1200	Injuries	2.8	120	373

[a] In a massive evacuation model 30%, 40%, and 30% are assumed with 1, 3 and 5 hours delay times, respectivley.

[b] Inner 3.6 km (two miles) is evacuated promptly, with sheltering in the 3.6 to 16 km (two to ten miles) annulus. If release occurs, shelterees in the plume deposition area would be located within four h. Relocation distance usually less than 2 km.

Note: Adapted from A Perspective on Emergency Planning, Risk and the Source Term Issue, J. Martin, United States Nuclear Regulatory Commission, 13 April (1983).

reductions in the maximum (or peak) number of early fatalities and fewer early radiation injuries. The graded response, which focuses the emergency workers on those who are truly at risk, is therefore far more efficient than a slow massive evacuation. These same NRC analyses calculated that, when combined with a very small reduction in the SST-1 source term (a factor of two, but not reducing the noble gas releases), zero early fatalities are expected when a graded response is fully used.

Several other useful observations can be made about the graded response. The inner, prompt action zone[4], encloses only 4% of the area of the whole EPZ (assumed to be 16 kilometers in radius in these cases). This inner area always contains the site, almost always a body of water, and usually has a low population density. As such, it is likely that less than 2% of the total EPZ population would ever require prompt evacuation in a graded response approach. The other 98% or more of the EPZ population would be asked to take shelter. Evacuees could be moved to a nearby shelter, such as a school or multistoried building, where the shielding is good. Therefore prompt evacuation involves moving just a small number of people a short distance.

[4] Local site characteristics would probably result in an inner zone that is not a perfect circle.

The NRC analyses and figure one are based on a short duration or "puff" release of radioactivity. A much more likely situation would be that a release would occur over an extended period of time, during which time wind directions are likely to shift.
This would mean that a larger area would be affected, but at a lower level of radiation. This would require that a somewhat greater number of sheltered people be relocated. On the other hand, the risk of early fatalities and early injuries would be smaller than the puff release case and more time would be available to implement relocation out of this somewhat larger area.

In the unlikely case where the plume did not meander, the plume width would be very narrow, with a center-to-edge distance of less than 2 km at the assumed 16 km radius of the EPZ, and even more narrow closer to the plant. Therefore relocation distances and times to cross the contaminated zone would be very short. Shelterees would only have to travel short distances, down streets they are familiar with, to reach safety.

A conservatism in these analyses is the assumption that the plume was cold and therefore non-buoyant. This assumption results in a higher calculated concentration of radioactive material, especially in the 3 or 4 km zone nearest the plant, than heated or buoyant plumes. However, accident analyses show that the class of very rare accidents which are characterized by prompt failure of the containment, are also likely to be very energetic. Prompt containment failure accidents, which tend to minimize the time to take emergency actions, have a compensating feature of having heated plumes which reduce radiation levels near the plant. Lower radiation levels enhance the use of sheltering, because it takes longer to accumulate a particular level of exposure. Preliminary, recent, studies of heated plumes indicate that the inner zone of a graded response should be even smaller than the assumed 3.2 kilometers (2 miles). This would result in <u>an even greater, or perhaps total, use of sheltering</u>. Thus, a sheltering response is particularly effective in dealing with severe accidents where the containment fails promptly.

The subject of evacuation warrants further discussion. An interesting analysis about evacuation was made several years ago. If a person evacuated radially away from the plant at walking speed, about 5 km/hr, he would not become an early fatality, even with an SST-1 release. This is so, provided that evacuation begins at or before the radioactive material enters the environment. This previous analysis is consistent with figure one and represents an evacuation response to the left of the leftmost evacuation line plotted in figure one. Therefore, an evacuation response can also result in an expected value of zero early fatalities providing evacuation begins earlier than the time of containment failure.

Three major conclusions can be reached:

1. Anyone who takes a timely emergency response, be it sheltering or evacuation, is very unlikely to become an early fatality.

   Virtually all of the early fatality risk is disproportionately associated with that small element of society that does not take a timely response. In the NRC massive evacuation analysis (See Table II) most of the risk is associated with that group of people who both delayed their evacuation for five hours and did not take continuous sheltering once the release of radioactive material had begun.

2. A graded response approach (almost all-sheltering) should be the basis for emergency planning.

   Since use of a graded response approach would be highly effective even in the more demanding prompt containment failure case, it should be the basis for emergency planning. Sheltering can be implemented very quickly in most instances.

   If a core damage accident did occur and containment failure were delayed, then evacuation of selected areas near the plant could be used. With the graded response as the framework for the emergency planning and recalling that low evacuation speeds are sufficient in the delayed containment failure case, then evacuation of these selected areas could be implemented on an *ad hoc* basis.

3. The graded response is simple, conservative, and more efficient than massive evacuations.

ЭКСПРЕСС-МЕТОДИКА ОПРЕДЕЛЕНИЯ РАДИАЦИОННЫХ ПОСЛЕДСТВИЙ АВАРИЙНЫХ СИТУАЦИЙ НА АТОМНЫХ СТАНЦИЯХ

Л.А. ИЛЬИН, О.А. ПАВЛОВСКИЙ
Институт биофизики Министерства здравоохранения СССР,
Москва,
Союз Советских Социалистических Республик

Abstract–Аннотация

HIGH SPEED METHODS OF DETERMINING THE RADIATION CONSEQUENCES OF ACCIDENT SITUATIONS AT NUCLEAR POWER PLANTS.

By means of computerized calculations of the radiation consequences of accidents of various degrees of severity at nuclear plants with water cooled, water moderated power reactors (WWER), high power vessel type reactors (RBMK) and fast neutron reactors (BN) a number of simple relationships have been obtained linking the basic radiation parameters within the radioactive plume zone after an accidental release (radioactivity of air, soil and vegetation, external and internal doses to humans) to the measured or calculated gamma radiation dose rates in the contaminated area. Practical application of the high speed methods is shown in the real case of eliminating the consequences of the Chernobyl accident.

ЭКСПРЕСС-МЕТОДИКА ОПРЕДЕЛЕНИЯ РАДИАЦИОННЫХ ПОСЛЕДСТВИЙ АВАРИЙНЫХ СИТУАЦИЙ НА АТОМНЫХ СТАНЦИЯХ.

На основе расчетов на ЭВМ радиационных последствий аварийных ситуаций различной степени тяжести на АЭС с реакторами ВВЭР, РБМК и БН получены ряд простых соотношений, связывающих основные параметры радиационной обстановки в зоне радиоактивного следа аварийного выброса (радиоактивность воздуха, почвы и растительности, дозы внешнего и внутреннего облучения людей) с измеренным или рассчитанным значением мощности дозы гамма-излучения на загрязненной территории. Показано практическое использование данной экспресс-методики в реальных условиях ликвидации последствий аварии на Чернобыльской АЭС.

Отечественная и зарубежная практика эксплуатации атомных станций показывает, что при их нормальной работе средние значения индивидуальной дозы облучения всего тела, а также наиболее значимых "критических" органов и тканей организма, обусловленные радиоактивными газоаэрозольными выбросами, для лиц, проживающих в несколько километрах от АЭС, не превышают единиц и даже долей процента от дозы естественного фона [1–3].

Наибольшую радиационную опасность для населения может представить авария на АЭС. Именно в этом случае возникает потенциальная опасность облучения населения высокими дозами и на обширных территориях. Это служит основной причиной широкораспространенной как у нас в стране, так и за рубежом практики размещения мощных АЭС в малонаселенной местности и на значительных расстояниях от крупных городов.

Однако в последнее время отчетливо проявляется тенденция приближения АЭС, а также АТЭЦ и АСТ к местам потребления. Поэтому повышаются требования к оценке возможной радиационной опасности при аварийных выбросах радиоактивности.

Очевидно, что с возрастанием масштаба аварии резко возрастает техническая сложность и, соответственно, стоимость защитных мероприятий по обеспечению безопасности населения. Поэтому в каждом конкретном проекте АЭС устанавливается перечень нарушений, радиационные последствия которых могут быть достаточно надежно локализованы предусматриваемыми техническими мерами. Такой класс аварий носит название "проектных аварий" (ПА). Под наиболее крупной аварией из этого класса – "максимальной проектной аварией" (МПА) – подразумевается "проектная авария с наиболее тяжелым исходным событием, устанавливаемым для каждого типа реакторов" [4]. В качестве примера можно показать, что для АЭС первого поколения (типа III- го блока НВАЭС) – МПА – это разрыв трубопровода диаметром до 100 мм, хотя на АЭС есть трубопроводы и больших размеров. Для АЭС второго поколения МПА значительно выше – разрыв основного трубопровода первого контура с одновременным истечением теплоносителя из обоих концов (ДУ-500 и ДУ-850 для АЭС с ВВЭР-440 и ВВЭР-1000 соответственно) [5].

Естественно, что при возникновении аварийной ситуации, превышающей по своим размерам МПА, защитные устройства АЭС могут оказаться неэффективными и последствия аварии выйдут за пределы, рассматриваемые в проекте. Поэтому в последнее время появилось мнение о необходимости анализа радиационных последствий значительно более крупных аварий – "гипотетических аварий" (ГА), которые не исключают возможности плавления активной зоны реактора [4, 6]. При этом под "максимальной гипотетической аварией" (МГА) подразумевается "гипотетическая авария, приводящая к максимально возможному выбросу радиоактивных веществ в окружающую среду при расплавлении твэлов и нарушении локализующих систем" [4]. Очевидно, что чем значительнее реально происшедшая авария будет превосходить проектную, тем ближе ее радиационные последствия могут оказаться к тем, которые оценены для МГА.

Мощность и радионуклидный состав выброса при аварийной ситуации зависят от характера аварии и конструктивных особенностей АЭС, а также от степени повреждения тех или иных защитных барьеров, создаваемых на пути радиоактивных веществ с АЭС во внешнюю среду. Проведенные нами оценки радионуклидного состава аварийных выбросов АЭС с реакторами ВВЭР, РБМК и БН показали (табл.I) [7], что при МГА, сопровождающейся разрушением всех защитных барьеров на пути распространения радиоактивных веществ, суммарная активность выброса продуктов деления может достичь 4×10^{19} Бк. При этом основной вклад в активность будут давать газообразные и летучие продукты ядерного деления, хотя будет выделено и значительное количество нелетучих радионуклидов.

Приведенные в табл. I данные целесообразно сравнить с реальными выбросами радиоактивных веществ при трех наиболее крупных авариях, происшедших на ядерных реакторах – в Уиндскейле, на АЭС "Три-майл-айленд-2" (TMI–2) [8] и на Чернобыльской АЭС [9] (табл. II). Из сравнения видно, что даже самая крупная из всех

ТАБЛИЦА I. РАДИОНУКЛИДНЫЙ СОСТАВ АВАРИЙНЫХ ВЫБРОСОВ АЭС

Класс аварии	Тип реактора на АЭС	Q, активность выброса, ЭБк	Относительный вклад в Q			
			РБГ	Иод	ДЖА	Актиниды
МПА	ВВЭР	0,12*	0,99	0,001	–	–
	РБМК	0,006	0,99	0,003	–	–
	БН	0,01	0,21	0,03	0,76	–
МГА	ВВЭР	44	0,39	0,53	0,08	–
	РБМК	19	0,73	0,12	0,15	–
	БН	41	0,24	0,33	0,30	0,13

Сокращения: РБГ – радиоактивные благородные газы; ДЖА – долгоживущие аэрозоли;
* Активность выброса нормирована на 1000 МВт (эл.).

ТАБЛИЦА II. РАДИОНУКЛИДНЫЙ СОСТАВ ВЫБРОСОВ ПРИ КРУПНЫХ АВАРИЯХ НА ЯДЕРНЫХ РЕАКТОРАХ

Авария на АЭС	Дата	Тип реактора на АЭС	Q, активность выброса, ЭБк	Относительный вклад в Q			
				РБГ	Иод	ДЖА	Актиниды
Уиндскейл	09.10.57	промышл.	0,014	0,92	0,055	0,035	–
TMI-2	28.03.79	PWR	0,093–0,48	1,00	1×10^{-5}	–	–
ЧАЭС-4	26.04.86	РБМК	3,5[a]	~0,48	0,23 (^{131}I – 0,08)	0,28	0,014 (^{239}Pu – $7{,}4 \cdot 10^{-6}$)

[a] Активность выброса пересчитана на 06.05.86.

аварий – авария на IV-м энергоблоке Чернобыльской АЭС (ЧАЭС-4) – по своим масштабам существенно ниже, чем аварийный выброс при МГА. Необходимо однако отметить, что специфика этой аварии привела к значительному обогащению выброшенной смеси тугоплавкими радионуклидами и трансурановыми элементами, по сравнению с общепринятыми сценариями развития аварий типа МГА.

Радиационные последствия аварии класса МПА рассмотрим на примере АЭС с ВВЭР-440. При такой аварии в окружающую среду выбрасываются в основном РБГ и небольшая часть изотопов иода, поэтому доза внешнего облучения человека обусловлена излучением облака выброса, а внутреннего облучения (щитовидной железы) – ингаляцией иода.

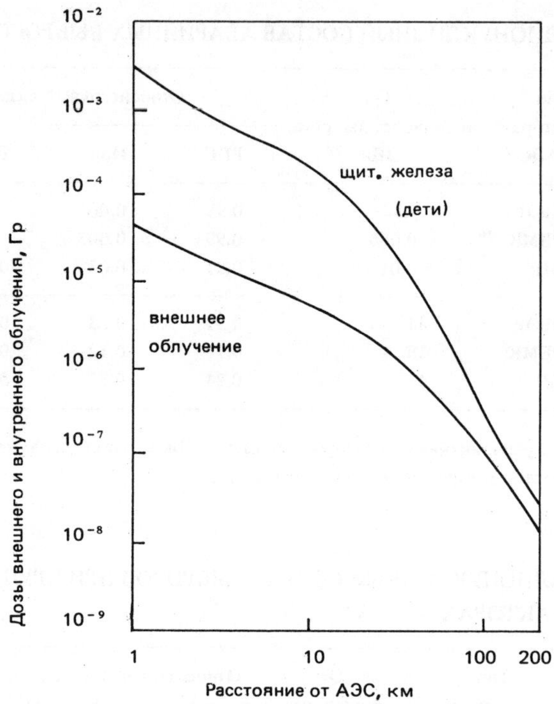

РИС. 1. Изменение с расстоянием дозы внешнего и внутреннего облучения человека за счет аварийного выброса при МПА на АЭС с ВВЭР-440.

Изменение с расстоянием от АЭС доз внешнего и внутреннего облучения организма человека, обусловленных выбросом активности при МПА на АЭС с ВВЭР-440, показано на рис. 1. Вычисления проводили по методике, изложенной в [10]. Из рисунка видно, что доза внутреннего облучения щитовидной железы при МПА почти на всех расстояниях от места выброса превышает дозу внешнего облучения, поэтому вклад внутреннего облучения в формирование суммарной дозы достигает на малых расстояниях от АЭС 50--60% и снижается до нескольких процентов на расстояниях, превышающих 50 км. Расчеты показывают, что при МПА АЭС с ВВЭР, РБМК и БН доза и внешнего, и внутреннего облучения человека на расстояниях больших 0,5 км от АЭС существенно ниже предела дозы аварийного облучения, установленного СП АЭС-79 – 0,1 Зв – внешнее облучение и 0,3 Зв – щитовидная железа детей. Таким образом, даже при неблагоприятных условиях распространения аварийного выброса, системы локализации, предусмотренные в проектах АЭС с реакторами разных типов, обеспечивают при МПА радиационную безопасность населения.

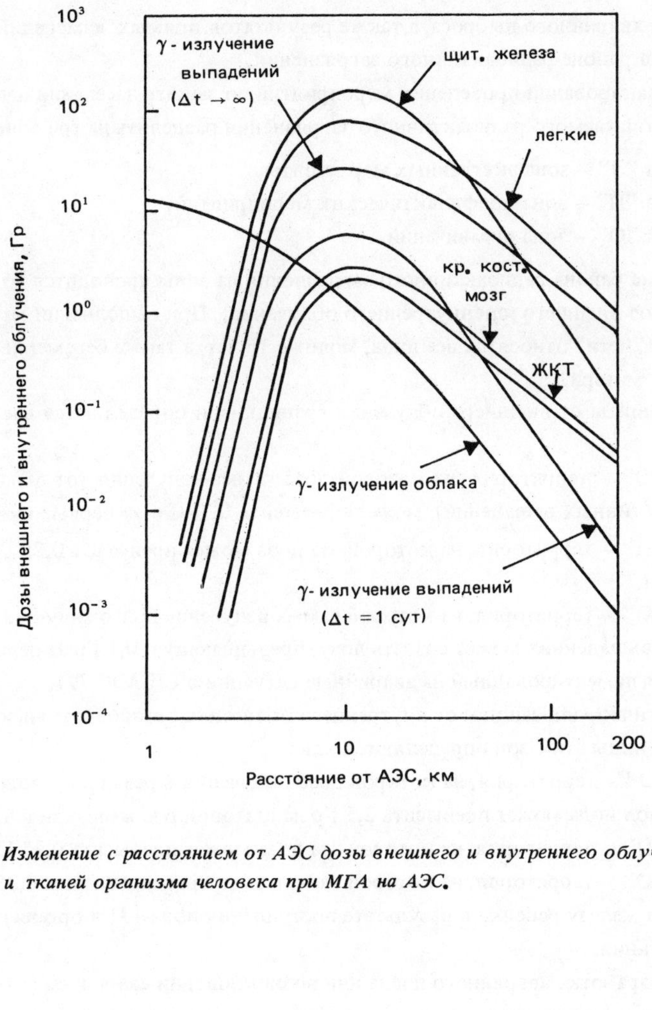

РИС. 2. Изменение с расстоянием от АЭС дозы внешнего и внутреннего облучения органов и тканей организма человека при МГА на АЭС.

Оценим теперь радиационную обстановку в районе размещения АЭС при МГА. Результаты расчетов доз внешнего и внутреннего облучения человека показывают (рис. 2), что наибольшую значимость при авариях такого масштаба на малых расстояниях от АЭС (до 3 км) играет гамма-излучение облака выброса, а на больших – возрастает значимость облучения гамма-излучением выпавших на местность радиоактивных продуктов и внутреннего облучения за счет ингаляции радионуклидов. Это значит, что при МПА существенную роль в последствиях аварийного облучения населения будут играть своевременно проведенные профилактические и защитные мероприятия [11].

Необходимость и достаточность проведения этих мероприятий определяется в каждом конкретном случае на основании анализа характеризующих аварию данных, экспресс-оценки с помощью излагаемой ниже методики возможных радиационных последствий аварийного выброса, а также результатов прямых измерений реальной обстановки в районе радиоактивного загрязнения.

При планировании проведения мероприятий по защите населения целесообразно район предполагаемого радиоактивного загрязнения разделить на три зоны:

- зона "Э" – зона экстренных мер защиты;
- зона "П" – зона профилактических мероприятий;
- зона "О" – зона ограничений.

Разграничение района радиоактивного загрязнения на зоны проводится по ожидаемым значениям доз внешнего или внутреннего облучения. При выполнении мер защиты к категории "дети" относятся все лица, моложе 16 лет, а также беременные женщины и кормящие матери.

Для защиты от внешнего облучения границы зон определяются следующим образом:

Зона "Э" – территория, на которой доза гамма-излучения (от облака выброса и от радиоактивных выпадений) может превысить 0,75 Гр за первые сутки.

Зона "П" – территория, на которой эта доза может превысить 0,25 Гр за неделю после аварии.

Зона "О" – территория, на которой гамма-излучение долгоживущих радионуклидов в выпадениях может создать дозу, превышающую 0,1 Гр за первый год (уровень, регламентированный на аварийные ситуации в СП АЭС-79).

Аналогично, для защиты от внутреннего облучения органов и тканей организма человека границы этих зон определяются как:

Зона "Э" – территория, на которой доза облучения в результате вдыхания радиоизотопов иода может превысить 2,5 Гр на щитовидную железу детей.

Зона "П" – территория, на которой эта доза может превысить 0,25 Гр.

Зона "О" – территория, на которой может быть превышена доза 0,25 Гр на щитовидную железу ребенка в результате поступления иода-131 в организм с продуктами питания.

При разработке аварийного плана или возникновении аварийной ситуации размеры этих зон определяются расчетным путем, исходя из типа постулируемой или происшедшей аварийной ситуации и погодных условий. Оценки показали, что из всего спектра возможных аварийных ситуаций на АЭС достаточно выбрать четыре типа аварий (i = 1,2,3, и 4), сопровождающихся выбросом радиоактивных веществ в окружающую среду:

- ПА – авария, сопровождающаяся истечением теплоносителя первого контура, восполняемым аварийными средствами подпитки. Выброс РБГ и иода-131 (менее 0,01% от суммарной активности выброса) – i = 1;
- МПА – авария, связанная с разрывом крупного трубопровода первого контура, приводящая к разгерметизации всех твэлов. Выброс РБГ и смеси изотопов иода (35% иода-131). Для АЭС с БН основную активность аэрозольной компоненты выброса дает натрий-24 [12] – i = 2;

– ГА – авария, связанная с плавлением 10% активной зоны и отказом всех локализующих систем. Выброс РБГ, изотопов иода и смеси других продуктов деления (12% активности смеси изотопов иода и 9% от активности всей аэрозольной компоненты выброса составляет иод-131) – $i = 3$;
– МГА – авария со 100% плавлением активной зоны при отказе всех локализующих систем. Выброс РБГ, изотопов иода и смеси других продуктов деления (иод-131 – 11% от активности смеси изотопов иода и 10% от активности всей аэрозольной компоненты выброса) – $i = 4$.

Рассматриваются также три типа погодных условий ($j = 1,2,3$) при аварийной ции ситуации на АЭС:

$j = 1$ – неустойчивая стратификация атмосферы, "н", характерная для очень солнечной летней погоды;
$j = 2$ – безразличная или нейтральная стратификация, "б", характерная для переменной облачности в течение дня, облачного дня или облачной ночи, а также дождливой погоды;
$j = 3$ – устойчивая стратификация, "у", характерная для ночи с переменной облачностью или ясной ночи, а также морозного зимнего дня. К этому же типу относятся случаи образования утренних и вечерних инверсий.

Большая серия специальных расчетов на ЭВМ показала, что в качестве основной исходной величины для разработки методики экспресс-оценки радиационных последствий аварийных ситуаций на АЭС наиболее удобно принять рассчитанное, в соответствии с разработанной методикой, или реально измеренное в момент t после аварии значение мощности дозы гамма-излучения над загрязненной территорией ($P_{i,j}(t)$). Для унификации последующих вычислений данное значение мощности дозы в точке, удаленной на расстояние x от АЭС, целесообразно привести к одному времени, например, к 24 часам после аварии ($P_{24,i,j}(x)$ в сГр/ч). В этом случае значения дозы внешнего облучения тела человека, находящегося на открытой местности в течение всего периода радиационного воздействия рассчитываются по формулам:

– внешнее облучение от облака выброса:

$$D_{обл.}(x) = K_{i,j} \cdot P_{24,i,j}(x), \text{ Гр} \quad (1)$$

– внешнее облучение от радиоактивных выпадений за период времени до момента t после аварии:

$$D_{вып.}(x,t) = K'_i(t) \cdot P_{24,i,j}(x), \text{ Гр} \quad (2)$$

Значения коэффициентов $K_{i,j}$ приведены в табл. III, а зависимость $K'_i(t)$ показано на рис. 3. Доза внутреннего облучения щитовидной железы человека за счет ингаляционного или перорального (с молоком) поступления радиоактивности в организм оценивается по формуле:

$$D_{щ.ж.}(x) = K''_i \cdot P_{24,i,j}(x), \text{ Гр} \quad (3)$$

ТАБЛИЦА III. ЗНАЧЕНИЯ КОЭФФИЦИЕНТОВ $K_{i,j}$

Категория погоды	$K_{i,j}$			
	i = 1	i = 2	i = 3	i = 4
"н", (j = 1)	13	0,4	1,6	0,7
"б", (j = 2)	18	0,5	1,8	0,8
"у", (j = 3)	120	1,9	3,3	2,0

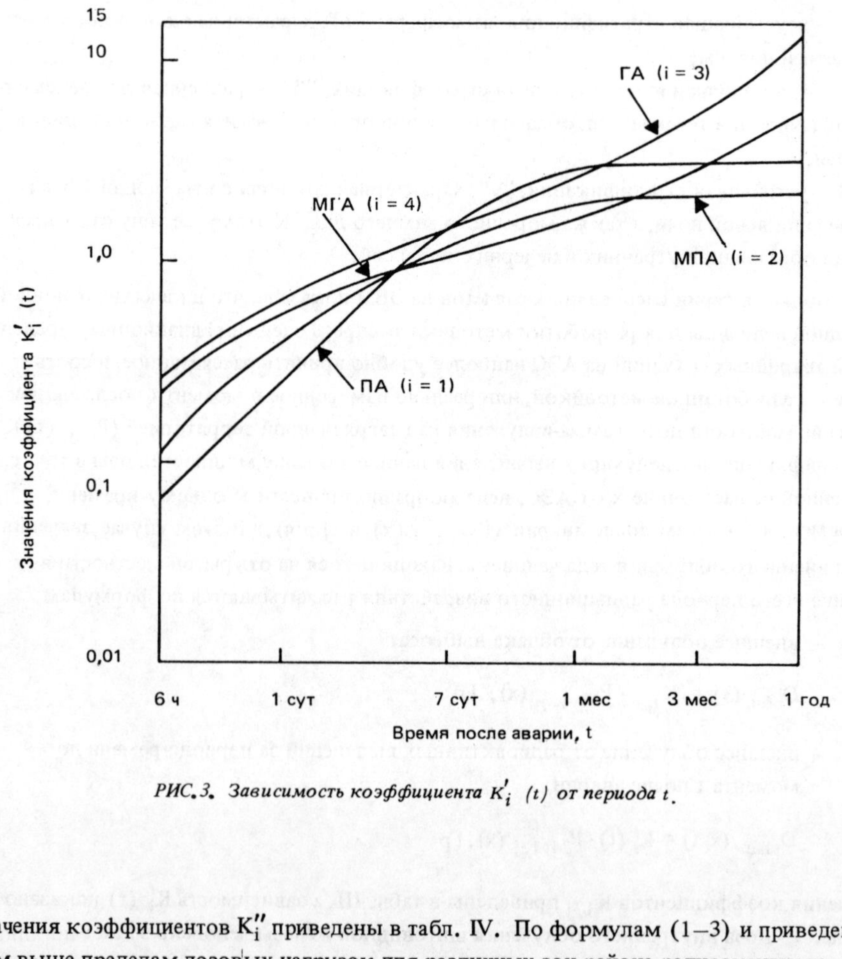

РИС. 3. Зависимость коэффициента K_i' (t) от периода t.

Значения коэффициентов K_i'' приведены в табл. IV. По формулам (1–3) и приведенным выше пределам дозовых нагрузок для различных зон района радиоактивного загрязнения могут быть рассчитаны размеры этих зон: максимальная протяженность по оси следа (L) и площадь зоны загрязнения (S). Анализ зависимости величин L и S от категории устойчивости атмосферы, высоты подъема облака выброса,

ТАБЛИЦА IV. ЗНАЧЕНИЯ КОЭФФИЦИЕНТОВ K_i''

Путь поступления активности	Категория облучаемых лиц	K_i''			
		i = 1	i = 2	i = 3	i = 4
Ингаляция	взрослые	0,27	0,17	0,08	0,11
	дети	0,46	0,29	0,13	0,18
Пероральный	дети	46	29	13	18

скорости ветра в момент аварии показал, что размеры зон "Э", "П", "О", особенно по внутреннему облучению щитовидной железы, максимальны при неустойчивой стратификации атмосферы. Зависимость L и S от высоты подъема облака выброса выражена слабо. На рис. 4 и 5 в качестве примера показаны корреляционные зависимости величин L и S от мощности дозы гамма-излучения в точке на оси следа на удалении 10 км от АЭС через 24 часа после аварии – $P_{24}(10)$ для безразличной стратификации атмосферы. Видно, что с достоверностью P = 0,981 площадь зоны загрязнения прямопропорциональна мощности дозы $P_{24}(10)$, а L пропорциональна корню квадратному из этой величины. Обобщение аналогичных данных для других вариантов расчетов позволило получить оценочные соотношения между S и $P_{24}(10)$:

$$S = a_j \cdot P_{24}(10), \text{ км}^2 \qquad (4)$$

где a_j – коэффициенты пропорциональности (табл.V). Длину L и полуширину (ΔY) радиоактивного следа можно оценить с помощью соотношений:

$$L_j = b_j' \sqrt{S_j}, \text{ км} \qquad (5)$$

$$\Delta Y_j = b_j'' \sqrt{S_j}, \text{ км} \qquad (6)$$

Значения коэффициентов b_j' и b_j'' приведены в табл. VI.

Были получены также пороговые значения мощности дозы через 24 часа после аварии ($P_{24}(x)$), позволяющие по результатам фактических измерений на местности или прогностическим оценкам отнести изучаемый район к той или иной зоне радиоактивного загрязнения (табл. VII). Кратность превышения реальной мощности дозы над приведенными в табл. VII пороговыми значениями позволяет оценить объем необходимых дополнительных мероприятий [11, 13] для снижения дозовых нагрузок на население до допустимых уровней.

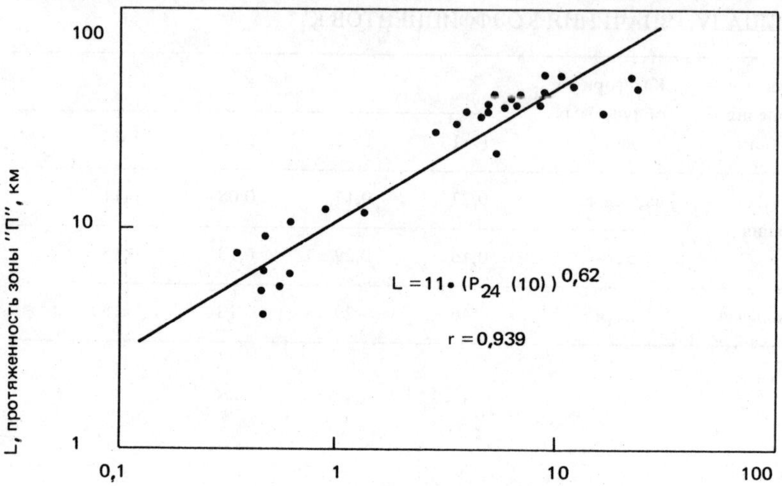

РИС. 4. Зависимость протяженности зоны "П" по внешнему облучению тела человека от мощности дозы на местности на оси следа на удалении 10 км от АЭС через 24 часа после аварии для случая категории устойчивости атмосферы "б".

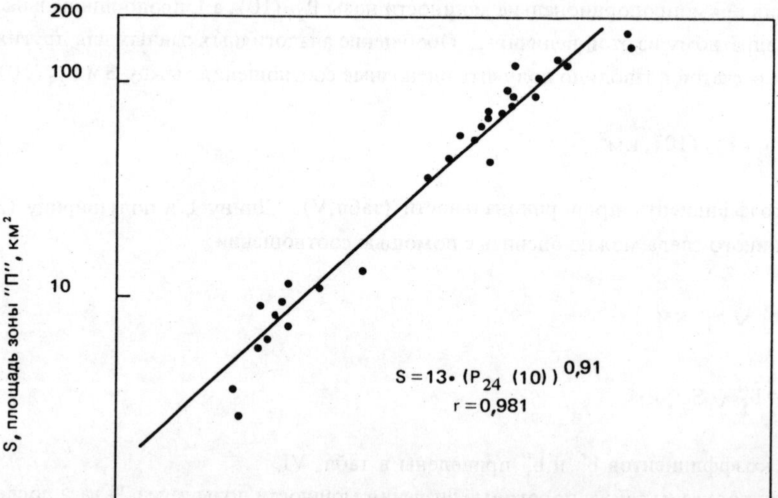

РИС. 5. Зависимость площади зоны "П" по внешнему облучению тела человека от мощности дозы на местности на оси следа на удалении 10 км от АЭС через 24 часа после аварии для случая категории устойчивости атмосферы "б".

ТАБЛИЦА V. ЗНАЧЕНИЯ КОЭФФИЦИЕНТОВ a_j

Зона	Предел дозы, Гр	Фактор радиационного воздействия	a_j		
			j = 1	j = 2	j = 3
"Э"	0,75	внешнее	4,6	2,7	4,3
	2,5	ингаляция (щит. железа)	210	100	90
"П"	0,25	внешнее	22	13	15
	0,25	ингаляция (щит. железа)	4000	1300	630
"О"	0,1	внешнее	150	80	130
	0,25	пероральное (щит. железа)	1×10^{6}	6×10^{5}	4×10^{5}

ТАБЛИЦА VI. ЗНАЧЕНИЯ КОЭФФИЦИЕНТОВ b_j' И b_j''

Коэффициент	j = 1	j = 2	j = 3
b_j'	2,3	3,8	4,0
b_j''	1,1	0,6	0,5

ТАБЛИЦА VII. ПОРОГОВЫЕ ЗНАЧЕНИЯ МОЩНОСТИ ДОЗЫ $P_{24,\,i}$ ДЛЯ ЗОНИРОВАНИЯ РАДИОАКТИВНОГО СЛЕДА, сГр/ч

Зона	Предел дозы, Гр	Фактор радиационного воздействия	$P_{24,\,i}$			
			i = 1	i = 2	i = 3	i = 4
"Э"	0,75	внешнее	–	–	2,2	3,2
	2,5	ингаляция (щит. железа)	0,05	0,09	0,19	0,14
"П"	0,25	внешнее	–	–	0,8	1,2
	0,25	ингаляция	0,005	0,009	0,019	0,014
"О"	0,1	все тело	0,21	0,53	0,11	0,24
	0,25	пероральное (щит. железа)	$5,4\times10^{-5}$	$8,5\times10^{-5}$	$1,9\times10^{-4}$	$1,4\times10^{-4}$

ТАБЛИЦА VIII. СООТНОШЕНИЕ МЕЖДУ $P_{\gamma, 24}$ И ВРЕМЕННЫМИ ИНТЕГРАЛАМИ КОНЦЕНТРАЦИИ ИОДА-131 В РАЗЛИЧНЫХ ОБЪЕКТАХ ОКРУЖАЮЩЕЙ СРЕДЫ

Интеграл концентрации	Величина интеграла концентрации
– для воздуха, ГБк·с/м³	0,06
– для поверхности почвы, ТБк·с/м²	1,0
– для пастбищной травы, ГБк·с/кг (сух. вес)	140
– для молока, ГБк·с/л	7

Следует подчеркнуть, что наиболее интенсивно дозы облучения людей будут формироваться в первые дни после аварии – 10–20% годовой дозы от выпадений при МГА создастся за первую неделю после аварии. Таким образом, необычайно возрастают требования к оперативному срабатыванию всех звеньев цепочки по быстрейшему введению в действие плана противоаварийных мероприятий по защите населения. Основные принципы организации этой работы на АЭС СССР изложены в [14].

Изложенная в докладе схема сразу после аварии на Чернобыльской АЭС была применена для оценки радиационной обстановки в районе размещения станции. Однако, поскольку аварийный реактор в течение сравнительно продолжительного времени (8–10 дней) оставался мощным источником выброса радиоактивных продуктов в атмосферу, картина загрязнения объектов окружающей среды как по уровню активности, так и по радионуклидному составу имела сложный характер ввиду изменяющихся во времени метеоусловий, высоты и интенсивности выброса [9]. В частности наблюдалось образование аномально высоких локальных загрязнений отдельных участков территории. Возникали также определенные трудности в установлении некоего типичного радионуклидного состава радиоактивных примесей в воздухе и на загрязненной территории. Так, например, содержание иода-131 в пробах воздуха и грунта варьировало от 8 до 40%, а цезия-137 от 1 до 20%. Указанное обстоятельство затрудняло прямое использование получаемых по описанной методике результатов, однако небольшие усовершенствования позволили откорректировать методику, и с точностью до 30–50% давать прогноз радиационной обстановки в населенных пунктах, намечаемых к эвакуации. Так как основной выброс радиоактивных веществ на зоны реактора прекратился к 6 мая 1986 г., то к этому моменту времени можно считать завершившимся и процесс формирования радиоактивного следа. В связи с этим было решено в качестве опорной величины использовать не $P_{24}(x)$, а $P_{\gamma, 15}$ – мощность дозы на местности на 15-ый день после начала аварии. С использованием этой величины были получены следующие соотношения между $P_{\gamma, 15}$ (в мкГр/ч) и дозой внешнего гамма-облучения от радиоактивного облака ($D_{обл.}$, сГр), дозой от радиоактивных выпадений за различное время после аварии ($D_{вып.}$, сГр), а также дозой внутреннего облучения щитовидной железы детей ($D_{щ. ж.}$, сЗв) за счет ингаляции и потребления загрязненного молока коров:

$$D_{\text{обл.}}\ (10\text{–}30\ \text{км}) = (0{,}028\text{–}0{,}007)\cdot P_{\gamma,15}$$
$$D_{\text{вып.}}\ (7\ \text{сут}) = 0{,}07\cdot P_{\gamma,15}$$
$$D_{\text{вып.}}\ (1\text{мес}) = 0{,}12\cdot P_{\gamma,15}$$
$$D_{\text{вып.}}\ (1\ \text{год}) = 0{,}25\cdot P_{\gamma,15}$$
$$D_{\text{вып.}}\ (50\ \text{лет}) = 0{,}8\cdot P_{\gamma,15}$$
$$D_{\text{щ.ж.}}\ (\text{инг}) = 1{,}0\cdot P_{\gamma,15}$$
$$D_{\text{щ.ж.}}\ (\text{пер}) = 100\cdot P_{\gamma,15}$$

Последнее значение относится к случаю, когда не проводится никакого ограничения на потребление загрязненного молока коров, что, естественно, возможно только в зонах с очень низкими уровнями загрязнения растительности иодом-131.

Были получены также простые соотношения, устанавливающие связь между мощностью дозы $P_{\gamma,\ 15}$ в мкГр/ч с различными характеристиками загрязнения объектов окружающей среды (табл. VIII). Эти соотношения в реальных условиях ликвидации последствий аварии на ЧАЭС позволили достаточно оперативно проводить как восстановление радиационной обстановки в первые дни после аварии, так и осуществлять прогноз ситуации на первые недели и месяцы после инцидента, исходя из полученной фактической информации о мощностях доз гамма-излучения на местности.

ЛИТЕРАТУРА

[1] UNITED NATIONS, Ionizing Radiation: Sources and Biological Effects Scientific Committee on the Effects of Atomic Radiation (Report to the General Assembly), (UNSCEAR), UN, New York (1982).

[2] БАБАЕВ Н.С., ДЕМИН В.Ф., ИЛЬИН Л.А. и др., Ядерная энергетика, человек и окружающая среда, Энергоатомиздат, Москва (1984).

[3] ВОРОБЬЕВ Е.И., ИЛЬИН Л.А., ТУРОВСКИЙ В.Д. и др., Радиационная безопасность АЭС в СССР, Атомная энергия, **54** 4 (1983) 227.

[4] Общие положения обеспечения безопасности атомных станций при проектировании, сооружении и эксплуатации (ОПБ-82), Сборник нормативных материалов по безопасности АЭС, Энергоатомиздат, I (1984) 80.

[5] ДОРОЩУК В.В., Ядерные реакторы на электростанциях, Атомиздат, Москва (1977).

[6] СИДОРЕНКО В.А., Вопросы безопасной работы реакторов ВВЭР, Атомиздат, Москва (1977) 216.

[7] ИЛЬИН Л.А., ПАВЛОВСКИЙ О.А., САЯПИН Н.П., "Оценка радиационных последствий аварийных ситуаций на АЭС и проблемы безопасности населения", Радиационная безопасность и защита АЭС, Энергоатомиздат, Москва, 8 (1984) 146.

[8] AMERICAN PHYSICAL SOCIETY STUDY GROUP, Radionuclide release from severe accidents at nuclear power plants, Rev. Mod. Phys. 57 3 (1985) 51–54.

[9] ГКАЭ СССР, "Авария на Чернобыльской АЭС и ее последствия", Информация, подготовленная для совещания экспертов МАГАТЭ (25–29 августа 1986 г., Вена), Москва, (1986).

[10] ПАВЛОВСКИЙ О.А., "Методы расчета дозовых нагрузок на население, проживающее в районе расположения АЭС", Радиационная безопасность и защита АЭС. Энергоатомиздат, Москва, 9 (1985) 95.

[11] Международное агентство по атомной энергии, Планирование защитных мер за пределами площадки в случае радиационных аварий на ядерных установках, Серия изданий по безопасности №55, МАГАТЭ, Вена (1981) 98.

[12] ПАВЛОВСКИЙ О.А., ПУРИМ Е.М., ЮРЧЕНКО Д.С., "Вопросы радиационной безопасности населения при эксплуатации АЭС с реакторами на быстрых нейтронах с жидкометаллическим теплоносителем", Радиационная безопасность и защита АЭС, Энергоатомиздат, Москва, 9 (1985) 120.

[13] ИЛЬИН Л.А., Основы защиты организма от воздействия радиоактивных веществ, Атомиздат, Москва (1977).

[14] НЕВСКИЙ В.П. и др., "Обеспечение радиационной безопасности персонала и населения во время возможных аварийных ситуаций на АЭС в СССР", Обеспечение радиационной безопасности при эксплуатации АЭС, Энергоатомиздат, Москва, 1 (1983) 49.

METHODS FOR EVALUATING THE RADIOLOGICAL CONSEQUENCES OF ACCIDENTAL RELEASES: TRENDS IN MODEL DEVELOPMENT INCLUDING THE IMPACT OF CHERNOBYL

W. NIXON, P.J. COOPER, G.C. MEGGITT
Safety and Reliability Directorate,
United Kingdom Atomic Energy Authority,
Culcheth, Warrington

S. NAIR
Berkeley Nuclear Laboratories,
Central Electricity Generating Board,
Berkeley, Gloucestershire

United Kingdom

Abstract

METHODS FOR EVALUATING THE RADIOLOGICAL CONSEQUENCES OF ACCIDENTAL RELEASES: TRENDS IN MODEL DEVELOPMENT INCLUDING THE IMPACT OF CHERNOBYL.

Broadly speaking, environmental consequence calculations may be classed as having applications in two areas: risk assessment (including probabilistic safety assessment, PSA) and emergency response. For each of these, the models used and research requirements differ somewhat; this is reflected in ongoing work programmes. In the PSA area, a wide variety of research topics are being pursued, ranging from atmospheric dispersion/deposition modelling, through various aspects of dosimetry to health effects calculations. Methods for evaluating the uncertainty associated with risk estimates are also being studied. In addition, a significant volume of work is devoted to the development of new, flexible consequence modelling codes, which are capable of easy modification and which are, thereby, able to capitalise on the ongoing research efforts. In the case of emergency response codes, a subdivision into short- (a few tens of km) and long-range may be made. Short-range models may benefit from the development of simple methods (possibly extensions of the Gaussian model) to take account of features such as coastal and topographical effects. Long-range models are likely to benefit from validation studies using the Chernobyl data as a benchmark. However, it is felt that very careful consideration should be given to the proposed application of any emergency response system, bearing in mind likely uncertainty levels, before deciding on the complexity of model required; in many cases, relatively simple approaches may suffice.

1 INTRODUCTION

Models for evaluating the environmental impact of atmospheric releases of radioactivity have many applications; these include risk assessment, evaluation of alternative plant design features, environmental impact assessment, emergency planning and response, development of criteria for the acceptability of risk and provision of focus for research needs [1]. These methods are currently the subject of much research activity and, in order to structure a discussion of this work, it is valuable to identify two general types of computer model or code:

i) Probabilistic Safety Assessment (PSA) Codes. Given a postulated release to the atmosphere, with an estimated frequency of occurrence, these methods perform atmospheric dispersion, dosimetry and consequence calculations for a range of possible weather sequences, each with an associated probability. This process is then repeated for each of a number of assessed releases. In this way a probability distribution of each consequence of interest (e.g. number of fatalities, area of land contaminated) is constructed, which characterises the risk associated with the plant under study [2].

ii) Emergency Response Codes. These methods perform atmospheric dispersion calculations for one weather sequence only, i.e. that prevailing at the time of an accident [3]. These "real-time" calculations are intended to provide information to aid the decision making process concerning, for example, which population groups (if any) should be evacuated. These methods can also be of help in training personnel in handling emergencies.

The models used by, and research requirements of, each of the above types of code differ somewhat. In view of this, the development trends in each are discussed in turn below.

2 TRENDS IN MODEL DEVELOPMENT - PSA CODES

Methods of this type have found fairly wide application. The first major work using probabilistic methods was the Reactor Safety Study (WASH-1400 [4]) in the USA; a further large study has since been published in that country (Aldrich et al [5]). Other milestones in this area are reflected in the studies performed in the Federal Republic of Germany [6] and for the Sizewell Public Inquiry in the UK [7].

In addition to these applications, an international model comparison exercise has been performed [8], organised under the auspices of the Nuclear Energy Agency of the Organisation for Economic Co-operation and Development. Over thirty organisations contributed to that study, which concluded that, while existing models provide a reasonable basis for estimating the radiological consequences of accidents, there were certain areas where model and data improvements would be useful. These included the prediction of external exposure from radioactive material deposited in urban environments, the modelling of atmospheric dispersion in complex terrain and the treatment of long duration releases (greater than about 10 hours).

Although PSA consequence codes have been fairly extensively used, it is widely recognised that there is a measure of uncertainty associated with their application. This, combined with the clear need to extend the applicability of existing methods to deal with new situations (e.g. to consider nuclear plant other than reactors), has led to much national and international research effort. This work is exemplified by the MARIA (Methods for Assessing the Radiological Impact of Accidents) project organised by the Commission of European Communities. This is a very extensive programme covering both experimental and theoretical investigations, in areas ranging from atmospheric transport and deposition in various conditions, through to dosimetry in the urban environment. Details of the first phase of this programme have been published [9].

The above research efforts were, of course, well established prior to the Chernobyl accident. Nevertheless, that event did provide some additional impetus. Existing consequence modelling techniques were successfully used by various countries to provide information on the nature of the spread of activity across the European continent and the likely radiological impacts. Furthermore, the monitoring programmes set up after the accident (involving measurement of deposition levels, concentrations in foodstuffs, etc) have provided a very large data base against which existing methods may be validated/ improved.

The results of the above studies will have an impact on the large computer codes used to perform PSA environmental consequence calculations. New models and data will require to be incorporated into their computational framework. Clearly, therefore, those codes which have been well designed and have a flexible, modular structure capable of easy modification, will be best placed to take advantage of the large body of research currently being performed. In view of this, some of the development work in the PSA consequence modelling field is geared towards code development, the principal aim being the production of very flexible, "user-friendly" computer systems, which have been subjected to rigorous quality assurance procedures.

The remainder of this discussion on PSA codes is devoted to a brief description of some of the more fundamental research areas in the field of consequence modelling, followed by a consideration of code development work. Note that both of the following sections cannot hope to be comprehensive; the principal aim is to provide an indication of development trends.

2.1 Fundamental research work

i) Atmospheric dispersion. Existing dispersion models in PSA codes tend to be based on the relatively simple Gaussian methods. The low computer costs associated with their application are a great advantage in the PSA context where many dispersion calculations are performed. Furthermore, studies using more complex techniques [10] have suggested that, considering the uncertainty associated with the end-points of PSA calculations, simple methods probably suffice. Nevertheless, there is still much scope for further development and work is currently underway or planned in many areas, for example:

Topographical effects. Some PSA applications involve areas with strong topographical features and the development of simple models will allow these effects to be included.

Gaussian puff modelling. Application of these techniques will allow improved simulation of the effects of long duration releases (greater than about 10 hours). Such releases are an increasingly common feature of source term analyses and their importance has clearly been demonstrated following the 10 day release from Chernobyl (see e.g. [11]).

Post-release processes. Processes such as aerosol agglomeration and steam condensation along with various chemical reactions which may occur during downwind transport, are not accounted for in existing codes. Some work is addressing this problem [12].

ii) Deposition modelling. Deposition calculations are a significant source of uncertainty in consequence calculations and much effort has been, and is continuing to be, expended in three general areas:

Dry deposition. Experimental and theoretical work is aimed at, for example, characterising deposition to different surface types, e.g. urban as compared to rural surfaces [13]. Other developments endeavour to take account of different physical and chemical forms of pollutant, e.g. different particle sizes. Some of this work is using the Chernobyl data base to characterise deposition [14].

Wet deposition. Of interest here is, for example, improved modelling of the spatial and temporal distribution of rainfall [15]; the known marked variations in rain-rate within a rainstorm are not as yet adequately accounted for in existing PSA techniques. (Such effects are, of course, principally of importance at the high consequence/low probability end of an estimated consequence distribution.)

The Chernobyl accident has clearly demonstrated the effects of rainfall.

Gravitational settling. Although large aerosol particles have not been a notable feature of most source term calculations to-date, this is likely to change as source term research progresses and consideration is given to a wider range of nuclear plant. Correspondingly, some research effort is being focussed on that area (e.g. [16]).

iii) Foodchain modelling. Developments in this field are largely aimed at dynamic models which also include the effects of seasonality (i.e. the timing of an accidental discharge relative to a crop harvest/growing season) [17,18]. These developments are being aided by the wide body of data available post-Chernobyl; comparison between model prediction and monitoring data should, in time, lead to better validated models.

iv) Dosimetry. Many aspects of dosimetry are currently being investigated; however, it is perhaps worth taking particular note of the significant effort being expended with the aim of better characterising external exposure in urban areas (see e.g. [19,20,21]). Particular areas of study include the shielding effect of various building types and the long term fate of deposited activity (often termed "weathering"). This latter topic includes use of the Chernobyl fall-out as a source of data.

v) Decontamination. Several experimental programmes are underway here, e.g. [22,23]. However, it is clear that the USSR experience, post-Chernobyl, will be valuable. Data on the effectiveness and cost of various decontamination procedures would be of value to the consequence modelling field. Perhaps of particular interest is the applicability (both in technical and financial terms) of methods over very large areas of contaminated land.

vi) Health effects. In this context it is worth noting that a large review has recently been completed within the USA [24]. In addition, experience within the USSR in dealing with the immediate aftermath of the Chernobyl accident, in combination with subsequent epidemiological studies (within certain groups of exposed people), should, in time, provide valuable additional information on the biological effects of radiation.

vii) Countermeasures. Most PSA consequence codes endeavour to simulate actions which may be taken to mitigate radiation exposure. Indeed, different countermeasures assumptions can have a marked impact on the estimated health effects of any release [8]. Clearly, therefore, PSA codes are likely to benefit from USSR experience post-Chernobyl on the effectiveness of, for example, evacuation and sheltering and the problems associated with introducing these measures.

viii) Uncertainty analysis. As has already been noted, PSA consequence calculations are subject to a measure of uncertainty. Correspondingly, methods are under development to allow detailed uncertainty analyses to be performed (see e.g. [25,26]). These methods allow simultaneous variations in many parameters to be considered, as opposed to single parameter sensitivity studies. The application of one such method, using Latin Hypercube sampling techniques, has recently been demonstrated [27].

2.2 Code development work

As has already been noted, new PSA environmental consequence codes are being developed to take best advantage of existing research programmes. These systems are being designed to be very flexible and are subject to rigorous quality assurance procedures. One such code is that developed in the USA by Sandia National Laboratories as part of the MELCOR programme [28]. Some details of that method, called MACCS (MELCOR Accident Consequence Code System), have been published elsewhere [29]. Another system is that being developed within the authors' own organisations, the Safety and Reliability Directorate of the UK Atomic Energy Authority and Berkeley Nuclear Laboratories of the Central Electricity Generating Board, UK, in collaboration with the UK National Radiological Protection Board. A very brief description of some aspects of this new system, called CONDOR, is given below, to provide some indication of the nature of these new codes. Further details may be found in [30].

A modular structure has been adopted for CONDOR to meet the above noted aim of producing a very flexible code capable of straightforward modification or extension. Details of the division into various modules may be found in [30]. Communication between modules is performed entirely through files in an agreed format; this promotes flexibility and enables modules to be developed quite independently of each other. The code is currently being produced using standard FORTRAN 77. To facilitate performing uncertainty analyses, the modules are being designed so that every variable may be changed by the user (i.e. there are no "hard-wired" variables).

Some of the more significant features of the modular structure are outlined below.

i) Extensive use is made of library data. i.e. to keep computational time to a minimum, modules are limited as much as possible to, say, multiplying pre-calculated library data (e.g. data on dose per unit intake from ingestion) with data passed from another module (e.g. time integrated concentrations). Much of the modelling complexity is, thereby, excluded from the coding.

ii) A temporal grid is included in the calculations. The details of the grid are determined by the time requirements of the countermeasures and health effects modelling. The grid provides information on the temporal distribution of dose and, in particular, allows long duration releases to be better modelled.

iii) A marine grid is included. This identifies specific volumes of sea from which seafoods are obtained.

iv) A unique surface type (e.g. sea, rural, urban) may be assigned to each spatial grid point. This not only enables changes in dispersion and deposition caused by the underlying surface to be accounted for, but also enables dose conversion factors and decontamination factors appropriate to each surface type to be selected.

v) Different chemical and physical (e.g. different particle sizes) forms of the source term are allowed. This information may be used for both dispersion/deposition and dosimetry calculations.

vi) The population at any grid point may be divided according to age and sex.

vii) The population at any grid point may be divided according to the countermeasures experienced.

Consistent with the above code structure, improved modelling techniques are being incorporated into the initial version of the system; these <u>include</u>:

i) Improved dry deposition modelling. Depending, inter alia, on the particle size range, dry deposition is being modelled using either the source depletion approximation or a modified tilted plume approximation (a gravitational settling model deduced using K-theory).

ii) Modelling of the exposure pathway involving external irradiation from activity deposited on skin and clothing.

iii) Use of a comprehensive library of dose conversion factors which take appropriate account of age and sex and, for the inhalation pathway, the physical and chemical form of the pollutant.

iv) Use of library data based on improved terrestrial foodchain modelling, particularly in respect of the dynamics of foodchain transfer and seasonal effects.

v) Estimation of the radiological impact of deposition of airborne activity over the sea.

vi) Modelling of the countermeasure enabling the decontamination of people to reduce doses from activity deposited on the skin and clothing.

Note that some of the above features are not likely to be used in some of the early applications of CONDOR; e.g. data may not be available to allow some aspects, like the effect of chemical form (and how it varies with time) on deposition, to be utilised. However, it must be appreciated that the code has been designed to last for some considerable time and to be able to incorporate the results of current and future researches. In addition, the inclusion of some of the above features will allow wide-ranging uncertainty/sensitivity analyses to be performed.

3 TRENDS IN MODEL DEVELOPMENT - EMERGENCY RESPONSE CODES

A number of systems have been developed to assist the personnel who have to make decisions on actions to be taken in actual emergencies [31]. The consequence modelling can be divided conveniently into "short-range" and "long-range", to reflect the different needs of the users and the different expectations we may have of the modelling. Each is discussed in turn below.

3.1 Short-range models

In an emergency which involves a large release to the atmosphere there may be an early need to take actions within a few kilometers of the incident location. These actions will include investigative monitoring in the environment and may include countermeasures such as issue of iodate tablets, sheltering or evacuation. It is therefore possible to set a probable upper-limit on the distance to which modelling is required in this context to a few tens of km.

The overall specification of an appropriate model is that it should take available information on the source term, combine this with meteorological data and produce predictions of airborne concentrations and ground deposition. These should then be converted, using appropriate dosimetric and environmental transfer factors, into the radiation doses which would be received from external (cloudshine and groundshine) and internal (by inhalation and, possibly, ingestion) sources. These predictions can then initially be used to decide upon local emergency monitoring strategies and, if considered appropriate, for early decisions on countermeasures.

This straightforward (if difficult) modelling aim must be combined with other requirements if the system is to be really useful. In the early stages of the incident it is likely that the nature of the source term will be poorly known. It is therefore essential that the system should be able to assess rapidly a range of possible source terms for scoping calculations. This factor will be reflected in the nature of the modelling.

Because of the likely uncertainties in the source term and other factors, the priority in an emergency is to obtain some environmental monitoring results, either by the deployment of survey teams or by means of automatic systems which telemeter results to a central point. The aim of the monitoring must be to obtain estimates of both source term magnitude and isotopic composition as well as values of relevant environmental levels at particular locations. The function of the modelling in these circumstances is to estimate the implications of particular monitoring results for the surrounding area. The system must therefore be capable of interpolating between and extrapolating from these limited data and this too imposes constraints on the type of modelling which can be undertaken.

The predictions of the model will be based on the weather conditions prevalent at the time of the release and this, as has been noted, distinguishes it from the models used for risk assessment purposes where many weather sequences are required to be analysed. However, in many circumstances the modelling will be required to address questions concerning the possible future development of the accident and its consequences. This implies the ability to investigate the possible consequences of changes in the meteorological conditions; there will be a need for scoping calculations to consider the effects of, for example, precipitation. The modelling procedure needs to be sufficiently flexible for such calculations to be performed quickly.

In addition to the above, the nature of the modelling must take account of the knowledge of the intended user. If models could be relied upon in all circumstances then this might not be such a significant factor. As it is, all dispersion models are fallible under certain circumstances and the interpreter of the model results must be aware of the limitations of the model.

There are thus a number of essentially non-technical constraints on the nature of the model: it must be flexible to allow rapid calculations for a variety of source terms and possible weather conditions; it must be able to accommodate monitoring results as they become available and it must be simple enough for its limitations to be appreciated by the user; in addition it needs to have an accuracy commensurate with the numerous uncertainties which exist in real emergencies.

These considerations suggest to the authors that the most appropriate dispersion model for short range predictions is that based on the Gaussian approach. The Gaussian model is reliable enough under most weather conditions for predictions out to the required distance of a few tens of km and is simple enough in calculational terms for implementation on a small computer. It is also simple enough in conceptual terms for a non-specialist user to appreciate how it works and the limitations on its performance.

There is no doubt that the Gaussian model has limitations. It cannot deal easily with conditions of severe wind shear but this is unlikely to be a real handicap in the short-range of interest. It is doubtful if it is applicable when the wind speed is very low but this is true of most models. Within the limits of our knowledge and when combined with a suitable sub-model the Gaussian model is capable of predicting ground deposition both by dry and wet mechanisms. However, two particular features of dispersion require detailed consideration from the point of view of model development: coastal effects and topography. Each is considered very briefly below.

In the United Kingdom most power plants and other major nuclear facilities are sited near the coast. The particular meteorological structures which can occur in such locations, including sea breezes, have been subjected to intensive study and methods have been developed to include these into the Gaussian framework [32,33]. As yet these Gaussian model adjustments have not (as far as is known) been incorporated into a working emergency prediction system. This is not a major task in computational terms but it remains to be seen how the additional complications can be dealt with in an emergency situation.

The question of topography is more difficult. While model modifications have long been in existence to predict the ground level concentration from a plume which passes over a hill, there is little evidence that much can be done to predict where the plume actually goes within a scheme compatible in complexity with the Gaussian model. In the United Kingdom this is not a severe problem because the majority of facilities are located in areas where the topography is simple. However, it is one which clearly should be addressed, perhaps with the help of specific experimental studies, where topography is complex.

3.2 Long-range modelling

As in the case of short-range modelling it is appropriate to ask what the purpose of long-range emergency response modelling is and what information will be available as the basis for calculations. The organisations most likely to be interested in the effects at greater distances are the Governmental bodies responsible for implementation of longer-term countermeasures, such as the interdiction of foodstuffs. Since the time available for such decisions is days rather than hours, it is to be expected that there will be adequate time to complete at least a preliminary monitoring programme, covering measurement of deposition levels and concentrations in milk for example. It then seems likely that these decisions would be based on measurements rather than predictions.

The most useful role for long-range dispersion models is, therefore, considered to be the prediction of the trajectory (or trajectories) of the released activity. This information may then be used to direct monitoring teams efficiently. The calculation of trajectories is, of course, a major undertaking both technically and organisationally, since it requires the collection and processing of substantial amounts of meteorological data to obtain a reliable result. The authors conclude that <u>detailed</u> dispersion calculations, which would be superimposed on a trajectory analysis and which can be very expensive to perform, may be of minimal value in this context, particularly when it is recalled that information on the magnitude and duration of the release is likely to be subject to much uncertainty.

Future work in this field may lead to better validated models, using, perhaps, the Chernobyl data base as a benchmark. However, it is felt that the above points concerning model complexity should be borne in mind prior to the adoption of a particular system for real-time use.

4 CONCLUSION

The foregoing discussion has illustrated the large body of research and development work currently underway in the field of environmental consequence modelling. Most of this work is, at present, aimed at improving methods used in the context of probabilistic safety assessment: many fundamental research programmes (some influenced by Chernobyl) are being performed and, to take advantage of this, new, flexible computer codes are being devised. (As an example of the latter, the joint SRD/CEGB/NRPB code, CONDOR, has been outlined above.) Effort is, however, being expended in the field of emergency response: methods are being developed and improved, aimed at predicting doses in real time, both for short range (up to a few tens of km) and long range. This area is set to expand further, particularly post-Chernobyl. Nevertheless, the authors are of the view that detailed consideration needs to be given to the applications envisaged for any real time system, taking due account of the uncertainty associated with the relevant calculations, before deciding on the level of modelling complexity required. It is likely that in many cases relatively simple, computationally inexpensive models may suffice.

Finally, it is perhaps worth commenting that while the general field of environmental consequence assessment will benefit from analyses of the data base arising from Chernobyl, the many ongoing research programmes which were, of course, underway before the accident, will have the greatest long-term impact on the consequence modelling field.

5 REFERENCES

[1] US NUCLEAR REGULATORY COMMISSION. "Environmental transport and consequence modelling", Chapter 9 of PRA Procedures Guide - A Guide to the Performance of Probabilistic Risk Assessments for Nuclear Power Plants, NUREG/CR-2300 (1983).

[2] NIXON W, COOPER P J, UNDERWOOD B Y, PECKOVER R S. Accident consequence analysis, Nucl. Energy 24 4(1985)229.

[3] MEGGITT G C. A review of computer methods for emergency response to atmospheric releases, Presented at Workshop on Real Time Computing of The Environmental Consequences of an Accidental Release to Atmosphere from a Nuclear Installation, Luxembourg, September 1985.

[4] US NUCLEAR REGULATORY COMMISSION. "Calculation of reactor accident consequences", Appendix VI of Reactor Safety Study, NUREG-75/014 (1975).

[5] ALDRICH, D C et al. "Technical Guidance for Siting Criteria Development", NUREG/CR-2239, SAND81-1549 (1982).

[6] BUNDESMINISTER FUER FORSCHUNG UND TECHNOLOGIE. "Unfallfolgenrechnung und Risikoergebnisse", Fachband 8 Deutsche Risikostudie Kernkraftwerke - Eine Untersuchung zu dem durch Stoerfaelle in Kernkraftwerken verursachten Risiko, Verlag TUV Rheinland, 1981.

[7] KELLY G N, CLARKE R H. "An Assessment of the Radiological Consequences of Releases from Degraded Core Accidents for the Sizewell PWR", NRPB-R137 London, HMSO (1982).

[8] ORGANISATION FOR ECONOMIC CO-OPERATION AND DEVELOPMENT. "International Comparison Study on Reactor Accident Consequence Modelling", Summary Report to CSNI by an NEA Group of Experts (1983).

[9] COMMISSION OF EUROPEAN COMMUNITIES. Workshop on Methods for Assessing the Off-Site Radiological Consequences of Nuclear Accidents (Proc. Luxembourg, 1985) (1985).

[10] WOODARD K. Sensitivity of consequence results to more sophisticated atmospheric dispersion and other modelling enhancements, presented at Workshop on Reactor Accident Off-Site Consequence Modelling Assessment and Application, Tokai, Japan, April 1984.

[11] MIKKELSEN T, LARSEN S E, THYKIER-NIELSEN S. "Description of the Risϕ puff diffusion model, Nuclear Technology 67 (1984)56.

[12] COOPER P J, UNDERWOOD B Y, BREARLEY I. Review of Specific Effects in Atmospheric Dispersion Calculations: The Impact of Source-Term Characteristics - and the Processes that Modify them Post Release - on Dry and Wet Deposition Rates, Commission of the European Communities Report EUR 8935EN Volume 3, 1985.

[13] ROED J. Dry deposition in rural and urban areas, presented at CEC Workshop on the Consequences of an Accidental Contamination of the Urban Environment, Risϕ, Denmark, June 1987.

[14] NICHOLSON K W. The deposition of radioactive particulate matter onto buildings, Presented at NATO/CCMS 16th Int. Tech. Mtg. on Air Pollution and its Applications, Lindau, FRG April 1987.

[15] APSIMON H M, SIMMS K L. Estimating the effect of rain and snow on potential reactor accident consequences, Nucl. Energy 254 (1986) 235.

[16] UNDERWOOD B Y. "The modelling of gravitational settling in assessment of the consequences of accidental releases of radioactivity", Workshop on Methods for Assessing the Off-Site Radiological Consequences of Nuclear Accidents (Proc. Luxembourg, 1985) (1985).

[17] LINSLEY G S, SIMMONDS J R, HAYWOOD S M. FOOD-MARC: The Foodchain Transfer Module in the Methodology for Assessing the Radiological Consequences of Accidental Releases, NRPB-M76(1982).

[18] NAIR S. Models for the Evaluation of Ingestion Doses from the Consumption of Terrestrial Foods Following an Atmospheric Radioactive Release, Central Electricity Generating Board Report RD/B/5200/N84 (1984).

[19] JACOP P, MECKBACH R. Shielding factors and external dose evaluation, Presented at CEC Workshop on Consequences of an Accidental Contamination of the Urban Environment, Risϕ, Denmark, June 1987.

[20] LE GRAND J, ROUX Y, PATAU J. Evaluation of doses and protection afforded by dwellings against atmospheric releases, Radiation Protection Dosimetry 11 1 (1985) 41.

[21] STEINHAUSLER F. The effect of fallout deposition on indoor gamma radiation levels in single family dwellings, Presented at CEC Workshop on Consequences of an Accidental Contamination of the Urban Environment, Risϕ, Denmark, June 1987.

[22] SANDALLS F J. Removal of radiocaesium from urban surfaces, Presented at CEC Workshop on Consequences of an Accidental Contamination of the Urban Environment, Risϕ, Denmark, June 1987.

[23] WARMING L. Decontamination of radioactivity deposited on an asphalt road by shot-blasting, Presented at CEC Workshop on Consequences of an Accidental Contamination of the Urban Environment, Risϕ, Denmark, June 1987.

[24] EVANS J S, MOELLER D W, COOPER D W. Health Effects Model for Nuclear Power Plant Accident Consequence Analysis, NUREG/CR-4214, SAND85-7185 (1985).

[25] FISHER F. Role and importance of uncertainty analysis, Workshop on Methods for Assessing the Off-Site Radiological Consequences of Nuclear Accidents (Proc. Luxembourg 1985) (1985).

[26] HOFER E. Probabilistic uncertainty analysis; methods and their application, Workshop on Methods for Assessing the Off-Site Radiological Consequences of Nuclear Accidents (Proc. Luxembourg, 1985) (1985).

[27] ALPERT D J, IMAN R L, HELTON J C, JOHNSON J D. A Demonstration Uncertainty/Sensitivity Analysis Using the

Health and Economic Consequence Model CRAC2, NUREG/CR-4199. SAND 84-1824 (1985).

[28] SPRUNG J L et al. Overview of the MELCOR risk code development program, Proceedings of the International Meeting on Light Water Reactor Severe Accident Evaluation, Cambridge, MA (1983).

[29] ALPERT D J et al. The MELCOR accident consequence code system (MACCS), Workshop on Methods for Assessing the Off-Site Radiological Consequences of Nuclear Accidents (Proc. Luxembourg, 1985) (1985).

[30] COOPER P J, BOARDMAN J, BREARLEY I R, EGAN M J, NAIR S, NIXON W, PONTING A C. A description of CONDOR - a new probabilistic consequence analysis code, Presented at the International Topical Conference on Probabilistic Safety Assessment and Risk Management, Zurich, Switzerland, September 1987.

[31] MEGGITT G C, FRASER G. Real time computing of the environmental consequences of an accidental release to atmosphere from a nuclear installation - summary of the Luxembourg Workshop, Proceedings of an International Symposium on Emergency Planning and Preparedness for Nuclear Facilities, Rome, November 1985.

[32] COOPER P J, NIXON W, GHOBADIAN A. A Modelling Study of Dispersion of Elevated Plumes at a Coastal Location During Onshore Flow, SRD R307 (1987).

[33] JONES J A. Models to Allow for the Effects of Coastal Sites, Plume Rise and Buildings on Dispersion of Radionuclides and Guidance on the Value of Deposition Velocity and Washout Coefficients, NRPB-R157 (1983).

ATMOSPHERIC RELEASES FROM SEVERE NUCLEAR ACCIDENTS: ENVIRONMENTAL TRANSPORT AND PATHWAYS TO MAN

Modelling of radiation doses to man from Chernobyl releases

L.R. ANSPAUGH
Lawrence Livermore National Laboratory,
Livermore

M. GOLDMAN
University of California,
Davis

R.J. CATLIN
Electric Power Research Institute,
Palo Alto

California, United States of America

Abstract

ATMOSPHERIC RELEASES FROM SEVERE NUCLEAR ACCIDENTS: ENVIRONMENTAL TRANSPORT AND PATHWAYS TO MAN: MODELLING OF RADIATION DOSES TO MAN FROM CHERNOBYL RELEASES.

The Chernobyl accident released a large amount of highly fractionated radioactive debris, including approximately 89 PBq of ^{137}Cs. The resulting collective dose commitment to the Northern Hemisphere via the pathways of external exposure and ingestion of radionuclides with food was calculated. In order to do this, a rural/urban model of external dose was developed using the PATHWAY model for ingestion. The results are a collective dose commitment of 630 000 person-Gy over the first year and 1 200 000 person-Gy over 50 years.

1. INTRODUCTION

The accident at Reactor 4 of the Chernobyl Nuclear Power Station resulted in the release of a very large amount of highly fractionated radioactive debris. Of the released material, it is now apparent that the more dosimetrically significant radionuclides are ^{137}Cs, ^{134}Cs, and ^{131}I. The approximate releases of these radionuclides were 89, 48, and 1700 PBq, respectively[1]; this calculation[2] is based upon the use of the PATRIC atmospheric transport model and measured concentrations of radionuclides in air in Europe and elsewhere throughout the Northern Hemisphere. These values are consistent with the integrated deposition of ^{137}Cs in the Northern Hemisphere[1] and with the 70 PBq release of ^{137}Cs estimated by Cambray et al.[3]. This amount of ^{137}Cs is about one-third of that estimated to be in the reactor core at the time of the accident and is about one-sixteenth of that released to date from all tests of nuclear weapons in the atmosphere.

The release from the stricken reactor took place in two distinct phases: the initial one on April 26, 1986, resulting from the initial accident and a second one several days later resulting from elevated temperatures within the remaining core[4]. Thus, the cloud of radioactive debris was very complex and did not follow a single path. Some regions in Eastern and Western Europe were impacted by significant deposition of fallout debris. For example, some locations in Bavaria actually experienced a deposition of ^{137}Cs per unit area that was roughly 10 times higher than the average deposition from fallout from nuclear weapons testing in the atmosphere. At other locations, such as the United States of America and Canada, the deposition was still measurable, but of minor magnitude compared to global fallout from weapons testing.

The purpose of the work reported here was to estimate a collective dose commitment for the Northern Hemisphere. To do this, we combined the results for the USSR reported by the Soviets[4,5] with our own calculations.

2. METHODS

Our method of calculation is simple and direct and depends for input on simple measurements of either external exposure rate or the deposition per unit area of ^{131}I or ^{137}Cs. In order to apply this general method, it is necessary to know the relative radionuclide mixture at some point in time and to calculate the relative mixture at other points in time. To define our reference radionuclide mixture, we used the measured values reported for Nurmijarvi, Finland, for a mean collection time of 66 hours post accident[6]. Many other reported measurements indicate similar relative mixtures[1]. With the use of such a reference radionuclide mixture, the measurement of any one radionuclide is sufficient to infer the presence of the others.

The inference of deposition of radionuclides from a measurement of external gamma-exposure rate is only slightly more difficult. In this case, the relative radionuclide concentration must be weighted by its efficacy of producing an external gamma-exposure-rate field per unit deposition. For this calculation, we used the conversion factors (exposure rate per unit areal deposition) published by Beck[7]. Further details are provided in [1].

2.1 External Dose

Once the ground deposition of the radionuclide mixture is established, external exposure in air is calculated by projecting the resulting exposure rate into the future for 50 years. This is easily done by using the standard decay relationships for the reference radionuclide mixture and by allowing for the decrease in exposure rate due to weathering. For rural environments, this weathering consists of vertical movement into the soil column. We used the standard concept that fallout radionuclides are distributed exponentially with depth[8]: short-lived (half life of less than 14 days) radionuclides with a relaxation depth of 0.16 g/cm^2, intermediate-lived (half life longer than 14 days, but less than 200 days) with 1.6 g/cm^2, and long-lived with a depth of 4.8 g/cm^2. The latter value is equivalent to an average depth of penetration of 3 cm, which is the standard value used by UNSCEAR for long-lived radionuclides[9].

For urban environments, horizontal movement occurs. Jacob[10] has reported that roughly half of the activity deposited within urban environments following the Chernobyl accident disappeared within a few days. We include this effect in our model with the assumption that half of the initially deposited material weathers with a half time of 7 days.

Additional factors are a conversion of 0.0087 Gy per R and 0.3 to convert from absorbed dose in air to organ dose including the effects of building shielding and occupancy. These same factors have been used by UNSCEAR[9], and are stated to be average values for the Northern Hemisphere. We have made the additional assumptions that 30% of each country's population lives in an urban environment where our urban runoff model applies and where building shielding and occupancy are expressed better by a factor of 0.15 to convert from dose in air to organ dose.

2.2 Internal Dose

The Chernobyl reactor accident happened at a time of transition from stored feed to pasture use for cows. Thus, it is appropriate to use a model that includes seasonal dependence of pasture use and growth and use of other crops such as fresh vegetables. For the calculations here, we used the PATHWAY model of Whicker and Kirchner[11] and their tabulated integrated intake values for an example exposure occurring on April 25 in the western USA. For this situation, radionuclide movement is calculated with the assumption that cows are not on pasture, but that they subsequently move onto pasture and/or derive about 20% of their dry matter intake from pasture or green chop beginning on May 1. This apparently corresponds to the actual situation in much of northern Europe at the time of the accident.

This model also incorporates a value of 0.39 m^2/kg for the normalized retention of fallout by vegetation. This value was measured for fallout deposited close to the Nevada Test Site under dry conditions and for relatively large particles. This value also appears to be appropriate numerically for Chernobyl fallout, which at farther distances consisted of smaller particles deposited primarily by rain[1].

The output of PATHWAY is integrated intake of a radionuclide per unit areal deposition. We then applied dose-conversion factors calculated by Ng[12], who used the ICRP methodology[13].

3. RESULTS

The results of our calculations combined with those reported by the Soviets for the USSR[4,5] are shown in Table I. The numbers in this table are the sums of the dose commitments via the external and the ingestion pathways. About half of the total dose commitment comes from the external exposure pathway; the radionuclide ^{137}Cs would contribute about two-thirds of

TABLE I. ESTIMATES OF COLLECTIVE AND AVERAGE INDIVIDUAL DOSE COMMITMENT FOR VARIOUS REGIONS

	Collective Dose Commitment (person-Gy)			
	1st Year		50 Year	
	Preferred Estimate	Average Individual Dose (mGy)	Preferred Estimate	Average Individual Dose (mGy)
USSR (European)	2.5E+05	3.3E+00	4.7E+05	6.1E+00
USSR (Asian)	5.9E+04		1.1E+05	
Europe (non-USSR)	3.1E+05	6.4E-01	5.8E+05	1.2E+00
Asia (non-USSR)	1.4E+04	7.6E-03	2.7E+04	1.4E-02
North America		2.4E-03		4.5E-03
United States	5.7E+02	2.4E-03	1.1E+03	4.6E-03
Canada	5.0E+01	2.1E-03	9.4E+01	3.9E-03
Northern Hemisphere	6.3E+05		1.2E+06	

the total from both pathways. We estimate that the total collective dose commitment for the first year following the accident was 630 000 person-Gy and that it will be 1 200 000 person-Gy over 50 years. Most of this will be experienced within the USSR and within the non-USSR part of Europe.

ACKNOWLEDGEMENT

Part of this work was performed under the auspices of the U.S. Department of Energy by the Lawrence Livermore National Laboratory under contract W-7405-Eng-48.

REFERENCES

[1] GOLDMAN, M., CATLIN, R.J., ANSPAUGH, L.R., et al., Health and Environmental Consequences of the Chernobyl Nuclear Power Plant Accident, U.S. Department of Energy Rep. DOE/ER-0332 (1987).

[2] GUDIKSEN, P.H., LANGE, R., Atmospheric dispersion modeling of radioactivity releases from the Chernobyl event, Nature (in press).

[3] CAMBRAY, R.S., CAWSE, P.A., GARLAND, J.A., et al., Observations on radioactivity from the Chernobyl accident, Nucl. Energy 26 2 (1987) 77.

[4] The Accident at the Chernobyl Nuclear Power Plant and Its Consequences, USSR State Committee on the Utilization of Atomic Energy, IAEA Translation (1986).

[5] Analysis of the Radiological Consequences of the Accident at the Chernobyl Nuclear Power Plant for the Population of the European Regions of the USSR, Ministry of Health of the USSR (WHO A40/INF.DOC/9, May 1987).

[6] PAAKKOLA, O., AALTONEN, H., ARVELA, H., et al., Second Interim Report, Radiation Situation in Finland from May 5 to May 16, 1986, Finnish Centre for Radiation and Nuclear Safety, Helsinki, Rep. STUK-B-VALO 45 (1986).

[7] BECK, H.L., Exposure Rate Conversion Factors for Radionuclides Deposited on the Ground, U.S. Department of Energy Environmental Measurements Laboratory, New York, Rep. EML-378 (1980).

[8] BECK, H.L., Environmental gamma radiation from deposited fission products, 1960-1964, Health Phys. 12 (1966) 313.

[9] Ionizing Radiation: Sources and Biological Effects, United Nations Scientific Committee on the Effects of Atomic Radiation, New York (1982).

[10] JACOB, P., Paper presented at the Workshop on the Radiological Consequences of Chernobyl, Commission of European Communities, Brussels, February 3–5, 1987.

[11] WHICKER, F.W., KIRCHNER, T.B., PATHWAY: A dynamic food-chain model to predict radionuclide ingestion after fallout deposition, Health Phys. 52 (1987) 717.

[12] NG, Y.C., Personal communication, Lawrence Livermore National Laboratory, Livermore, Calif. (1987).

[13] Limits for Intakes of Radionuclides by Workers, International Commission on Radiological Protection, ICRP Publication 30, Part 1, Pergamon Press, New York (1979).

THE SECTOR OF AGRO-INDUSTRIAL PRODUCTION: POST-ACCIDENT RADIOLOGICAL CONSEQUENCES AND THE BASIC PROTECTIVE MEASURES

N.A. KORNEEV, A.P. POVALYAEV, R.M. ALEKSAKHIN,
L.I. PANTELEEV, A.N. RATNIKOV, S.V. KRUGLOV,
N.I. SANZHAROVA, N.N. ISAMOV, A.N. SIROTKIN
USSR State Agro-Industrial Committee (Gosagroprom),
Moscow,
Union of Soviet Socialist Republics

Abstract

THE SECTOR OF AGRO-INDUSTRIAL PRODUCTION: POST-ACCIDENT RADIOLOGICAL CONSEQUENCES AND THE BASIC PROTECTIVE MEASURES.

The accident at the Chernobyl nuclear power station led to fallout of radioactive substances on agricultural land. In this connection, the USSR State Agro-Industrial Committee, Gosagroprom, faced a number of tasks concerning evaluation of the radiological situation in the sector of agro-industrial production, organization of radiation monitoring of agricultural products and implementation of a series of agricultural ameliorative measures to reduce the radionuclide contamination of foodstuffs. The study of the radiological situation revealed a complex dynamics of the emergency release of radioactivity and inhomogeneous composition of the fallout on the contamination zone. Early autumn fallout of radionuclides caused aerial contamination of the leaves and stems of a number of agricultural crops — winter, natural and sown perennial grasses. The other types of plant were contaminated mainly through the soil and through wind lift. After deposition the bulk of the radionuclides were concentrated in the top soil. The availability of radionuclides for root assimilation depends on the proportion between the forms in which they occur in soils, which varies considerably in soils of different types. Application of amendments changes the transfer of radionuclides to agricultural crops. In animal husbandry two principal areas of activity are of the greatest interest: evaluation of the degree of contamination of products (first of all, milk and meat) and determination of the physiological condition of livestock in the immediate neighbourhood of the power station. An evaluation has been made of the organizational and agricultural ameliorative measures which were applied to ensure stable agro-industrial production in the contaminated area.

1. INTRODUCTION

The accident at the Chernobyl nuclear power station resulted in fallout of radioactive substances on agricultural land, and the products of the agro-industrial complex containing radionuclides became a source of additional exposure for the population. While the contribution of the exposure associated with the consumption of the locally grown foodstuffs to the total exposure dose at different stages of the post-accident period may vary, this pathway was, on the whole, sufficiently significant so as to predetermine the importance of the protection measures which are being implemented in the sector of agro-industrial production in the overall system of activities designed to counteract the consequences of the accident and to ensure the radiation safety of the population. Another important factor should be emphasized — the pathway of human exposure associated with the consumption of contaminated agricultural products is the one which is most susceptible to control among the other sources of additional exposure (especially in comparison with external exposure). Therefore, reduction of the total dose burden for the population depends to a considerable extent on the possibilities of limiting the internal exposure dose due to the consumption of foodstuffs.

The basic tasks of the USSR State Agro-Industrial Committee, Gosagroprom, in connection with controlling the consequences of the Chernobyl accident were:

(a) To evaluate the radiological situation in the agro-industrial production sector in the area affected by the accident;
(b) To implement a complex of emergency measures designed to preclude the consumption of a number of agricultural products (especially milk) containing iodine radionuclides;
(c) To evaluate the characteristics of contamination of agricultural land and products and to organize radiation monitoring thereof;
(d) To predict the transfer of radionuclides to agricultural products and to organize the agro-industrial complex so as to ensure minimal transfer of radionuclides to plant and animal products in the second and subsequent years after the accident;
(e) To put into effect a complex of agricultural ameliorative measures with a view to reducing the uptake of radionuclides in agricultural products in the long term.

From the radiological standpoint, two features of the accident situation at the Chernobyl power station should be noted which are relevant to agriculture:

(1) The release of radioactive substances occurred at the end of April and the beginning of May, when the main field activities (sowing and planting) in this region had been completed and stabled livestock had been put out to pasture;
(2) The soil cover of the contamination zone consists of light textured sandy and argillo-arenaceous sward-podzolic acid soils with a low content of nutrients, a low sorption capacity and flushing regime (wooded regions) and this predetermines sufficiently intensive incorporation of radionuclides in the agricultural chains of migration (chiefly from soil to plant).

2. RADIOLOGICAL SITUATION IN THE AGRO-INDUSTRIAL PRODUCTION SECTOR IN THE REGION OF THE ACCIDENT

The density of contamination of arable land was evaluated by vehicle borne survey using generally accepted techniques. All activities on the assessment of the

TABLE I. RADIONUCLIDE COMPOSITION OF ARABLE SOILS OUTSIDE THE 30 km ZONE ON 15 JUNE 1986
(% of total gamma emitting nuclides: main radionuclides are shown)

Sector	Zr-95 +Nb-95	Ru-103	Ru-106	I-131	Cs-134	Cs-137	Ba-140	Ce-141	Ce-144
North	15.5	23.7	27.0	1.8	5.2	12.3	2.1	4.7	6.0
	8.2	24.5	13.6	2.0	13.2	30.0	3.2	1.8	3.4
	0.8	24.4	—	4.2	14.6	43.7	4.4	0.9	2.0
South	28.6	11.4	5.5	2.6	11.2	15.9	5.7	6.3	12.3
	33.1	10.4	6.0	0.5	4.5	8.1	19.6	4.8	13.0
West	17.6	9.0	16.2	2.7	10.3	16.7	14.0	4.3	9.2
East	2.9	22.8	4.2	5.1	7.7	24.4	9.2	8.4	8.3

Note: In the case of the northern and southern sectors respectively, three and two series of data for regions at different distances from the power station are given.

radiological situation in the agricultural sector carried out by Gosagroprom were co-ordinated with those on the assessment of the radiation situation by the USSR State Committee on Hydrometeorology.

The nuclide composition of gamma emitters in soil and plant samples was determined by semiconductor gamma spectrometry. The radiochemical determination of ^{89}Sr and ^{90}Sr was performed by standard methodologies. The content of radionuclides in soil was measured from 10 May till the end of 1986. In soil samples the following radionuclides were identified: ^{95}Zr, ^{95}Nb, ^{103}Ru, ^{106}Ru, ^{131}I, ^{134}Cs, ^{137}Cs, ^{140}Ba, ^{140}La, ^{141}Ce and ^{144}Ce (Table I).

The complex dynamics of radionuclide release from the damaged reactor unit and the sharp change in the meteorological conditions in this period were responsible for the inhomogeneous nuclide composition of the fallout on the contamination zone. In the northern sector we note a change in radionuclide composition in favour of volatile nuclides: ^{131}I, ^{134}Cs and ^{137}Cs and in the south in favour of the high melting ones: ^{95}Zr, ^{95}Nb, ^{141}Ce and ^{144}Ce. There was probably a change in the radionuclide composition also in the other directions considered although no clear trend was noticed in these data.

Figure 1 shows straight lines indicating a correlation between the content of ^{95}Zr and ^{137}Cs and that of ^{144}Ce, the data being expressed as a percentage of the total activity of gamma emitting nuclides. The $^{95}Zr/^{144}Ce$ concentration ratio remained constant in a wide range of contamination density and on 15 June 1986 had a value of 0.35. The $^{137}Cs/^{144}Ce$ concentration ratio in the same contamination

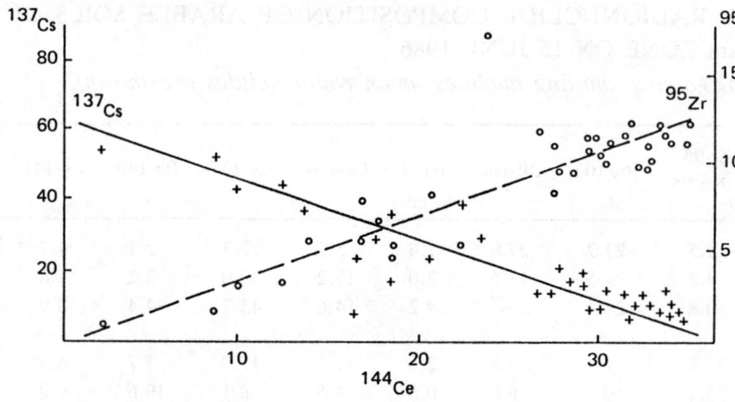

FIG. 1. Correlation between the content of Cs-137 and Zr-95 (%) and that of Ce-144 (%) in the soil.

TABLE II. PARAMETERS OF THE REGRESSION EQUATIONS DESCRIBING THE RELATION BETWEEN THE CONTENTS OF Sr-90, Cs-137 AND Ce-144 IN SOILS ON 15 JUNE 1986

Parameter	A_{Sr-90}/A_{Ce-144}			A_{Sr-90}/A_{Cs-137}		
	0.01–0.05	0.05–0.15	>0.15	0.01–0.05	0.05–0.15	0.15
a	0.03 ± 0.003	0.10 ± 0.005	0.26 ± 0.03	0.77 ± 0.05	0.94 ± 0.03	1.10 ± 0.07
b	−0.08 ± 0.09	−0.28 ± 0.16	0.07 ± 0.08	−1.55 ± 0.06	−1.06 ± 0.05	0.53 ± 0.05
g	0.81	0.95	0.88	0.89	0.93	0.93
p	44	41	32	76	52	36

density range decreased from 6 to 0.66, indicating volatile ^{137}Cs enrichment in the fallout with increasing distance from the power station. The change in the radionuclide concentration in the fallout as a function of direction and distance from the release source makes it impossible to use the average data on radionuclide composition to obtain correct estimates of the radionuclide contamination density of the area on the basis of measurement of the total activity or dose of gamma radiation in the locality. For this reason, it is necessary to carry out a substantially greater volume of direct gamma spectrometric measurements of radionuclide content. It is especially laborious to determine ^{90}Sr content in environmental bodies; we therefore attempted to find correlations between the ^{90}Sr content and the quantity of any gamma emitting nuclide which can be determined with sufficient ease and reliability by gamma spectrometry. A linear regressive relationship was established for the ^{90}Sr–^{144}Ce pair:

$$A_{^{90}Sr} = aA_{^{144}Ce} + b$$

and an exponential dependence for the ^{90}Sr–^{137}Cs pair:

$$\lg A_{^{90}Sr} = a \lg A_{^{137}Cs} + b$$

In the contamination zone we can distinguish three groups of region differing in the regression equation parameters shown in Table II.

3. AERIAL UPTAKE OF RADIONUCLIDES BY AGRICULTURAL PLANTS

The early spring fallout of radionuclides after the accident caused aerial contamination of the leaves and stems of agricultural plants — winter, natural and perennial sown grasses.[1] In other types of plant the contamination was mainly through soil and wind lift. The content of radionuclides taken up by the leaves and stems depended on their total mass and a number of other parameters characterizing the interaction between the radioactive impurity and plants. The contamination was highest in the case of natural herbage and perennial sown grasses and somewhat lower for winter rye. The lowest accumulation of radionuclides was noted in crops which had not suffered from direct (aerial) contamination, for example, barley, oats and maize (Table III). The differences in radionuclide accumulation in plants are due to the difference in the rates of initial interception. The starting date for calculations was taken as D+15. In this period the bulk of the fallout radionuclides were contained in the soil and turf — on an average up to 60–80%. Natural herbage retained 30–45% of the deposited amount of gamma emitting radionuclides, the interception of ^{137}Cs being about 30%. These values were somewhat lower in the case

[1] Seasons refer to the Northern Hemisphere.

TABLE III. LEAF AND STEM CONTAMINATION IN PLANTS ON 10 JUNE 1986

Crop	Calculated contamination of plants; n $[10^{-8}$ (Ci/kg)/(Ci/km^2)]^a
Natural grasses	10.0–16.2
Perennial sown grasses	1.5–22.5
Winter rye	1.1– 7.5
Barley	0.3– 0.5
Oats	0.2– 0.3
Maize	0.4– 2.0

^{a} 1 Ci = 3.70 × 10^{10} Bq.

of perennial sown grasses: 20–40% for the total gamma emitting nuclides and 7–15% for ^{137}Cs. For winter rye the initial interception of the total gamma emitting nuclides varied from 10 to 30% and that of ^{137}Cs from 7 to 20%. The range of variation in these figures for one and the same crop depends both on the crop status in the fallout period and on the location of the field in relation to the contamination source.

During the vegetative period the radionuclide concentration in plants decreases as a result of radionuclide decay, growth in the biomass and cleaning of the above ground part of plants by the action of meteorological factors. The most intensive fall in the contamination level of vegetation occurs in the first period after fallout. By the beginning of June the leaves and stems of winter rye retained no more than 10–20% of the content of total gamma emitting nuclides on D+15 and by harvest time this value fell to 5–10%. For herbage these values were somewhat lower: 9–15% and 1–6%, respectively. The first period of semicleaning for leaves and stems of winter rye was about 15 days. Later the second effective period of semicleaning of plants was more extended: about 50 days. In the case of herbage these values were 10–12 and 23–30 days, respectively. The sharp decline in the contamination level of plants in the first period was caused mainly by the decay of the short lived nuclides and fall in their contribution to the total activity. The decisive influence of the decay of the short lived nuclides in the first period is confirmed if we consider the dynamics of ^{137}Cs contamination of vegetation. By the beginning of June the leaves and stems of winter rye contained up to 50–70% of the ^{137}Cs from the fallout and by harvest time up to 30–40%. These values were somewhat lower for herbage: 15–40% and 5–10%, respectively. Thus, the ^{137}Cs concentration fell much more slowly than that of the total gamma emitting nuclides.

Substantial differences were observed between the contamination values for the first harvest of grasses which had suffered aerial contamination and those for the second harvest of grasses contaminated predominantly by radionuclide uptake from turf and soil and also by wind lift. The hay of the first harvest contained 3–6 times more gamma emitting nuclides and 2–3 times more ^{137}Cs than the hay of the second harvest.

Reploughing, followed by sowing of mixed grass crops, led to a reduction in the contamination of the herbage of hay fields and pasture. The contamination level of the herbage of a vetch–oat mixture was lower by a factor of 8–10 than that of herbage from undisturbed land. The bulk of the radionuclides remain during the entire vegetative period in the near ground part of the crops. Thus, in the second harvest of grasses up to 90% of the total radionuclides and more than 60% of ^{137}Cs were concentrated in the lower 0–5 cm of the herbage. In this manner, in undisturbed hay fields the radionuclide uptake by plants through their basal part may make an additional contribution to the contamination of herbage both in the first year after fallout and in the subsequent period.

4. DISTRIBUTION AND FORMS OF ^{137}Cs IN SOILS

The inclusion of radionuclides in the agricultural chains of migration in the second and subsequent post-accident years will be determined, first of all, by the availability of the radionuclides for root assimilation from soil, i.e. it will depend on

TABLE IV. DISTRIBUTION OF RADIONUCLIDES IN THE PROFILE OF SWARD-PODZOLIC ARGILLO-ARENACEOUS SOIL

Type of land	Depth (cm)	Per cent of total content in 0–20 cm layer					
		1986-05-08		1986-06-10		1986-08-05	
		Σ_γ	Cs-137	Σ_γ	Cs-137	Σ_γ	Cs-137
Perennial	0– 2	98.0	99.5	97.3	99.1	79.8	75.6
Sown	2– 5	1.5	0.45	2.1	0.8	9.1	12.9
Grass	5–10	0.5	0.005	0.6	0.1	8.6	6.1
Field	10–20	—	—	—	—	2.7	5.4
Maize	0–02	99.8	99.6	99.3	97.3	61.9	57.6
field	2– 5	0.16	0.37	0.68	1.8	33.5	33.8
	5–10	0.04	0.03	0.02	0.9	4.1	6.2
	10–20	—	—	—	—	0.5	2.4

TABLE V. COEFFICIENTS OF ACCUMULATION OF Cs-137 IN OATS, WHEAT, MUSTARD AND MAIZE

Variants of experiment	Oats		Wheat		Mustard		Maize leaves and stems
	Grain	Straw	Grain	Straw	Grain	Straw	
1	2	3	4	5	6	7	8
Sward-podzolic argillo-arenaceous silty sand							
Without fertilizer (control)	0.042	0.11	0.030	0.08	0.18	0.24	0.07
$N_{70}P_{70}K_{70}$ (background)	0.027	0.07	0.024	0.05	0.14	0.19	0.06
Background + $1H_hCaCO_3$	0.030	0.08	0.021	0.05	0.15	0.20	0.05
Background + $2H_hCaCO_3$	0.024	0.06	0.018	0.04	0.14	0.18	0.10
Background + K_{140}	0.012	0.03	0.018	0.04	0.13	0.17	0.08
Background + P_{140}	0.027	0.07	0.015	0.05	0.16	0.22	0.07
Background + $P_{140}K_{140}$	0.015	0.04	0.009	0.02	0.17	0.23	0.07
Background + zeolite 5 t/ha	0.033	0.08	0.012	0.03	0.26	0.35	0.08
Background + zeolite 10 t/ha	0.039	0.10	0.012	0.03	0.20	0.25	0.09
Background + $1H_hCaCO_3$ + zeolite 10 t/ha	0.030	0.08	0.027	0.07	0.16	0.22	0.12
Background + $1H_hCaCO_3$ K_{140} + P_{140} + zeolite 10 t/ha	0.027	0.08	0.021	0.05	0.14	0.20	0.07
Sward-podzolic sandy loam with a higher sand fraction							
Without fertilizer (control)	—	—	0.12	0.28	0.66	0.90	—
$N_{70}P_{70}K_{70}$ (background)	—	—	0.10	0.24	0.48	0.64	—
Background + $1H_hCaCO_3$	—	—	0.11	0.27	0.69	0.93	—
Background + zeolite 10 t/ha	—	—	0.11	0.27	0.42	0.72	—
Background + $1H_hCaCO_3$ + zeolite 10 t/ha	—	—	0.17	0.36	0.57	0.78	—

TABLE V. (cont.)

Variants of experiment	Oats		Wheat		Mustard		Maize leaves and stems
	Grain	Straw	Grain	Straw	Grain	Straw	
1	2	3	4	5	6	7	8
Sward-podzolic light loam with a higher sand fraction							
Without fertilizer (control)	—	—	0.15	0.36	0.78	1.10	—
$N_{70}P_{70}K_{70}$ (background)	—	—	0.14	0.33	0.84	1.14	—
Background + $1H_hCaCO_3$	—	—	0.14	0.33	0.57	0.81	—
Background + zeolite 10 t/ha	—	—	0.16	0.36	0.81	1.11	—
Background + $1H_hCaCO_3$ + zeolite 10 t/ha	—	—	0.17	0.39	0.54	0.72	—
Sward-podzolic argillo-arenaceous silty sand							
Without fertilizer (control)	—	—	0.024	0.063	—	—	0.10
$N_{70}P_{70}K_{70}$ (background)	—	—	0.024	0.063	—	—	0.08
Background + $1H_hCaCO_3$	—	—	0.024	0.057	—	—	0.07
Background + K_{140}	—	—	0.015	0.042	—	—	—
Background + zeolite 10 t/ha	—	—	0.027	0.069	—	—	0.13
Background + $1H_hCaCO_3$ +zeolite 10 t/ha	—	—	0.027	0.069	—	—	0.10

their distribution in the soil profile and on their forms. Immediately after fallout the bulk of the radionuclides were concentrated in the upper 0–2 cm layer of the soil. During the vegetative period there was some redistribution in the 0–5 cm layer (Table IV). Reploughing of the soil led to a uniform distribution of ^{137}Cs in the arable layer of the soil: 0–5 cm, 31.7%, 5–10 cm, 37.7% and 10–15 cm, 30.6%.

By the beginning of October 1986 a considerable part of ^{137}Cs which had fallen on the soil was in the non-exchangeable state. The content of the different forms of ^{137}Cs in soils of different types varied. The quantity of water soluble and exchangeable ^{137}Cs together varied from 16 to 36% in the soils studied. The content of strongly bound ^{137}Cs (not extractable by 3NHCl) varied from 36 to 71%. The low quantity of ^{137}Cs in the strongly bound form in the soils of the western sector (less than 5%) is evidently attributable to the predominance there of sandy soils with a very small content of the clay fraction.

The results of many years' study in stationary experiments (1977–1984) at the All-Union Scientific Research Institute of Agricultural Radiology of Gosagroprom showed that the quantity of exchangeable ^{137}Cs in the arable layer of sward-podzolic argillo-arenaceous soil decreased by a factor of two, after eight years of residence in the soil, i.e. there was no sharp fall with time in the amount of ^{137}Cs available to plants. It should be noted, however, that after eight years of interaction with soil a considerable part of ^{137}Cs (70%) changes into the strongly bound form, the proportion of the fixed ^{137}Cs increasing by a factor of 1.5 with time.

5. ACCUMULATION OF RADIONUCLIDES IN PLANTS FROM SOIL

As has already been pointed out, in the mixture of radionuclides contained in the fallout on the contaminated area the most important from the standpoint of agricultural radiology is ^{137}Cs (and to some extent during the first 1–2 years, ^{134}Cs). With a view to evaluating the inclusion of ^{137}Cs in the soil–plant chain vegetative experiments were carried out on four of the most representative sward-podzolic soils from the northern sector of the contaminated territory.

The quantity of soil in each vessel was 1–5 kg.

As numerous studies have shown, among the most effective methods of reducing ^{137}Cs uptake from soil by plants in the case of low fertility light textured acid soils are the application of mineral fertilizers (especially potassium fertilizers, considering the antagonism during the assimilation of ^{137}Cs and its chemical analogue potassium by root systems) and the liming of soils. Moreover, there are data indicating a decrease in the transfer of radionuclides (including ^{137}Cs) to plants in light textured sward-podzolic soils when zeolite type clay minerals are added. In this case, the soil becomes enriched in finely dispersed fractions of clay minerals with a high exchange capacity, ensuring fixation of ^{137}Cs and its conversion into a state which is not easily assimilable by plants.

On the whole, addition of lime, mineral fertilizers and zeolites generally causes a reduction in the accumulation of ^{137}Cs in plants (Table V). The most effective reduction results from the addition to the soil of large amounts of potassium fertilizers (the greatest decrease was observed in the grain of oats in a soil of the sward-podzolic argillo-arenaceous silty sand type — by a factor of up to 3.5). On an average, in the case of grain, leaves and stems in all the plants studied, higher doses of potassium fertilizers lowered the transfer of this radionuclide to plants by a factor of up to 1.5.

The plants' response, in the form of their altered ^{137}Cs assimilation from the soil, to the addition of lime and mineral fertilizers and also to that of zeolites, was not observed in all the soils investigated. Thus, in sward-podzolic sandy loam with a higher sand fraction no decrease in ^{137}Cs transfer to plants was noted after liming and addition of fertilizers and zeolite.

The addition of zeolites was effective only in one of the four soil types studied and only for particular crops: the greatest reduction in ^{137}Cs accumulation was detected in wheat (by a factor 2.5 in grain). In a number of variants the application of zeolite increased ^{137}Cs transfer to plants.

The effectiveness of the agricultural ameliorative measures depended on the biological characteristics of plants and the time of interaction of the amendments with soil. Thus, in the case of mustard (grain and straw) and maize (leaves and stems), the addition of lime, fertilizers and zeolite did not substantially lower ^{137}Cs transfer to plants. The accumulation of ^{137}Cs, expressed as the ratio of radionuclide concentration in plants and soil respectively, in mustard grain in a soil of the sward-podzolic argillo-arenaceous silty sand type varied in the 0.13–0.26 range,the figures for the grain of wheat and oats being 0.009–0.030 and 0.012–0.042, respectively. In the same crop ^{137}Cs accumulation depends on soil properties and can vary by a factor of up to 4–5. It is to be emphasized that the effectiveness of adding zeolites from the standpoint of reduction of ^{137}Cs transfer to plants should be evaluated after a prolonged period of interaction between zeolites and the solid phase of soil.

6. ASPECTS OF RADIONUCLIDE MIGRATION AND BIOLOGICAL ACTION OF RELEVANCE TO ANIMAL HUSBANDRY

In animal husbandry, two main activities are of the greatest interest: assessment of the degree of product contamination and determination of the physiological condition of livestock in the immediate neighbourhood of the nuclear power station.

In the aerial contamination of pastures and hay fields, the most important were the nuclides of iodine and caesium as potential contaminants of animal products. Moreover, it is well known that the significance of radioiodine lies in its accumulation in the thyroid. The content of these radionuclides in the grass in the neighbourhood of the nuclear power plant during the first few days was considerable.

The high solubility of the radioactive substances deposited from the fallout was conducive to the absorption of ^{131}I and ^{137}Cs from the gastrointestinal tract and their transfer to milk and meat. In this case, the concentration of radioiodine in milk was, on an average, 10 times higher than that of radiocaesium. The transfer factors for the iodine and caesium radionuclides from diet to milk essentially agree with the data obtained earlier (Table VI).

The physiological condition was studied in 400 cattle, 5 sheep and 15 horses, which were pastured near (10–15 km) the nuclear power station till the end of the

TABLE VI. TRANSFER OF RADIONUCLIDES FROM DIET TO MILK (% OF DAILY DIET IN 1 L OF MILK)

Radionuclide	Source of entry of radionuclides into the body	Transfer factor	
		Main part of the area	Individual sectors
I-131	Aqueous solution	1.0	1.0
	Feed	0.96	0.86–1.05
Cs-137	Aqueous solution	1.0	1.0
	Feed	0.94	0.86–1.23

TABLE VII. EXPOSURE DOSE RATE IN THE REGION OF THE THYROID AND STOMACH IN LIVESTOCK ON 7 MAY 1986 (%)

Type of livestock	Region of dosimetry	
	Thyroid	Xiphoid process of the sternum
Cattle: adult	100.0 ± 13.3	7.0 ± 0.9
Cattle: new born calves	44.9 ± 2.3	3.9 ± 0.4
Horses	30.5 ± 8.0	9.8 ± 0.8
Sheep	88.7 ± 16.0	5.5 ± 0.7

TABLE VIII. DYNAMICS OF THYROID HORMONES IN CATTLE (ng/mL)

Hormone	Time of study				Control
	Jul.	Nov.	Jan.	Feb.	
T_3	1.24 ± 0.07	1.11 ± 0.16	0.70 ± 0.16	0.78 ± 0.16	1.60 ± 0.11
T_4	42.99 ± 2.76	29.42 ± 3.91	27.14 ± 9.17	43.65 ± 4.08	59.8 ± 6.60

TABLE IX. RESULTS OF IMMUNOLOGICAL STUDIES OF BLOOD SERUM IN LIVESTOCK

Type	Time of study				Control
	May	Jul.	Nov.	Feb.	
			Leucocytes (thousands)		
Cattle	6.8 ± 0.4	7.3 ± 0.4	8.4 ± 0.7	6.7 ± 0.5	4.5 ± 12.0
Horses	—	—	—	8.3 ± 1.1	7.0 − 12.0
Sheep	—	—	—	7.5 ± 0.9	6.0 − 14.0
			Bactericidal action (%)		
Cattle	82.9 ± 1.7	70.1 ±1.0	66.4 ± 2.3	64.6 ± 0.9	84.0 ± 0.5
Horses	59.7 ± 1.0	—	59.2 ± 4.2	51.0 ± 2.5	60.5 ± 1.5
Sheep	70.0 ± 0.1	—	65.9 ± 1.7	50.6 ± 3.3	78.9 ± 1.4
			Lysozyme (%)		
Cattle	40.4 ± 2.3	—	40.4 ± 2.9	45.6 ± 2.8	48.4 ± 6.9
Horses	44.9 ± 7.1	—	34.0 ± 5.9	35.8 ± 4.1	51.0 ± 3.0
Sheep	38.7 ± 3.1	—	38.7 ± 3.1	39.5 ± 1.7	37.5 ± 1.3
			Beta haemolytic activity (%)		
Cattle	15.9 ± 0.8	16.4 ± 1.1	18.8 ± 2.2	27.5 ± 0.9	16.2 ± 0.7
Horses	58.5 ± 2.5	—	—	68.8 ± 4.2	60.6 ± 1.2
Sheep	21.2 ± 1.3	—	—	18.0 ± 0.6	20.0 ± 1.5

pasturing season. The content of radionuclides in the body of the animals was evaluated from the gamma radiation level in the thyroid and in the gastrointestinal tract, which was recorded by dosimetric instruments (Table VII).

Different species of livestock exhibited different doses in the thyroid region, and this was evidently associated with the mass of the organ and the measurement geometry. This is proven by the close values of dose in cattle and considerably lower but similar values in horses (in the latter the dose in the thyroid region was lower by a factor of 2–3 than in cattle). The doses in new born calves up to the age of three days was lower by a factor of up to two than in their mothers. Dose measurements in the region of the xiphoid process of the sternum in order to determine the concentration of radionuclides in the gastrointestinal tract showed lower levels differing from the radionuclide concentration in the thyroid by a factor of 10–50.

The radiation effect from incorporated radionuclides did not produce any clinical symptoms and was not reflected in the productivity of the animals. At the same time, there were cases of change in the functional status of the thyroid. Thyroid hypofunction phenomena were confirmed by the results of radioimmunoassay (Table VIII). There were no significant changes in the haematological indices. No outbreak of infectious and invasive diseases was recorded nor any increase of cases of non-contagious diseases. The indices of non-specific immunity in blood also varied within measurement errors (Table IX).

Thus, the results of radiological studies showed that in the whole of the area affected by radioactive contamination there were no grounds to expect or apprehend changes in the economically useful qualities of livestock owing to the radiation factor.

7. ORGANIZATIONAL MEASURES IN THE GOSAGROPROM SYSTEM WITH A VIEW TO ELIMINATING THE CONSEQUENCES OF THE ACCIDENT

On the basis of the radiation situation, as it became established, and the predictions about the transfer of radionuclides through the agricultural chains and accumulation of radioactive substances in products in the different post-accident periods, a number of organizational and agricultural ameliorative measures were implemented during the first year after the accident in the contaminated agricultural land. These measures were designed to obtain agro-industrial products fit for consumption in accordance with the standards on the maximum permissible content of radionuclides. In addition the following steps were taken in order to limit the consumption of certain types of product:

(a) Removal from the diet of the population, in the first period, of milk and some types of green vegetables containing high quantities of radionuclides (chiefly ^{131}I);

(b) Reprocessing of milk with high radionuclide concentrations into fuel oil;

(c) A complex of agricultural ameliorative measures in farming (liming of acid soils, addition of high doses of mineral fertilizers, especially potassium fertilizers, and addition of zeolites to soil);

(d) Radical improvement of natural meadows and expansion of areas occupied by sown perennial grasses to ensure sharp reduction of radionuclides in feeds and subsequently in animal products;

(e) To obtain meat satisfying the standards on ^{137}Cs content, it was suggested that before slaughter animals should be given feeds with a relatively low content of radionuclides;

(f) Over a limited area the profile of agro-industrial production has been changed with a view to obtaining products fit for consumption in accordance with the standards on the maximum permissible content of radionuclides.

During 1986–1987 Gosagroprom put into effect a system of radiation monitoring covering all agricultural products obtained in the area affected by the accident at the Chernobyl nuclear power station. All the foregoing measures made it possible to ensure stable functioning of the agro-industrial complex over the whole area contaminated by radioactive substances as a result of the accident.

BIBLIOGRAPHY
(in Russian)

ALEKSAKHIN, R.M., et al., Agrochemistry of caesium-137 and its accumulation in agricultural plants, Agrokhimiya **2** (1977) 129.

Information on the accident at the Chernobyl nuclear power station and the consequences thereof compiled for the IAEA, At. Ehnerg. **61** 5 (1986) 301.

ISRAEL, Yu.A., et al., Radioactive contamination of the natural environment in the zone affected by the accident at the Chernobyl nuclear power station, Meteorol. Gidrol. **2** (1987) 5.

KORNEEV, N.A., et al., Reduction of Radioactivity in Vegetation and Animal Products, Kolos, Moscow (1977).

MAREJ, A.N., et al., Global Fallout of Caesium-137 and Man, Atomizdat, Moscow (1974).

MOISEEV, I.T., et al., Accumulation of ^{137}Cs in agricultural plants in leached chernozem, Agrokhimiya **9** (1972) 122.

SIROTKIN, A.N., ''Uptake of fission products in the body of agricultural animals and transfer of radionuclides to animal products'', Radiobiology and Radioecology of Agricultural Animals (ANNENKOV, B.N., DIBOBES, I.K., ALEKSAKHIN, R.M., Eds), Atomizdat, Moscow (1973).

YUDINTSEVA, E.V., GULYAKIN, I.V., Agrochemistry of Strontium and Caesium Isotopes, Atomizdat, Moscow (1968).

ENVIRONMENTAL CONSEQUENCES OF CHERNOBYL IN WESTERN EUROPE

G.A.M. WEBB, M.E. MORREY
National Radiological Protection Board,
Chilton, Didcot, Oxfordshire,
United Kingdom

Abstract

ENVIRONMENTAL CONSEQUENCES OF CHERNOBYL IN WESTERN EUROPE.

The accident at Chernobyl in April 1986 spread radioactive contamination throughout Europe. Using measurements of activity concentrations in the air, on the ground and in a range of foods, estimates of the doses received by the populations of western Europe have been made. For the European Community (EC) countries, the average individual effective doses accruing from the first year following the accident are estimated to range from less than a microsievert in Portugal to a few hundred microsieverts in the Federal Republic of Germany, Italy and Greece. Elsewhere in western Europe, similar average individual effective doses have been estimated, up to a value of about 700 μSv for Austria. These values can be compared with the average effective dose of about 2000 μSv which the populations of western Europe receive each year from natural radiation. The restrictions placed on consumption of some foods are estimated to have been effective in reducing doses to the most exposed individuals by a factor of two in some countries. Throughout the EC, however, countermeasures are estimated to have reduced the collective effective dose by about 5%. The health impact on western Europe from the accident at Chernobyl can be assessed from the collective effective doses and the collective thyroid doses estimated together with the appropriate risk factors. For the EC the collective effective dose estimate is about 80 000 man Sv; considering western Europe as a whole the collective dose would be a few tens of percent higher. Using the risk factor recommended by United Nations Scientific Committee on the Effects of Atomic Radiation, the theoretical number of extra cancers predicted in western Europe due to the Chernobyl accident is of the order of 1000. This can be compared with the number of fatalities which would be expected over the next fifty years in western Europe as a result of radiation from natural sources, which is over half a million.

1. INTRODUCTION

The accident at the Chernobyl Unit 4 reactor in the USSR, in April 1986, resulted in large quantities of radioactive materials being released into the atmosphere over a period of several days. During this time a complex set of meteorological conditions occurred which dispersed and deposited these materials over large areas of Europe. Levels of radioactivity in the environment were monitored by national authorities, and in some countries countermeasures were introduced in order to reduce the radiation exposure of individuals in the population.

In this paper we review the environmental contamination measurements in member states of the European Community (EC) and some other countries in western Europe which were available in the first few months following this accident. From these we make a preliminary assessment of individual and population doses for each country, estimate the resulting health impact and indicate the effects of various preventive measures instigated by member states in terms of the reductions in both individual and population exposure which they produced. Finally, we compare these preliminary dose estimates with the results of some whole body monitoring programmes which are now becoming available; generally these indicate that the theoretical estimates of intakes tend to be pessimistic.

The dose estimates given in this paper are mainly based on the results of a study carried out by The UK National Radiological Protection Board (NRPB) for the Commission of the European Communities(1). Dose estimates provided by individual European countries and published in a recent Nuclear Energy Agency (NEA) report are also used to provide perspective(2).

2. ENVIRONMENTAL CONTAMINATION IN WESTERN EUROPE

The accident at the Chernobyl Unit 4 reactor resulted in two explosions and a subsequent fire which released large quantities of radioactive material into the atmosphere. Most of this release took place over a period of about ten days and was characterised by two peaks in the release rate, on the first day (26th April) and on the tenth day (5th May). During this time a complex and varying set of meteorological conditions developed over Europe, which dispersed the cloud over a very wide area. Subsequently, the release continued for many weeks at a much reduced rate until the reactor was finally sealed some five months later.

Figure 1 shows estimates of the areas of land covered by the main body of the cloud at various times during the main release. On Saturday 26th April, the day of the accident, only areas to the north of Chernobyl were affected by the cloud, but by the end of the first week the area of contaminated air extended from north-western Europe to south-eastern Europe; of the western European countries, only Spain and Portugal remained relatively unaffected.

As the cloud moved it deposited radioactive material on the ground. The extent of this deposition was very uneven in Europe, both between and within countries. This was due in part to the varying distances of each country from Chernobyl, but more importantly to the development of a frontal system which gave rise to a pattern of localised heavy rainfall over

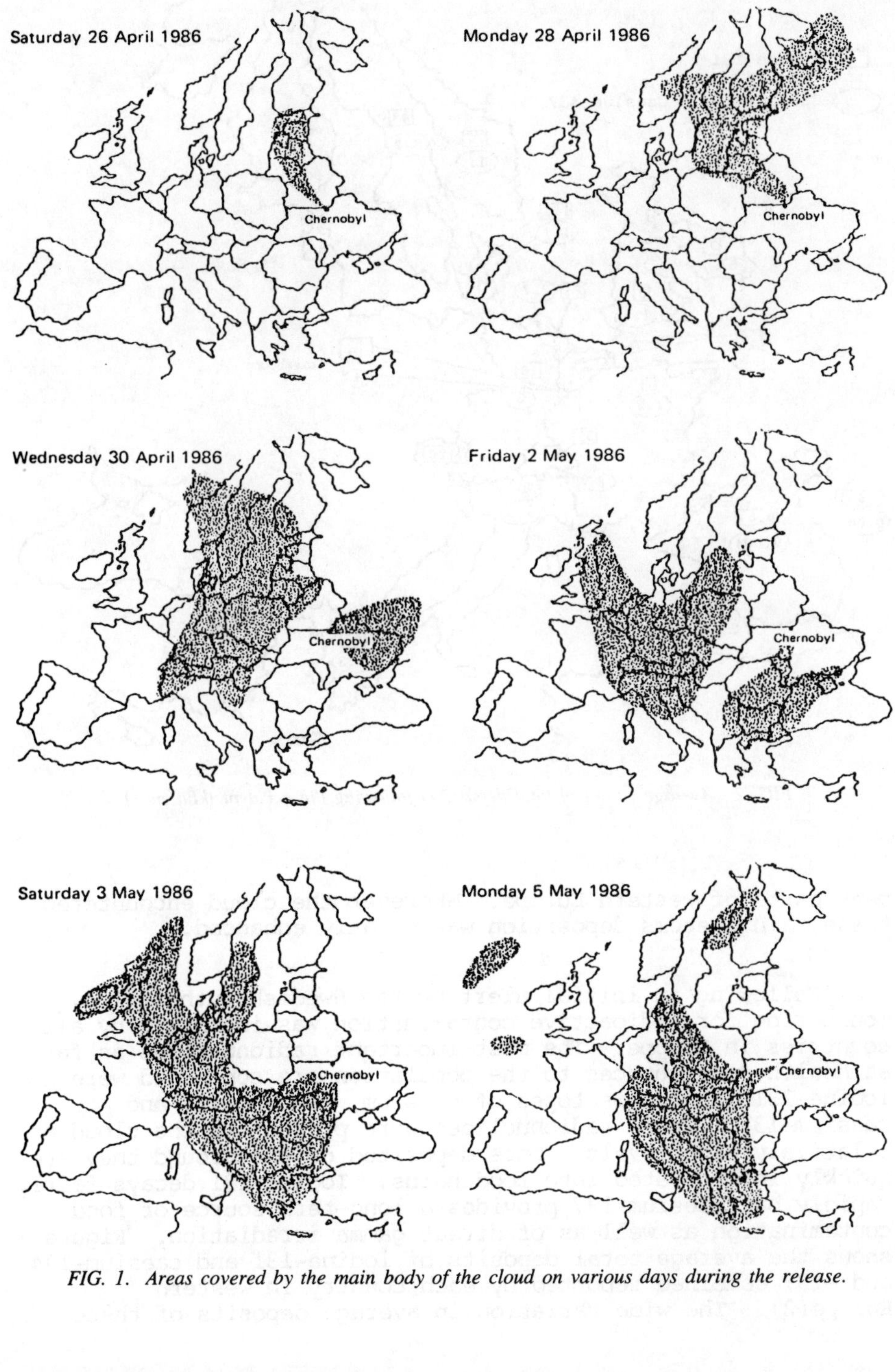

FIG. 1. Areas covered by the main body of the cloud on various days during the release.

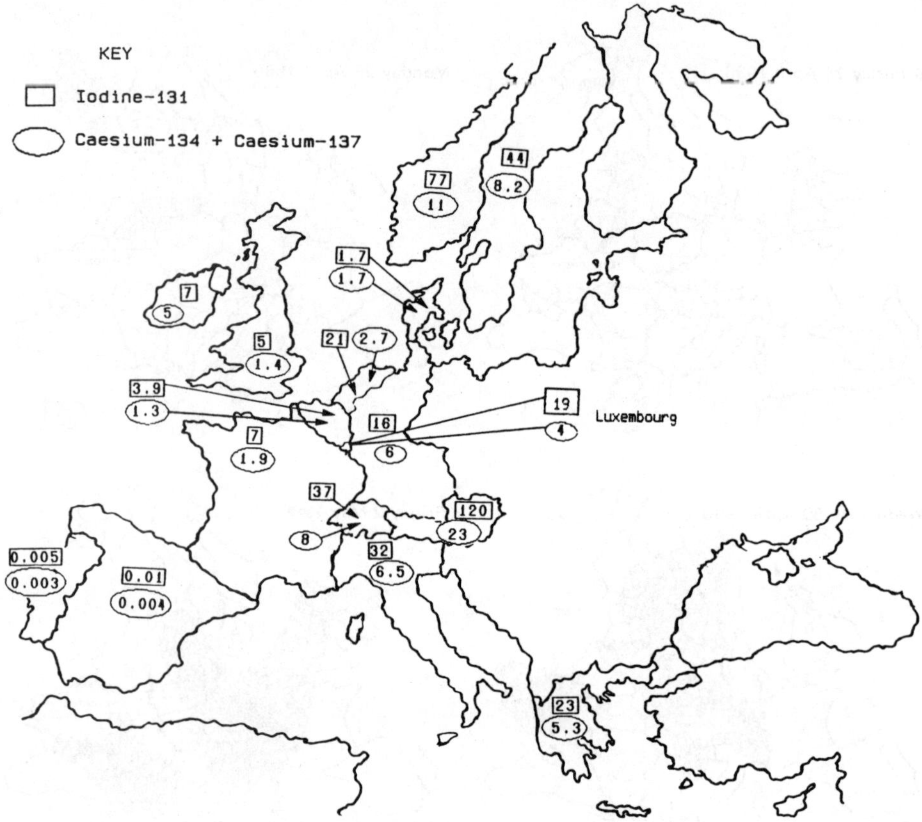

FIG. 2. Average values of total deposition following the accident ($kBq \cdot m^{-2}$).

many parts of western Europe. Wherever the cloud encountered these rain systems deposition was greatly enhanced.

Following an initial alert by the Swedish authorities, monitoring for radioactive contamination was increased by all countries in Europe. The most important radionuclides as far as health consequences to the population are concerned were iodine-131 and two isotopes of caesium, caesium-134 and caesium-137. These radionuclides were present in the cloud at relatively high levels. Once deposited on the ground they are quickly incorporated into foodchains. Iodine-131 decays fairly rapidly but caesium-137 provides a long-term source of food contamination as well as of direct gamma irradiation. Figure 2 shows the average total deposits of iodine-131 and caesium-134 and -137 combined reported by each country in western Europe(2). The wide variation in average deposits of these radionuclides is clear, ranging from very low values in Portugal to values in Austria of over 100 kBq m^{-2} for iodine-131 and about 20 kBq m^{-2} for caesium. Generally those countries which experienced relatively high deposits of iodine-131 also recorded high deposits of caesium. The factor by which the maximum recorded total deposition densities exceeded the average values varied between countries; Federal Republic of Germany, Italy, Scandinavia and the UK reported the greatest regional variation in deposition, of an order of magnitude or more, whereas several of the north European countries (Belgium, Denmark, the Netherlands and Luxembourg) received more uniform contamination(2).

Once deposited the radionuclides became incorporated into foods, both directly into plants and, via plants, into animals. The activity concentration levels measured in foods partly reflect the levels of deposit and partly the agricultural practices of each country, including any countermeasures taken to prevent contamination entering foods; for example, in some countries the diet of dairy cattle grazing out-of-doors was supplemented with stored feed. The ranges of activity concentration levels measured in different countries were very wide. For many foods the lower values were reported as below detection limits, whilst the highest concentrations, particularly of iodine-131, varied between countries by several orders of magnitude. For example, peak activity concentrations of iodine-131 in milk ranged from a few Bq l^{-1} in Portugal to nearly 20 000 Bq l^{-1} in parts of Federal Republic of Germany; in green vegetables the peak activity concentrations of caesium-137 exceeded 1000 Bq kg^{-1} in Federal Republic of Germany, Greece and Italy(1).

3. THE DOSE ASSESSMENT PROCEDURE

Following the release of radionuclides to atmosphere people will be irradiated via a number of different routes. Those pathways which are likely to give rise to significant radiation exposure are external irradiation of the whole body from the cloud and from deposited material, and internal irradiation from inhalation of the material in the cloud and from ingestion of contaminated foods. Account was taken of each of these dose pathways for the dose estimates given in this paper. Inhalation doses and doses from external gamma irradiation by the cloud have been estimated using measured activity concentrations in air. Doses from external gamma irradiation from material deposited on the ground have been estimated using measured dose rates and levels of deposition of important radionuclides. Doses from ingestion have been estimated using both deposition levels and measured activity concentrations in foods. In calculating these doses, account has been taken of the countermeasures introduced in some countries.

3.1 Estimation of individual dose

The radiation dose to any individual depends on many factors, including the fraction of their time spent outdoors, their breathing rate and diet, their body size and metabolism. Thus for any given measured level of contamination, there can be a relatively wide range of individual doses. In order to give some idea of the likely range, doses from the internal irradiation pathways have been estimated for three representative age-groups, a 1 year old infant, a 10 year old child and an adult, using intakes of air, water and various foods appropriate for each age-group, and taking into account the body-size and metabolic characteristics of each group. The variation with age of external irradiation doses is small and is not considered here. For each age-group doses have been evaluated for an average member of the population, and separately for a "critical" member of the population, that is, a theoretical person with habits which lead to higher than average exposure from all of the pathways considered. The doses estimated for the theoretical critical individual are intended to give an indication of the possible upper end of the range of doses in the population.

3.2 Estimation of collective dose

It is useful to have a measure of the total radiation dose to a whole population. The quantity used to express this total is the collective dose obtained by adding all the individual doses in the exposed population. In this paper we calculate both collective effective doses and collective thyroid doses, and evaluate them separately for the populations of each country.

4. RESULTS OF THE DOSE ASSESSMENT

In carrying out theoretical dose assessments a number of judgements and assumptions are required. These give rise to uncertainties in the dose estimates that are made. Many EC countries have made their own detailed theoretical dose assessments(2) and inevitably there are differences in the dose estimates made, particularly those for critical groups. However, the theoretical estimates of average doses are within factors of two or three of each other and we consider this to be indicative of the reliability of the results.

4.1 Average individual dose

The average effective doses to adult individuals in each EC country from exposure in both the first year and over the first fifty years following the Chernobyl accident are shown in Table I. The average doses in the first year range from 0.2 μSv in Portugal to 150–300 μSv in the Federal Republic of Germany, Italy and Greece. In most cases, consumption of food and external irradiation from deposited material are the major contributors to doses in the first year. In the case of food consumption, the average doses in the first year are mainly due to caesium-134 and caesium-137, which together contribute between about 65% and 90% of the dose, the remainder being due to iodine-131. As can be seen from Table I, exposure in the first year contributes between 45% and 75% of the total dose over 50 years. The range of doses listed in Table I can be compared with other published dose estimates; generally average adult effective doses throughout western Europe are estimated to lie within this range, although the average dose received in Austria is likely to have been slightly higher at about 700 μSv(2).

The average effective doses may be placed in some perspective by comparing them with the effective dose received each year from natural background radiation. In Europe, average annual individual doses from natural background are in the range 1000–3000 μSv. Thus average doses to individuals over the 50 years following the Chernobyl accident will be less than those received in one year from natural background radiation.

Average doses to the thyroids of individuals in the first year after the accident are shown in Table II. In general, the pattern of results is the same as that for effective doses, but there are small differences due to variations in the amounts of iodine and caesium radioisotopes deposited in countries. Again, other published dose estimates indicate a similar range of average thyroid doses for most western European countries, but doses in excess of 1000 μSv have been estimated for Austria, Federal Republic of Germany, Greece, Italy and Switzerland(2).

4.2 Doses to the critical group

Effective doses to the theoretical critical group in the first year after the accident are also shown in Table I. Differences between countries are partly a reflection of the different intake rates assumed, but mainly result from differences in environmental concentrations. The critical group effective doses range from 4 μSv in Portugal to about

TABLE I. AVERAGE ADULT AND THEORETICAL CRITICAL GROUP EFFECTIVE DOSES IN EC COUNTRIES[1]

Country	Effective Dose (μSv)					
	Critical Group[2]		Average Adult[3]			
	1st year[4]		1st year[4]		50 years[5]	
Belgium	430	(I)	44		82	
Denmark	400	(I)	55		95	
France	970	(I)	39	(21-110)	76	(39-220)
FRG	1900	(I)	150	(46-380)	360	(85-950)
Greece	1300	(C/A)	300		530	
Ireland	520	(A)	97		170	
Italy	1100	(A)	160	(120-200)	310	(210-400)
Luxembourg	400	(I)	47		86	
Netherlands	380	(A)	60		100	
Portugal	3.6	(I)	0.2		0.3	
Spain	70	(I)	0.9	(0.2-3.9)	1.2	(0.3-4.7)
UK	840	(I)	32	(20-190)	46	(25-270)

Notes

1. The results are given to two significant figures only to avoid rounding errors in subsequent calculations; this degree of accuracy is not implied.

2. The age group forming the critical group is indicated in brackets: I = 1 year old infant, C = 10 year old child, A = adult.

3. Where deposition levels varied considerably within a country, average doses were calculated for separate regions. The principal value given is the average adult effective dose for the whole country, the values in brackets indicate the range of average doses calculated within the country.

4. The doses arising from intakes and external exposures during the year May 1986 - April 1987.

5. The doses arising from intakes and external exposures during the 50 years following the accident.

TABLE II. AVERAGE AND THEORETICAL CRITICAL GROUP THYROID DOSES IN EC COUNTRIES IN THE FIRST YEAR[1,2]

Country	Critical Group[3]	Thyroid Dose (μSv)					
		Adult		Average Child[4]		Infant[5]	
Belgium	10 000	290		630		2500	
Denmark	9 300	450		1000		2700	
France	23 000	280	(200- 590)	560	(400-1200)	1500	(1100-3200)
FRG	30 000	540	(280-1100)	1100	(630-2100)	4000	(2400-7400)
Greece	6 500	530		1100		2700	
Ireland	6 100	370		770		2000	
Italy	4 700	810	(550- 910)	950	(730-1000)	800	(560-860)
Luxembourg	9 200	370		830		2500	
Netherlands	5 500	370		730		1500	
Portugal	50	0.7		1.3		4.6	
Spain	2 000	6.8	(0.9-30)	15	(1.7-66)	52	(5.1-240)
UK	11 000	200	(130-640)	430	(250-1300)	1100	(490-3800)

Notes

1. The results are given to two significant figures only to avoid rounding errors in subsequent calculations; this degree of accuracy is not implied.
2. Where deposition levels varied considerably within a country, average doses were calculated for separate regions. The principal value given is the average adult effective dose for the whole country, the values in brackets indicate the range of average doses calculated within the country.
3. The critical group is 1 year old infants for all countries except Italy where the critical group is 10 year old children.
4. Doses estimated for a child aged 10 years.
5. Doses estimated for an infant aged 1 year.

1000–2000 μSv in Italy, Greece and the Federal Republic of Germany. The critical group of individuals consists of 1 year old infants, 10 year old children or adults depending on the country, but generally there is little difference in the doses predicted for the three age groups. The most important exposure pathway in all cases is ingestion of contaminated food. For infants milk is generally the most important food but for the other age groups milk, lamb, green vegetables, grain and fruit all make significant contributions to the critical group doses.

Doses to the thyroids of the theoretical critical group in the first year after the accident are shown in Table II. These doses range from about 50 μSv to 30 000 μSv, and are dominated by radioisotopes of iodine, taken into the body via milk.

These theoretical critical group doses can be compared with the effective dose received each year due to natural background radiation (1000–3000 μSv). Although doses to particular organs, especially the thyroid, may be greater, in no country is the theoretical effective dose estimated for critical groups much greater than that received in one year from natural background, and in most cases it is considerably lower.

4.3 Collective doses

Estimates of the collective effective and collective thyroid doses to the populations of EC countries integrated over all time (ie the collective dose commitments), are given in Table III. The collective effective dose commitment to the whole population of the EC is estimated to be 78 000 man Sv. The highest collective effective doses are those to the populations of the Federal Republic of Germany (30 000 man Sv) and Italy (27 000 man Sv), because contamination levels in these countries were relatively high and their populations are relatively large. The main pathways contributing to the collective effective dose are consumption of contaminated food and external irradiation from deposited material. In the first year food consumption is the dominant pathway, but in later years external irradiation from deposited material increases in relative importance. For both pathways, caesium-137 is the most important radionuclide. The value of 78 000 man Sv received over all time from the Chernobyl accident can be compared with the collective dose of about 500 000 man Sv which is received by the populations of EC countries every year from natural sources. For western European countries outside the EC an additional 10 000 man Sv has been estimated for the collective dose from the first year(2); the total collective dose is likely to be about twice this value.

TABLE III. COLLECTIVE DOSES IN EC COUNTRIES[1]

Countries	Collective Effective Dose Commitment (man Sv)	Collective Thyroid Dose Commitment (man Sv)
Belgium	940	3 100
Denmark	1 100	2 700
France	5 600	19 000
FRG	30 000	52 000
Greece	8 500	11 000
Ireland	950	2 200
Italy	27 000	67 000
Luxembourg	42	150
Netherlands	1 200	5 400
Portugal	2	8
Spain	57	340
UK	3 000	12 000
Total	78 000	170 000

Note

1. The results are given to two significant figures only to avoid rounding errors in subsequent calculations; this degree of accuracy is not implied.

The total collective thyroid dose to the EC population is estimated to be 170 000 man Sv. The dominant radionuclide in this instance is iodine-131 and since this has a relatively short radioactive half-life (8 days), a large fraction of the dose had been received within a few weeks of the accident. For thyroid doses, ingestion is the most important pathway. For western European countries outside the EC an additional 30 000 man Sv has been estimated for the collective thyroid dose from the first year(2); the total collective thyroid dose is likely to be some tens of percent higher.

4.4 The effect of countermeasures

The introduction of countermeasures will have the greatest percentage impact on the doses to the critical groups in the populations. Estimates have been made of the percentage of the effective dose potentially received in the first year which has been saved by the countermeasures. The largest calculated reductions in effective dose are 60% and 50% for Greece and Italy, where restrictions on the consumption of milk and green vegetables were introduced. Significant reductions in the individual effective doses, in the range 20-40% are also estimated for the Federal Republic of Germany, the Netherlands and the UK. Outside the EC countries, countermeasures are also estimated to have had a significant impact on the doses received by the most exposed individuals. In Austria the estimated effective dose saving to the critical group ranged from about 60% for adults to 80% for infants(2).

The calculated percentage reductions in thyroid doses to theoretical critical groups are generally greater than the reductions in the effective dose and reach 90% in Greece and Italy. This is because the thyroid doses are mainly due to iodine-131 and are received in a relatively short time. In most cases, countermeasures introduced for short periods were effective in reducing the thyroid dose from iodine-131 but had less influence on the effective doses which are mainly due to the caesium radioisotopes and which will be delivered over longer time periods.

The effect of countermeasures on the collective dose was less than the reduction estimated for the most exposed individuals. Throughout the EC countermeasures are calculated to have reduced the collective effective dose by about 5% and the collective thyroid dose by about 25%.

4.5 Comparison with whole body monitoring results

Some countries have recently published the results of whole body monitoring programmes(2,3,4). Comparing these with the corresponding theoretical dose estimates indicates that most of the theoretical estimates are within the ranges of uncertainty expected (factors of two or three). Where significant differences between the theoretical dose estimates and the whole body monitoring results occur, the whole body monitoring results are the lower values. This is probably because the theoretical calculations have taken no account of the consumption of tinned and frozen food which would have been prepared before the accident and so would have been uncontaminated. It may also result from the assumption made in our assessment when estimating doses to individuals in the more highly contaminated areas of each country that all their food was obtained from the region where they lived whereas a large proportion of their food would have originated in less contaminated areas.

5. THE HEALTH IMPACT ON THE POPULATION OF THE EUROPEAN COMMUNITY

5.1 Risk estimates

If individuals are exposed to very high levels of radiation, serious injury and even death can occur within a matter of weeks. The radiation doses received by even the critical individuals in the EC as a result of the Chernobyl accident were well below those at which these early, acute effects of radiation would be manifested. In predicting the health impact on the EC population it is therefore only necessary to consider the incidence of longer term health effects. In this paper we estimate the total number of cancer fatalities which may occur in the exposed population and, because of the importance of iodine-131 concentrating in the thyroid gland, the incidence of fatal and non-fatal thyroid cancers.

The fundamental processes by which cancer is induced by radiation are not fully understood, but a greater incidence of cancers has been observed in various groups of people who were exposed to high doses of radiation years previously, for example the survivors of the atomic bombings in Japan. Studies of such population groups can be used to derive "risk factors" which express the predicted excess numbers of various cancer types in a group of exposed individuals that may result from the radiation doses received by that group. The United Nations Scientific Committee on the Effects of Atomic Radiation (UNSCEAR) reviews all the information on risk factors periodically, and has provided estimates of risk factors for cancers as a result of irradiation of various tissues at moderately high doses. Based on these factors and interpolating between the observed effects and zero, an excess of 100 thyroid cancers would be expected in a population receiving a collective dose to the thyroid of 10 000 man Sv, of these 5 would be expected to be fatal. In a population receiving a collective effective dose of 10 000 man Sv the total number of cancer fatalities expected would be 125. We have used these risk factors, which are based on a worldwide scientific consensus, in this study.

5.2 Results

The collective doses, both effective and thyroid, have been presented in Table III for each member state of the EC. Using the risk factors given above, estimates can be made of the extra numbers of non-fatal and fatal thyroid cancers (from collective thyroid doses) and total numbers of fatal cancers (from collective effective doses) which would be statistically expected in the EC following the Chernobyl accident. In this way, the expected number of additional thyroid cancers occurring within EC countries due to Chernobyl is estimated to be some two thousand, of which about 100 are expected to result in fatality. The number of additional fatalities from cancers of all types due to Chernobyl is expected to be in the region of a thousand. These extra cancers are predicted to occur spread out in time over a few decades following the accident. Considering western Europe as a whole, these statistical estimates would be a few tens of percent higher.

The estimates need to be seen against the background of cancers that would occur in the population even if the Chernobyl release had not happened. Members of the EC population are subject to continuous irradiation from natural radiation sources, for example, cosmic rays, terrestrial radiation, and irradiation from naturally-occurring materials in the diet. It is therefore possible to make an estimate, on the same basis, of the cancer fatalities which are expected to result from natural irradiation. A typical annual effective dose from natural sources in EC countries is about 2000 μSv; this corresponds to approximately eight thousand cancer fatalities per year in the EC due to natural radiation sources, or nearly half a million over a fifty year period. This is nearly five hundred times greater than the excess cancer fatalities predicted to result from the Chernobyl accident.

ACKNOWLEDGEMENTS

The authors would like to thank the CEC and NEA for permission to quote the results given in this paper. Part of the work presented was also funded by the CEC.

REFERENCES

[1] MORREY, M., BROWN, J., WILLIAMS, J.A., CRICK, M.J., SIMMONDS, J.R., HILL, M.D., A Preliminary Assessment of the Radiological Impact of the Chernobyl Reactor Accident on the Population of the European Community (Jan. 1987), EUR Rep. CEC, Brussels (in preparation).

[2] ORGANISATION FOR ECONOMIC CO-OPERATION AND DEVELOPMENT, The impact of the Chernobyl accident in OECD member countries, Committee on Radiation Protection and Public Health, OECD, Paris (in press).

[3] FRY, F.A., Chernobyl reactor accident: the impact on the United Kingdom. Brit. J. Radiol. (in press).

[4] WATSON, W.S., Human caesium-134, caesium-137 levels in Scotland after Chernobyl, Nature (London) **323** (Oct. 1986).

PROJECTED GLOBAL HEALTH IMPACTS FROM SEVERE NUCLEAR ACCIDENTS: CONVERSION OF PROJECTED DOSES TO RISKS ON A GLOBAL SCALE
Experience from Chernobyl releases

R.J. CATLIN
Electric Power Research Institute,
Palo Alto

M. GOLDMAN
University of California,
Davis

L.R. ANSPAUGH
Lawrence Livermore National Laboratory,
Livermore

California, United States of America

Abstract

PROJECTED GLOBAL HEALTH IMPACTS FROM SEVERE NUCLEAR ACCIDENTS: CONVERSION OF PROJECTED DOSES TO RISKS ON A GLOBAL SCALE: EXPERIENCE FROM CHERNOBYL RELEASES.

A recent evaluation by the authors and their colleagues of possible health effects arising from Chernobyl releases has just been published by the U.S. Department of Energy. New models were used to evaluate external exposure and other environmental dose pathways. Estimates of projected collective dose and average individual dose commitments from these releases were made for various regions. These results were combined with those of the Soviet scientists to provide collective dose commitments for the Northern Hemisphere. Consideration was given to the possible effectiveness of protective actions taken by various countries to reduce projected doses to their populations. Although some preliminary data indicate possible mean reductions of about 25% in total collective doses over the first year, and of about 55% in collective dose to the thyroid, no corrections were made to these dose estimates because of the variable nature of the data. A new combined set of dose-effect models recently published by the United States Nuclear Regulatory Commission was then applied to estimate the ranges of possible future additional health effects due to the Chernobyl accident. In this method possible health effects are estimated on an individual site basis and the results are then summed. Both absolute and relative risk projection models are used. By use of these methods, ''best'' estimates of possible additional health effects were projected for the Northern Hemisphere as follows: 1) over the next 50 years, up to 28 thousand radiation-induced fatal cancers, compared to an expected 600 million cancer deaths from natural or spontaneous causes; 2) over the next year, up to 700 additional cases of severe mental retardation, compared to a normal expectation of 340 thousand cases; and 3) in the first generation, up to 1.9 thousand radiation-induced genetic disorders, compared to 180 million naturally-occurring cases. The possibility of zero health effects at very low doses and dose rates cannot be excluded. Owing to the very large numbers of naturally-occurring health effects, it is unlikely that any additional health effects will be demonstrable except, perhaps, for the more highly exposed population in the immediate vicinity of Chernobyl.

1. INTRODUCTION

The assessment of possible health impacts from the Chernobyl accident presented in this paper is based on a recent evaluation by the authors and their colleagues recently published by the U.S. Department of Energy (USDOE) [1]. New models were used to evaluate external exposure and other environmental dose pathways, as reported in a companion paper by Anspaugh, et al., at this Conference [2]. Estimates of projected collective dose and average individual dose commitments from Chernobyl releases were made for various regions. These results were then combined with those of the Soviets [3,4] to provide the projected collective dose commitments over the first year and for 50 years for the Northern Hemisphere, as summarized in Figure 1 and used for the calculations in this paper. The principal contributions to dose are from external sources and ingestion of radionuclides, in approximately equal amounts. As shown in Figure 1, about half of the total projected dose commitment occurs in the first year.

2. EFFECTIVENESS OF PROTECTIVE ACTIONS (COUNTERMEASURES)

Following the Chernobyl accident, various protective actions were taken by various nations in Europe and Asia to reduce possible exposures to external sources of radiation or to intake of radionuclides through the foodchain. The need for protective actions at various times and places was governed by the patterns of environmental dispersion of the releases, and the concern of health authorities over local conditions, in particular during early times when knowledge of such dispersion was scanty or absent. Much credit must be given to Soviet and neighboring authorities for early actions such as sheltering, evacuation, and use of potassium iodide and other controls to minimize uptake of radioiodine through the milk chain.

Thus, protective actions were initiated, modified, and terminated at different times in different places and, in some cases, were applied on a local or regional rather than national basis. Moreover, little normalization of protective action guide levels has occurred among nations, as illustrated in Figure 2 in which peak ^{131}I concentrations in cows milk are compared to the action/emergency protective action guide (PAG) levels adopted by various countries [1]. The peak concentrations of ^{131}I generally occurred in milk from farm cows. Milk from dairy cows had lower concentrations, while peak concentrations in sheep and goat milk were found in the thousands of bequerels per liter as shown in Figure 3 [1]. Unlike radioiodine, the foodstuff levels of ^{137}Cs are more persistent and may entail protective actions extending for several years.

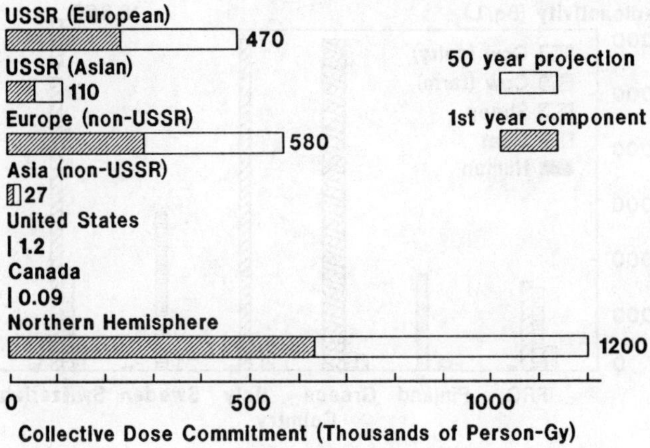

FIG. 1. Total collective dose commitment projected for the Northern Hemisphere.

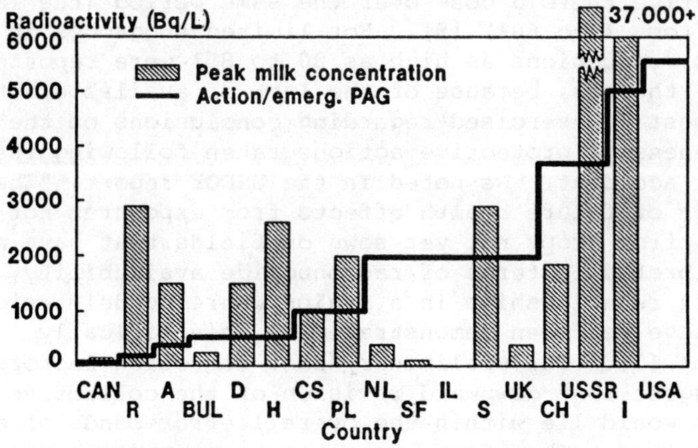

CAN	Canada	CS	Czechoslovakia
R	Romania	PL	Poland
A	Austria	NL	Netherlands
BUL	Bulgaria	SF	Finland
D	Germany, Fed. Rep. of	IL	Israel
H	Hungary	S	Sweden
		I	Italy

FIG. 2. Peak concentrations of ^{131}I in cows' milk: comparison with protective action guides (critical group: infant).

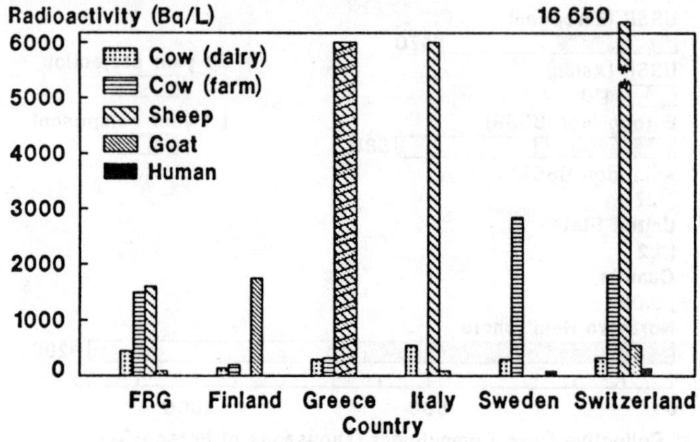

FIG. 3. Peak ^{131}I concentration in different milk. (Human milk peak of 4 Bq/L in Italy and Finland.)

In terms of effectiveness of such actions for general populations, one such study now nearing completion indicates a mean reduction of about 25% in total collective dose over the first year (with a range from 1 to 50%), and about 55% reduction in collective thyroid dose over the same period (the latter ranging from 1 to 60%) [5]. For limited subgroups in a few locations, reductions as high as 80 to 85% were reported.

Nevertheless, because of the lack of available data, caution must be exercised regarding conclusions on the overall effectiveness of protective actions taken following the Chernobyl accident. As noted in the USDOE report, "The projection of future health effects from exposures not yet received, from crops not yet sown on fields that have not yet been calibrated in terms of radionuclide availability, and by using risk relationships in a region where no deleterious health effects have yet been demonstrated is scientifically difficult" [1]. The preliminary dose reduction factors given above suggest that downward revision of the collective dose estimates would lie within the overall error bands of a factor of two or less. Therefore, no attempt was made in the USDOE report to adjust exposure predictions on the basis of protective actions taken except where such adjustments were taken into account in national summaries.

3. DOSE-EFFECT PROJECTION MODELS

Projected health consequences in various populations of the Soviet Union from exposure to radioactivity released at Chernobyl have been derived from the Soviet reports [3,4].

TABLE I.
CHERNOBYL HEALTH EFFECTS PROJECTIONS (USSR 1987)

Group at Risk	*Source*	*Excess Cancer Deaths*	*Spontaneous Cancer Deaths*	*% Increase*
Chernobyl Evacuees (135,000)	External	270	~17,000	1.6%
European/USSR (74.5 million)	External	2 850	~9.5 million	0.03%
European/USSR (74.5 million)	I-131	150	15 000	1%
European/USSR (74.5 million)	Internal	3 800	~9.5 million	0.04%
	Totals	~7 000	~9.5 million	[0.04% Avg.]

Projections of cancer risks, both spontaneous and radiation-induced, were given for two populations: 1) about 135,000 evacuees from the Chernobyl area (out to 30 km); and 2) about 75 million persons in the European part of the USSR. These projections are given in Table I. Based on Soviet data, the lifetime cancer mortality risk coefficient used appears to be on the order of 1.4 - 1.7 cancer deaths per 100 person-sievert, based on a nonthreshold absolute risk, linear dose-effect projection model consonant with that of the ICRP [6] which uses a lifetime mortality risk factor for radiation-induced cancers of 1.7 x 10-2 per sievert as an average for both sexes and all ages.

The radiation exposures following the Chernobyl accident were protracted and received at progressively decreasing low dose rates. As noted in the USDOE report, the 135 000 evacuees received their exposures over a period of days even though some 24,000 persons in that group were exposed to somewhat elevated dose rates [1]. Only low-LET radiation is involved. Knowledge of dose-effect relationships in human populations is limited largely to persons exposed to high doses at high dose rates. Interpolation of risk coefficients from these high dose levels at which health effects are observed in humans down to zero has generally been done on a linear basis to assess an upper limit of risk.

Although many studies have been documented, the analysis of atom-bomb survivors is the major one underlying estimates of cancer mortality risk. The lifetime fatal cancer risk is on the order of 1 x 10-2 per gray, as a first approximation. The dosimetry is being reevaluated and the assigned radiation doses will likely be lowered by a factor of up to two, which would increase the risk estimate up to 2 x 10-2 per gray [1]. A value of 2 x 10-2 per gray has also been calculated by Charles and Lindop using "all" data except that from the Japanese [7].

In its 1980 report, the BEIR Committee of the U.S. National Academy of Sciences/National Research Council considered the effect of dose rate on dose-response relationships [8]. For low-LET radiation human data on chronic exposures at low dose rates is limited. Experimental data in animals give strong indications that a given dose of low-LET radiation will produce fewer effects at low dose rates than at high dose rates. A reduction factor of two to three has been considered in the 1980 BEIR report, based on both human and animal data. For single exposures to doses not exceeding 0.2 gray delivered at any dose rate, and for any total dose delivered at a rate not exceeding 0.05 gray per year, the dose-effect reduction factors are likely to be between two and ten [8].

Various models and assumptions may be used for estimating possible latent health effects in exposed populations. For example, annual doses of 10 microsievert or a lifetime dose of about 500 microsievert would be unlikely to produce additional risk. Therefore, a major portion of the collective dose in the Northern Hemisphere might produce no additional radiogenic risk from Chernobyl fallout. There are no direct data that confirm that a few random ionizations in tissue cause fatal cancers, as observed in NCRP Report No. 64 [9]. Moreover, the BEIR Committee reports of both 1980 and 1972 noted that for low dose and dose rates, the likelihood of zero deleterious health effects is not precluded [8,10].

Although the Soviet projections of additional mortality appear to use risk coefficients for an absolute risk, linear dose-effect projection model as set forth by both the ICRP [6] and in the 1980 BEIR report [8], such an approach does not take into account the higher risks that might be obtained by use of relative risk projection models. The BEIR report provided risk coefficients for both absolute and relative risk projection models generally without indication of preference.

Newer epidemiological studies over the last decade have resulted in three reports issued by the US National Institutes of Health [11], the US Nuclear Regulatory Commission (NRC) [12], and the United Nations Scientific Committee on the Effects of Atomic Radiation [13] that assign both absolute and relative risk projection models and dose-response models on a site specific basis. Overall risk is obtained by the summation of risk estimates for each primary site. The reports are internally consistent on model assignments, as shown in Table II. The models and parameters reported in this paper are taken from the NRC report that was prepared by the Harvard University School of Public Health and Sandia National Laboratories. In this report

TABLE II.
COMPARISON OF RISK PROJECTION MODELS

Site of Interest	Projection Model* NIH	NUREG/CR-4214**	UNSCEAR (1986)
Leukemia	A	A-A-A	A
Bone Cancer	A	A-A-A	A
Breast Cancer (F)	R	R-A-A	R
Lung Cancer	R	R-R-A	A or R
Ca of thyroid	R	A-A-A	A
Ca of GI tract	R	R-R-A	—
Other Ca	R	R-R-A	—

* A = Absolute, R = Relative model
** Model order: Upper-best-lower estimate (bottom = 0)

(NUREG/CR-4214), risk projection models and coefficients are given for upper, central and lower estimates of the top risk range. For the low doses and dose rates concerned, a zero risk model is compatible with each set of estimators. The central or "best" estimate is considered to be the most realistic by the authors of the USDOE report [1], with risk coefficients as given in Table IIIa. Dose rate and total dose reduction factors for these coefficients are given in Table IIIb. The overall risk coefficient for lifetime cancer mortality for the general populace under review is about 2.3 x 10-2 per sievert. For the 8 to 15 week-old fetus, the projected additional incidence of severe mental retardation in the first year is 40% per gray, and for genetic ill health, the risk coefficient is 30 x 10-4 effects per gray in the first generation.

4. PROJECTIONS OF FUTURE HEALTH EFFECTS

These risk coefficients were applied to the collective dose estimates presented by Anspaugh, et al. [2], as summarized in Figure 1, to produce "best" estimates of possible additional health effects due to Chernobyl releases for various regions in the Northern Hemisphere, as given in Table IV. Thus, for the Northern Hemisphere, up to 28 thousand additional radiation-induced fatal cancers have been projected, compared to 600 million cancer deaths expected from natural of spontaneous cause over the next 50 years. About 700 additional cases of severe mental retardation are projected, above the 340 thousand cases expected in the first year after the Chernobyl releases. And in

TABLE IIIA. RADIOGENIC RISK MODELS (NUREG/CR-4214, "BEST" ESTIMATE)

Cancer Site or Effect	*Risk Coefficient*
Leukemia	$2.24/10^4$ PY-Gy × F
Bone	$0.1/10^4$ PY-Gy × F
Breast	45%/Gy
Lung	18%/Gy × F
GI tract	39%/Gy × F
Thyroid	$0.25/10^4$ PY-Gy (age < 20)
(external)	$0.125/10^4$ PY-Gy (age ≥ 20)
(^{131}I)	[1/3 of above values]
Other ca.	20%/Gy × F
In utero ca.	$21/10^4$ PY-Gy (to fetus)
Severe mental retardation	40%/Gy (incidence) (8- to 15-wk-old fetus)
Genetic ill health	$30/10^4$ Gy (effects per generation)

TABLE IIIB. REDUCTION FACTOR (F) FOR RADIOGENIC RISK MODELS

Dose Rates	*F*
< 0.05 Gy/day	0.3
≥ 0.05 Gy/day	0.30 + 0.47 D(Gy)

Total Dose D(Gy)	*F*
0.01	0.30
0.10	0.35
0.25	0.42
0.50	0.54
0.75	0.65
1.00	0.77
1.25	0.89
≥ 1.50	1.00

TABLE IV.
PROJECTED CHERNOBYL HEALTH IMPACTS
Estimates of Spontaneous and Radiation-Induced Effects*

	Fatal Cancer		*Severe Mental Retard.*		*Genetic Disorders*	
	Natural (10^3)	*Rad-Induced[a] (10^3)*	*Natural (10^3)*	*Rad-Induced[a] (10^3)*	*Natural (10^3)*	*Rad-Induced[a] (10^3)*
USSR evacuees	17	0.41	0.013	0.018	6.9	0.06
European USSR	9 400	11	7.2	0.27	3 900	0.75
Asian USSR	28 000	2.5	22	0.06	12 000	0.18
Europe (other)	72 000	13	38	0.33	21 000	0.93
Asia (other)	450 000	0.6	240	—	130 000	0.04
United States	41 000	0.02	22	—	12 000	—
Northern Hemisphere	~600 000	28	340	0.07	180 000	1.9

[a]The possibility of zero health effects at very low doses and dose rates cannot be excluded.

TABLE V.
COMPARISON OF MODEL PROJECTIONS
Excess Cancer Mortality in European USSR

		Risk Model		
		*NUREG/CR-4214**	*BEIR (1980)**	
Population at Risk	*USSR Estimate (derived)*	*Best Estimate*	*REL-LQ*	*ABS-LQ*
Chernobyl Evacuees (135,000)	270	410	360	120
European/USSR (75 million)	6 800	10 300	8 400	3 300
Total	7 000	10 700	8,800	3 500

* Possibility of zero health effects is not precluded

the first generation, up to 1.9 thousand cases of genetic ill health are estimated, compared to the natural expectation of about 180 million cases. About 50% of the additional cases are projected to occur within the Soviet Union, 80% of which would appear within the European portion. The possibility of zero health effects at the very low doses and dose rates generally encountered cannot be excluded.

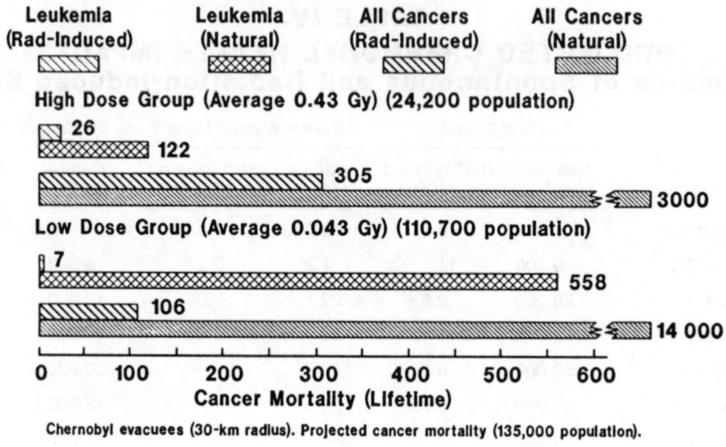

FIG. 4. Estimates of possible effects in the Chernobyl evacuees.

Estimates of cancer mortality derived for the European portion of the USSR are in general agreement with those obtained by use of the NRC "best" estimate model and the BEIR report relative risk, linear-quadratic dose-effect model, as shown in Table V. Due to the very large numbers of health effects that occur from natural or spontaneous causes, it is unlikely that any additional effects will be demonstrable except, perhaps, for the more highly exposed population in the immediate vicinity of Chernobyl. Of this latter group, about 24 thousand persons are reported to have received an average individual absorbed dose of about 0.43 gray, as shown in Figure 4. Some effects, such as myeloid leukemia, may be seen by epidemiological evaluation of this group over the next 30 years if the risk factors and reported levels of dose are valid.

5. UNCERTAINTIES AND LIMITS OF RISK ESTIMATION

Uncertainties in these projections arise because the quantitative risk assessment models used for low dose and dose-rate exposures are based on extrapolations from observations of high dose and dose-rate experiences. Refinements of the models occur as newer data become available. The "best" (or central) and upper estimate values are less likely to underestimate true risk because large errors in this direction would become apparent in epidemiological and laboratory studies. Radiation-induced cancer risk models all use assumptions that are not proven but are generally compatible with observations. Caution

TABLE VI.
PROJECTED FATAL CANCER RISKS (LIFETIME)
Northern Hemisphere

Population at Risk: 3.5 Billion Persons (3.5 E + 09)	
Cancers from natural or spontaneous causes	~600 000 000
— Including cancers from natural background radiation	Up to 6 000 000 (1%)
— Plus cancers from nuclear tests	Up to 100 000 (0.02%)
— Plus cancers from Chernobyl Releases	0 up to ~28 000 0.004%

must be taken also in applying risk factors derived from one population to another. Some of the differences between national and regional cancer statistics may be due to varying age distributions or to competing risks. Others may be due to differences in reporting, in population diet, or in environmental factors. Still others may arise from true genetic variations.

The radiation dosimetry for the various exposed populations has been assessed and reported only to an initial degree. In some instances, several years of observation will be required to determine future exposures from Chernobyl radionuclide releases passing through the foodchain, and the effects of mitigating actions to reduce dose are incompletely documented and evaluated. As indicated previously, protective actions already taken could reduce projected health effects by a factor perhaps as high as 2 in some instances.

6. SUMMARY

In summary, it appears that the number of fatal cancers in the Northern Hemisphere due to Chernobyl could possibly increase by as much as 28 thousand or 0.004% over the next 50 years, as shown in Table VI. This may be compared to the expected 600 million cancer deaths from natural or spontaneous causes, including up to 6 million fatalities or 1% due to natural background radiation, plus up to 100 thousand cancer deaths or 0.02% from worldwide nuclear tests. A possible increment of 700 cases of severe mental retardation of children irradiated as fetuses in 1986 could also appear in the more heavily irradiated population groups in Europe, an increase of about 0.2% over the number of spontaneous cases. Finally, a genetic disorder increment in the first generation could range up to about a 0.001% increase. This projection is highly speculative, being based on extrapolations from animal data. Such effects have not been demonstrated in humans. In any event, the projected increment would not be expected to be detected even within the more highly exposed populations in the general area of Chernobyl.

REFERENCES

[1] GOLDMAN, M., et al., Health and Environmental Consequences of the Chernobyl Nuclear Power Plant Accident, Rep. DOE/ER-0332, United States Department of Energy, Washington, DC (1987).

[2] ANSPAUGH, L.R., GOLDMAN, M., CATLIN, R.J., "Atmospheric releases from severe nuclear accidents: Environmental transport and pathways to man: Modelling of radiation doses to man from Chernobyl releases", Paper IAEA-CN-48/274, these Proceedings.

[3] USSR STATE COMMITTEE ON THE UTILIZATION OF ATOMIC ENERGY, The Accident at the Chernobyl Nuclear Power Plant and its Consequences, information compiled for the IAEA Experts Meeting, Vienna, 1986.

[4] USSR MINISTRY OF HEALTH, Analysis of the Radiological Consequences of the Accident at the Chernobyl Nuclear Power Plant for the Population of the European Regions of the USSR, WHO Info. Doc. A40/INF.DOC/9, May (1987).

[5] NUCLEAR ENERGY AGENCY OF THE OECD, The Radiological Impact of the Chernobyl Accident in OECD Countries, (Proc. OECD/NEA, Paris 1987) (to be published).

[6] INTERNATIONAL COMMISSION ON RADIOLOGICAL PROTECTION, Recommendations of the International Commission on Radiological Protection. Publication 26, Annals of the ICRP, Pergamon Press, Oxford and New York (1977).

[7] CHARLES, M.W., LINDOP, P.J., J. Soc. Radiol. Protect. **1** (1981) 15–19.

[8] UNITED STATES NATIONAL ACADEMY OF SCIENCIES/NATIONAL RESEARCH COUNCIL, Committee on the Biological Effects of Ionizing Radiation, the Effects on Populations of Exposure to Low Levels of Ionizing Radiation, Rep. BEIR 3, National Academy Press, Washington, DC (1980).

[9] NATIONAL COUNCIL ON RADIATION PROTECTION AND MEASUREMENTS, Influence of Dose and its Distribution in Time on Dose-Response Relationships for Low-LET Radiation, Rep. 64, NCRP, Washington, DC (1980).

[10] UNITED STATES NATIONAL ACADEMY OF SCIENCES/NATIONAL RESEARCH COUNCIL, Committee on the Biological Effects of Ionizing Radiation, the Effects on Populations of Exposure to Low Levels of Ionizing Radiation, Rep. 1, BEIR 1, National Academy Press, Washington, DC (1972).

[11] UNITED STATES NATIONAL INSTITUTES OF HEALTH, Report on the National Institutes of Health Ad Hoc Working Group to Develop Radioepidemiological Tables, Rep. NIH 85-2748, US Public Health Service, Washington, DC (1985).

[12] UNITED STATES NUCLEAR REGULATORY COMMISSION, Health Effects Model for Nuclear Power Plant Accident Consequence Analysis, NRC Contractor Rep. NUREG/CR-4214 (SAND-85-7185), US Govt. Printing Office, Washington, DC (1985).

[13] UNITED NATIONS, Genetic and Somatic Effects of Ionizing Radiation 1986, Scientific Committee on the Effects of Atomic Radiation (UNSCEAR), UN, New York (1986).

SAFETY IMPROVEMENTS IN DESIGN AND OPERATION

Technical Sessions 3.3A and 3.3B

Chairman (Technical Session 3.3A)

C. SENNIS
Italy

Chairman (Technical Session 3.3B)

M. COPIC
Yugoslavia

IAEA-CN-48/226

РАЗВИТИЕ ТРЕБОВАНИЙ ПО БЕЗОПАСНОСТИ И СИСТЕМЫ ГОСУДАРСТВЕННОГО НАДЗОРА КАК ОСНОВЫ БЕЗОПАСНОГО РАЗВИТИЯ ЯДЕРНОЙ ЭНЕРГЕТИКИ

П.П. АЛЕКСАШИН, А.М. БУКРИНСКИЙ,
О.М. КОВАЛЕВИЧ, В.М. МАЛЫШЕВ,
В.А. СИДОРЕНКО
Государственный комитет СССР по надзору за безопасным ведением работ в атомной энергетике,
Москва,
Союз Советских Социалистических Республик

Abstract–Аннотация

DEVELOPMENT OF SAFETY REQUIREMENTS AND SYSTEM OF STATE SUPERVISION AS BASES FOR THE SAFE DEVELOPMENT OF NUCLEAR POWER.

The paper evaluates the situation with regard to the development of nuclear power in the USSR following the accident at the Chernobyl nuclear power plant. It is stated that the development of nuclear power in the USSR will continue with due regard to the lessons learnt from the Chernobyl tragedy. In the present circumstances renewed and increased attention is being paid to safety, which even before was of prime importance in the development of nuclear power. This attention is focused on increasing the quality of operation, adopting the necessary technical measures at existing plants, reviewing and improving safety requirements and achieving more efficient and better adapted functioning of the system of State supervision. Specific measures are described which have been undertaken or are being undertaken to improve the safety of nuclear power plants with RBMK reactors and plants with WWER-1000 reactors which are likely to serve as the basis for the further development of nuclear power in the USSR. Consideration is given to reactors with an increased level of safety.

РАЗВИТИЕ ТРЕБОВАНИЙ ПО БЕЗОПАСНОСТИ И СИСТЕМЫ ГОСУДАРСТВЕННОГО НАДЗОРА КАК ОСНОВЫ БЕЗОПАСНОГО РАЗВИТИЯ ЯДЕРНОЙ ЭНЕРГЕТИКИ.

В докладе оценивается ситуация с развитием атомной энергетики в СССР после аварии на Чернобыльской АЭС. Излагается позиция, что развитие атомной энергетики в СССР будет продолжаться с учетом всех уроков Чернобыльской трагедии. Безопасность, и ранее стоявшая во главе угла развития атомной энергетики, в нынешних условиях требует нового повышенного внимания. Это внимание сосредотачивается на повышение качества эксплуатации и проведения необходимых технических мероприятий на действующих станциях, пересмотру и совершенствованию требований по безопасности, более четкому и налаженному функционированию системы государственного надзора. Излагаются конкретные меры, предпринятые и предпринимаемые по повышению безопасности АЭС с реакторами РБМК и АЭС с реакторами ВВЭР-1000, которые должны стать основой дальнейшего развития атомной энергетики в СССР. Приводятся соображения относительно реакторов с повышенной безопасностью.

Тяжелая авария, которая произошла в Советском Союзе на Чернобыльской атомной электростанции в апреле 1986 года, так же как авария на американской АЭС "Три-майл-айленд-II" в 1979 г., еще раз со всей остротой поставила проблему обеспечения безопасности АЭС, как главную проблему, определяющую возможность дальнейшего развития атомной энергетики. Все страны, развивающие атомную энергетику, в свете происшедших событий анализируют свои подходы к решению этой проблемы, уточняют и реализуют необходимые изменения в этих подходах. Советский Союз также переоценивает достигнутый уровень безопасности и предпринимает энергичные меры по совершенствованию государственного надзора за проектированием, строительством и эксплуатацией ее объектов, как важной составной части всей системы обеспечения безопасности.

В Советском Союзе на 1 мая 1987 г. действуют 42 энергоблока общей установленной мощностью 30 110 МВт. К 2000 г. планируется значительное увеличение установленной мощности станций на ядерном топливе. Может ли Советский Союз, оценивая последствия аварии на Чернобыльской АЭС, отказаться от дальнейшего развития атомной энергетики? Анализ показывает, что, несмотря на наличие в Советском Союзе больших запасов других источников энергии, развитию атомной энергетики в европейской части Советского Союза нет разумной альтернативы. Отказ от развития атомной энергетики потребовал бы отказаться от осуществления многих социальных и социально-экономических программ развития страны. Поэтому атомная энергетика должна развиваться, но безопасность ее развития должна быть существенно повышена.

Подходы к обеспечению безопасности атомной энергетики постоянно совершенствовались по мере ее развития. Если на ранних стадиях атомной энергетики основное внимание уделялось обеспечению надежной и безопасной работы компонентов и оборудования АЭС, то в дальнейшем все большее внимание стало уделяться использованию на станциях технических средств для ограничения последствий возможных отказов. Авария на АЭС "Три-майл-айленд-II" побудила мировое сообщество подвести итоги достигнутого уровня безопасности АЭС на международной конференции, которая состоялась в Стокгольме 20–24 октября 1980 г. В советском докладе, представленном на этой конференции [1], подробно были рассмотрены все аспекты безопасности, связанные с развитием атомной энергетики в нашей стране. С тех пор в СССР была введена в действие вторая редакция нормативного документа по безопасности "Основные положения обеспечения безопасности атомных станций при проектировании, сооружении и эксплуатации" (ОПБ-82), образован и функционирует с 1984 г. единый государственный орган надзора за безопасностью в атомной энергетике – Государственный комитет СССР по надзору за безопасным ведением работ в атомной энергетике (Госатомэнергонадзор СССР), выведены из эксплуатации первенцы атомной энергетики – первые блоки Нововоронежской и Белоярской АЭС, разработаны и реализуются меры по повышению безопасности других энергоблоков ранних разработок.

В ОПБ-82, так же как и в его редакции, разработанной в 1971 г., в качестве основного принципа обеспечения безопасности был принят принцип предотвращения тяжелых аварий при постулированных исходных событиях, в т.ч. при мгновенном разрыве главного циркуляционного трубопровода первого контура. Техническим средствам, которые необходимо предусматривать в реакторных установках с водяным теплоносителем, должно предотвращаться массовое расплавление тепловыделяющих элементов, а системы локализации аварий должны быть рассчитаны на максимальную проектную аварию с полной разгерметизацией оболочек тепловыделяющих элементов. Количественно определен еще один барьер защиты от радиационных последствий аварий на АЭС – удаление атомных станций от населенных пунктов с численностью населения более 100 тысяч человек.

Для атомных станций теплоснабжения, которые должны располагаться вблизи населенных пунктов, регламентируется более жесткий подход к техническим средствам обеспечения безопасности. Здесь при разработке мер по обеспечению безопасности должна учитываться возможность повреждения корпуса реактора.

Чернобыльская авария потребовала критического подхода как к отдельным требованиям так и ко всему комплексу требований по обеспечению безопасности атомных станций.

Подводя итоги анализа аварии на Чернобыльской АЭС, международная консультативная группа по ядерной безопасности (INSAG) пришла к выводу, что Чернобыльская авария не обнаружила каких-либо новых явлений, которые бы не были известны раньше, и что в целом она не поставила под сомнение те подходы к обеспечению безопасности АЭС, которые были выработаны предыдущей практикой и которые с небольшими непринципиальными различиями получили распространение во всех странах, развивающих атомную энергетику. Вместе с тем огромный масштаб ущерба, причиненного аварией, и особенно вызванное ею психологическое воздействие на широкие круги общественности требуют более внимательного осмысления того, что же произошло в Чернобыле.

Четвертый блок Чернобыльской АЭС считался удовлетворяющим современным требованиям безопасности, тем не менее на нем произошла авария, вероятность которой предполагалась столь малой, что допускалось не считаться с ней. И хотя непосредственной причиной аварии явились многочисленные нарушения эксплуатационным персоналом технологического регламента, которых не должно было быть,и их возможность проектом не рассматривалась, они все-таки были, и авария произошла. Отсюда возникает вопрос: можно ли полностью доверять оценкам малых вероятностей, которые всегда базируются на вполне детерминированных и ограниченных по количеству условиях? Можно ли быть уверенным, что все условия, которые могут иметь место в действительности, приняты во внимание? Опыт Чернобыля показывает, что нет. С другой стороны, малая вероятность какого-либо события вовсе не означает, что оно не может осуществиться завтра же. Но если это событие сопряжено с огромным ущербом, то очень слабым утешением является то, что следующее подобное событие может произойти лишь в очень отдаленном будущем. Поскольку закладываемые в основу проекта постулированные исходные события всегда выбирались на основе оценки малых вероятностей, то с учетом сказанного выше этим и определяются недостатки такого подхода.

На Чернобыльской АЭС произошла худшая из реактивностных аварий – неконтролируемый разгон реактора. Если в ходе такой аварии происходит значительное выделение энергии, то в результате разрушения ядерного топлива могут сложиться благоприятные условия для мощного парового взрыва, что и имело место на АЭС в Чернобыле, а еще раньше было известно из подобного же развития событий на американской экспериментальной установке SL-I. Взрыв на Чернобыльской АЭС разрушил защитное здание вокруг реактора, которое, правда, не было рассчитано на повышение давления. Вместе с тем имеются обоснованные сомнения в том, что можно построить вокруг реактора здание, способное выдержать силу подобного взрыва. В связи с этим возникает вывод о том, что возможность подобных взрывов должна быть в принципе исключена. Надежно это может быть обеспечено лишь с помощью внутренних обратных связей, т.е. за счет обеспечения соответствующих физических характеристик активной зоны реактора. В этой связи следует отметить, что ОПБ-82 допускали в некоторых эксплуатационных режимах иметь положительный быстрый мощностной коэффициент реактивности, если в проекте обеспечена и обоснована безопасность реактора в стационарных, нестационарных и аварийных режимах, т.е. если проявление этого положительного коэффициента реактивности компенсируется соответствующими защитными техническими устройствами. Однако опыт Чернобыльской аварии показал недостаточную эффективность такого требования ОПБ-82. В связи с этим возникает вопрос о целесообразности выделения принципиального различия между физическими и техническими принципами защиты. Физические принципы реализуются за счет внутренних обратных связей и являются практически абсолютно надежными. Технические же принципы защиты обеспечиваются с помощью соответствующих технических устройств, которые с определенной вероятностью могут отказать. Поэтому, если дело касается необходимости исключить в принципе возможность какого-либо события, то решать эту задачу необходимо опираясь лишь на физические принципы защиты. Изменение соответствующего требования уже внесено в ОПБ-82. Необходимые первоочередные изменения внесены также в конструкцию эксплуатируемых в настоящее время реакторов РБМК.

Устранены те свойства реактора, которые способствовали опасному развитию аварии: недопустимо высокое значение положительного коэффициента реактивности и недостаточные эффективности органов аварийной остановки реактора в нерегламентных режимах работы.

В качестве первоочередных технических мер реализованы:

- перестановка на действующих АС с реакторами РБМК концевых выключателей так, чтобы в крайнем положении все стержни были погружены в активную зону на глубину 0,7 м, что повысило скоростную эффективность защиты и устраняет возможность повышения размножающих свойств активной зоны в нижней части при движении стержня вниз;
- одновременно увеличено количество неизвлекаемых органов регулирования реактора во всех его режимах работы;
- установлено 80 поглощающих стержней, постоянно находящихся в активной зоне, что снижает паровой эффект реактивности.

В настоящее время на действующих 13 энергоблоках с реактором РБМК осуществляются дополнительные технические меры по повышению безопасности – улучшение физических характеристик активных зон (введение топлива с обогащением 2,4%) и увеличение быстродействия аварийной защиты. На достраиваемых энергоблоках с реакторами РБМК эти мероприятия будут реализованы до ввода блоков в эксплуататацию.

Использование реакторов РБМК в дальнейших планах развития атомной энергетики не предусматривается.

Программа развития атомной энергетики в СССР до 2000 г. в основном будет базироваться на АЭС с реакторами ВВЭР-1000, головной блок серии которых сооружен на Запорожской АЭС. По своим характеристикам и оснащенности средствами безопасности этот блок соответствует уровню, принятому в большинстве стран, развивающих атомную энергетику. Выполненный после Чернобыльской аварии анализ показал, что такое развитие событий, какое имело место при этой аварии, на энергоблоках данного типа практически невозможно. Тем не менее, как уже отмечалось выше, Чернобыльская авария поставила на повестку дня вопрос о необходимости повышения уровня безопасности и энергоблоков данного типа. Здесь речь идет прежде всего о мерах повышения безопасности на случай тяжелой, запроектной аварии, приводящей в худшем случае к полному расплавлению ядерного топлива. В странах мира после аварии на "Три-майл-айленд-II" авариям этого типа уделяется все более пристальное внимание. Проведено большое количество исследований процессов при таких авариях, которые обеспечили возможность определения последствий. В Советском Союзе вводится практика рассмотрения тяжелых аварий, которые не исключаются предусмотренными средствами защиты и приводящих, например, к расплавлению ядерного топлива. Для новых поколений блоков с целью локализации и ограничения выхода в окружающую среду радиоактивных продуктов при таких тяжелых авариях прорабатываются соответствующие технические средства. Конечный уровень требований по защите окружающей среды от радиоактивных загрязнений может быть различным для постулированной максимальной проектной аварии и для постулированного состояния тяжелой аварии. В последнем случае эти требования должны ограничивать зону вокруг атомной электростанции, в которой потребуется применять экстренные меры по защите населения на случай аварии.

В целом Чернобыльская авария показала, что с возможностью тяжелых аварий необходимо считаться, если они не исключены защитой, построенной на физических принципах.

Выводы из уроков Чернобыльской аварии были бы неполными, если бы остались не отмеченными недостатки в деятельности органов надзора за безопасным ведением работ на атомных электростанциях.

На упоминавшейся международной конференции по проблемам безопасности в Стокгольме большое внимание было уделено проблеме "человек–машина", поскольку в ходе развития аварии на "Три-майл-айленд-II" выявилась необходимость изменения установившейся к тому времени в западных странах концепции оператора, когда при управлении техническими устройствами предпочтение отдавалось средствам автоматизации, а действия человека заранее программировались и строго ограничивались инструкциями. Когда на "Три-майл-айленд-II" возникла ситуация, не предусмотренная инструкциями, то операторы не сумели правильно оценить состояние реактора и их действия усугубили развитие аварии. На конференции отмечалась необходимость более глубокой подготовки операторов. Они должны хорошо разбираться в процессах, чтобы более правильно действовать в непредвиденных случаях. Таким образом, в концепции "человек–машина" на этой конференции наметилась трансформация в сторону повышения роли личности оператора, большего доверия к нему. Именно такой подход всегда отличал концепцию оператора в СССР. В Советском Союзе операторы пульта управления являются, как правило, специалистами с высшим образованием, имеющими обширную инженерно-физическую подготовку, разбирающимися в основах физических процессов установки и изучившими оборудование и приборы. Как показал опыт эксплуатации АЭС в СССР, операторы, как правило, хорошо справлялись со сложными ситуациями, выходящими за рамки инструкций. Вместе с тем опыт аварии на Чернобыльской АЭС показал, что и в таком подходе к оператору, какой был принят в Советском Союзе, имеются недостатки. Высокая эрудиция и глубокие знания оператором физических процессов придают ему излишнюю самоуверенность и смелость. Эта смелость психологически подкрепляется сравнительно благополучной эксплуатацией в течение длительного времени. В нестандартных условиях, когда на IV блоке Чернобыльской АЭС проводились процедуры, связанные с испытанием оборудования, и когда протекание испытаний отклонилось от запланированного, операторами самостоятельно был принят ряд необоснованных рискованных решений. Мотивация этих решений с точки зрения операторов определялась интересами дела, которое им было поручено, однако, их последствия оказались неожиданными. Чернобыльская авария показала, что слишком большая самостоятельность оператора так же опасна, как и ее отсутствие. Здесь необходим компромисс, опирающийся на более четкое и строгое выделение допустимых границ самостоятельности оператора. Необходимо лучше организовывать контроль за строгим соблюдением этих границ. Это явная недоработка регулирующих органов. Высокая требовательность инспекторов к должностным лицам подконтрольных объектов атомной энергетики – важнейшее условие, создающее предпосылки их безопасной эксплуатации.

Так как причинами вывода реактора РБМК-1000 в аварийное состояние явились многочисленные нарушения персоналом АС регламентов и инструкций по эксплуатации, были предприняты организационные меры по укреплению технологической дисциплины и качества эксплуатации. Прежде всего по результатам анализа причин аварии уточнены и усилены регламентные и организационные ограничения. Это дало возможность, опираясь на знание причин происшедшего, сохранить в эксплуатации другие аналогичные блоки (кроме блоков Чернобыльской АЭС, которые были остановлены).

Проведена переподготовка персонала и сдача экзаменов, введены в действие более строгие правила и нормы по эксплуатации. В частности запрещено проведение испытаний и экспериментов без обоснования безопасности реакторной установки и получения разрешения на их проведение от инспектора Госатомэнергонадзора СССР.

Накопленный опыт развития атомной энергетики показал, что государственный надзор за безопасным ведением работ в этой области является исключительно важным звеном обеспечения ее безопасности развития. Поэтому в СССР в 1984 г. вместо нескольких существовавших тогда разрозненных органов надзора был образован единый независимый орган — Государственный комитет СССР по надзору за безопасным ведением работ в атомной энергетике (Госатомэнергонадзор СССР). Чернобыльская авария потребовала проанализировать эффективность деятельности этого государственного органа и предпринять энергичные шаги по ее повышению. Сам факт возникновения такой аварии говорит о том, что в этом была настоятельная необходимость.

После своего создания Госатомэнергонадзор СССР основное внимание уделял вопросам надзора за соблюдением действовавших в атомной энергетике норм и правил при проектировании, изготовлении оборудования и эксплуатации объектов атомной энергетики.

Для решения этих задач был сформирован необходимый центральный аппарат в г. Москве и региональные органы надзора, постоянно контролирующие деятельность предприятий и учреждений по указанным направлениям. Следует отметить, что вопросам разработки требований к уровню безопасности, научной обоснованности представляемых разработчиками доказательств безопасности принимаемых решений, вопросам аттестации и качества подготовки специалистов на первом этапе деятельности Госатомэнергонадзора СССР уделялось меньшее внимание. В результате анализа деятельности Госатомэнергонадзора СССР в связи с уроками Чернобыльской аварии произведены существенные изменения как в структуре его органов, так и в направлениях их деятельности.

На Госатомэнергонадзор СССР в настоящее время возложена серьезная задача разработки совместно с вновь образованным Министерством атомной энергетики СССР и Госкомитетом по использованию атомной энергии СССР требований к безопасности объектов атомной энергетики и к эффективности разрабатываемых для ее обеспечения технических средств. Госатомэнергонадзор СССР организует и контролирует разработку норм и правил в атомной энергетике, содержащих эти требования.

Одним из важных направлений в деятельности Госатомэнергонадзора СССР при осуществлении надзора за соблюдением правил, норм и инструкций по безопасности в атомной энергетике является контрольно-профилактическая работа, т.е. деятельность по предупреждению нарушений нормативно-технических документов на всех этапах создания и эксплуатации объектов атомной энергетики.

С этой целью на передний план выдвигается:

1. Деятельность по выдаче разрешений на право проектирования, строительства, монтажа, этапов ввода АС в эксплуатацию и эксплуатацию объектов атомной энергетики, а также на конструирование, изготовление и ремонт подконтрольного оборудования. Основная задача этого вида деятельности — обеспечить полную готовность предприятий, организаций к качественному выполнению работ.
2. Развитие и повышение ответственности внутриведомственного надзора на объектах атомной энергетики, служб технического контроля при сооружении АС; изготовлении, монтаже и ремонте оборудования объектов атомной энергетики.
3. Повышение эффективности надзора за подготовкой, аттестацией и переподготовкой персонала АС, а также инженерно-технических работников всех подконтрольных

организаций, предприятий и объектов, предварительная проверка готовности учебных центров и подразделений для качественного обучения и аттестации.

4. Переориентация проведения обследований и проверок на ранние стадии работ с целью своевременного предупреждения нарушений правил, норм и инструкций по безопасности. Улучшение подготовки к проведению проверок и их качества путем разработки детализированных методических пособий, обучения инспекторского состава, расширения его прав; организация контроля за деятельностью инспекций и оказание им необходимой помощи.

5. Повышение требовательности к руководителям подконтрольных организаций и предприятий, к должностным лицам за соблюдением правил, норм и инструкций по безопасности. Ни один факт нарушения не должен оставаться без последствий.

Для научного обеспечения деятельности по надзору за безопасным ведением работ в атомной энергетике при Госатомэнергонадзоре СССР создается Научно-технический центр.

Усилились и в дальнейшем будут расширяться связи Госатомэнергонадзора СССР с МАГАТЭ и с национальными органами по безопасности атомной энергетике в других странах.

Вопрос о необходимости использования принципов, обеспечивающих реакторной установке повышенную безопасность, ставился еще на Стокгольмской конференции в октябре 1980 г. К этим принципам относятся уже упоминавшиеся физические принципы обеспечения защиты, такие как саморегулируемость и внутренняя самозащищенность, низкие энергонапряжения и нагрузки, естественная циркуляция и применение пассивных защитных и локализующих устройств. В СССР в наибольшей степени в соответствии с этими принципами развивалась реакторная установка АСТ-500, предназначенная для атомных станций теплоснабжения. Две такие станции сооружаются в Советском Союзе для городов Горький и Воронеж. Тем не менее Чернобыльская авария потребовала проведения дополнительных защитных мероприятий и для реакторов этого типа.

Тяжелая авария на любой АЭС, как показала Чернобыльская авария, безусловно, с большой вероятностью, может воздействовать на население соседних стран. А учитывая, что сегодня отсутствует приемлемая альтернатива широкому развитию атомной энергетике в мире, ее безопасное развитие будет в большой степени определяться развитием доверия, широким обменом опытом и технологией, сосредоточением совместных усилий на наиболее проблемных направлениях.

Именно такое развитие должна получить выдвинутая СССР "Программа создания международного режима безопасного развития ядерной энергетики".

ЛИТЕРАТУРА

[1] СИДОРЕНКО В.А., КОВАЛЕВИЧ О.М., Ат. Энерг. **50** (1987) 93.

THE ROLE OF THE INTERNATIONAL ATOMIC ENERGY AGENCY SAFETY MISSIONS IN UPGRADING PLANT SAFETY AT KRŠKO NUCLEAR POWER PLANT

D. FERETIĆ
University of Zagreb,
Zagreb

Ž. PAVLOVIĆ
Krško Nuclear Power Plant,
Krško

Yugoslavia

Abstract

THE ROLE OF THE INTERNATIONAL ATOMIC ENERGY AGENCY SAFETY MISSIONS IN UPGRADING PLANT SAFETY AT KRŠKO NUCLEAR POWER PLANT.

The role of the International Atomic Energy Agency (IAEA) safety missions in upgrading plant safety at Krško NPP is defined. Krško NPP is a two loop PWR plant designed by Westinghouse. The safety missions, following the invitation of the Yugoslavian Government, visited the plant on numerous occasions from the early construction phase to full power operation. The purpose of inviting the missions was to provide a second independent international viewpoint on various subjects related to safety features of plant design, equipment characteristics, utility organization during construction and operation, quality control and quality assurance programmes and plant commissioning. The main activities of the IAEA missions during plant construction were related to the analysis of the preliminary and final safety analysis reports, especially where specific site conditions deviate from the standard United States Regulatory Guides (for example site meteorology and related environmental dose calculations) or where new design features have been introduced (such as 16×16 fuel design or emergency core cooling systems for two loop plants). Special consideration has been given to plant safety systems upgrading, particularly since the Three Mile Island accident. Mission advice on this subject was very valuable because it has provided the utility and licensing body with independent engineering judgements on the merit of particular hardware changes. In addition, the missions spent considerable time in analysing the role, functions and training of regulatory body personnel, the functions of utility organizations, especially in the areas related to safety (function of the quality assurance department and future organization for plant operation). Following the start of plant commercial operation the IAEA was invited to send the Operational Safety Review Team and the Assessment of Safety Significant Events Team. The paper describes the outcome of these missions. Finally, some general comments are given about the benefits and the possible areas of improvement through using the services of IAEA safety missions.

1. INTRODUCTION

Krško NPP is the first Yugoslavian NPP. The plant is a two loop PWR with 632 MW net output and was built by Westinghouse. One of the principal utility approaches to plant safety was the request that the plant be licensed in the United States of America as the country of origin. All codes and standards valid in the USA at the date of the contract were to be implemented in the plant design. It should be pointed out that the utility organization and the regulatory body in developing countries encounter difficulties in verifying that these commitments are fully implemented.

It might be possible to use the services of a consultant, preferably from the supplier's country, and/or to seek international assistance through the International Atomic Energy Agency (IAEA). Krško NPP did both.

In order to establish regular and systematic co-operation with the IAEA, Yugoslavia signed a trilateral agreement with the USA and the IAEA covering guarantees for nuclear fuel supply and implementation of safeguards measures. At the same time the plant has achieved the status of an IAEA project, which ensured IAEA assistance at various stages of project completion, from evaluation of the preliminary safety analysis report to the evaluation of plant safety status prior to startup. The most important IAEA safety mission results and recommendations are summarized briefly in this paper.

2. IAEA SAFETY MISSIONS DURING THE DESIGN AND CONSTRUCTION PHASE

Evaluation of the Preliminary Safety Analysis Report

The mission (April 1976) included experts from the IAEA, United States Nuclear Regulatory Commission (NRC) and Spain. This mission, despite limited time available for evaluation of the preliminary safety analysis report (PSAR), has produced a number of valuable recommendations which influenced plant design and licensing activities. Some of the recommendations were:

(1) Additional documentation needed for in-depth plant safety evaluations, such as:

- Westinghouse topical reports (including proprietary reports) which are applicable to the review of the nuclear facility.
- All NRC dockets on the reference plants and NRC standard review plants.
- Safety reports prepared by the NRC for two loop nuclear facilities.
- Applicable safety codes and standards.

(2) Necessity to perform additional analysis, for example:

— Recalculation of radiation doses on plant exclusion area boundary (500 m from reactor centreline) with realistic site meteorological conditions in order to assure the compliance with the requirements of 10 CFR 100.
— Re-analyse reactor vessel supports for asymmetric loads following loss of coolant accident (LOCA) conditions.
— Request from reactor vendor more data and analysis relative to nuclear fuel design and performance. This is necessary because Krško NPP is the first PWR to apply 16 × 16 fuel elements.
— Demonstrate the independence of 380 kV and 110 kV external power supply.
— Analyse the possibility of redesigning the spent fuel pit racks to increase the fuel pool capacity (originally designed for 4/3 core).
— Calculate the radioactive effluents from the plant on the basis of 0.12%, instead of 0.05%, fuel cladding defect value to demonstrate conformance with 10 CFR 50 App. I requirements.
— Calculate the accident iodine doses to demonstrate conformance with 10 CFR 100 guidelines.
— Consider probabilistic based analysis of plant systems (particularly auxiliary feedwater system and residual heat removal system) to meet the requirements of the WASH-1270 report.
— The final safety analysis report should contain the technical specifications in the new format developed for Westinghouse PWR plants.
— The quality assurance activity should be effected by an independent group which reports directly to the utility director.
— Demonstrate that the emergency core cooling system (ECCS) analysis meets the acceptance criteria (10 CFR paragraph 50.46) and that the limiting core parameter value (2.32) is adequate.
— Analyse inadvertent valve motion of power operated valves as part of single failure criteria requirements.
— The pre-operational tests should be based on United States Regulatory Guide 1.79. The containment sump recirculation test should be reanalysed to secure minimum pump head due to vortex formation.
— Analysis of water level following a LOCA in containment and verification that electrical components of ECCS valves will not be submerged.

(3) Safety systems upgrading

— Installation of loose part monitoring system
— Seismic qualification of sump monitoring system
— Automatic feature to back up the operator's action to achieve a proper recirculation mode following a LOCA.

Review of Final Safety Analysis Report

The review of final safety analysis report (FSAR) was submitted to IAEA upon completion. The IAEA mission (April 1980) presented some major comments on this report. Since at that time the first NRC recommendations and ground rules related to actions after the TMI accident were known, the mission also presented some viewpoints on these actions.

Apart from presenting a number of comments and remarks on various FSAR chapters, mainly connected with estimates of radioactive releases, environmental dose calculations and potential hazard from chlorine release from a nearby paper mill factory the mission submitted a number of recommendations which have influenced substantially activities in the area of plant safety. Some of these recommendations were:

- — Domestic capability to perform plant safety analysis to be established and supported on a long term basis. The transients and accidents presented by the vendor in FSAR Chapter 15 to be confirmed and the analysis to be extended to stable plant conditions.
- — Apart from the deterministic type of safety analysis, probabilistic studies establishing the reliability of safety related systems also to be carried out.
- — Quality assurance organization to be shifted from construction to startup and plant operation.
- — Training on a plant specific simulator, when available, to be organized for the Krško NPP operator in addition to the training and retraining on a generic four loop PWR plant simulator.
- — A corresponding post-TMI 2 action plan to be elaborated on the basis of the lessons learned from that accident as given in the latest revision of NRC's NUREG 0660 [1].

3. PRE-CRITICALITY SAFETY MISSIONS

In the period shortly before the first criticality (which was achieved in September 1981) a number of IAEA safety missions was invited by the licensing authorities mainly to analyse specific pending issues, such as (1) site meteorological data to be used in environmental dose calculations; (2) function of regulatory body and on-site inspection during plant startup and operation, specially in view of the fact that the responsibility for plant licensing is divided between two neighbouring republics, Slovenia and Croatia; (3) review of plant technical specifications with up to date practice for similar plants.

The Krško site is unique from the point of view of site meteorology. It has a high frequency of low wind speeds (about one third of the time wind speed is less then 1 m/s), persistent thick fogs and 'stagnation' diffusion conditions. The question to be solved was the application of standard IAEA Safety Guides for such conditions.

After thorough analysis of all calculations and measurements the IAEA expert opinion was that the FSAR application of standard regulatory diffusion models is adequate at Krško to ensure proper safety conditions.

With respect to the regulatory body functions, the IAEA mission recommended an increase in the number of staff of regulatory body and safety inspectors and further use of the assistance of IAEA expert groups for the regulatory activities. The help for regulatory body activities should also come from local specialized institutes that could perform the safety analyses needed in the regulatory process.

Plant technical specifications — FSAR, Chapter 16 of standard format — have been reviewed. (The document was originally prepared by Westinghouse using the old format.) The review and recommendations reflected the newest viewpoints in the USA on specific chapters and paragraphs.

Emergency Planning

In early 1981 the IAEA was requested by the regulatory authorities to provide assistance on appropriate emergency planning arrangements in support of Krško NPP. This was in response to a circular letter from the IAEA which offered assistance upon request. Such assistance was essential because of lack of experience in this field.

The expert report addressed the following subjects:

— Emergency planning;
— Emergency preparedness;
— Evaluation of emergency plans.

Based on the recommendations of this mission a full scale emergency drill was executed the following year in the area surrounding the plant.

TMI Action Plan of Krško NPP

At the time of the TMI accident Krško NPP was in the startup phase. Following the request of the licensing authorities Krško NPP has developed a TMI action plan based on NRC requirements as defined in NUREG 0737 [2]. This plan included 53 tasks. After careful study and investigation carried out by the utility, independent experts and foreign consultants some doubts appeared about timely fulfilment of all requirements.

Krško NPP then requested the help of an IAEA safety mission to discuss the merits and background of each TMI task and define feasible actions. After joint engineering judgement and meetings held with licensing authorities it was concluded that preference should be given to preventive actions (i.e. actions most important to prevent core melt) and to the majority of software actions (analysis, training, procedures, organization).

Through this approach priority has been given to actions such as emergency power to pressurizer heaters, direct indication of relief and safety valve position, subcooling margin monitor, auxiliary feedwater system initiation and flow rate indications, etc., and to the actions related to improved safety analysis and probabilistic risk assessment studies, improved operator training and examinations, introduction of shift technical adviser, technical support centre, control of operational activities and procedures related to control room activities.

The second priority was given to tasks requiring installation of reactor head vent, reactor vessel level measurement, high level radiation measurement, post-accident sampling, etc., which are intended to be applied in post-accident conditions.

The help of the IAEA mission to Krško NPP in determining the proper selection of the most important post-TMI tasks was essential.

4. IAEA SAFETY MISSIONS DURING PLANT OPERATIONS

Upon the request of the Yugoslavian Government the IAEA's Operational Safety Review Team mission visited Krško NPP in February 1984, i.e. about 18 months after start of plant commercial operation. The team consisted of eight members from the IAEA and operational plants (from Brazil, Federal Republic of Germany, Switzerland, Sweden and the United Kingdom). The review covered the following areas:

— Training and qualifications
— Operations
— Technical support
— Maintenance
— Radiation protection
— Chemistry
— Emergency planning.

Since the reviewed areas are extensive, the mission audits resulted in a considerable number of remarks, observations and recommendations. The recommendations, for example, included the necessity to upgrade the FSAR to reflect the latest changes in plant organization, the functions of the Krško safety committee, operations in the quality assurance department, upgrading of some part of the plant manual, functions of safety inspections, functions of the plant safety engineer and the Independent Safety Engineering Group, changes in controlled access area, radiation protection functions, equipment, etc.

Many of these recommendations derived from sound engineering judgement and were not unexpected by a relatively experienced plant staff in plant startup and operation. However, the recommendations were useful because they were not only of a formal nature but reflected actual experience from other plants and because they confirmed the priorities for some specific actions.

In order efficiently to implement these recommendations it is important that the mission defines the most important and essential points which are also accepted as being important by the plant staff and plant management. The main benefit of this mission was the opportunity for the plant engineers to exchange viewpoints with the mission members on possible improvements of various specific areas of plant operational activities. This is especially important since Krško NPP does not belong to any formal international nuclear operating group. The licensing authorities and the public were also interested to hear the mission's opinion on the plant operational safety.

4.1. IAEA's Assessment of Safety Significant Events Team mission to Krško NPP

Krško NPP was the first NPP to receive the recently established IAEA's Assessment of Safety Significant Events Team mission at the end of 1986. The purpose of the mission was to analyse in detail the safety significant events and determine the remedial actions.

Upon the request of the Yugoslavian Government the analyses of the mission were devoted to the following events, experienced by Krško NPP during the first years of operation:

- — Loss of second off-site power source;
- — Inadvertent depressurization of the reactor coolant system;
- — Leakage of the inner seal ring on reactor vessel flange;
- — Containment pressure increase;
- — Incorrect operation of certain protective parts in diesel generator sets;
- — Number of safety related events reported to the regulatory body in 1983, 1984 and 1985 (deviations and incidents).

The assessments focused on human and technical aspects of each topic. Conclusions of these assessments were intended to provide Krško NPP staff and the regulatory body with an independent opinion on the appropriateness of corrective actions already implemented and on possible additional corrective actions to prevent the occurrence of similar events. The team had prepared a guide together with a questionnaire analysing precursory elements of each event by establishing a failure sequence. The purpose of such an approach was to identify the direct cause among six predetermined categories of deficiencies:

- — Design, manufacturing or installation deficiencies related to equipment performance;
- — Work control, deficiencies related to personnel performance.

All available facts were analysed to develop a logical sequence leading from the direct cause to the last consequence observed which actually led to the event. Contributory elements to the direct cause were analysed through establishment of a 'weakness tree' leading to the identification of root causes among three predetermined categories of deficiencies:

— Surveillance of the equipment;
— Surveillance of man–machine interface (operator aids, tools and procedures);
— Surveillance of personnel qualifications.

For each event the lesson learned had to be defined and specific corrective actions undertaken.

5. CONCLUSIONS

The role of IAEA safety missions to upgrade plant safety in helping both the utility and the regulatory body was substantial. This is especially the case of the Krško NPP. Krško is the only NPP in Yugoslavia with relatively modest domestic support for plant operation.

The role of IAEA missions to upgrade plant safety is reflected mainly as follows.

Overall nuclear safety regulations have not been issued in most developing countries in which first NPPs are being constructed. IAEA expert groups were invited for work in special areas, such as site survey, site investigation, design, construction, startup and operation, assist in interpretation of the contents of the regulations on individual investigation tasks or application of specific procedures used during construction and issued in the country of the supplier of nuclear technology.

The missions give support both to the company and the regulatory body, which can then act promptly by means of issuing the specific authorizations and permits. The regulatory body relies upon the objectivity of the members of the expert mission.

The 'turnkey' contract is the usual form of contract for the construction of the first NPP. In such a contract there is no obligation for the supplier of the NPP to develop also the corresponding infrastructure support which is indispensable for the future work of the plant operator. The members of a mission are experienced experts who have constructed NPPs and who have successfully passed through this initial period in their own countries. Many things which seem to be trifles at the beginning become very complex problems later. If due attention is paid to problems at the beginning, the implementation of the nuclear power programme will be smooth and no omissions will be made.

The utility constructing its first nuclear plant can benefit considerably by learning from the experience of mission members. Since the mission members are from various countries this experience represents a cross-section of international practice in a specific field. The personal contacts established with mission members allow the continuing exchange of information also after the mission period.

The mission, by detecting weak points in plant operations and recommending adequate changes, contributes directly to the improvement of plant safety and plant performance.

The regulatory body obtains advice and support for specific regulatory decisions.

The plant management can more easily obtain support from the utility organizations and from governmental authorities if the request is based on recommendations from the mission.

The public is more reassured on plant safety if the mission confirms that the plant is operated in accordance with international safety standards.

It is likely that the vendor will pay more attention to the plant design and requests from the utility, especially in relation to plant safety, if the activities are constantly exposed to international surveillance.

Some difficulties encountered by the utility in connection with the mission are the following:

The plant staff, usually overloaded with day to day duties, sometimes find it difficult to dedicate their full attention to the mission activities on the site, especially if the mission stay is long.

The plant staff often consider the mission audit of operation activities as part of the licensing process (the findings are normally submitted to the local licensing authorities) and they are reluctant to discuss freely all the problems connected with the plant.

The basis for mission evaluation is usually not founded on known and pre-established criteria but on the experience and background of individual mission members. This may create different evaluations and recommendations in various fields examined by the mission.

The main prerequisites for a successful mission are:

(1) Good preparation: The plant staff and mission members must be prepared for the mission. All documents necessary for the mission should be identified in advance and be available on time.
(2) Proper timing: The mission should arrive on site when the problems to be considered are clear and the necessity for their resolution recognized by the counterparts.

REFERENCES

[1] UNITED STATES NUCLEAR REGULATORY COMMISSION, NRC Action Plan Developed as a Result of the TMI-2 Accident, Vol. 2, Rep. NUREG-0660-V-2, USNRC, Washington, DC (1980).

[2] UNITED STATES NUCLEAR REGULATORY COMMISSION, Clarification of TMI Action Plan Requirements, Technical Rep. NUREG-0737, Office of Nuclear Reactor Regulation, USNRC, Washington, DC (1980).

ENSEIGNEMENTS TIRES ET ACTIONS CORRECTIVES DECIDEES APRES L'INCIDENT SURVENU A BUGEY 5 (14 AVRIL 1984)

J.-M. GANDIT
Institut de protection et de sûreté nucléaire,
Commissariat à l'énergie atomique,
Fontenay-aux-Roses, France

Abstract–Résumé

LESSONS LEARNT AND CORRECTIVE ACTION DECIDED ON FOLLOWING THE BUGEY-5 INCIDENT ON 14 APRIL 1984.

On the basis of an example, the paper shows how lessons are drawn from a significant incident for the purpose of improving the safety of the unit affected, and subsequently, of adapting the corrective measures to other reactors. The example taken is the incident which occurred at the Bugey-5 unit on 14 April 1984. The procedure followed for the analysis is mentioned, and the causes and actual and potential consequences are analysed from the point of view of safety. The thinking derived from the analysis shows, in particular, that the failures due to slow deterioration, not foreseen in the design, give rise to uncertainties in operation as well as to difficulties in interpreting alarm information from the control room. This may lead to a complex situation making control of the unit a very critical matter. The lessons drawn and the corrective measures applied to units at the different levels of the French pressurized water reactor park are described. The paper stresses the generic nature of the lessons drawn and shows that the slow deterioration of the parameters, especially as regards control sources, is a mode of failure which cannot be analysed. It is a limit of safety analysis. The purpose of the corrective measures applied to the French PWRs is to protect them against uncertainties in the operation of electrical systems. Obviously, they can be applied to all types of reactors.

ENSEIGNEMENTS TIRES ET ACTIONS CORRECTIVES DECIDEES APRES L'INCIDENT SURVENU A BUGEY 5 (14 AVRIL 1984).

A partir d'un exemple, on montre comment les leçons sont tirées d'un incident significatif en vue d'augmenter la sûreté de la tranche affectée et, par suite, d'adapter les mesures correctives à d'autres réacteurs. L'exemple pris est l'incident de la tranche Bugey 5 du 14 avril 1984. La démarche suivie pour l'analyse est rappelée, les causes puis les conséquences réelles et potentielles du point de vue de la sûreté sont analysées. Les éléments de réflexion tirés de l'analyse montrent en particulier que les défaillances par dégradation lente, non prévues à la conception, induisent des modes de fonctionnement aléatoires ainsi que des difficultés d'interprétation des informations d'alarme de la salle de commande. Elles peuvent amener à une situation complexe rendant la conduite de la tranche très délicate. Les enseignements tirés ainsi que les mesures correctives appliquées aux tranches des différents paliers du parc des réacteurs à eau sous pression français sont exposés. Ce mémoire souligne le caractère générique des enseignements tirés et montre que la dégradation lente des paramètres, surtout dans le domaine des sources de contrôle-commande, est un mode de défaillance qui ne peut pas être analysé. C'est une limite de l'analyse de sûreté.

Les mesures correctives mises en place sur les REP français ont pour objectif de se prémunir contre les fonctionnements aléatoires des systèmes électriques. Elles peuvent être appliquées bien évidemment à tous les types de réacteurs.

1. INTRODUCTION

L'incident survenu à Bugey 5 le 14 avril 1984 est sans aucun doute le plus grave depuis le démarrage des tranches à eau sous pression du parc français. Il est caractérisé par son aspect un peu spectaculaire (perte des sources de puissance externes ainsi que de la source interne de la voie A) faisant suite à la défaillance de la source de contrôle-commande de la voie A; il aurait pu évoluer vers un accident grave pour lequel aucune disposition particulière n'était encore opérationnelle sur le site si une défaillance avait affecté l'autre voie.

Les agents de conduite ont très rapidement compris ce qui se passait et, malgré l'absence de consignes de conduite appropriées à la situation, ont très bien maîtrisé l'incident.

Les conséquences réelles sont faibles; les conséquences potentielles, bien plus importantes, montrent que la perte des systèmes redondants est possible.

Ce mémoire, après un rappel des événements concernant plus spécialement les sources électriques, rappel fait pour mieux comprendre l'origine et la portée des actions correctives, met l'accent sur les conséquences réelles et potentielles de l'incident du point du vue de la sûreté et souligne les enseignements génériques applicables aux tranches du parc REP français. Les enseignements sont, dans leur intégralité, applicables à d'autres types de réacteurs.

2. DEMARCHE SUIVIE POUR L'ANALYSE

L'analyse approfondie des incidents jugés significatifs est décidée suivant des critères d'identification en rapport avec l'importance qu'attachent les analystes de sûreté à certains événements, définis par des critères informels tels que les événements non enveloppés par une situation prise en compte à la conception.

Le principe d'effectuer une analyse approfondie étant décidé, celle-ci est réalisée selon la démarche suivante:

1) Au départ, il faut disposer de renseignements précis quant à l'état initial de l'installation, le déroulement des événements et l'état normal de la fonction de sûreté en cause. C'est l'exploitant qui fournit, dans un compte rendu rédigé dans les deux mois qui suivent l'incident, toutes ces informations.

2) Sur cette base, l'analyse est entamée, les compléments d'information nécessaires étant demandés à la centrale ou aux services centraux de l'exploitant.

L'analyse tient compte des incidents connus du même type, des incidents précurseurs, de la réaction des opérateurs, des erreurs humaines éventuelles, des défaillances matérielles. Le faisceau des causes, les conséquences réelles et potentielles sur la sûreté sont examinés.

3) Les éléments de réflexion dégagés (anomalies de conception, défaut d'application des consignes, insuffisance dans les spécifications techniques d'exploitation) permettent de formuler des recommandations qui forment en partie la conclusion du rapport. Le rapport est discuté avec l'exploitant avant sa transmission aux autorités de sûreté. Cette discussion finale avec l'exploitant ne limite pas l'autonomie de l'analyste de sûreté et elle permet une meilleure qualité de l'analyse.

Les analystes ne proposent pas de solution technique, ceci restant de la seule responsabilité de l'exploitant aussi bien pour l'étude que pour la mise en œuvre des modifications décidées.

L'incident décrit dans le chapitre suivant a été traité suivant la démarche qui vient d'être énoncée. Les mesures correctives appliquées ont été étudiées et mises en œuvre par l'exploitant.

3. RAPPEL SUCCINCT DES EVENEMENTS

La tranche fonctionnait en puissance. Après la défaillance de la régulation du redresseur en service, celui-ci s'est arrêté. Dès lors, les circuits de contrôle-commande de la voie A n'étaient alimentés que par la batterie d'accumulateurs au plomb dont la décharge commençait[1].

La batterie d'accumulateurs débitant seule dans les circuits, il s'ensuivit une baisse lente de la tension du tableau de distribution.

L'alarme signalant en salle de commande la «tension faible» du tableau est apparue lors du dépassement du seuil préréglé. Cependant, celle-ci étant constituée par le regroupement de plusieurs informations (défaut d'isolement, déclenchement d'un disjoncteur de départ), elle a été interprétée comme la réapparition d'un défaut d'isolement non identifié.

Après plusieurs heures de décharge lente, la tension du tableau est descendue à une valeur très faible. C'est alors que des désordres sont apparus dans le fonctionnement des voyants de la salle de commande. Ceci a incité un agent de

---

[1] Une source importante pour la sûreté, dite sans coupure, comporte une batterie d'accumulateurs et deux redresseurs. Le jeu de barres de distribution du tableau est alimenté par le redresseur en service qui maintient la charge de la batterie (charge en floating). En cas de panne du redresseur en service, le second redresseur est mis en service manuellement et le redresseur en panne est découplé du jeu de barres. Si aucun des deux redresseurs ne peut être alimenté par le réseau, c'est la batterie qui alimente seule le jeu de barres de distribution avec cependant un temps limité de fonctionnement.

conduite à se rendre dans les locaux électriques. Sur place, l'agent s'est rendu compte de la situation et, avant qu'il n'ait eu le temps d'enclencher le redresseur de secours, le disjoncteur d'arrêt d'urgence du réacteur de la voie A s'est ouvert par relâchement de la bobine à manque de tension.

L'enchaînement des événements qui se sont alors succédés montre que tout contrôle des transitoires sur les sources électriques de puissance de la tranche était impossible.

La première source touchée bien que n'étant affectée d'aucune perturbation, la source externe principale a été séparée de la tranche. Le disjoncteur de ligne s'est ouvert, après le déclenchement de la turbine, car l'interrupteur enclencheur ne s'est pas ouvert et le contacteur d'excitation de l'alternateur est resté fermé. Le turbo-alternateur est resté couplé sur les jeux de barres de la tranche.

Du fait du ralentissement de la turbine, la tension et la fréquence fournies se dégradaient. Le seuil de tension faible a été atteint trois minutes après la séparation du réseau. Ce seuil étant activé, il y a eu un début de changement de source externe.

Les jeux de barres de puissance ont été séparés du groupe, mais ils n'ont pas pu être couplés sur la source externe auxiliaire (deuxième source touchée).

Le tableau secouru de la voie A, mis hors tension, n'a pas pu être réalimenté par le groupe diesel de secours correspondant.

Il faut nécessairement une tension de contrôle-commande pour qu'il démarre et se couple sur le jeu de barres (la troisième source étant hors service).

Le tableau secouru de la voie B a été normalement réalimenté en puissance par le groupe de secours.

Pendant les phases de ralentissement du groupe turbo-alternateur, l'onduleur du groupe de protection n° 1 a déclenché (quatrième source touchée de façon indirecte).

La tranche est restée environ 50 minutes seulement alimentée en puissance par la voie de sûreté B.

Un retour vers une situation plus saine a été amorcé après que la manoeuvre manuelle des disjoncteurs d'alimentation des tableaux 6,6 kV de tranche eut été faite. La source auxiliaire couplée a permis une remise en service de la source de contrôle-commande de la voie A, donc une maîtrise de la situation.

Les répercussions sur le circuit primaire principal des désordres apparus dans la distribution électrique sont nombreuses.

Les pompes primaires étant arrêtées, il n'y avait plus d'aspersion normale dans le pressuriseur. L'aspersion auxiliaire ne fonctionnait plus (manque de tension de contrôle-commande). La pression régnant dans le circuit primaire principal qui fonctionnait en thermosiphon était écrêtée par l'ouverture répétée des vannes de décharge dont seule, à la fin, celle de la voie B pouvait encore s'ouvrir. La température du fluide primaire toujours lisible en salle de commande a décru rapidement après que les agents de conduite eurent ouvert manuellement les vannes de contournement de la vapeur à l'atmosphère (circuit secondaire). Le refroidissement du combustible a toujours été assuré par l'alimentation en eau de secours des générateurs de vapeur. Il n'y a pas eu de rejet radioactif à l'extérieur de l'enceinte de confinement. La barrière thermique des joints des pompes primaires a toujours été refroidie sauf pendant quelque temps lors des changements de source externe et l'injection d'eau aux joints des pompes primaires a été assurée. De faibles fuites ont été mises en évidence aux joints des pompes primaires. Après la remise en place des joints, les fuites ont été résorbées.

4. CAUSES DE L'INCIDENT

A l'origine de l'incident, deux causes sont retenues. A chacune de ces causes correspondent des conséquences pour la sûreté.

La première cause, dans l'ordre chronologique, est la défaillance de la régulation du redresseur en service suivie de l'arrêt de celui-ci sans que la panne soit signalée. Cette cause est bénigne en elle-même; cependant elle prend des proportions importantes à partir du moment où l'arrêt du redresseur n'est pas signalé en salle de commande, les dispositifs prévus n'étant pas à même de le faire.

La seconde cause, dont l'apparition est bien repérée, est technique, mais elle se situe à l'interface homme-machine ce qui, tout comme pour la première, aggrave son caractère banal. Il y a eu apparition d'une alarme lorsque, la tension de la source étant suffisamment dégradée, le seuil a été sollicité. Malheureusement, l'alarme étant un regroupement d'informations d'importances inégales pour la sûreté, elle a été mal interprétée.

C'est là que se situe la défaillance humaine: il y a eu mauvaise interprétation d'une information.

5. CONSEQUENCES DU POINT DE VUE DE LA SURETE

Les conséquences de l'incident de dégradation de la tension de la source de contrôle-commande de la voie A sont examinées en séparant les conséquences réelles des conséquences potentielles. Les conséquences sont analysées aussi bien sur le plan technique que sur le plan de l'application des procédures incidentelles.

5.1. Conséquences réelles

Celle qu'il faut retenir en premier lieu est le mode commun de défaillance induit par le fonctionnement de la source avec une tension dégradée, en dehors des limites admises. Ce mode de fonctionnement produit des défaillances en chaîne de façon directe (source externes de puissance) ou indirecte (source de protection).

La seconde conséquence est que les moyens d'action normaux depuis la salle de commande sont diminués, voire inutilisables. Seuls certains ordres sont exécutés: ceux de la voie B de façon sûre, quelquefois ceux de la voie A mais avec de nombreux aléas de commutation.

Par exemple, les disjoncteurs de la distribution de puissance électrique n'ont pas répondu aux ordres émis par les automatismes en voie A; par contre, les pompes de l'eau de secours du circuit secondaire ont quand même fonctionné sur demande émise par les capteurs de la voie A.

La troisième conséquence se rapporte à la conduite de la tranche. Les informations logiques et analogiques fournies aux opérateurs et alimentées par les sources défaillantes ont également été perdues. La difficulté qui s'ensuivait était que certaines informations présentes en salle de commande ne reflétaient pas la réalité.

Il était difficile de séparer les bonnes informations de celles qui étaient douteuses.

La dernière conséquence concerne l'utilisation des procédures incidentelles. Jusqu'à l'occurrence de cet incident, les défaillances admises de sources étaient limitées à une seule, les procédures existantes ne tenant pas compte de cas, plus rares il est vrai, de perte multiple de sources. Le jour de l'incident, aucune aide n'a pu être fournie aux opérateurs par les procédures existantes. Il n'est pas possible d'utiliser plusieurs procédures incidentelles simultanément. Malgré le développement des procédures, il subsiste encore des séquences incidentelles non identifiées et l'incident montre que les opérateurs ne sont pas suffisamment armés pour faire face à certains types de défaillances concernant les sources électriques, en particulier les défaillances partielles et multiples.

5.2. Conséquences potentielles

Comme cela a déjà été écrit en introduction, une seule défaillance sur la voie B (voie de sûreté) aurait suffi pour que la tranche se trouvât hors du domaine de dimensionnement.

Le jour de l'incident, aucune disposition particulière prévue n'était opérationnelle sur le site.

On rappelle que, pour faire face à une situation de perte totale des sources électriques de puissance, les dispositions actuelles comportent:

– une réalimentation en eau des joints des pompes primaires;

– une réalimantation électrique d'une partie de l'appareillage de contrôle-commande (suivi des paramètres concernant le cœur);

– une réalimentation en air comprimé par des bouteilles externes de certains circuits pneumatiques;

– enfin des opérations sont prévues pour rétablir une alimentation en 6,6 kV de la tranche soit par:

- une tranche îlotée du site ou d'un site voisin ou d'un groupe hydraulique proche,
- une turbine à gaz dont la mise en service demande plus de temps,
- un groupe diesel de secours d'une autre tranche.

6. ENSEIGNEMENTS TIRES

L'incident, analysé tant du point de vue des causes que des conséquences sur les systèmes électriques, comporte de nombreuses facettes.

Les enseignements qui ont été tirés ont eu des répercussions sur tous les paliers (900 et 1300 MWe) des REP français. Ceci s'est traduit par des modifications techniques et par la création de procédures de conduite des tranches en régime incidentel.

Les enseignements sont suffisamment généraux pour avoir également une application sur toutes les autres filières de réacteurs existantes.

Avant de montrer la concrétisation des enseignements tirés, on exposera les éléments de réflexion qui ne sont pas nouveaux d'ailleurs mais qui s'appuient sur des faits bien réels qu'on avait tendance jusque-là à classer dans les événements hypothétiques. Ensuite, les mesures correctives techniques mises en œuvre en vue de réduire les causes puis celles qui ont pour but de minimiser les conséquences de ce type d'incident seront succinctement décrites. Enfin, les nouvelles procédures incidentelles seront abordées.

6.1. Eléments de réflexion

L'analyse a mis en lumière trois éléments principaux de réflexion:

1) *L'importance d'une bonne organisation des alarmes* (regroupements, couleurs des voyants, emplacements): il s'agit là du premier élément de l'interface homme–machine dont l'analyse n'a pas été suffisamment poussée après l'accident de TMI.

2) *La possibilité de défaillances non franches:* jusqu'à présent, toutes les analyses de sûreté, qu'elles soient déterministes ou probabilistes, ne prennent en compte que les défaillances complètes des équipements à l'exclusion des modes de fonctionnement dégradés ou aléatoires. La prise en compte de tels fonctionnements n'est pas, en l'état actuel des moyens d'analyse, possible. Tout doit donc être mis en oeuvre pour éviter ce mode de défaillance.

3) *Les interdépendances qui existent entre les différentes sources électriques ont été mises en évidence:* la complexité des relations entre elles fait que la défaillance de l'une peut simultanément affecter les moyens d'action et les informations fournies aux opérateurs, ce qui rend le contrôle de telles situations

particulièrement délicat. Malgré le développement des procédures incidentelles utilisables individuellement, l'incident a montré que les opérateurs de la salle de commande ne sont pas suffisamment armés pour faire face à certaines situations telles que les défaillances multiples. L'effort entrepris depuis TMI dans le domaine des procédures incidentelles doit être complété par un effort semblable dans le domaine des sources électriques de puissance et de contrôle-commande.

6.2. Mesures correctives décidées pour éliminer les causes

Il n'est pas admissible qu'un matériel important pour la sûreté cesse de fonctionner pour cause de panne sans que les agents de conduite en soient informés.

Pour surveiller le fonctionnement des redresseurs des sources sans coupure, la solution appliquée consiste à surveiller le sens de circulation du courant dans la batterie d'accumulateurs associée au redresseur. Si le sens du courant indique que la batterie d'accumulateurs est en charge, la source fonctionne dans les conditions normales. Si le sens s'inverse, le système de surveillance est sollicité et provoque l'apparition d'une alarme, indiquant le début de décharge de la batterie, en salle de commande.

Cette mesure corrective complète l'alarme de tension basse existante. L'alarme du début de décharge prévient très tôt les agents de conduite de l'apparition d'une anomalie de fonctionnement, avant que la batterie n'ait perdu une grande partie de sa capacité.

Cette mesure corrective est appliquée à toutes les sources sans coupure des tranches de tous les paliers REP.

L'exploitant s'est en outre rendu compte de l'effet néfaste de certains regroupements d'informations pour constituer des alarmes. Plus particulièrement, pour les sources sans coupure, un réaménagement a été effectué. Il a consisté à séparer les alarmes concernant la tension des autres alarmes, comme celle du défaut d'isolement. L'alarme de début de décharge de la batterie d'accumulateurs est également séparée des autres alarmes.

6.3. Mesures correctives décidées pour minimiser les conséquences

Comme l'enchaînement des événements qui se sont succédés au cours de l'incident le montre, la défaillance non franche d'un système électrique provoque des fonctionnements aléatoires dans d'autres systèmes, les systèmes servis en particulier. La conduite de la tranche dans ces conditions présente des difficultés importantes. La garantie que les pannes non franches ne sont pas possibles ne pouvant être fournie, l'exploitant a décidé, en cas de symptôme déclaré laissant préjuger que des risques de défaillance non franche sont grands, d'isoler la source douteuse de façon délibérée en provoquant une coupure franche afin de conduire la tranche avec un procédure incidentelle connue et qualifiée.

Il y a deux méthodes pour réaliser cette coupure: soit manuellement sur ordre, soit par un automatisme présentant toutes les garanties de fiabilité et de redondance.

La première solution a été adoptée pour la plupart des sources de contrôle-commande. Auparavant, la puissance de la tranche est ramenée au minimum technique.

La seconde solution a été adoptée pour quelques sources jugées plus importantes. Elle est en particulier adoptée pour la source à l'origine de l'incident de Bugey 5.

6.4. Mesures correctives appliquées aux procédures

Dans ce domaine les actions correctives effectivement opérationnelles sont moins nombreuses. La réflexion dans laquelle s'est engagé l'exploitant n'est pas encore terminée; cependant, les objectifs qu'il s'est assignés sont précisés et il est prévu qu'un début de concrétisation ait lieu avant la fin de cette année.

L'exploitant a engagé des actions à deux niveaux qui se complètent. Le premier niveau concerne le choix de la procédure incidentelle sur des critères d'alarmes présentes en salle de commande. C'est le document d'entrée en consigne. Le second niveau concerne la conduite de la tranche en situation incidentelle de perte multiple de sources électriques.

6.4.1. Document d'entrée en consigne (DEC)

Ce document est une nouvelle consigne servant uniquement au diagnostic de l'incident en cours et renvoyant à la procédure incidentelle adaptée à la situation.

L'opérateur éprouve souvent de grandes difficultés dans le choix de la bonne procédure de conduite lorsqu'un nombre important d'alarmes apparaît. Le DEC est un guide basé sur l'allumage d'alarmes singularisées par un repère collé sur la verrine. Il y a environ 70 alarmes repérées qui sont les critères d'entrée dans les procédures. Les critères sont classés par ordre d'importance ce qui, automatiquement, réalise une hiérarchisation des consignes de conduite incidentelles. La hiérarchisation des procédures conduit naturellement à considérer certaines comme plus importantes que d'autres parce qu'elles les enveloppent, ou se rapportent à des incidents plus graves qui demandent des actions plus urgentes.

Pratiquement, il résulte des principes énoncés que le document présente les procédures classées entre elles par un organigramme qui n'a qu'une seule entrée et autant de sorties que de procédures.

Le cheminement dans l'organigramme est organisé de telle manière que chaque test réalisé demande une réponse «oui» ou «non».

Si la réponse est oui, cela identifie la procédure nécessaire. Si la réponse est non, il faut continuer le cheminement.

La première réponse positive trouvée est la bonne, et il ne faut pas poursuivre car le diagnostic est fait. Même si certains tests suivants peuvent être validés, ils conduisent nécessairement à des consignes incidentelles ne traitant que partiellement de l'incident.

Dans des situations plus complexes où le document d'entrée en consigne ne permet de valider aucune procédure (cas des pertes partielles ou multiples de sources par exemple), la tranche se trouve hors du domaine de validité du document d'entrée ou des consignes. On pénètre là dans des situations plus rares qui font l'objet du paragraphe suivant.

6.4.2. Pertes multiples de sources

Pour ce niveau d'action sur les procédures, la réalisation effective des améliorations n'a pas encore abouti. Les objectifs et les délais que s'est fixés l'exploitant sont néanmoins connus.

Le problème posé comporte deux volets: la conduite de la tranche en cas de perte multiple de sources et les précautions à prendre pour la remise en service après réparation.

Pour le premier volet, l'étude en cours est engagée sur la base des événements précurseurs de défaillances de cause commune sur les alimentations électriques ainsi que des incidents connus et analysés. Les cas retenus détermineront la forme des améliorations de procédures de conduite.

Le second volet est à un stade plus avancé. Les procédures incidentelles comportent dorénavant un chapitre supplémentaire explicitant les actions à exécuter suivant l'état de la chaudière avant la remise en service des sources.

Les précautions à prendre ne concernent que les sources de contrôle-commande. Il faut principalement s'assurer que les organes de commande manuels sont dans une position correspondant à la position réelle de l'actionneur.

Pour les sources de puissance, la remise en service des tableaux de distribution peut se faire sans précaution particulière car en cas de manque de tension les actionneurs sont automatiquement délestés, passé un certain délai, pour justement éviter toute remise en service intempestive.

7. CONCLUSION

L'analyse d'un incident comportant des situations non prévues à la conception ayant des conséquences potentielles importantes est riche d'enseignements.

Entre ce qui avait été admis à la conception, c'est-à-dire la défaillance brutale d'un matériel, et ce qui avait été postulé par les analystes de sûreté, c'est-à-dire la perte totale de systèmes redondants, il existe toutes sortes de défaillances comme la perte partielle et la dégradation qui n'ont pas été analysées. L'analyse de la dégradation et, en corollaire, celle des fonctionnements aléatoires des matériels, est d'ailleurs impossible.

C'est un des aspects des enseignements tirés de l'incident, valable pour tout réacteur.

Les mesures correctives effectivement appliquées sur les tranches REP françaises ont toutes pour objectif d'éviter le fonctionnement de la tranche avec une source de contrôle-commande dégradée.

Il est même préférable de mettre manuellement hors service, passé un certain seuil, une source dont la tension se dégrade que de la conserver en service au risque d'induire des fonctionnements aléatoires de certaines parties des actionneurs de la tranche.

L'incident analysé est donc très bénéfique pour la sûreté tant du point de vue des éléments de réflexion mis en lumière que des mesures correctives originales apportées. Leur caractère générique a été montré et cet aspect dégagé par l'analyse montre, à l'évidence, qu'ils sont applicables à tous les types de réacteurs.

SAFETY ASPECTS OF THE AGEING AND MAINTENANCE OF NUCLEAR POWER PLANTS: SYMPOSIUM OVERVIEW

S. NOVAK
Division of Nuclear Safety,
International Atomic Energy Agency

M. PODEST
Division of Nuclear Power,
International Atomic Energy Agency

Vienna

Abstract

SAFETY ASPECTS OF THE AGEING AND MAINTENANCE OF NUCLEAR POWER PLANTS: SYMPOSIUM OVERVIEW.

Some NPPs are approaching or are even beyond the end of their planned lifetime: 25–40 years of operation. In the year 2000 approximately 160 plants will be 25 years old, and 69 will be 30 years old or more. The question of the safe operation of rather 'old' plants and the question of NPP life extension as an alternative to decommissioning will thus become of growing importance worldwide. Therefore, the International Atomic Energy Agency (IAEA) organized an international symposium, held in Vienna from 29 June to 3 July 1987, to provide a forum for the exchange of technical information on Member States' approaches to and activities regarding ageing phenomena in NPPs and for the exchange of technical and economic information on approaches to plant life extension. It was generally agreed at the symposium that much more concern should be devoted to this issue. During a panel discussion the IAEA's activities in international co-operation were emphasized and the active role of the IAEA in helping to solve ageing related problems of NPP operational safety and lifetime extension policies was commended.

1. INTRODUCTION

For more than 30 years NPPs have been supplying electrical energy to national grids and 397 NPPs in 26 countries with a total electrical capacity of 273 795 MW had produced about 1514 TW·h(e) by the end of 1986.

Some of these NPPs are already approaching or are even beyond the end of their planned lifetime, which, for most designs, was estimated at 25–30 years of operation. Whether or not measures should be taken to extend this lifetime by refurbishment of the worn or obsolete components is becoming an important economic question in a number of countries. However, long before the end of design lifetime is reached, age related processes can potentially affect the safety of NPPs in operation, if appropriate measures are not taken.

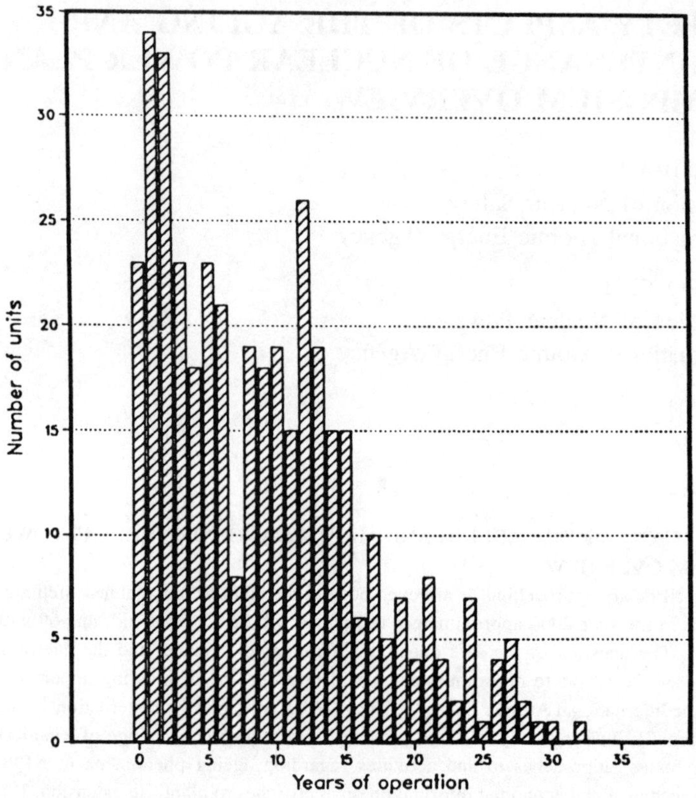

FIG. 1. Histogram of years of operation of reactors in Member States.

TABLE I. NUMBER OF NPPs HAVING REACHED 25 OR 30 YEARS OF OPERATION WITHIN THE PERIOD 1990–2000

Years of operation	Year										
	1990	1991	1992	1993	1994	1995	1996	1997	1998	1999	2000
25	37	41	48	53	63	69	84	99	118	144	160
30	14	15	22	24	29	37	41	48	53	63	69

Figure 1 shows a histogram of years of operation of reactors in Member States.

The number of NPPs which will reach the age of 25 or 30 years during the period between 1990 and 2000 is shown in Table I (taken from the IAEA Power Reactor Information System).

It can be seen that whilst in 1990 37 plants will be 25 years old and 14 will be 30 years old, by the year 2000 the figures will increase to 160 and 69, respectively. The question of the safe operation of rather 'old' plants and the question of NPP life extension as an alternative to decommissioning will therefore be of growing importance worldwide.

The IAEA, in accordance with requests from Member States, has commenced activities with regard to questions of ageing and life extension. Following the Technical Committee meetings on Safety Aspects of NPP Ageing, the IAEA international symposium on Safety Aspects of the Ageing and Maintenance of Nuclear Power Plants, held in Vienna from 29 June to 3 July 1987, was organized to provide a forum for the exchange of technical information on Member States' approaches to activities regarding ageing phenomena in NPPs and for the exchange of technical and economic information on approaches to plant life extension.

The symposium was the first organized by the IAEA on this subject; therefore it was directed at a broad spectrum of participants: technical and managerial staff engaged in NPP operation, regulatory body staff, consulting and architect–engineering organizations, vendor technical and management staff and NPP technical and managerial staff involved in maintenance activities.

2. TECHNICAL SESSIONS

The topics of the symposium were presented and discussed in six technical sessions which are reported here. The proceedings of the symposium were published in early 1988 [1].

Session 1: General Safety Aspects of Ageing

The main purpose of this session was to give participants a broad overview of the NPP ageing issue.

Three papers were presented, covering general problems of NPP ageing. One of the papers from the USA gave a complex view of ageing, which affects all reactor structures, systems and components, having the potential to increase risk to public health and safety if its effects are not controlled. The ageing processes showed that owing to the broad spectrum of material and designs and different operational conditions, there are a variety of factors which can lead to the degradation of the functional capability of equipment. Another paper surveyed ageing failures of light water reactor safety systems and components in the USA. The systems/components combinations most affected by ageing are (a) valves and pumps in normally operating systems, (b) class 1E diesel subsystem components in harsh environments, (c) heat exchangers in the main feedwater systems and (d) stand-by liquid control system pumps.

All papers accentuated how the NPP could be affected if degradation of key components or structures is not detected before loss of functional capability and if timely corrective actions are not taken. Emphasis was laid on the necessity to understand how the ageing process may change the likelihood of component failures in systems which are designed to mitigate transients and accidents, leading to reduced safety margins, and how age degradation can cause such events to be initiated and reduce defence-in-depth against catastrophic failures.

The United States Nuclear Regulatory Commission (USNRC) has initiated a complex programme for Nuclear Plant Ageing Research (NPAR) to achieve an understanding of nuclear plant ageing, its potential effects on safety, and methods for its detection and mitigation. This is designed to provide a basis for discussion of safety and regulatory issues and supporting regulatory decisions on issues.

The NPAR programme involves (a) risk oriented identification and selection of components and systems for which assessments of the impact of ageing on safety performance are to be conducted; (b) review of design base safety margins, qualification testing, operating experience, experts' opinions, methods for surveillance, inspection, monitoring and maintenance, leading to the development of recommendations for in-depth engineering studies; (c) engineering studies including verification of inspection, surveillance, monitoring and maintenance methods, in situ examinations and collection of data from operating equipment, post-service examinations and tests of naturally aged equipment, and cost–benefit analyses.

Session 2: Conceptual Approach to the Safety Aspects of NPP Ageing

Of the three papers presented in this session, two explained the regulatory body approach to ageing problems. The following were presented as important tools:

— Programmes for periodic assessment based on broad operational data collection;
— Programmes for preventive maintenance;
— Continuous monitoring of NPP components;
— Feedback from incident analyses;
— Individual and component qualification;
— Personnel education and training;
— External control and quality assurance.

Quantitative relationships between ageing failure data and risk were described. A procedure is under development in the USA for quantifying NPP risk from ageing related component failure data. Within the procedure, a preliminary set of reliability models has been developed which relates ageing component failure rates to ageing mechanism parameters which are estimable from ageing related failure data. The models are used to estimate the age dependent plant risk, such as core melt frequency, to give the dominant contributors to ageing plant risks, and to identify sensitivity and uncertainty contributors.

The most straightforward ageing model which is utilized in the ageing risk quantification procedure is the linear failure rate model. The linear ageing failure rate is derived from phenomenological considerations by which the ageing rates are related to ageing mechanism attributes. Conditions are defined under which a linear ageing rate may be applied to components and structures and non-linear models are developed where needed.

The above ageing risk quantification procedure has been applied to evaluate the implications of ageing failure data on component and system unavailabilities. The application of the procedure is presently being extended to details of testing and maintenance and to the evaluation of more detailed plant models including more comprehensive treatment of structure failures.

Session 3: Methods for Critical Components and Systems Determination

Eight papers presented methods for determination of critical components and systems. One of them described the French qualification programmes, the purpose of which is to demonstrate which safety related mechanical and electrical equipment is capable of carrying out its functions under accident conditions where required. To a great extent the accelerated ageing phase makes it possible to determine critical components in relation to the life of the equipment; however, in certain cases the methods of artificial ageing cannot fully represent real conditions and a policy of in-service surveillance should be defined for such components. It is quite clear that even if the ageing phases of qualification tests make possible the identification of a number of critical components and pieces of equipment, an approach of this kind should be supplemented, firstly, by a policy of surveillance and inspection of equipment throughout the life of the power plant and secondly, by a preventive maintenance programme.

A few papers indicated the importance of collecting and evaluating operational data. The advantage of a computerized data collection system which incorporates inspection, surveillance and maintenance control and evaluation of critical component life was presented in a paper from Czechoslovakia. For electrical equipment, the role of operational feedback in developing test specifications was described in a paper from France which illustrated commonly used equipment such as condensers, relays and medium voltage motors. In a paper from the USA the use of naturally aged NPP components for examination and testing was demonstrated at the Shippingport Atomic Power Station now undergoing decommissioning. This power station is a major source of components and specimens for ageing evaluations. Specific aspects of optimization of operating conditions in 'old' Magnox reactors were also presented in a paper from the United Kingdom. A simple probabilistic model was used to provide advice to operators on how to optimize lifetime outputs. In another paper (from Canada), an overview was given of methods for the detection and mitigation of ageing effects before they affect plant safety and production reliability. Regulatory audits of these methods, to ensure that an acceptable level of plant safety is maintained by the NPP licensees, were also described. The following methods were discussed: a preventive maintenance programme, significant event reporting systems, and a reliability based assessment of performance of safety related systems.

Present and future developments include revisions of current monitoring, testing and inspection methods to ensure that appropriate time variant parameters (capable of revealing ageing degradation before loss of functional capability) are monitored, and reviews of the effectiveness of existing maintenance programmes and methods in mitigating ageing and service wear effects.

Session 4: Methods and Provisions to Handle Ageing Phenomena during NPP Operation

Ten papers were presented in this session. Operational condition evaluation methods to determine the ageing of motor operated valves were the subject of two papers. One paper also presented evaluation criteria for valves based on electrical current and the measurement of water flow changes. Reactor pressure vessel (RPV) radiation damage monitoring was reported on by participants from Finland and France. The French study concluded that there will be no problems within the planned life of their RPVs, whereas specific measures had to be taken to cope with the embrittlement of RPVs in Finland, including the improved surveillance programme and the modified fuel loading pattern.

Application of technical diagnostics to trace system ageing was discussed in the paper from the German Democratic Republic and an intention was stressed to consider NPP ageing in a more detailed way than has been the case up to now. By periodically taking the essential parameters of different operational ages by means of trend analysis and applying suitable models to ageing, it should be possible to trace system degradation and to estimate the remaining lifetime. Technical diagnostics have further aspects with respect to ageing. On the one hand early detection of failures permits safe operation, which minimizes ageing. On the other hand, ageing is not restricted to technical degradation; it also concerns safety that can be improved by systematically including the technical diagnostics in plant monitoring. In any case the surveillance of parameters over a period of some decades demands a reliable surveillance system, reproducible performance of analysis, careful storage of data, etc. These problems and their possible solutions were the main subjects discussed.

The importance of water chemistry monitoring in prevention of ageing related corrosion phenomena was emphasized in another paper (from Finland). Corrosion phenomena, particularly stress corrosion cracking and corrosion fatigue, are caused by environment, material and stress. The ageing of NPPs causes changes in materials properties and can also cause differences in the stress states of structures. Because corrosion attacks are more likely and crack growth rates faster in aged than in new materials, environmental control becomes more important during plant ageing. Almost all corrosion phenomena in NPPs can be prevented or at least moderated by water chemistry control or by the change of water chemistry. Successful water chemistry control requires regular and continuous monitoring of such water chemistry parameters as dissolved oxygen content, pH, conductivity and impurity contents.

The results of long term studies on the behaviour of cables were presented in another paper (from France). Priority was given to gaining knowledge of the mechanisms of cable ageing under environmental conditions specific to the reactor building of PWR power plants, since this equipment, composed partly of polymers sensitive to temperature and ionizing radiation, is difficult to replace inside the reactor building. Furthermore, synergetic phenomena were studied and estimations were made of the period over which a cable operates satisfactorily.

One of the US papers showed that the current question facing the industry could be simply stated as 'How can an acceptable level of safety and availability be maintained throughout the operational life of a nuclear power plant'? To aid in answering these questions a two-bin model is proposed to identify ageing research needs, guide technological advancement, and assure timely incorporation of research products into plant operations. The birst bin, considered tactical or short term planning, includes those current daily activities carried out with different degrees of success in all operating units. Such activities as normal operating practices, surveillance tests, and preventive and corrective maintenance can be considered here. The second bin is much more long term and can be termed as strategic planning. This planning must review repair/replace/upgrade strategies, requirements for spare parts, and trade-offs to optimize preventive and corrective maintenance. It is here that planning for the next ten years of successful operation takes place, reliability practices and enhanced plant performance monitoring are considered, ageing problems are diagnosed and research needs established. The general issues associated with and the complexities involved in long range, strategic planning and their relationship with day to day, tactical plant operation decisions were discussed.

Session 5: General Aspects of NPP Life Extension

Six papers were presented in this session, four of them dealing with nuclear power plant life extension (NUPLEX) programmes in Japan, France and the USA, and of the two remaining US papers, one gave an analysis of the economic impact of NUPLEX and the other was more plant oriented.

The Japanese eight year light water reactor power plant life extension programme, as a part of the long term nuclear energy development plan, was started in 1985 at the Japan Power Engineering and Inspection Corporation (JAPEIC) and under the sponsorship of the Ministry of International Trade and Industry. The main reasons for starting this programme were economic ones because the cost of decommissioning a power reactor of 1000 MW(e) was estimated as 30×10^9 yen (about US $200 million). Therefore, upgrading of LWRs so as to have enhanced reliability as well as greater economy and higher usage factors would be desirable. The eight year programme can be divided into three phases of which the first one had largely been completed by early 1987. Phase II will last for another five years, being more oriented to experimental work on life estimation and development of suitable devices for replacement or repair of critical equipment, whereas the last phase will deal with overall final estimation and conclusions.

The French approach to the NUPLEX programme is based on the use of the French standardized PWRs which, in the year 1986, produced more than 70% of the electricity in France. At the end of 1985 Electricité de France initiated a plant life time project which comprised a four year work programme. The first stage, to make a full inventory of the data available in order to predict the ageing of sensitive NPP equipment and to keep track of it with a view to prevention or remedy, has already been concluded. A preliminary analysis of ways in which such equipment may be damaged shows that radiation embrittlement of the pressure vessel and fatigue of main components of the steam supply system will not limit the life of these sensitive parts, except in the very long term. Provided the effort called for is made, PWR units may have a very long lifetime.

The United States Department of Energy (USDOE) research programme on NUPLEX, started in 1985, is based on the fact that life extension will be an urgent issue for many US NPP owners within the next ten years. At present there are about 100 NPPs licensed to operate in the USA. Because of the lead time required to construct replacement power plants, many owners, and the nation as a whole, will have to make decisions about life extension of NPPs in the mid- to late 1990s. The bulk of the methodology and technology needed for informed decision making is encompassed in the above NUPLEX research programme. USDOE research involvement and support are foreseen for all long term activities in this area in order to continue the pilot plant studies, to develop generic conclusions, resolve technical issues and to develop better decision making tools. It may be concluded that this NUPLEX programme is the most developed worldwide.

The NUPLEX Steering Committee was established in the USA in late 1985 to help to develop the technology required for owners to evaluate life extension, to justify decisions to the USNRC and to facilitate a stable and predictable licence renewal process, which should be on a generic basis wherever possible. In the light of the needs of the earliest licensed reactors in the USA, a licence renewal process should be in place as soon after 1993 as possible. The basic premise of the NUPLEX Five Year Plan is that life for reactors may continue beyond an age of forty years. The role of the NUPLEX Steering Committee will be to advocate the timely establishment of a licence renewal process by the USNRC, to sponsor the necessary regulatory studies, and to speak for the nuclear utilities on proposed USNRC policy and rules on licence renewal.

The systematic evaluation of BWR operating experience in the USA over the past twenty years provides the best single source of information on plant life extension. In both the fatigue as well as the stress corrosion cracking phenomena laboratory tests, the BWR field data can show the life limiting mechanisms, and also indicate methods for preventing, mitigating and monitoring the effects of age related degradation. Based on this experience, suggestions have been made for lengthening the service life of the BWRs.

An economic assessment of the costs and benefits of NUPLEX for the USA as a whole under widely varying economic assumptions was the subject of study in another paper from the USA. Two preliminary engineering estimates indicate that NUPLEX refurbishment would cost about \$150–300/kW (1986 dollars), thus enabling the refurbished plant to operate for an additional 30 rather than 20 years as assumed previously. This is, of course, a much more favourable assessment of NUPLEX. The new assessment shows also that NUPLEX could save future US electricity consumers almost US \$390 × 10^9 (1986 dollars), over twice as much as previous estimates, even in comparison with the so-called NEWCOAL plants (relatively cheap coal and new coal utilization technologies).

Session 6: Technological Aspects of NPP Life Extension

Five papers were presented in this session; with the exception of the first, which was an overview of the OECD/NEA Symposium on Plant Life Extension, all others were oriented to the technological aspects, assessing the residual life of some critical components with a view to NUPLEX.

The overview summarized discussions and conclusions from the symposium organized by OECD/NEA in conjunction with the IAEA and held at the OECD headquarters in Paris in February 1987. This symposium, attended by 72 delegates from 14 of the OECD and the IAEA Member States, reviewed the experience in design and maintenance practices relevant to NPP life extension, identified life limiting factors and discussed the viability of major refurbishment as an alternative to decommissioning and replacement. Participants at this meeting — utility operators as well as manufacturers as opposed to government regulators — gave emphasis to the economic rather than safety aspects of NUPLEX.

Residual life assessment of major PWR components was performed by the Idaho National Engineering Laboratory, USA, supported by the USNRC. The main criterion used in selecting these components is their role in preventing the release of fission products to the public. The stressors, possible degradation sites and mechanisms, potential failure modes, and currently used methods for non-destructive examination were discussed for five PWR components: pressure vessels, containment structures, reactor coolant piping, steam generators, and RPV supports. Important technical issues related to life extension of these components were identified, and several recommendations were provided.

Another part of the complex US NUPLEX programme performed for the USDOE by the Sandia National Laboratories aims to develop predictive performance monitoring and reliability/safety technology to enhance the operational lives of

NPPs. Probabilistic risk assessment (PRA) techniques were initially used to help identify components critical to safe operation. These components were then evaluated for degradation and failure mechanisms; equipment was identified for which it is important to maintain initial reliability levels and for which improvements are necessary. The work will continue with the commercial nuclear industry to provide proven technologies for improved plant performance.

The nominal licensed lives of NPPs in Japan have been set at 30 years (for PWRs) to 40 years (for BWRs) and after about 20 years of operating history, more than 40 NPPs are either in operation or in construction. It is therefore easily understandable that a life extension programme has been started in Japan, encompassing among other aims material engineering studies of ageing phenomena in structural materials of LWRs. Regarding the most critical component in plant life, i.e. the RPV, some conclusions based on the research work were drawn regarding the quality of the steels used. Practical engineering procedures for the evaluation of ageing and the associated life assessment are under development, including the scientific basis for accelerated tests and monitoring methodology for determination of the degree of thermal ageing in well defined testing under reasonably simulated service conditions.

The French regulatory requirements concerning the primary circuit of water reactors were recapitulated in a paper describing a procedure for keeping account of operating transients occurring in the primary circuits of water reactors. The procedures adopted by the operator encompass the transient detection, identification and characterization. The procedure affords a means of checking that operating transients remain generally within the design hypotheses. In practice this monitoring has been found to be a valuable tool for incorporating feedback from experience. The accounting, in combination with other methods, can provide an accurate and differentiated indicator for keeping track of the ageing of the reactor units.

3. PANEL DISCUSSIONS

Three panel discussions were organized on the following subjects:

(1) Ageing and Safety Relationship;
(2) Operators' Activities to Prevent Undesired Consequences of Ageing Phenomena during NPP Operation;
(3) International Co-operation and Future IAEA Activity in NPP Ageing and Plant Life Extension.

At the first panel discussion some issues concerning ageing of active and passive components were presented. It was emphasized that the contribution of ageing plant components to risk (despite the possibility of replacement the risk can be high) must be carefully considered. Current databases at NPPs and systems are not adequate for handling ageing phenomena and for evaluating the risk on the basis of operational data. These database systems should be improved to collect more detailed data on failures, repairs and maintenance. It was emphasized that ageing is not an issue concerning only the future, but rather one that also concerns the present in all NPPs, including those which are not very 'old'. This issue should also be included in research programmes. Current methods of testing, monitoring and maintaining of NPPs are not adequate for coping with the ageing issue. Quantification of the ageing impact is very important to assure the safe operation of NPPs and to supply proven data to the regulatory bodies. Prevention rather than corrective action should be used to manage ageing, and relevant activity should concentrate on the most sensitive parts of NPPs.

One of the most important operator activities is the systematic identification of the ageing degradation of components and systems important to safety and reliability. The degradation mechanisms must be understood to track the level of degradation and to measure the exact amount of deterioration. Measures to initiate the mitigation of ageing and corrective action should be implemented as soon as possible.

Adequate record keeping and better characterization of operational modes, transients and events should be implemented at NPPs. A great amount of information has been accumulated from NPP operation, but its complicated handling limits the use of this information in evaluating ageing effects.

One of the most important measures for managing ageing phenomena in NPP operation is an adequate maintenance programme based on detailed data analyses with the purpose of replacing items before they fail. Examples of some measures to decrease the number of reactor scrams and to implement low leakage core configurations were considered in the discussions to be useful measures for handling NPP ageing.

The third panel discussion concentrated on the IAEA role in international co-operation on NPP ageing and life extension and considered: methodological aspects; exchange of experience in monitoring, surveillance and testing methods and results; co-ordination of special research programmes, etc. Several detailed proposals for future activities were put forward for consideration by the IAEA.

4. CONCLUSIONS

During the symposium, which was attended by 141 participants from 31 countries and three international organizations, 35 papers and four extended synopses of posters were presented.

It was generally agreed that ageing is a serious issue.

Nuclear power plant equipment ageing, and the consequent failure of components, could have a strong impact on both plant availability and safety if degradation of key components, systems and structures is not detected before the loss of functional capability and if appropriate corrective measures are not taken. Ageing processes also alter the probability of component failures in systems which are designed to mitigate transients and accidents and thereby reduce safety margins, possibly causing transients and accidents to be initiated. Ageing degradation could represent a potentially significant contribution to common cause failure of safety related equipment during transients and accidents.

Coping with the ageing process of NPPs requires a systematic approach in analysing these phenomena. Experimental and theoretical methods should be developed to evaluate its impact on NPP performance. Effective methods of inspection, surveillance and monitoring should be implemented during operation to evaluate the 'qualified life' of the components, systems and structures. This should be the basis for their timely and effective maintenance, repair and replacement, with particular attention to systems and components important to safety.

There is an important relationship between safety and ageing and the availability and reliability of NPPs. With the high investment costs involved in NPPs and with worldwide attention focused on nuclear safety, the safe and undisturbed operation of existing NPPs is a major concern. Owing to capital cost development, extension of the lifetime of existing NPPs beyond their design life may be economically justifiable even if significant capital expenditures are required to assure nuclear safety and reasonable availability. Technical, economic and regulatory lessons have to be learned before decisions can be taken concerning plant life extension, but NPP life extension in the planning of nuclear power systems development had not been tabled for discussion during the symposium.

Although nuclear plant ageing could have an impact on the efficiency of electric power generation, the concern is that plant safety could be affected if degradation of key components or structures is not detected before the loss of functional capability, and timely corrective actions are not taken.

It is very important to achieve an understanding of NPP ageing, its potential effects on safety and methods for its detection and mitigation. A complex programme for handling ageing phenomena exists in the USA. In some countries (Canada, France, Finland and Japan) elements of such complex programmes exist and these were presented at the symposium. Other countries presented only partially carried out activities to handle ageing phenomena (surveillance, preventive maintenance, operation data collection and evaluation, examination results in the ageing of NPP items, etc.).

Close interconnection between plant ageing and plant life extension was demonstrated. Quantification of ageing impact is very important to assure the safe operation of NPPs and to supply proven data to regulatory bodies. NPP lifetime improvement and licence extension represent an important policy issue that must be resolved to help assure an adequate supply of safe, reliable and economical electrical energy.

The symposium was the first organized by the IAEA on this subject. It successfully fulfilled its purpose. It was generally agreed that much more time should be devoted to the issues of NPP ageing and NPP life extension. During the panel discussion the role of the IAEA in international co-operation was emphasized and the importance of the NPP ageing issue and life extension in future IAEA operational safety and reliability programmes was stressed.

REFERENCE

[1] Safety Aspects of the Ageing and Maintenance of Nuclear Power Plants (Proc. Symp. Vienna, 1987), IAEA, Vienna (1988).

THE MAN-MACHINE INTERACTION IN REACTOR SAFETY

T.L. WATERS, N.J. HOLLOWAY
UKAEA Safety and Reliability Directorate,
Culcheth, Warrington,
United Kingdom

Abstract

THE MAN-MACHINE INTERACTION IN REACTOR SAFETY.

The paper describes some of the particular problems involved in the optimisation of man-machine interaction in systems as complex as those found in the nuclear industry. It identifies three activities necessary to achieve this: the application of standard ergonomics principles, the comprehensive and systematic analysis of the operators task and the application of specific defences against operator error. Work which is being undertaken by the UKAEA to encourage the wider use and understanding of techniques within the first two areas is described and the results of in-house activities are provided. Currently, the effectiveness of the results in the third area depends upon the prediction of operator behaviour. While this is possible in proceduralised tasks it is not considered appropriate for non-proceduralised tasks where there are a large number of different plans and actions possible. Instead, it is argued that an alternative systems led deterministic approach is required. This approach is outlined and the role of the assessment of operator feedback is described. The limitations of the quantification of human error probabilities in both routine and transient conditions are discussed and recent projects involving the application of different methods are contrasted. Finally, an approach is outlined which attempts to integrate the stages of task analysis, error analysis and error reduction. The method is intended to enable engineers with limited human factors knowledge to analyse systematically error possibilities in proceduralised tasks.

1 INTRODUCTION

The occurrence of serious accidents attributable in part to operator errors has focused attention on the need to address the possibilities of such errors in system and ergonomic design and analysis, and, if possible, to validate the defences of future reactor systems against unplanned or mistakenly planned operator actions and associated errors.

The man-machine interface in reactor control and safety systems is the interface at which human thoughts are translated into system actions. Its design, and the validation of its design, must therefore be addressed if we are to have reasonable assurances that human errors during incidents will not negate the carefully laid plans for reactor safety represented in the safety equipment designs and operating procedures.

Classical ergonomic principles for the optimisation of human performance were developed on systems which (relative to those in existence today) were simple, with only a small number of controls and displays. In such simple situations the achievement of good ergonomic design was relatively easy, providing the potential for high levels of human reliability. Additionally the range of man-machine interaction meant that the number of human errors which were possible was relatively low, these could be predicted and if necessary, special measures taken to prevent their occurrence or aid recovery.

In larger and more complicated systems this traditional approach became more difficult. While the application of the same ergonomic principles undoubtedly contributed to the improvement of human reliability, the situation was complicated by the increasing complexity. Individual aspects of the operator's task could no longer be considered in isolation because the effect of complicated interdependencies became significant, and methods of task analysis were developed to aid the assessment. The range of opportunities for error also increased as the system became larger, making the process of prediction more difficult and extremely time consuming. At this point the contribution of systems analysis became necessary to identify those errors which had a significant impact upon the system and which were worthy of further detailed ergonomics analysis.

In very large and complex systems, such as nuclear power plants, the analysis is complicated further. While it is still important to apply standard ergonomics principles to achieve a good basic level of design, and also to apply task analysis techniques, the scope for operator error is now so wide that it is practically impossible to predict precisely and comprehensively how operators will react in all possible situations. For this reason in complex situations a comprehensive defence against operator error cannot be made purely on the basis of an ergonomics driven approach. In such situations it is necessary to bound the ergonomics analysis by first performing a comprehensive systems analysis to identify those combinations of events which are potentially hazardous and restrict the detailed ergonomics assessment to these.

This paper will describe the work which SRD has performed to improve the quality of the man-machine interface in complex systems by the three measures described above and also discuss the benefits which can be obtained through the quantification of human reliability.

2 THE APPLICATION OF HUMAN FACTORS KNOWLEDGE

The past fifty years has seen much research in the areas of human factors and the man-machine interface. As a result there is a vast amount of both specific and general information which if applied would do much to improve operator performance. Unfortunately much of this information is often difficult to obtain and rarely in a form which the designer can use. As a result, standards of the man-machine interface and operator performance are often less than optimal. For example, many of the problems which occurred at the TMI-2 incident could have been avoided if only simple ergonomics principles had been followed.

The UKAEA has recognised the need to make this information more easily accessible and understandable to designers. As a result two complementary projects have been undertaken in an attempt to improve this situation.

2.1 Guide to reducing human error in process operation

The UKAEA has published a 'short guide' [1] intended for use by designers and assessors and divided into five sections:

1 The Operator/Process Interface
2 Procedures
3 Work Place and Working Environment
4 Training
5 Task Design and Job Organisation.

Each section comprises a number of questions, which act as a simple checklist to cover the major ergonomics design principles.

A 'long guide' is shortly to be published which will provide background information, illustrations and case studies on each of the areas listed above.

2.2 Ergonomics information database

As a complement to the long and short guides the UKAEA have produced a prototype database containing a limited amount of human factors information [2]. The database comprises two parts, one part is a traditional relational database with a very simple structure. A large amount of information can be stored in this form and additional information can easily be added. This system takes full advantage of the computer's speed and specific information can be found quickly and easily by using a word search facility. The other part of the database comprises a hierarchical tree structure. The user works through a series of menus until the required detail of information is found. Careful design of the structure makes it possible to educate and alert the user to important ergonomics principles as the tree is descended. The feasibility study has proved to be successful and it is hoped that further development work will be possible in the future.

3 THE ANALYSIS OF THE OPERATOR'S TASK

3.1 Guidelines on task analysis techniques

A comprehensive understanding of the operator's task is a prerequisite to effective ergonomics analysis. The aim of this being to systematically analyse the match between the demands of the task and equipment with the capabilities of the operator. Mismatches and the presence of other factors likely to degrade performance can then be identified and steps can be taken to improve the situation.

The types of tasks and situations vary greatly, for example some may involve mainly heavy manual tasks in severe environments, others may require fine manual dexterity whereas yet others may involve cognition such as diagnosis and decision making. Additionally, constraints may affect the type of analysis that is possible, for instance, the analysis of a task for a process which is still at the conceptual design stage will require a different approach from that for an existing task which is readily observable. There are many methods available to the ergonomist though unfortunately few techniques are adequately documented and there is little guidance for the non-specialist.

The importance of task analysis is widely acknowledged and the UKAEA is currently supporting a group of experts, whose objective is the publication of a guide of task analysis techniques. This will enable non-specialists to choose and apply the most effective technique given the task and situation characteristics and the various factors constraining the analytical resources.

3.2 A case study

A case study has recently been completed at one of the Authority's small research reactors [3]. The study involved the analysis of the operators' task during a loss of coolant accident. Information was obtained from a study of emergency operating procedures, discussions with the operators within the control room and a table top simulation exercise. A number of ergonomic issues emerged from the study concerning among other things: the display of alarms, the high workload of the operator and a distinct difference in strategies employed by different operators to control the situation (FIG 1). A number

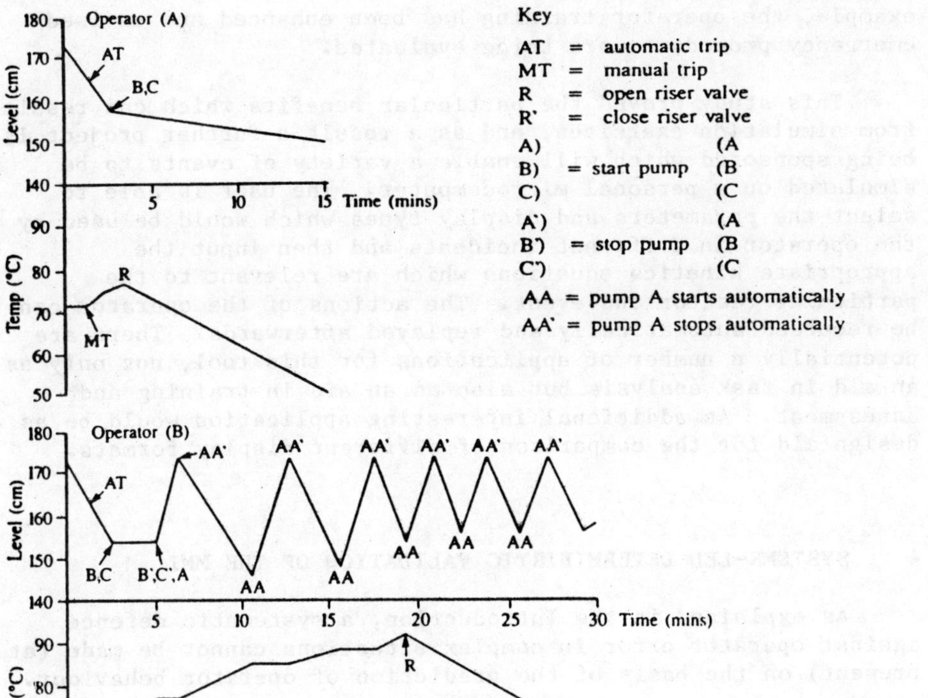

FIG. 1. Different post-loss of coolant accident operator responses (simulation exercise, Harwell reactors).

of practical benefits have resulted from the study. For example, the operator training has been enhanced and revised emergency procedures are being evaluated.

This study proved the particular benefits which can result from simulation exercises, and as a result a further project is being sponsored which will enable a variety of events to be simulated on a personal microcomputer. The user is able to select the parameters and display types which would be used by the operator in different incidents and then input the appropriate kinetics equations which are relevant to the particular reactor and event. The actions of the operator can be recorded automatically and replayed afterwards. There are potentially a number of applications for this tool, not only as an aid in task analysis but also as an aid in training and assessment. An additional interesting application would be as a design aid for the comparison of different display formats.

4 SYSTEMS-LED DETERMINISTIC VALIDATION OF THE MMI

As explained in the Introduction, a systematic defence against operator error in complex situations cannot be made (at present) on the basis of the prediction of operator behaviour.

To overcome this problem work is being undertaken at SRD to investigate the feasibility of employing systems analysis as a means towards the deterministic validation of the man-machine interface [4]. The method recognises and makes explicit the importance of operator recovery towards system success. The systems analysis which underlies the method is used at the outset to identify safety functions which are necessary to achieve system safety for various initiating events. (Events which would in any case be considered within the normal safety analysis.) However, whereas currently certain combinations of safety system failures are considered incredible, this proposed approach makes no such assumption. This is because the operator, having made an incorrect diagnosis, could intervene and cause a combination of events which could not reasonably occur as a result of hardware failures alone. In these situations one critical redeeming factor is the quality of the feedback which is given to the operator regarding the state of the plant.

Ergonomics analysis of the MMI is performed to assess the quality of this feedback because this determines directly the likelihood that the operator will perceive that the system is in an unsafe condition and make a suitable recovery. The criteria of feedback include the strength, the clarity, the timing and

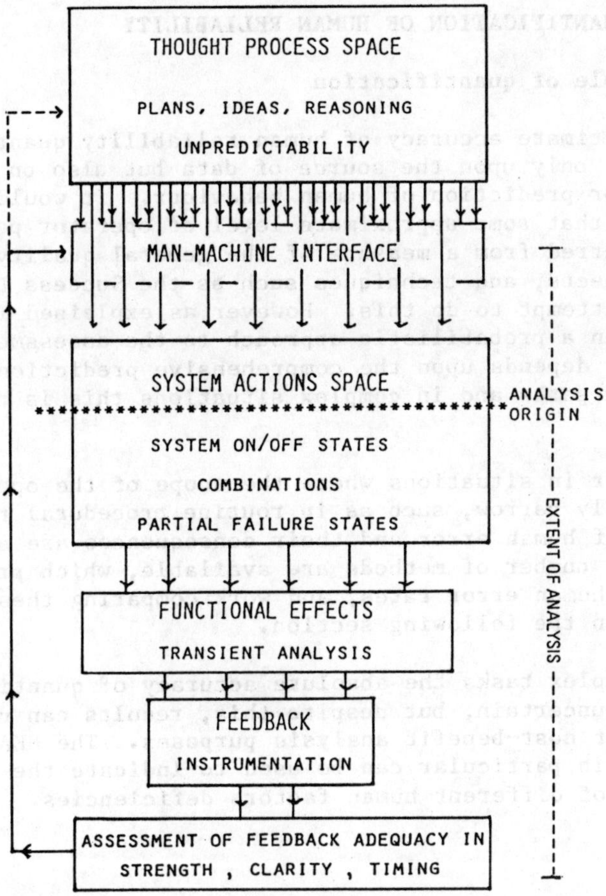

FIG. 2. Scheme for deterministic analysis of operator errors and feedback.

the existence or otherwise of recovery advice. The quality of the displays, the existence of alarm suppression, prioritisation etc and the provision of a safety parameter display system would all be important factors in this assessment. If feasibility studies are successful, then it is envisaged that future developments of this approach could include whole system applications and probabilistic assessment. FIG 2 illustrates the above concepts in which the analysis originates in the MMI and proceeds to an assessment of the feedback.

5 THE QUANTIFICATION OF HUMAN RELIABILITY

5.1 The role of quantification

The ultimate accuracy of human reliability quantification depends not only upon the source of data but also on the modelling or prediction of human behaviour. It would seem reasonable that some approximate level of operator performance can be inferred from a measure of the general quality of human factors aspects, and techniques such as the Success Likelihood Index [5] attempt to do this. However as explained in the Introduction a probabilistic approach to the assessment of human reliability depends upon the comprehensive prediction of all human error paths and in complex situations this is practically impossible.

However in situations where the scope of the operators task is relatively narrow, such as in routine procedural tasks, then the types of human error and their consequences are easier to predict. A number of methods are available, which provide sources of human error rates, and work comparing these will be described in the following section.

In complex tasks the absolute accuracy of quantification methods is uncertain, but despite this, results can usefully be employed for cost-benefit analysis purposes. The HEART method [6] in particular can be used to indicate the relative importance of different human factors deficiencies.

5.2 Guidelines in the use of quantification methods

The dearth of 'real' and usable human error data has resulted in a plethora of different approaches and methods for the estimation of human error probabilities. The UKAEA is involved in three different and complementary projects which will provide information to the non-ergonomist regarding the limitations of human reliability assessment and also provide guidelines to help in their application.

The first project followed from the table-top simulation exercise mentioned in Section 3.2. An attempt was made to quantify the probability that the operator would fail to act correctly during the incident. This was done using six well known techniques and the results compared.

The second project involves the production of a guide describing a number of case studies in which different methods have been applied in a variety of industries.

The third project is the JRC Ispra organised Human Factors Reliability Benchmark exercise in which a number of organisations throughout Europe and USA have applied human reliability quantification methods in two tasks - a proceduralised test routine and a non-proceduralised situation during a transient.

The three projects are either completed or near completion and their comparison will prove particularly informative. Quantification in the first project was predominantly performed by one analyst whereas the third project involved many different analysts working independently. Whereas these projects involved the analysis of the same event, the second project describes the analysis of different events in different industries, thus providing information regarding the applicability of different methods for different situations.

6 SYSTEMATIC HUMAN ERROR REDUCTION AND PREDICTION APPROACH (SHERPA)

The previous sections show that an ergonomics assessment is comprised of a number of different stages. The number of approaches possible at each stage make it difficult for the non-specialist to tackle assessments confidently. Therefore, particularly until the projects described earlier have been completed, there is a need for an integrated scheme to aid assessors. The UKAEA has recently sponsored work [7] which has resulted in a framework which contains modules of task analysis, error analysis, error quantification and error reduction through procedural, equipment and training modifications. The method is primarily intended for procedural applications for which recovery from error is a particularly important consideration. The approach has been applied in this context on a Non-Destructive Testing calibration procedure and proved particularly helpful in the identification of points in the procedure where extra checks should be added.

7 CONCLUSION

The importance of the man-machine interface with respect to system safety and reliability is now widely accepted. Unfortunately information regarding the techniques available to assess and improve the MMI has been confined to a small and highly specialised group of people. If high standards of ergonomics are to be achieved throughout the industry it is essential that this information is put into a form which is easily accessible and usable. The UKAEA is currently attempting to do this in the areas of task analysis techniques, and human reliability assessment methods and has also produced guidelines containing both general principles and specific data which if used properly will reduce operator error. An integrated approach (SHERPA) has also been developed with which the non-specialist can assess human factors issues in proceduralised tasks. Applications of some of the methods described above have succeeded in identifying problem areas in operating plant and practical benefits have accrued as a result.

The quantification of human reliability is considered to be generally useful for purposes of cost benefit analysis, but for non-proceduralised tasks a systems led deterministic approach is considered more appropriate, particularly when it is desired to establish an identifiable line of defence against human error.

REFERENCES

[1] UNITED KINGDOM ATOMIC ENERGY AUTHORITY, Guide to Reducing Human Error in Process Operation, Rep. SRD R-347, UKAEA, Warrington, UK (Feb. 1985).

[2] WATERS, T.L., Human Factors Research — A Feasibility Study on a Database of Human factors Information and Ergonomics Knowledge, Rep. for the Commission of the European Communities, UKAEA, Warrington, UK (Oct. 1985).

[3] WATERS, T.L., ''Human reliability assessment. A case study involving the analysis of operator actions during a loss of coolant accident'', Proc. Reliability '87 Conf. Birmingham, UK (1987).

[4] HOLLOWAY, N.J., WATERS, T.L., ''The assessment of the efficiency of system feedback provided to the operator and its role in probabilistic safety analysis'', The Human factor Information Feedback in Nuclear Power: Implications of Operating Experience on System Analysis, Design and Operation (Proc. IAEA/NPPCI Specialists' Mtg, Risø, Denmark, 1987), Risø National Lab. (1987) Session 4.

[5] EMBREY, D.E., The Use of Performance Shaping Factors and Quantified Expert Judgement in the Evaluation of Human Reliability: An Initial Appraisal, Rep. NUREG/CR-2986, United States Nuclear Regulatory Commission, Washington, DC (1983).

[6] WILLIAMS, J.C., ''HEART — A proposed method for assessing and reducing human error'' (9th Symp. on Advances in Reliability Technology, Bradford Univ.), Bradford Univ., UK (1986).

[7] MURGATROYD, R.A., EMBREY, D., BALLARD, G., TAIT, J., ''The reliability of ultrasonic inspection'', Non-Destructive Evaluation in the Nuclear Industry (Proc. ANS Top. Mtg Orlando, FL, 1986), American Nuclear Society, Hinsdale, IL (1986).

AMELIORATION DE LA SURETE PAR LES ENSEIGNEMENTS TIRES D'AUTRES SOURCES QUE LES INCIDENTS D'EXPLOITATION

D.B. LÉVY, H.E. WALLARD, J.-R. JARDEL
Service central de sûreté des installations nucléaires,
Ministère de l'industrie, des P. et T. et du tourisme,
Paris

A. CERNES
Institut de protection et de sûreté nucléaire,
Commissariat à l'énergie atomique,
Fontenay-aux-Roses

France

Abstract–Résumé

IMPROVEMENT IN SAFETY BASED ON EXPERIENCE FROM SOURCES OTHER THAN OPERATIONAL INCIDENTS.

The feedback of experience and the improvement of facilities on the basis of incidents is completely formalized in France. However, the feedback of experience should also be based on other information, often available earlier than that from incidents. This information is mainly of three types: that collected during inspection visits by safety authorities, that emerging from checkups, tests and appraisals made during construction and operation, and, lastly, that obtained from studies in greater depth or from an improvement in knowledge. This approach supplements the feedback originating with analyses of incidents, especially in the following areas: more accurate and more effective knowledge of the state of the facility; better knowledge of the state of the materials and detection of manufacturing faults or faults due to operation; and finally, a more concrete approach to operational methods and the human factor, and therefore rapid perception of a possible discrepancy in these methods. Generally speaking, the effort here is to show how safety can be improved through the analysis of events more 'subtle' than significant incidents.

AMELIORATION DE LA SURETE PAR LES ENSEIGNEMENTS TIRES D'AUTRES SOURCES QUE LES INCIDENTS D'EXPLOITATION.

Le retour d'expérience et l'amélioration des installations à partir des incidents est tout à fait formalisé en France. Mais le retour d'expérience doit également se faire à partir d'autres informations, souvent plus précoces que les incidents. Ces informations sont essentiellement de trois types: les informations recueillies lors des visites de surveillance par les autorités de sûreté, celles issues des contrôles, essais et expertises effectués à la construction et en fonctionnement, et, enfin, les informations acquises grâce à des études plus approfondies ou à une amélioration des connaissances. Cette approche complète le retour d'expérience issu de l'analyse des incidents en particulier dans les domaines suivants: connaissance plus précise et plus effective de l'état de l'installation, meilleure connaissance également de l'état des matériels et détection des défauts de fabrication ou des défauts engendrés par l'exploitation, enfin, approche souvent plus concrète des méthodes d'exploitation et du facteur humain, et donc perception rapide d'une dérive éventuelle dans ces méthodes. De manière générale, on essaie ici de montrer comment la sûreté peut être améliorée à partir de l'analyse d'événements plus «fins» que les incidents significatifs.

1. INTRODUCTION

Le retour d'expérience sur les réacteurs à eau sous pression à partir de l'analyse des incidents significatifs est tout à fait formalisé en France et un autre mémoire est présenté sur ce sujet (vol. II, IAEA-CN-48/123). Mais le retour d'expérience doit également se faire à partir d'autres informations, plus précoces que les incidents. Il est clair, en effet, qu'un des rôle principaux de la sûreté est d'anticiper et de prévoir tous les incidents. Le rapport de sûreté est rédigé à la conception dans ce but et il convient, durant l'exploitation, de devancer les incidents par une analyse permanente permettant de déceler toute source potentielle d'incidents non prévue à la conception.

Le but principal de cette analyse est de s'assurer que l'exploitation est conforme aux règles prévues à la conception et que l'état des matériels est également satisfaisant. Ceci peut se faire par deux voies: d'une part, à travers les informations obtenues lors des visites de surveillance effectuées par les autorités de sûreté, et, d'autre part, à travers celles qui résultent du suivi des activités de maintenance et des essais périodiques.

Cette approche complète les enseignements tirés de l'analyse des incidents, en particulier dans les domaines suivants: connaissance plus précise et plus effective de l'état de l'installation, meilleure connaissance également de l'état des matériels et détection des défauts de fabrication ou des défauts engendrés par l'exploitation et, enfin, approche souvent plus concrète des méthodes d'exploitation et du facteur humain, et donc perception rapide d'une dérive éventuelle dans ces méthodes.

Le chef du Service central de sûreté des installations nucléaires (SCSIN) a également estimé utile de compléter les visites de surveillance effectuées à sa demande par la création d'un groupe de travail sur la formation dans les centrales REP. La formation est en effet essentielle, tant dans le domaine de la prévention des accidents que dans celui de leur gestion. Ce point sera donc évoqué dans ce mémoire.

2. LES VISITES DE SURVEILLANCE

Les visites de surveillance constituent un élément indispensable des actions menées par les autorités de sûreté. Au moyen des contacts directs qu'elles permettent avec les exploitants, dans les bureaux d'études, les usines, les chantiers et les installations nucléaires, elles donnent de façon privilégiée la mesure des problèmes de sûreté et constituent une source d'information précieuse.

C'est grâce à elles que le SCSIN peut s'assurer que les dispositions imposées par les textes réglementaires sont respectées par les exploitants et que les installations sont construites et exploitées conformément aux rapports de sûreté qui les décrivent.

Le SCSIN établit tous les semestres, en liaison avec les inspecteurs et ses appuis techniques, un programme de visites de surveillance. Ce programme vise, d'une part, à assurer une surveillance homogène sur l'ensemble des installations nucléaires, et, d'autre part, à porter l'accent sur des aspects techniques particuliers (protection contre l'incendie, confinement, maintenance, contrôle des fabrications, etc.).

En outre, à la suite de certains incidents significatifs, le SCSIN fait effectuer, si cela est jugé utile, une visite de surveillance de l'installation concernée.

2.1. Généralités

Les visites de surveillance ont principalement pour objectif de vérifier le respect des dispositions réglementaires et des prescriptions imposées ou approuvées par les autorités de sûreté: décrets d'autorisation de création, prescriptions accompagnant les autorisations de mise en service, règles générales d'exploitation ainsi que les règles que se fixe l'exploitant (rapport de sûreté, normes et cahiers des charges divers, consignes, etc.), ou les engagements qu'il prend.

Compte tenu de la complexité des problèmes et des équipements rencontrés dans la plupart des installations nucléaires, les visites de surveillance portent souvent plus particulièrement sur un ou plusieurs thèmes.

De plus, compte tenu de la taille et de la complexité des installations nucléaires, il est exclu que l'inspecteur examine de façon systématique l'ensemble des locaux, l'ensemble des circuits et l'ensemble des problèmes posés par une installation nucléaire, ou, en d'autres termes, se substitue à l'exploitant. Seul un contrôle par sondage est effectué.

Ces visites de surveillance permettent donc en particulier de juger de la qualité de la construction ou de l'exploitation. Dans ce cadre, toute dérive est le plus souvent rapidement décelée, et il est alors immédiatement demandé à l'exploitant d'y porter remède. Elles permettent également de détecter des événements qui n'auraient pas été jugés significatifs par l'exploitant, donc de compléter efficacement le retour d'expérience par les incidents significatifs. Elles permettent enfin de recueillir des informations supplémentaires sur les incidents suffisamment rapidement pour que les incidents soient encore dans la mémoire des personnes concernées.

Les problèmes soulevés par ces visites de surveillance peuvent être de deux natures, ponctuelle ou générique. Dans le premier cas, le service local compétent du Ministère de l'industrie donne les suites nécessaires à l'inspection, et demande en particulier à l'exploitant de remédier aux dérives constatées. Dans le deuxième cas, ou quand certaines anomalies ponctuelles se reproduisent sur plusieurs sites, les suites sont données au niveau national par le SCSIN. A cet égard, le fait que la plupart des centrales nucléaires françaises soient du même type permet à l'exploitant de réagir de manière plus efficace lorsque sont détectées des dérives.

Nous ne développerons pas ici les suites qui ont pu être données à des problèmes ponctuels; par contre, nous nous attacherons à donner des exemples d'actions engagées au niveau national à partir de visites de surveillance.

2.2. Organisation de la qualité

A la suite de constatations effectuées, au cours de visites de surveillance, en matière d'organisation de la qualité, différentes actions ont été menées pour demander à l'exploitant d'améliorer cette organisation dans les domaines qui concernent la sûreté. Il a été jugé nécessaire de compléter ces actions par un arrêté relatif à la qualité de la conception de la construction et de l'exploitation des installations nucléaires paru le 10 août 1984. Cet arrêté a pour principal but de définir un cadre minimum aux dispositions que l'exploitant de toute centrale nucléaire doit prendre pour viser à obtenir et maintenir une qualité de son installation et des conditions de son exploitation suffisante pour en garantir la sûreté. Cet arrêté a pu être en grande partie rédigé grâce aux constatations effectuées lors des visites de surveillance.

Les dispositions techniques de l'arrêté sont, pour la plupart, la codification sous une forme réglementaire de la pratique des industries intervenant dans le domaine nucléaire. Il apparaît nécessaire de figer cette pratique afin de s'assurer que les actions nécessaires sont menées par l'exploitant dans le domaine de la qualité.

Une des principales conséquences de la publication de cet arrêté a été l'obligation pour Electricité de France (EDF) de rédiger un certain nombre de notes d'organisation relatives à l'exploitation des centrales et de mettre en place un système d'agrément des prestataires.

On voit donc que les problèmes ponctuels constatés durant les visites de surveillance ont conduit à mettre en place, au niveau national, une structure susceptible de résoudre ces problèmes et d'éviter qu'ils ne se renouvellent. Cette structure apparaît aujourd'hui indispensable à une exploitation et une construction satisfaisantes des centrales nucléaires. A partir des dérives constatées lors de visites de surveillance, on peut donc considérer qu'a été construite une structure destinée à prévenir les accidents.

Notons enfin que des visites de surveillance sont maintenant organisées afin de vérifier la bonne application des notes d'organisation écrites par EDF. Ces visites, ainsi que l'analyse qui doit être faite des notes d'organisation, devraient permettre d'améliorer encore celles-ci.

2.3. Protection contre l'incendie

La protection contre l'incendie fait l'objet de dispositions constructives prévues dès la conception. Il a été malheureusement constaté, à la suite de visites de surveillance, que certains problèmes subsistaient tant dans le domaine de la construction que dans celui de l'exploitation. La protection contre l'incendie implique en effet un suivi tout particulier lors de la construction, afin de s'assurer par exemple que la séparation des voies électriques est effectuée conformément aux règles en vigueur. Ce suivi tout particulier est souvent difficile à réaliser sur un chantier aussi complexe que celui des centrales nucléaires. La complexité des locaux demande également des connaissances parfaites de la part des exploitants, afin que ceux-ci soient en mesure de détecter et d'éteindre tout début d'incendie dans les meilleurs délais. Cette complexité peut également induire des erreurs lors de la conception, erreur par exemple dans la séparation de locaux.

Ces trois types de problèmes ne peuvent être décelés qu'une fois la construction largement entamée ou au cours de l'exploitation. Ils ne sont en particulier pas détectables par une analyse, ni par un quelconque retour d'expérience à partir d'incidents. Les visites de surveillance ont, par contre, permis de cerner les problèmes réels et d'engager les mesures correctives nécessaires.

La chronologie dans la gestion de ce type de problèmes a été la suivante en France:

— réalisation d'une série d'inspections sur les diverses installations REP;

— demandes effectuées au niveau national par le SCSIN auprès d'EDF;

— analyse des mesures proposées par EDF, et visites de surveillance effectuées sur les mesures déjà prises;

— nouvelles demandes effectuées par le SCSIN après cette analyse et ces visites de surveillance.

Le travail engagé a été tout à fait considérable: il a consisté en la définition plus précise de règles de conception et de construction relatives à l'incendie, en la mise en œuvre de modifications pour améliorer la conception, et en la remise en état d'un certain nombre de matériels. Ces mesures concernent à la fois l'amélioration de la protection, l'amélioration de la détection et des moyens d'extinction, ainsi que celle de la préparation du personnel.

Il est clair que les conclusions issues des visites de surveillance ont joué un rôle fondamental dans la définition des mesures à entreprendre, et que, dans ce domaine, la conjonction de l'analyse de sûreté d'une part, et de visites de surveillance d'autre part, a pu montrer toute son efficacité.

3. LE SUIVI DE LA DISPONIBILITE DES MATERIELS

Les spécifications techniques d'exploitation imposent, en France comme dans de nombreux pays, le cadre dans lequel doit fonctionner la centrale nucléaire, en définissant notamment les limites de disponibilité des matériels et les valeurs de fonctionnement. Les dépassements de spécifications techniques font l'objet de déclarations d'incidents, et sont donc intégrés dans le retour d'expérience par les incidents significatifs. Il convient néanmoins de s'assurer que les matériels requis par les spécifications techniques sont effectivement disponibles, et donc d'anticiper les éventuelles défaillances de matériels.

La réponse à la question «Comment peut-on s'assurer qu'un matériel est disponible?» est triple:

— pour ce qui est des matériels qui peuvent être testés, il semble nécessaire de procéder à des essais périodiques;

— pour les autres types de matériels, il convient de définir un programme de maintenance adapté afin que tout défaut soit détecté;

— enfin, pour confirmer les dispositions prises ci-dessus, il importe que toute défaillance de matériel fasse l'objet d'une analyse et que les conséquences en soient tirées.

Le troisième point fait l'objet du retour d'expérience par les incidents significatifs; par contre, les deux premiers permettent d'améliorer notablement la sûreté sans qu'il soit fait appel à un incident.

Avant de développer ce qui a été fait en France dans le domaine des essais périodiques et de la maintenance, soulignons dès à présent le rôle primordial joué dans cette approche par la standardisation des réacteurs français. Il est clair que les structures mises en place tant dans le domaine des essais périodiques que de la maintenance n'ont pu l'être que grâce aux moyens qu'il est possible d'engager pour un parc important de réacteurs identiques.

3.1. Les essais périodiques

Les essais périodiques ont pour but de vérifier que tous les matériels importants pour la sûreté qui peuvent être testés sont en mesure d'assurer leurs fonctions dans toutes les situations de fonctionnement. Ils permettent en outre de détecter les dérives de paramètres significatives d'un vieillissement des matériels. Ils permettent donc, de manière générale, d'améliorer la sûreté en anticipant une éventuelle défaillance.

Pour que les essais périodiques soient correctement réalisés, il est nécessaire, d'une part, que les règles d'essais, et notamment la liste des matériels testés et les critères de sûreté, soient correctement définis, et, d'autre part, que l'organisation mise en place pour réaliser ces essais périodiques soit suffisamment fiable. Les autorités de sûreté françaises ont tout particulièrement veillé à ce que ces deux facteurs soient réunis.

Pour l'élaboration des programmes d'essais périodiques, les autorités de sûreté ont défini avec EDF une méthode équivalente à celle qui avait été établie pour les essais de démarrage:

— la liste des systèmes importants est définie par EDF et approuvée par les autorités de sûreté;

— l'ensemble des systèmes importants pour la sûreté donne ainsi lieu à la rédaction de notes d'analyses et d'essais périodiques;

— ces notes et ces règles sont analysées puis, après d'éventuelles corrections, approuvées par le SCSIN;

— enfin un tableau récapitulatif des principaux critères et de la périodicité est établi et intégré dans les règles générales d'exploitation.

Cette méthode a permis d'assurer une analyse exhaustive et approfondie des essais périodiques, donc une définition satisfaisante de ceux-ci et des critères à respecter; elle permet en conséquence de déceler a priori toute dérive significative d'une dégradation de matériels.

La réalisation des essais périodiques donne lieu, quant à elle, à la rédaction d'une note d'organisation, conformément à l'arrêté relatif à la qualité publié en 1984. Cette note d'organisation précise comment doivent être organisés, réalisés puis analysés tous les essais périodiques effectués sur les matériels importants pour la sûreté. Elle est un des maillons indispensables de la qualité en exploitation. Comme mentionné précédemment, l'application est vérifiée par sondage grâce aux inspections effectuées par les inspecteurs du SCSIN.

La méthode mise en place pour définir puis réaliser les essais périodiques permet de s'assurer que la qualité de ces essais est satisfaisante et qu'ils permettent donc de déceler des dérives importantes sur les matériels qui peuvent être testés. Nous citerons comme exemple les essais effectués sur les batteries. Ces essais ont permis de montrer que les batteries de secours des tableaux basse-tension se dégradaient au bout d'environ six ans, et qu'il convenait donc de les changer de manière préventive.

On voit donc que les essais périodiques sont indispensables pour la détection de la dégradation de matériels et qu'ils permettent d'éviter que certains incidents soient aggravés ou créés.

3.2. Les programmes de base de maintenance préventive

Les essais périodiques permettent de connaître l'état des matériels qui peuvent être testés. Il importe également que l'état des matériels statiques notamment soit connu. La réglementation française prévoit que les programmes de travaux effectués lors des arrêts de tranches doivent être approuvés par les autorités de sûreté. Ces programmes de travaux sont aujourd'hui définis, en ce qui concerne la maintenance, à partir de programmes nationaux de maintenance préventive élaborées par les services centraux d'EDF. Ces programmes ont pu être définis grâce à l'énorme expérience acquise par EDF tant sur les centrales thermiques classiques que sur les centrales nucléaires. Chaque année, EDF dispose en effet d'une expérience supplémentaire de plus de 40 années-réacteur. Les centrales doivent, sauf justifications, appliquer ces programmes de maintenance. De plus, à la fin de chaque arrêt de tranche, le chef de la centrale concernée doit transmettre aux autorités de sûreté le bilan des activités principales effectuées lors de l'arrêt de tranche, ainsi que celui des anomalies rencontrées. Ainsi, par l'approbation initiale du programme de travaux et par les résultats qui lui sont transmis à la suite de celui-ci, le SCSIN est-il en mesure de connaître l'état réel de la tranche.

Ces programmes sont bien évidemment définis de telle sorte que toute dégradation de matériel soit décelée aussi rapidement que possible. Ils sont également remis à jour dès qu'une anomalie nouvelle est décelée et qu'il apparaît nécessaire d'effectuer des contrôles différents ou plus fréquents. Grâce à l'ampleur des travaux engagés et au résultat considérable acquis chaque année sur une quarantaine de tranches similaires, on peut considérer que l'état exact des matériels des centrales nucléaires est connu de manière satisfaisante tant par l'exploitant que par les autorités de sûreté. Ceci doit permettre d'éviter toute dégradation nocive de matériels.

Ce type de contrôles permet par exemple de connaître chaque année l'état réel des générateurs de vapeur, mais il permet également de déceler d'autres défauts jusqu'alors non connus. Ainsi, suite notamment à un incident découvert sur une tranche de la centrale de Bugey, mais dont les origines semblent spécifiques à cette tranche, un programme de contrôle important des portions de tuyauteries comprises entre deux clapets et appelées «bras morts» a-t-il été entrepris. Ces «bras morts» semblent en effet être l'objet de phénomènes thermohydrauliques complexes. Le programme mis en œuvre a permis de déceler en 1987 une fissure, attribuée à ce phénomène de «bras morts», sur une tranche de la centrale de Blayais. Dans un tel cas, les contrôles sont de nouveau renforcés, et l'analyse de la portion de tuyauterie incriminée est bien sûr en cours.

En conclusion, grâce aux essais périodiques et au programme de maintenance mis en œuvre sur les tranches nucléaires françaises, on peut estimer que l'exploitant effectue un bilan de santé permanent de ces centrales, ce qui lui permet d'éviter, et ceci a priori, toute dégradation de matériels qui pourrait s'avérer nuisible.

4. LA FORMATION

Les structures mises en place pour tirer le meilleur profit des visites de surveillanc, de la maintenance ou des essais périodiques sont aujourd'hui clairement définies. Les problèmes de formation sont par contre beaucoup plus délicats à cerner. EDF a, depuis le début du programme électronucléaire, mis en place des moyens importants pour assurer une formation satisfaisante de ses personnels. Il est clair qu'une telle formation permet, grâce aux meilleures connaissances des matériels ou de l'installation qu'auront les exploitants ou les concepteurs, de prévenir certains incidents, de déceler des anomalies et de mieux prendre en compte tout ce qui concerne la sûreté.

4.1. Les actions de formation au sein d'Electricité de France

Electricité de France a été amené à définir au niveau national un plan guide de formation établi pour chaque fonction par la division chargée de la formation au sein d'EDF. Sur chaque centrale, ce plan guide est complété par un programme local d'adaptation professionnelle. De plus, chaque agent possède un plan individuel de formation qui doit comporter la liste des formations qu'il a reçues. C'est en fonction de ces actions que doit normalement être définie son habilitation. Ces habilitations ne couvrent pas uniquement le domaine de la conduite mais, également par exemple, celui de la maintenance.

Ces deux niveaux de formation, national et local, doivent permettre une certaine souplesse grâce au programme local d'adaptation professionnelle, et ils ont également permis à Electricité de France de disposer de moyens lourds de formation tels que les simulateurs, et de créer des centres de formation où sont dispensés les principaux cours généraux.

Le propos de ce mémoire n'est pas de décrire dans sa totalité la formation mais de montrer, sur des exemples précis, comment elle peut améliorer de manière notable la sûreté par le retour d'experience, indépendamment de tout incident.

4.2. Un exemple: la rupture de tube de générateur de vapeur

L'exemple de l'intervention de l'opérateur dans l'accident de rupture de générateur de vapeur a été choisi parce qu'il montre tant l'influence de la bonne formation des opérateurs sur la réussite de la procédure que la détection de problèmes liés à cette procédure découverts au cours d'exercices sur simulateurs.

En cas de rupture de tube de générateur de vapeur, le risque principal est le remplissage en eau du générateur de vapeur concerné, ce qui peut conduire à l'ouverture et éventuellement au blocage des soupages disposées sur le circuit secondaire. Cette ouverture et ce blocage peuvent amener des rejets radioactifs. Un des buts essentiels de la procédure accidentelle mise en place est donc d'éviter autant que possible le remplissage en eau du générateur de vapeur.

Une telle procédure est particulièrement délicate, et elle mérite donc qu'un entraînement tout particulier soit suivi par les opérateurs. Dans le cadre de la formation et du recyclage que suivent tous les opérateurs de centrales nucléaires, la formation sur simulateurs joue une rôle prépondérant. Ainsi, les opérateurs, parfois en équipe constituée, exécutent-ils sur simulateurs de nombreuses procédures accidentelles. Dans le cas de la procédure utilisée en cas d'accident de rupture de tube de générateur de vapeur (RTGV), cette formation s'avère indispensable en raison de la difficulté de la procédure. Elle permet aujourd'hui de s'assurer que les opérateurs de centrales nucléaires sont à même d'appliquer la procédure dans un temps suffisamment bref pour éviter autant que faire se peut le remplissage du générateur de vapeur.

De même, les temps constatés lors des exercices sur simulateurs ont permis aux concepteurs non seulement d'améliorer la procédure accidentelle, mais également d'étudier l'accident de RTGV dans des conditions représentatives de son déroulement en ce qui concerne la réaction de l'opérateur. Dans un tel cas, le temps de réaction devant être connu à environ 5 min près, le simulateur s'est avéré doublement efficace, aussi bien pour former des opérateurs comme nous venons de le voir que pour connaître leur comportement.

D'autres exemples auraient pu être cités de modifications effectuées sur les centrales, comme le pilotage en mode «gris» où des actions de formation ont pu permettre que les opérateurs acquièrent une bonne connaissance de la modification avant que celle-ci soit mise en place définitivement sur le site. Il ne paraît pas utile de décrire plus longuement tous les avantages du point de vue de la sûreté que procure la formation. Au paragraphe 4.3. seront uniquement décrites certaines actions engagées par le SCSIN pour suivre la formation du personnel sur les tranches REP.

4.3. Les actions du SCSIN

En 1980, à la suite de l'accident de Three Mile Island, le Ministre de l'industrie avait constitué un groupe d'experts chargé d'examiner les conditions de formation des personnels appelés à participer à la conduite des réacteurs ou à la contrôler. Les principales conclusions de ce groupe avaient porté essentiellement sur la création d'un ingénieur dit de sûreté radioprotection, et sur la formation pratique des opérateurs tant par l'utilisation de simulateurs que par celle de l'expérience tirée des incidents. L'ingénieur de sûreté radioprotection est maintenant en place sur toutes les centrales REP, ses principales fonctions étant de compléter l'équipe de conduite lors des situations post-accidentelles, et, en situation normale, d'assurer une partie de la formation des équipes de conduite.

En 1985, un deuxième groupe de travail a été créé qui avait pour but de faire le point sur les actions de formation engagées depuis 1980. Les principales conclusions de ce groupe sont la nécessité, pour parfaire encore le niveau général de formation, de réaliser certaines actions dans les domaines suivants:

— formalisation des programmes locaux d'adaptation professionnelle;
— définition plus précise des habilitations dans le domaine de la maintenance;
— meilleure prise en compte par les opérateurs du retour d'expérience.

De tels groupes de travail permettent, par les visites qu'ils effectuent après des exploitants sur un domaine précis, de mieux percevoir les problèmes réels et de déceler les améliorations qui pourraient être apportées sur des sujets aussi vastes que la formation. Il s'agit donc de fait d'un retour d'expérience direct.

5. CONCLUSION

De manière générale, le retour d'expérience consiste à améliorer la sûreté des tranches à partir de symptômes révélateurs de dégradations.

Les incidents significatifs sont les dégradations directement visibles, mais d'autres structures permettent de percevoir des dégradations moins importantes ou plus difficiles à percevoir.

Ainsi en est-il des visites de surveillance qui permettent de cerner les dérives éventuelles dans l'exploitation des tranches, des «bilans de santé» qui font découvrir toute dégradation même faible de matériel, ou des actions telles que les groupes de travail constitués sur la formation qui montrent, sur un sujet proche du facteur humain très délicat à percevoir, les améliorations qui peuvent être nécessaires.

Ces structures sont complémentaires de la structure formalisée, inévitablement moins apte à percevoir des dérives plus fines, du retour d'expérience par les incidents significatifs.

СИСТЕМА КОНТРОЛЯ И ДИАГНОСТИКИ РЕЖИМОВ РАБОТЫ ЭНЕРГОБЛОКА АЭС

А.А. АБАГЯН, В.М. ДМИТРИЕВ, Л.А. КЛЕБАНОВ, А.Е.
А.Е. КРОШИЛИН, Е.П. ЛАРИН, С.К. МОРОЗОВ
Министерство атомной энергетики СССР,
Москва,
Союз Советских Социалистических Республик

Abstract—Аннотация

SYSTEM FOR MONITORING AND DIAGNOSING THE OPERATING MODES OF A POWER UNIT AT A NUCLEAR POWER PLANT.

The authors discuss prospective diagnostic systems for identifying accident modes of operation and determining equipment defects as early as possible. A description is given of the structure, the tasks which need to be solved and the organization of the work involved in establishing one of these systems in the USSR: "a system for monitoring operating modes of power units at nuclear power plants". The system checks that the sensors are working properly, carries out a vibroacoustic test on the equipment, analyses the noises associated with technological parameters, identifies accident modes of operation, monitors non-steady-state operating modes and the remaining lifetime of the equipment and records non-steady-state and accident operating modes of the power unit. The authors examine the algorithms used to identify accident modes. These algorithms are based on a comparison of the accident mode with modes from the library of standard non-steady-state modes. The standard non-steady-state modes are obtained from analysis of experimental data and by using a detailed dynamic model of a power unit. In establishing these algorithms, maximum use is made of the experience of qualified expert technologists. This made it possible to create concise and effective identification methods.

СИСТЕМА КОНТРОЛЯ И ДИАГНОСТИКИ РЕЖИМОВ РАБОТЫ ЭНЕРГОБЛОКА АЭС.

В докладе обсуждены перспективные системы диагностики, направленные на идентификацию аварийных режимов и определение неисправности оборудования на возможно более ранней стадии. Изложены структура, задачи, которые должны решаться, и организация работ по созданию в СССР одной из таких систем: "системы контроля режимов работы энергоблоков АЭС". Система осуществляет контроль исправности датчиков, виброакустический контроль оборудования, анализ шумов технологических параметров, идентификацию аварийных режимов, контроль нестационарных режимов и остаточного ресурса оборудования, регистрацию нестационарных и аварийных режимов работы энергоблока. Рассмотрены алгоритмы используемые при идентификации аварийных режимов, которые основаны на сравнении аварийного режима с режимами из библиотеки эталонных нестационарных режимов. Эталонные нестационарные режимы получаются из анализа экспериментальных данных и с использованием подробной динамической модели энергоблока. При создании этих алгоритмов максимально учтен опыт квалифицированных экспертов-технологов, что позволило создать компактные и эффективные методики идентификации.

Развитие ядерной энергетики в СССР неразрывно связано с повышением требований к безопасности и надежности работы АЭС, как в стационарных, так и в переходных и аварийных режимах.

Это приводит к интенсивному развитию различных диагностических систем и методов диагностики. Условно системы диагностики энергоблоков АЭС можно разбить на две тесно связанные группы в соответствии с решаемыми задачами: диагностика оборудования остановленного энергоблока и оперативная диагностика (рис.1).

Отличительной особенностью оперативной диагностики является ее тесная связь с технологическим процессом производства электроэнергии и системами АСУ ТП АЭС.

Опыт показывает, что полная автоматизация управления энергоблоком (по крайней мере в настоящее время) невозможна. Могут возникнуть ситуации, при которых для предотвращения серьезных аварий необходимо вмешательство оперативного персонала. В этом случае важное значение имеет оперативная диагностика персоналом АЭС нестационарных процессов, происходящих в оборудовании АЭС, позволяющая определить первопричину, вызвавшую аварийный режим, оценить сложившуюся на АЭС ситуацию и наметить оптимальную стратегию ликвидации аварии.

Традиционные пути развития систем оперативной диагностики АЭС во всех странах до последнего времени заключались в увеличении детализации контроля за счет непрерывного увеличения числа контролируемых параметров. Это привело к тому, что в настоящее время контролируются показания порядка 10^4 датчиков. Такой большой объем информации, доступной определенному персоналу, затрудняет процесс принятия правильного решения в сжатые сроки.

Проведенный в последние годы в США анализ аварийных ситуаций, имевших место на реакторных установках, показал, что в подавляющем большинстве случаев причинами аварий являлись ошибочные действия обслуживающего персонала, а не автоматов.

Кроме того, традиционный подход позволяет выявлять аварийные процессы, которые уже интенсивно развиваются.

Из вышесказанного следует, что дальнейшее развитие оперативных систем контроля и диагностики должно идти по пути идентификации аварийных режимов на возможно более ранней стадии, максимальной автоматизации и обработки информации с помощью ЭВМ и представления ее в наиболее информативном и удобном виде. Именно в этом направлении развиваются системы оперативной диагностики в различных научных центрах СССР, в том числе и во ВНИИАЭС. Разработка и создание новых эффективных систем диагностики основаны на использовании достаточно сложных математических методов. Это естественно требует существенно более сложных алгоритмов обработки информации и, следовательно, использования вычислительной техники с достаточно широкими возможностями.

С другой стороны сложные алгоритмы обработки предполагают тщательную отладку и проверку в режиме опытно-промышленной эксплуатации перед внедрением в состав штатных систем АСУ. Этим они существенно отличаются от традиционных контрольно-сигнальных алгоритмов существующих систем АСУ. Для доведения разрабатываемых систем до уровня штатных в СССР на одной из АЭС создан экспериментальный полигон, на котором проводится опытно-промышленная эксплуатация систем

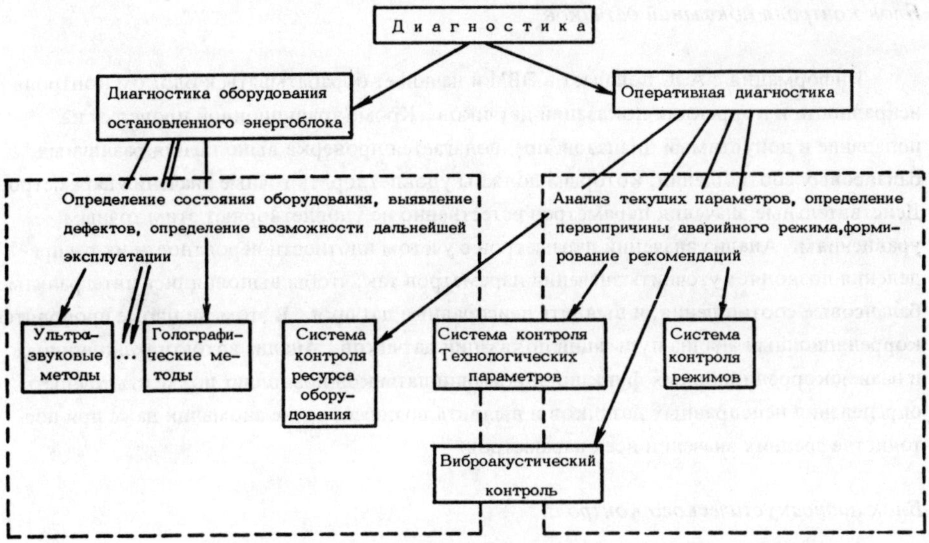

РИС. 1. Условная схема систем диагностики энергоблоков.

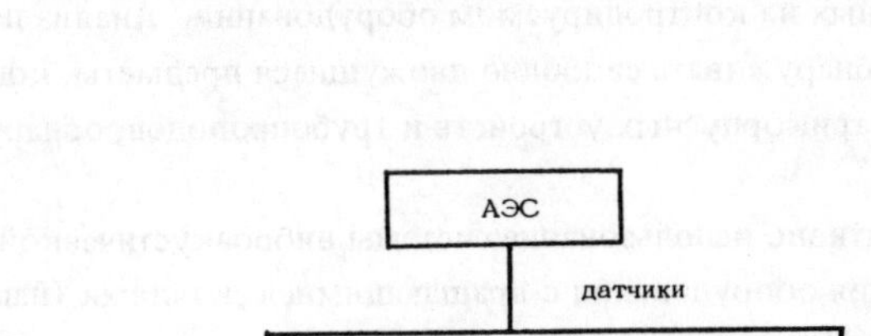

РИС. 2. Функциональная структура системы контроля режимов и ресурса.

контроля и диагностики. Одной из таких систем является "Экспериментальная система контроля режимов работы энергоблока и ресурса оборудования АЭС", функциональная структура которой приведена на рис. 2. Ниже приводится краткое описание всех основных блоков, входящих в систему.

Блок контроля показаний датчиков

Информация с АЭС попадает в ЭВМ и начинает обрабатываться блоком контроля исправности и коррекции показаний датчиков. Кроме традиционной проверки на попадание в допустимый диапазон, предполагается проверка выполнения различных балансовых соотношений, которым должны удовлетворять точные значения параметров. Действительные значения параметров естественно не удовлетворяют этим точным уравнениям. Анализ значений параметров с учетом плотности вероятности их распределения позволяет уточнить значения параметров так, чтобы выполнялись интегральные балансовые соотношения, и выявить неисправные датчики. В этом же блоке проводится корреляционный анализ пульсаций показаний датчиков. Анализ автокорреляционных и взаимокорреляционных функций показаний датчиков позволяет повысить точность определения неисправных датчиков и выявить возникновение аномалий даже при постоянстве средних значений всех параметров.

Блок виброакустического контроля

Диагностическими сигналами являются сигналы акустических и вибрационных датчиков, установленных на контролируемом оборудовании. Анализ показаний этих датчиков позволяет обнаруживать свободно движущиеся предметы, контролировать уровень вибраций внутрикорпусных устройств и трубопроводов, обнаруживать протечки теплоносителя.

Особенно эффективно использование системы виброакустической диагностики для контроля состояния оборудования с вращающимися деталями (например ГЦН, ТПН, турбина и т.д.). Накоплен положительный опыт использования виброакустической информации не только для определения работоспособности такого оборудования, но и для определения фактического его ресурса. Это позволяет обоснованно, по фактическому ресурсу, не разбирая оборудования, определять оптимальные сроки профилактического осмотра и ремонта такого оборудования, что существенно увеличивает надежность его работы и уменьшает затраты на ремонт. Для этого изучаются характеристики и изменение виброакустического сигнала в процессе эксплуатации оборудования. Создается атлас виброакустических сигналов, в зависимости от остаточного ресурса оборудования. В дальнейшем, сравнивая текущий виброакустический сигнал с эталонными, делается заключение о фактическом ресурсе оборудования.

Блок системы контроля шумов

Диагностическими сигналами являются шумы нейтронного потока, пульсации давления, температуры и других технологических параметров. Система позволяет по шумам контролировать технологические процессы. Имеется положительный опыт (в частности в ГДР) использования шумов нейтронного потока для определения закипания в активной зоне, что позволяет избежать опасных разрушений тепловыделяющих элементов. Включение этого блока в разрабатываемую систему позволяет так же обнаруживать недопустимые перемещения шахты реактора и измерять расход теплоносителя.

Блок идентификации аварийных режимов

В этом блоке на основе сравнительного анализа информации о реально протекающем процессе с эталонной информацией, полученной средствами математического моделирования, и на основании априорных знаний о процессе оперативно идентифицируется аварийный режим и определяется первопричина возникновения аварии. Более подробно структура этого блока будет описана ниже.

Блок контроля нестационарных режимов

Осуществляется проверка различных регламентных ограничений, экстраполяция изменения параметров и, если параметры приближаются к опасному пределу, вырабатывается предупредительное сообщение. Кроме того, контролируется медленный "квазистационарный" дрейф параметров (в течение суток, недель), трудно контролируемый оперативным персоналом.

Все перечисленные системы обеспечивают обнаружение неисправности в оборудовании на ранней стадии.

Последний входящий в систему блок предназначен для диагностики такого состояния оборудования, когда вероятность аварий недопустимо возрастает.

Эта задача решается *системой контроля фактического ресурса оборудования*.

Исходной информацией для этой системы являются показания штатных датчиков и результаты работы системы контроля режимов. Система во всех нестационарных режимах контролирует напряжения, возникающие в оборудовании, и рассчитывает фактический остаточный ресурс оборудования и вероятность возникновения неисправностей в оборудовании. Это позволяет определить самые опасные, с точки зрения аварий, узлы оборудования, определить вероятность этих аварий. Система использует стандартные, используемые в проекте методики, но фактические характеристики нестационарных и аварийных режимов. Это позволяет обеспечить любую, наперед заданную, вероятность безаварийной работы оборудования и определить оптимальные сроки планово-профилактических ремонтов и замены оборудования.

Кроме перечисленных функциональных блоков в описываемую систему входят еще два блока, обеспечивающих *формирование диагностических сообщений, диалог оперативного персонала с системой и регистрацию аварийных режимов*.

При разработке этих блоков особое внимание уделялось вопросу информативности и наглядности представления результатов, удобству и простоте диалога оперативного персонала с системой.

На рис. 3. приведен состав технических средств, используемых в описываемой системе диагностики. Центральным элементом технических средств являются две высоко производительные мини ЭВМ повышенной надежности, одна из которых устанавливается во ВНИИАЭС, а вторая – на полигоне. Повышенная надежность

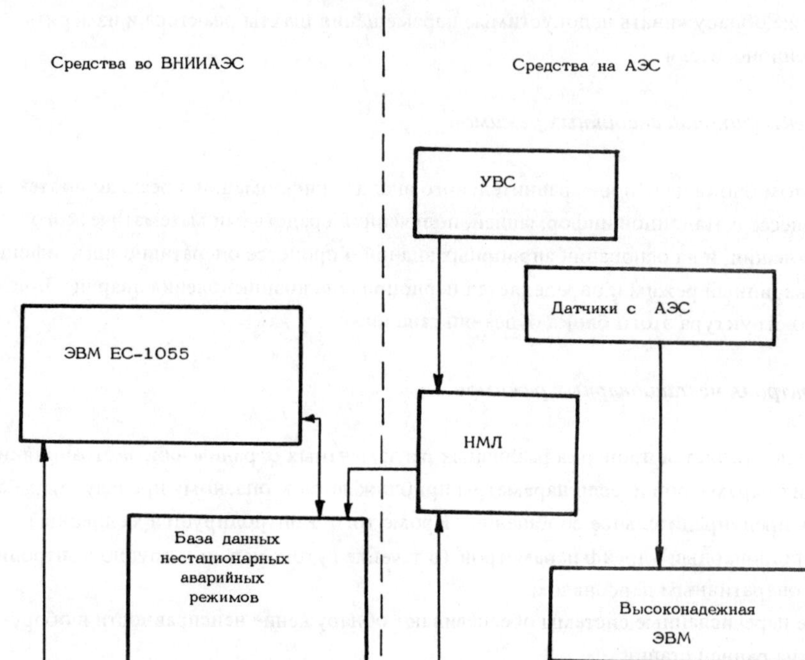

РИС. 3. Технические средства системы контроля режимов и ресурс.

является существенным требованием для ЭВМ, участвующих в работе систем диагностики с непосредственным выходом на оперативный персонал. Используемая многопроцессорная ЭВМ имеет два синхронно работающих центральных процессора с аппаратной проверкой результатов каждой операции и память с избыточной информацией в каждом слове, что позволяет исправлять случайные сбои в памяти. Такие характеристики технических средств позволяют свести к минимуму вероятность ошибочных диагностических сообщений, вызванных сбоями оборудования.

Кроме того, повышенная надежность ЭВМ существенна для обеспечения бесперебойной работы диагностической системы в возможно тяжелых (в аварийных режимах)

РИС. 4. Структура программного обеспечения во ВНИИАЭС.

условиях. Существенная часть программного обеспечения при создании математического обеспечения экспериментальной системы размещается на ВЦ во ВНИИАЭС, где производится отладка всего программного обеспечения. Для этой отладки принципальное значение имеет база данных о нестационарных и аварийных режимах работы АЭС. Эта база данных пополняется через упомянутую выше систему регистрации нестационарных и аварийных режимов. До начала работ по созданию системы контроля и диагностики был разработан механизм заполнения базы данных с помощью штатных средств регистрации аварийных режимов имеющихся в составе АСУ АЭС. Заполнение базы данных началось с момента проведения пусконаладочных испытаний I блоков. Специально созданные программные средства обеспечивают расшифровку стандартной информации, поступающей со штатных систем АСУ АЭС, и содержат диалоговые средства представления этой информации в удобном графическом виде.

Остановимся подробнее на одном из наиболее сложных с математической и алгоритмической точек зрения блоке – блоке идентификации аварийных режимов. На рис.4 показана структура размещаемого во ВНИИАЭС программного обеспечения системы контроля режимов и ресурса. Центральное место в составе этого программного обеспечения имеет математическая модель физических процессов и прочностных характеристик энергоблока ВВЭР-1000. Эта модель является основным инструментом при уточненном анализе аварийных режимов на АЭС, обеспечивает необходимой информацией блок уточнения анализа фактического ресурса и разработку методик анализа аварийных и нестационарных режимов. Адекватность математической модели

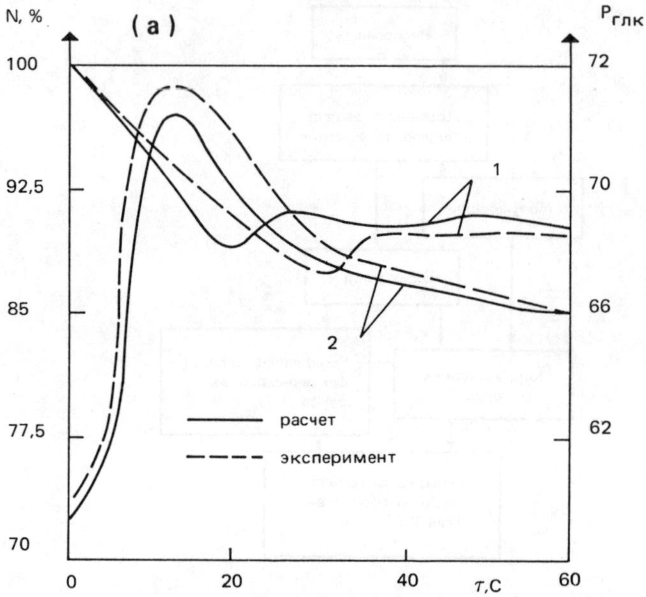

1. Изменение нейтронной мощности реактора в режиме закрытия СРК ТГ.
2. Изменение давления в ГПК в режиме закрытия СРК ТГ.

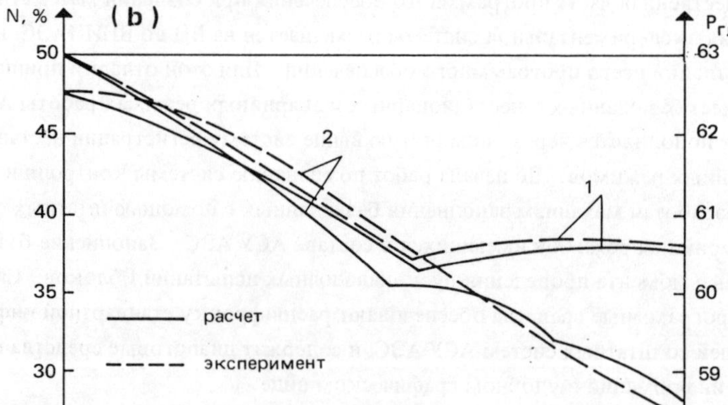

1. Изменение нейтронной мощности реактора в режиме отключения одного ГЦН из четырех.
2. Изменение давления в ГПК в режиме отключения одного ГЦН из четырех.

РИС. 5. Результаты расчетных и экспериментальных исследований двух сложных нестационарных режимов работы АЭС.

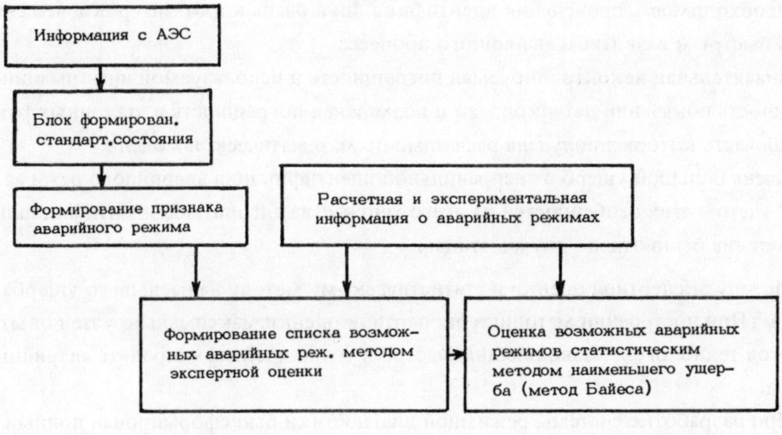

РИС. 6. Структура системы идентификации аварийных режимов АЭС с ВВЭР-1000, размещаемой на полигоне.

обеспечивается постоянным анализом поступающей в базу данных экспериментальной информации и необходимым ее уточнением. Используемый программный комплекс моделирует работу всего основного оборудования АЭС с ВВЭР-1000. Большой имеющийся опыт эксплуатации модели показал высокую точность и эффективность. В качестве иллюстрации на рис. 5 приведены результаты расчетных и экспериментальных исследований двух достаточно сложных нестационарных режимов работы АЭС. Структура размещенной на полигоне системы идентификации аварийных режимов показана на рис. 6. В блоке формирования стационарного состояния производится учет медленного дрейфа режимных параметров на АЭС, и адаптация методик диагностики аварийных режимов к текущему состоянию энергоблока.

Блок формирования признака аварийного режима производит анализ текущих показаний параметров, а сравнение отклонений с характеристиками шума датчиков определяет, когда изменения показаний являются случайным выбросом, а когда эти изменения с наперед заданной вероятностью являются результатом аварийных режимов.

Блок расчетной и экспериментальной информации в аварийных режимах содержит эталонную информацию о аварийных режимах, сравнение с которой позволяет определить тип аварийного режима.

Проблема идентификации текущего аварийного режима является типичной задачей теории распознавания образов. Особенностями задачи распознавания аварийного режима на АЭС являются:

1. Необходимость обработать большой объем информации: число возможных типов аварийных режимов составляет 10^2, а число датчиков 10^3 и даже после отсеивания неинформативных датчиков составляет 10^2.

2. Необходимость проведения идентификации в очень короткие сроки, что связано с очень быстрым развитием аварийного процесса.
3. Значительная неконтролируемая погрешность в используемой информации: как погрешность показаний датчиков, так и возможная погрешность в эталонных режимах, большая часть которых получена расчетами по математической модели.
4. Очень большой ущерб от неправильной идентификации аварийного режима.

С учетом этих особенностей из всего множества алгоритмов идентификации предпочтение было отдано двум методам.

1. Методу экспертной оценки и статистическому методу наименьшего ущерба (метод Байеса). При построении методики экспертной оценки максимально учтен опыт экспертов-технологов, позволяющий быстро и точно идентифицировать аварийные режимы.

При разработке системы режимной диагностики был сформирован полный перечень всех возможных типов аварийных режимов. Перечни, составленные экспертами СССР и Финляндии, согласуются, что указывает на отсутствие большого произвола при формировании такого перечня. Затем, после внимательного анализа экспертами всех аварийных режимов, были выбраны наиболее характерные параметры для каждого режима. Авторы не ограничивали экспертов в числе таких характерных параметров, однако их число оказалось очень небольшим и составляло 4–8 параметров в зависимости от режима (для разных режимов это свои параметры). Это позволило резко сократить объем анализируемой информации и повысить быстродействие программ.

Непосредственное сравнение расчетных значений характерных параметров с экспериментальными невозможно из-за неточности информации. В качестве примера на рис. 7 приведены две кривые, отличающиеся одна от другой небольшим растяжением временной оси. Видно, что в точках 1, 2 значения параметров отличаются даже знаками. Для решения возникающей проблемы был проведен анализ рассуждений экспертов при идентификации режимов, который позволил выделить всего шесть типов поведения режимных параметров: рост, падение, макс., мин., резкий провал или возрастание со стабилизацией на новом уровне. Эти довольно грубые типы поведения нечувствительны относительно различных погрешностей. Конкретные характеристики этих типов поведения (скорость роста и т.д.) брались из блока информации о типах аварии. Введение такого грубого представления о поведении параметров кроме повышения помехоустойчивости методики позволило существенно повысить ее быстродействие, так как каждый тип поведения определяется не более чем по трем точкам, что позволяет увеличить время цикла опроса датчиков и уменьшить объем экспериментальной информации.

Метод определения принадлежности поведения параметра к заданному типу строился так же с учетом алгоритма действий человека с привлечением понятий теории нечетких множеств. Было введено понятие "похожести" поведения параметра на эталонное. Количественное значение "похожести" задавалось функцией принадлежности. Типичный вид функции принадлежности показан на рис. 8. Значение параметра $P^*_{\text{расч.}}$ дает точное эталонное значение параметра. Ширина 2Δ функции принадлежности

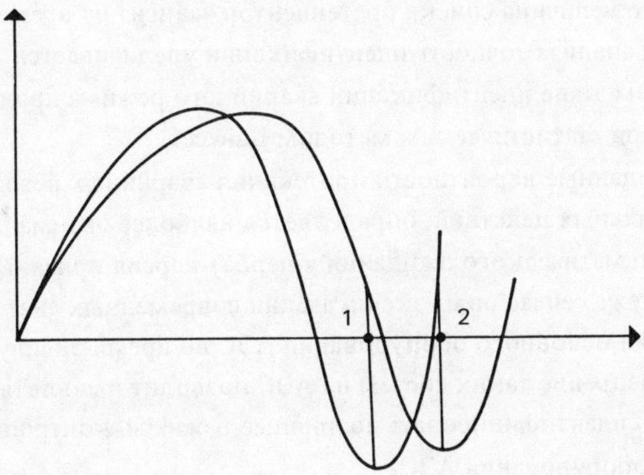

РИС. 7. Пример сравнения расчетных значений характерных параметров с экспериментальными.

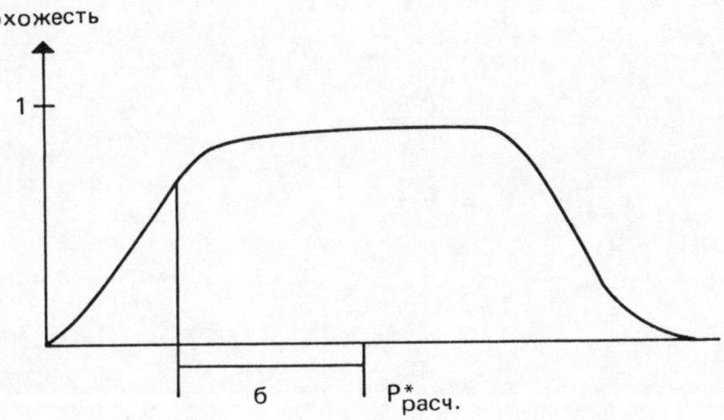

РИС. 8. Типичный вид функции принадлежности.

определялась с учетом погрешностей: датчика, эталонного значения, определения времени начала аварийного режима и других погрешностей.

Вид функции принадлежности и методика экспертной оценки аварийных режимов были выбраны такими, чтобы вероятность ошибочной идентификации была меньше наперед заданной (с учетом дисперсии сигналов и ошибки эталонных режимов).

Проведенные расчеты показали, что вплоть до погрешности в 1% по давлению и температуре и 3% по уровням и расходам блок экспертной оценки позволяет однозначно идентифицировать почти все режимы (список возможных аварийных режимов содержит мало режимов, один–два). До погрешности 2% и 6% список претендентов содержит похожие режимы. При больших погрешностях блок экспертной идентификации дает очень большие списки претендентов.

Очевидно, что величина списка претендентов зависит от времени анализа. С увеличением времени анализа точность идентификации увеличивается.

На следующем этапе идентификации аварийного режима производится анализ списка претендентов статистическим методом Байеса.

Используя заданные вероятности протекания аварийных режимов и величины ущерба от неправильных действий, определяется наиболее оптимальная (в смысле минимального математического ожидания ущерба) версия причины аварийного режима.

Имеющийся уже сейчас опыт эксплуатации современных систем диагностики и контроля состояния основного оборудования АЭС во время эксплуатации позволяет утверждать, что внедрение таких систем на АЭС позволит выявлять аварии на ранней стадии, оперативно идентифицировать аварийные режимы, контролировать фактический ресурс основного оборудования АЭС.

Все это позволит повысить безопасность и экономичность АЭС.

L'AIDE INFORMATIQUE AUX OPERATEURS

B. MÉCLOT
Service de la production thermique,
Electricité de France,
Paris

M. APPEL
Direction de l'équipement,
Electricité de France,
Villeurbanne

France

Abstract–Résumé

DATA-PROCESSING ASSISTANCE TO OPERATORS.

The principles of control for the PWR units of Electricité de France have been defined in the case of the first units of the 900 MW level since 1973. Since then, although these general principles have been maintained, a steady evolution in control has taken place; on the one hand, operating experience and safety analyses have given rise to new requirements, such as the safety panel, on the other, on the occasion of the passage to the 1300 MW level, the adoption of programmable automatic control devices to replace electromagnetic circuitry for almost the whole of the system, represented an improved technological development. However, it became apparent that the present structure, in which considerable room was left for information and control by the operator, presented certain limitations. For the new N4 level of 1400 MW, the first unit of which is that of Chooz B in the Ardennes (anticipated entry into service in 1991), it was decided, as early as 1981, to initiate a study involving the organization of the control room and the man–machine interface; this project has now materialized in the form of a control room with data-processing. The study of incidents which have taken place at nuclear power plants, and especially that at Three Mile Island, has shown the need for improving control in accident situations and for employing means of reducing human error, in particular, by ensuring a redundancy of diagnostic information and by enabling operators to understand what is happening and to take correct action in all circumstances: this is the purpose of the safety panel installed in the 900 MW units.

L'AIDE INFORMATIQUE AUX OPERATEURS.

Les principes de contrôle-commande des tranches REP d'Electricité de France ont été définis pour les premières tranches du palier 900 MW depuis 1973. Depuis, si ces principes généraux ont été conservés, une évolution constante du contrôle-commande a eu lieu; d'une part, l'expérience d'exploitation et les analyses de sûreté ont fait apparaître de nouveaux besoins, comme le panneau de sûreté; d'autre part, à l'occasion du passage au palier 1300 MW, l'adoption pour la quasi-totalité du système d'automates programmables en remplacement du relayage électromagnétique a marqué une importante évolution technologique. Toutefois, il est apparu que la structure actuelle, dans laquelle une large place est laissée aux moyens conventionnels pour l'information et la commande par l'opérateur, présentait certaines limitations. Pour le nouveau palier N4 de 1400 MW, dont la première tranche est celle de Chooz B, dans les Ardennes (mise en service prévue en 1991), il a donc été décidé d'engager, dès 1981, une étude remettant en cause l'organisation de la salle de commande et l'interface homme-machine: ce projet est maintenant concrétisé sous la forme d'une salle de commande informatisée. L'étude des incidents survenus sur les centrales nucléaires, et particulièrement de celui de TMI, a montré la nécessité d'améliorer la conduite en situation accidentelle et de mettre en œuvre des moyens visant à minimiser les erreurs humaines, notamment en assurant une redondance du diagnostic et en permettant aux opérateurs de comprendre ce qui se passe et d'agir correctement en toutes circonstances: c'est l'objectif visé par le panneau de sûreté mis en place sur les tranches de 900 MW.

1. INTRODUCTION

C'est en 1973 que furent définis les principes de conduite des tranches du palier REP 900 MW, ainsi que l'architecture et la technologie du contrôle-commande. Compte tenu des choix faits en matière de technologie (relayage électromagnétique) et du manque d'expérience d'exploitation des tranches REP en matière de conduite post-accidentelle, les aides informatiques prévues à l'époque furent extrêmement réduites. Seul un journal de bord avec historique, ainsi que l'état de toutes les informations analogiques et logiques des systèmes, étaient disponibles en salle de commande sur écran.

Dans cet exposé sont présentés les moyens d'aide aux opérateurs mettant en oeuvre des moyens informatiques importants:

– les panneaux de sûreté des tranches des paliers 900 MW et 1300 MW;
– la salle de commande informatisée des tranches du palier 1400 MW.

2. LE PANNEAU DE SURETE

Le panneau de sûreté constitue un des éléments de la réflexion menée à la suite de l'accident de TMI par Electricité de France (EDF) pour améliorer la conduite des tranches REP.

Il a pour objectif d'aider les opérateurs, et d'une manière générale tous les intervenants, en cas de situation incidentelle. A ce titre, il est intégré à l'organisation de la conduite en régime perturbé.

Dans cette éventualité, trois groupes d'acteurs interviennent:

1) Les opérateurs chargés d'appliquer les consignes: ils effectuent toutes les opérations de conduite et agissent en quatre étapes:

– détection des événements;
– confirmation des actions automatiques;
– diagnostic de l'événement et choix de la consigne à appliquer;
– application des consignes de conduite.

2) L'ingénieur de sûreté et de radioprotection (ISR), appelé en salle de commande sur détection d'un événement: par l'utilisation d'une consigne

indépendante de celle des opérateurs, il surveille l'évolution des paramètres de la chaudière. Sur des critères précis d'état physique de la chaudière, il peut:

– demander aux opérateurs d'effectuer des actions complémentaires;
– demander aux opérateurs d'abandonner la consigne en cours d'application et d'appliquer une consigne d'urgence dont le but est de sauvgarder le cœur et de limiter les rejets radioactifs par tous les moyens disponibles.

3) Une équipe de crise, composée de spécialistes, que le chef de centrale peut réunir en quelques heures: ces spécialistes se répartissent en:

– une équipe de crise locale qui rejoint la centrale accidentée, et
– une équipe de crise nationale qui se réunit à Paris.

Pour remplir son rôle d'outil d'aide à la conduite, ou à la réflexion, le panneau de sûreté est constitué des éléments suivants:

– un ensemble informatique composé de deux ordinateurs d'acquisition des informations fonctionnant en normal-secours et d'un ordinateur de traitement et de présentation de l'information vers trois postes de dialogue pour chacun des trois acteurs;
– un microprocesseur qui acquiert directement des températures et des pressions pour présenter aux opérateurs la température maximale en sortie du cœur et la marge à l'ébullition d'un manière redondante à l'équipement informatique.

Les informations traitées par le panneau de sûreté sont présentées:

– aux opérateurs par deux écrans et un clavier;
– à l'ISR par un écran et un clavier;
– à l'équipe locale de crise dans la salle réservée par un écran et un clavier.

Les centres nationaux de crise sont reliés directement aux panneaux de sûreté de tranche.

Les fonctions disponibles ont été conçues en même temps que les nouvelles consignes de conduite et sont cohérentes avec elles.

2.1. La détection des événements

Le but est d'indiquer aux opérateurs la cause des ordres ayant initié les actions de sûreté et le bon déroulement des actions automatiques.

L'image apparaît automatiquement sur le poste «opérateurs». Elle présente:

– le libellé des paramètres physiques ayant déclenché l'action de protection et l'heure d'apparition;
– la (ou les) action(s) de sûreté initiée(s) sur chacune des voies;
– le compte rendu bon ou mauvais d'exécution de chaque action de protection.

Cette image est mise à jour au fur et à mesure que les protections successives sont sollicitées.

2.2. La confirmation des actions automatiques

Dans cette deuxième phase, les opérateurs doivent vérifier que les actions automatiques demandées se sont bien effectuées.

En cas de compte rendu incorrect, ils peuvent appeler sur écran des images synoptiques des systèmes de sûreté. Ces images présentent:

- la position des organes;
- leur discordance éventuelle par rapport à une position de référence;
- leur disponibilité (c'est-à-dire s'ils sont susceptibles d'être télécommandés);
- l'état de toutes les sources électriques de la tranche.

2.3. L'aide au diagnostic: choix d'une consigne

Une fonction particulière présentée sous forme de logigramme aide les opérateurs à choisir la consigne à appliquer après injection de sécurité.

Une fonction analogue, en cours de développement actuellement, aidera l'opérateur à choisir la consigne à appliquer pour toute situation incidentelle.

2.4. L'aide à l'application des consignes

Toutes les consignes à appliquer après incident ou accident ont été réécrites; en particulier, la conduite de l'injection de sécurité fait appel à des logigrammes et diagrammes.

Le panneau de sûreté aide les opérateurs en présentant des images animées dans les mêmes plans, sur lesquelles le point de fonctionnement est matérialisé, avec son historique sur 10 minutes:

- niveau du pressuriseur en fonction de la marge à l'ébullition pour la conduite de l'injection de sécurité;
- pression primaire en fonction de la température pour le refroidissement ou le réchauffage de la chaudière avec ou sans pompe primaire.

Le panneau de sûreté aide également à l'application de consignes de surveillance, par exemple le bon refroidissement du coeur par le système de réfrigération à l'arrêt.

2.5. L'aide à la surveillance de la tranche

Courant 1988, le panneau de sûreté sera équipé d'une fonction de surveillance de la tranche, reprenant les règles de conception de la consigne de surveillance de l'ISR.

Cette fonction réalisera d'une manière complètement indépendante un véritable contrôle du procédé.

2.6. L'aide aux équipes de crise (spécialistes)

Les spécialistes volontairement tenus hors de la salle de commande peuvent suivre l'évolution de la situation par consultation:

– des synoptiques donnant l'état des circuits (images utilisées par les opérateurs pour la confirmation des actions automatiques);

– d'une série d'images regroupant l'évolution des paramètres physiques caractéristiques des grandes fonctions de sûreté:

– contrôle de la puissance nucléaire,
– évacuation de la chaleur du coeur,
– évacuation de la chaleur pour le secondaire,
– masse du réfrigérant primaire,
– intégrité du confinement.

Une transmission en temps réel de ces mêmes informations vers le centre national de crise est en service.

Le panneau de sûreté 900 MW a été expérimenté et évalué sur simulateur en 1984 et 1985.

Début 1986, toutes les tranches CP1 et CP2 (34 configurations) et les trois centres de formation EDF (4 configurations) ont été équipés de panneaux de sûreté.

Le fonctionnement de ce système fait l'objet d'un suivi particulier et les deux premières années d'exploitation se soldent par un bilan très positif.

Un effort semblable est consacré aux tranches 1300 MW qui seront équipées d'un système comparable à partir de 1988.

3. LA SALLE DE COMMANDE INFORMATISEE DES TRANCHES DU PALIER REP 1400 MW

La mise en évidence de besoins nouveaux en matière d'aide à l'opérateur a conduit, comme nous l'avons vu précédemment, à implanter dans le système de contrôle-commande et en salle de commande des moyens nouveaux. Cette implantation n'a pu se faire que par une juxtaposition de moyens et non pas par une intégration. Il en est résulté:

– un coût de mise en œuvre élevé,
– une mauvaise optimisation des performances,
– des difficultés d'utilisation en salle de commande.

Il apparut donc nécessaire à EDF d'engager, dès 1981, une étude remettant en cause complètement l'organisation de la salle de commande et l'interface homme–machine et répondant aux objectifs suivants:

– toute information fournie aux opérateurs doit être fiable et pertinente;

– cette information doit pouvoir être élaborée, tant dans sa forme (information synthétique combinant plusieurs paramètres de la tranche) que dans sa présentation (courbe, barographe, abaque);

– les actions que doit entreprendre l'opérateur, tant pour la conduite normale que pour les actions correctives (fiches d'alarmes, procédures incidentelles ou accidentelles), doivent pouvoir être étroitement intégrées aux informations nécessaires pour effectuer ces actions;

– il est nécessaire que les opérations de maintenance effectuées sur l'installation (consignation) soient systématiquement portées à la connaissance de l'opérateur et que ce dernier puisse aisément signaler au service de maintenance les interventions à effectuer;

– les capacités d'analyses, a posteriori, doivent être accrues et le système de traitement correspondant doit être intégré dans l'architecture informatique du site.

3.1. Dispositions fonctionnelles retenues

Pour satisfaire cet ensemble de besoins, on a retenu les dispositions fonctionnelles suivantes:

3.1.1. Traitement de validation

Toute information venant de l'installation est systématiquement validée au niveau acquisition (par exemple, alimentation capteur) puis, lorsqu'elle fait l'objet d'une alarme, au niveau fonctionnel (par exemple, validation pression par pompe enclenchée) et au niveau situation (adaptation de la présentation de l'alarme, caractérisant sa gravité, à la situation de la tranche).

3.1.2. Présentation élaborée des informations

L'ensemble des informations nécessaires ou utiles à la conduite peut être fourni aux opérateurs sous forme d'images intégrant des synoptiques, des barregraphes, des courbes, des abaques.

Tout objet présenté sur image (un objet pouvant être une vanne, un moteur, un capteur, un organe de réglage, etc.) est identifié par un certain nombre d'attributs (par exemple, une vanne, son état ouvert ou fermé, sa disponibilité, son état «défaut», «essai», «consigné», etc.) qui, pour un objet donné, sont toujours les mêmes, de façon à avoir une représentation homogène et cohérente quelle que soit la partie de l'installation concernée.

Des vues synthétiques des différents systèmes de la tranche permettent aux opérateurs d'appréhender rapidement l'état global d'un système, ou de la tranche.

3.1.3. Etroite imbrication des moyens de conduite

En désignant un actionneur, un organe de commande ou de réglage sur une image, un opérateur peut provoquer une action; cette étroite imbrication des commandes et des informations permet de fournir à l'opérateur un traitement interactif des alarmes (accès direct aux actions correctives depuis les fiches d'alarme) et des procédures de conduite.

3.1.4. Cohérence entre actions de conduite et de maintenance

Le système de conduite et le système de maintenance sont interconnectés; l'opérateur a l'indication des états consignés sur ses images; il peut informer automatiquement le service de maintenance lorsqu'il a identifié un capteur ou un matériel défaillant.

3.1.5. Traitements d'aide à la conduite et d'analyse a posteriori

Ces traitements, déjà implantés sur les calculateurs de tranche des paliers 900 et 1300 MW (panneau de sûreté), sont totalement intégrés dans le système de conduite; en outre, les transferts vers les systèmes de site et les centre nationaux de crise sont prévus dès la conception.

3.1.6. Prise en compte des procédures de conduite par état en situation accidentelle

Les procédures de conduite par états, en cours de mise en oeuvre partielle par EDF sur les tranches 1300 MW pour les situations accidentelles, seront intégrées dans le système de conduite de N4.

3.2. Architecture du système de contrôle-commande

Pour satisfaire aux fonctionnalités précédemment définies, la structure du contrôle-commande retenue est une architecture à deux niveaux (fig. 1):

1) Un niveau 1, assurant l'interface avec l'installation (acquisition d'information–émission des ordres) chargé:

– des fonctions automatiques de protection,

– des automatismes non séquentiels (commandes groupées, verrouillage, automatismes de réglage),

– des traitements de validation d'acquisition et de validation fonctionnelle.

Le niveau 1 est réalisé par une nouvelle génération de systèmes programmés de même type que ceux actuellement utilisés sur les tranches 1300 MW.

2) Un niveau 2, relié au niveau 1, et assurant l'interface homme–machine, les traitements de présentation d'information, de validation de situation, d'archivage, d'aide à l'opérateur, les dialogues de commande, les séquences, etc.).

Le niveau 2 a la structure suivante:

– un système informatique de conduite, redondé, composé de plusieurs calculateurs et multimicroprocesseurs, effectue l'ensemble des traitements et dialogues nécessaires pour permettre la conduite de la tranche, en toute circonstance, depuis des postes opérateurs «informatisés»;

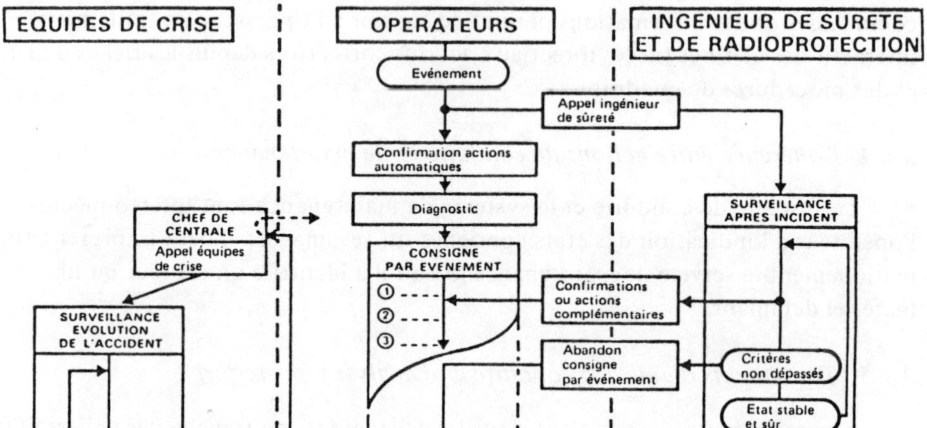

FIG. 1. Architecture du système de contrôle-commande.

– les calculateurs composant ce système sont reliés, entre eux et avec le niveau 1, par des bus rapides redondés au «standard» ETHERNET; ils peuvent échanger des informations avec le système informatisé d'aide à la gestion des consignations de matériels, avec le calculateur de gestion de site, et avec les centres nationaux de crise;

– un panneau auxiliaire doté de moyens indépendants du système informatique de niveau 2 (c'est-à-dire directement relié au niveau 1) et permettant, lors d'une perte de ce dernier, de ramener la tranche dans un état sûr et, en situation accidentelle, d'appliquer les procédures de conduite par états.

3.3. Organisation de la salle de commande

La salle de commande (fig. 2) est constituée:

– d'un synoptique mural de grandes dimensions, représentant une vue simplifiée des circuits de la tranche, où sont visualisés les valeurs des principaux paramètres et les états des principaux actionneurs et circuits; ce synoptique est destiné à fournir aux opérateurs, en permanence, une vue globale de l'ensemble de l'installation;

FIG. 2. Salle de commande.

FIG. 3. Poste opérateur.

– de trois postes opérateurs identiques comportant chacun (fig. 3):
i) des moyens de présentation des informations constitués de trois écrans couleur graphiques (700 X 1000 points) dits de conduite et de quatre écrans couleur semi-graphiques affectés à la présentation des alarmes;
ii) des moyens de dialogue constitués de deux claviers de fonction (dialogues → commandes, validation, dialogues → alarmes et procédures), de trois écrans monochromes sensitifs (gestion de postes, commande, compte rendu des actions), d'une boule roulante et d'un clavier alphanumérique et d'un afficheur.

Deux de ces postes constituent les postes de conduite; ils sont complétés par des équipements de communications (téléphone, etc.) et de quelques boutons-poussoirs pour les actions urgentes (arrêt d'urgence, injection de sécurité, etc.).

Le troisième poste, appelé poste d'observation, a ses possibilités de commande inhibées lorsque les deux autres postes sont opérationnels; il permet alors uniquement l'accès aux informations. Il peut être utilisé comme moyen de commande en secours, si l'un des deux autres postes est en panne.

Un panneau dit annexe supporte les commandes et l'information d'installations de surveillance (contrôle des accès, incendie, etc.).

3.4. Développements en cours

La première tranche REP 1400 MW doit être mise en service en 1991 et l'on vise à ce que le système de contrôle-commande, avec ses moyens informatiques, soit en partie opérationnel fin 1988.

Pour valider cette structure de salle de commande, nouvelle pour EDF, un simulateur à échelle 1 a été réalisé; il permet, sur un modèle de 1300 MW, de s'assurer que la conduite d'une tranche REP est possible en toute situation avec les moyens prévus et de mettre au point les traitements, dialogues, images et les méthodes de gestion des données. La campagne d'évaluation qui a été engagée en mai 1987 semble montrer la bonne adéquation de ce type de salle de commande aux objectifs visés. Cette campagne d'évaluation doit être terminée fin 1987.

4. CONCLUSION

Les opérateurs des tranches nucléaires à eau sous pression disposent ainsi d'aides à la conduite qui, notamment pour la conduite accidentelle, respectent les mêmes principes de base et sont conformes à l'organisation des équipes de conduite à Electricité de France. Au fur et à mesure de l'arrivée de nouveaux paliers techniques, ces outils d'aide sont de plus en plus intégrés au système général de contrôle-commande.

DEVELOPING SAFETY AT LOVIISA NUCLEAR POWER STATION

K. SJÖBLOM
Imatran Voima Oy (IVO),
Loviisa Nuclear Power Station,
Loviisa, Finland

Abstract

DEVELOPING SAFETY AT (LOVIISA) NUCLEAR POWER STATION.

The Loviisa nuclear power station (NPS) consists of two 445 MW PWR units. The main supplier was Atomenergoexport of the Soviet Union. The differences between the Finnish and Soviet safety philosophy resulted in a unique plant concept. During the 17 reactor-years of operation the high level of safety has not been affected by any incident. Process transients have led to a reactor scram only four times during the history of Loviisa NPS. The safety of Loviisa NPS has been further promoted by back-fitting and improved operational practice. The Chernobyl accident led to no major immediate plant or procedural changes because of the different nature of Chernobyl and Loviisa nuclear power plants. However, the Chernobyl accident underlined the importance of some special areas and of severe accidents in general. Reactor pressure vessel rupture, a total loss of redundant safety features or core melt have not been included in the original design basis. However, severe accidents have been widely studied and mitigating measures have been provided. This development work was intensified since Chernobyl. Various containment pressure relief concepts are being investigated. In order to ensure decay heat removal, even after the total loss of the turbine hall, an extra independent emergency feedwater system with diesel generators of its own will be constructed. In order to improve the reliability of the back-up electricity supply a special transmission line from a nearby hydro power plant directly to the diesel-backed buses has been constructed. A probabilistic safety assessment study is being conducted. The integrity of the reactor pressure vessel, taking into account the embrittlement due to neutron radiation, rapid accidental cooling or pressurization has been the subject of extensive studies. It is expected that the lifetime of Loviisa NPS will be determined by the aging of the reactor pressure vessels; until now no reason seems to prevent the operation of the plant up to the design age of 30 years.

1. COMBINATION OF SOVIET TECHNOLOGY AND WESTERN SAFETY PHILOSOPHY

The Loviisa nuclear power station (NPS) consists of two 445 MW PWR units that were commissioned in 1977 and 1980. The main supplier was Atomenergoexport (AEE) of the Soviet Union which delivered the main components: reactor pressure vessels, fuel, steam generators, turbines and generators. Also companies in the United States of America, Switzerland, Finland and the Federal Republic of Germany participated in deliveries and construction work. The utility, Imatran Voima Oy (IVO), participated widely in the plant design and in that way IVO developed engineering expertise of its own that has been valuable during the operation. The domestic share of the project cost was two thirds of the total.

The requirements of the Finnish nuclear safety authorities were based on western contacts; e.g. the 55 general design criteria for nuclear power plants (10 CFR 50, App. A) established by U.S. nuclear safety authorities in 1972 were applied to the design of Loviisa NPS. The differences between the Finnish and Soviet safety requirements resulted in a unique plant concept.

Loviisa has ice condenser containments. The redundancy of the emergency cooling pumps, diesel generators and most other active components which have a safety role is 4 x 100 %, and the redundancy of passive components is at least 2 x 100 %. The safety functions have been designed to fulfill the single failure criterion. The reactor protection system, which scrams the reactor, and the plant protection system, which activates other significant safety functions like emergency cooling, containment isolation, emergency feedwater injection and diesel generators, are totally separate. Less significant, but more frequent safety functions are activated by other plant instrumentation. The reactor scram system has 2 x 2/3 logic and the plant protection system 2/4 logic.

2. SAFETY EXPERIENCE

During the 17 reactor-years of operation the high level of safety has not been affected by any incident. Transients at Loviisa are relatively slow due to, amongst other factors, large water masses in the primary and secondary circuits. The safety margins of the core are relatively large and only a relatively small fraction of the events causes a scram. At Loviisa both units have two turbines; tripping only one turbine does not normally cause a reactor scram. The reactor scrams are much less frequent than turbine trips.

The average occurrence rate of unintentional scrams at Loviisa is 1.3 per reactor year. Process transients have led to a reactor scram only four times during the history of the Loviisa nuclear power plant. Learning from experience, modifications and other improvements have contributed to a significant decrease in the occurrence rates of scrams and other transients since the first years of operation as shown in table 1.

The safety of Loviisa NPS has been further promoted by backfitting and improved operational practice. The safety related modifications have been implemented with minimum impact on the electricity production. The cumulative capacity factors are 79 % and 84 % for the two units, respectively. The safety issues that have limited the plant operation the most have been the investigation of defects in steam generator primary collector welds (the annual outage was extended by several months in 1980), the checking and modification of anchor bolted pipe sup-

TABLE 1. TRIPS (UNINTENTIONAL) AT LOVIISA 1 AND 2

	Reactor trips		Turbine trips	
Year	L01	L02	L01	L02
1977	9			25
1978	2		5	
1979	0		5	
1980	2	5	2	5
1981	0	1	3	4
1982	0	1	1	2
1983	0	1	1	0
1984	0	0	1	1
1985	0	0	1	1
1986	0	0	2	0
1987 (until Oct.)	-	-	-	-

ports (three weeks shutdown), insufficient information concerning the safety limits of the AEE fuel (92 % power limitation for three years). All of these safety issues originated from reasons other than plant incidents.

3. THREE MILE ISLAND (TMI) RELATED ACTIONS

After the Three Mile Island accident, in order to eliminate water seals developed during small leakages of the primary circuit, connection lines between hot and cold legs by-passing steam generators and including also drain lines were installed in three of the six primary loops. Several alternatives for a (new) reactor pressure vessel (RPV) water level measurement system were studied but installing them was regarded as impractical; instead, a reactor coolant saturation margin monitoring system, based on the measurement of outlet pressure and temperature, was introduced; the saturation margin system serves as an indicator of adequate water level in the core.

The hydrogen generation in water/zirconium reaction during a LOCA was studied according to a revised design base, and hydrogen burns in the containment were analyzed. Because the design pressure of the ice condenser containments is relatively low (1.7 bar) the hydrogen problem was taken seriously. New hydrogen concentration monitors were developed and installed. About 70 glow plugs were placed in the containment in order to avoid hydrogen explosions by igniting the hydrogen already at lower concentrations. Gas removal from the RPV head was enabled.

New high range radiation monitors were installed in the containment and in the stack. Also sampling from the containment atmosphere and reactor water during an accident was made possible.

Introducing a shift technical advisor (STA) practice has been considered. Because the operators at Loviisa have a relatively high base education and extensive plant specific training, STAs were not regarded as necessary. However, recently a "safety engineer on duty" -practice has been introduced. The safety engineers on duty are chosen from experienced shift supervisors and operation engineers. The safety engineer on duty always has to be able to enter the power station within 30 minutes after being called by operators.

4. CHERNOBYL RELATED ACTIONS

After studying the Chernobyl accident the nuclear safety authority (STUK) and the utility considered no immediate major plant or procedural changes necessary, because of the different nature of Chernobyl and Loviisa nuclear power plants. However, the Chernobyl accident underlined the importance of some special areas and of severe accidents in general. According to the request of STUK the following actions were taken:

The potential and consequences of unintentional reactivity insertions taking account of various control rod positions, injection of unborated water, reactor cooling, changes in the specific weight of reactor coolant and illegal operations have been re-evaluated without discovering any new sequences leading to reactivity accidents. The study covered different operation modes in various stages of the loading period. The operators have been given extra training in reactor physics and reactivity transients will be included in the simulator training.

The adequacy of administrative procedures has been evaluated; the shift supervisors were asked about experiences in conducting test programs, by-passing the reactor or plant protection systems and how the plant procedures have been followed. The situations in which automatic protection or electricity supply could be inappropriately disconnected were studied. It was discovered that during some periodical tests the disconnection of direct current supplies caused no alarm to the control room; this led to modification to ensure the awareness of control room operators of the disconnection of dc supplies.

The display of safety parameters and functions in the control room will be reviewed in connection with further development of the emergency procedures. A new safety parameter display system is being designed.

5. SEVERE ACCIDENT POLICY

Reactor pressure vessel rupture or a total loss of redundant safety features or core melt had not been included in the original design basis accidents. However, severe accidents have been widely studied and mitigating measures have been provided at Loviisa like at many other nuclear power stations. This development work has been intensified after Chernobyl.

The following severe accident sequences are being studied:

- The rupture of a steam generator collector and failure of operator action;
- Total loss of normal and emergency feedwater;
- Total loss of electricity supply and breaking seals of primary coolant pumps;
- A small LOCA and failure of high pressure safety injection;
- A large LOCA together with a failure of both high and low pressure emergency cooling.

Because the (ice condenser) containment can probably maintain its integrity at markedly higher pressures than the design pressure of 1.7 bar, the realistic pressure tolerance of the containment and the pressure loads in core damage accident sequences, including production of noncondensable gases in corium/concrete reactions, are being studied. On the basis of the severe accident analyses the design base of the potential containment pressure relief system will be defined. Various containment overpressurization concepts shall be investigated. Also the alternative possibility to depressurize the containment by spraying the steel containment from the outside has been studied. The phenomena connected with the rapid overpressurization of the containment will be carefully studied in order to ensure that the pressure relief system would not increase the environmental risk. The decision concerning a new containment depressurization system will be made when the studies are completed.

Those safety systems that should be operable even during severe accidents are defined, and their capability to withstand the prevailing environmental accident conditions within the containment is being studied. The need to have battery back-up electricity supply to the hydrogen ignition system, and the suitability of a potential new catalytic hydrogen burning system are being evaluated. The on-site radiation levels and the accessibility of vital plant areas are being re-examined.

The emergency preparedness is tested and the emergency preparedness organization is trained in annual emergency preparedness exercises. The scale of emergency preparedness exercises has been enlarged to encompass, in addition to the power plant organization, the utility headquarters, the nuclear safety authorities (STUK) and the local off-site rescue authorities; a total of nearly 100 persons participated. The on-site full-scale plant simulator has been made use of in preparing and conducting emergency preparedness exercises.

6. THE ASSURANCE OF DECAY HEAT REMOVAL

At Loviisa decay heat is removed through steam generators to the secondary circuit. At first the steam generators produce steam but later on when decay heat is reduced the steam generators are filled with water. For this reason there are more systems of safety grade in the secondary circuit than usual. Many of them have been located in the turbine building.

The fire protection of the turbine hall has been identified as an important safety issue, especially after a recent fire in the turbine hall of a coal-fired power plant in Helsinki. Potential fires in the turbine hall have been studied with modern computer codes. The area containing feedwater tanks and secondary circuit safety valves will be separated from the turbine hall by a new fire wall. In order to ensure decay heat removal even after the total loss of the turbine hall, an extra, independent emergency feed water system with diesel generators of its own will be constructed.

7. EXTRA BACK-UP ELECTRICITY SUPPLY

At Loviisa both units have four diesel generators. However, because there are only two totally independent redundancies in the emergency core cooling systems the single failure criteria is not totally fulfilled after a failure of a diesel generator and so the maximum time to repair a diesel generator during plant operation is limited to three days. In order to increase the reliability of the back-up electricity supply, a special transmission line from a nearby hydro power plant directly to the diesel-backed buses has been constructed. After commissioning this new link the utility will apply to prolong the maximum time allowed for repairing a diesel generator.

8. FINAL SAFETY ANALYSIS REPORT (FSAR)

During the plant construction phase when the FSAR was compiled as a part of the application for the operation licence (OL) the FSAR-concept was new to the main supplier and to the utility, and the regulations concerning FSAR were still at a development stage. Since then IVO has gained much more experience and information concerning the plant and its safety. The goals set up for FSAR have been raised; the main parts of the system reports have been revised, and the numerous plant modifications and safety analyses made since commissioning have contributed to the need of developing FSAR. It has been noticed that a proper and detailed documentation of the design bases and safety evaluations is extremely important in designing and evaluating plant modifications and in conducting new safety analyses; the need of new safety analyses seems to be continuous. In the next few years the emphasis of the FSAR work will concentrate on accident analyses and on the continuous general updating of FSAR.

9. PROBABILISTIC SAFETY ANALYSIS

A PSA study of level "1 and 1/2" is being conducted. Level "1 and 1/2" means analyzing the probabilities of core melts and containment by-passing. After the Chernobyl accident, environmental risk became a special point of interest and recently it was decided to continue the study to levels 2 and 3, which means estimating also the probabilities and environmental consequences of various release categories; the study will also be continued to cover external events as well. A systematic and covering risk study helps in coordinating various safety projects, and the results of a risk study are useful when decisions are made for new modifications. After Chernobyl the environmental risk may easily be overemphasized and the results of the levels 2 and 3 might be beneficial when the insurance terms are being negotiated.

10. THE EMBRITTLEMENT AND INTEGRITY OF THE REACTOR PRESSURE VESSEL

The RPV is the most critical component when considering the safe lifetime of the Loviisa NPS. At the core region its wall is exposed to neutron irradiation, which causes embrittlement of the RPV material. The integrity of the reactor pressure vessel, taking into account the embrittlement due to neutron radiation, rapid accidental cooling, and pressurization, has been the subject of extensive studies. Numerous material and thermohydraulic experiments as well as strength and process analyses have been conducted.

The inservice inspection program of the reactor pressure vessel has been relatively extensive, including among others ultrasonic and eddy-current examinations of the inside wall and the cladding. The embrittlement rate of the steel is followed by surveillance specimens taken from the reactor. Several cladding samples have been taken by grinding to estimate the neutron flux distribution. The Loviisa 1&2 are also participating in the international surveillance testing program coordinated by the IAEA by irradiating "Round Robin" specimens in the reactor (French and Japanese steel).

Because the embrittlement rate of the RPV material at the core region seemed to exceed design estimates, the outermost fuel elements have been replaced by dummy elements. The neutron flux to the RPV has been further reduced by introducing a so called "low leakage loading pattern" according to which new fuel elements are placed in the middle of the core and the older ones nearer the RPV walls. By changing the power distribution in the core and increasing the average linear power and maximum burnup of the fuel, a decrease in the thermal power of the reactor and a shortening of the loading period have been avoided.

The reduced fracture toughness can endanger the integrity of the RPV during overcooling transients that potentially can result in pressurized thermal shock (PTS). Because the RPV vessel failure is not included in the deterministic design basis accidents of the plant and because the earlier deterministic brittle fracture analyses of the RPV that employed postulated transients might describe neither representative nor limiting sequences, the risk of PTS to the RPV was studied probabilistically. The expectance value of the frequency of a rupture of LO1 reactor pressure vessel was estimated to be 4E-6/ reactor-year using conservative values extrapolated to the situation after 30 years of operation. The results of this study will be re-evaluated in 1989, in the light of the findings of the PSA study mentioned in chapter 9. A corresponding study for Loviisa 2 resulted in a risk estimate 3E-6/ reactor-year (extrapolated to the situation after 40 years of operation). A rupture of RPV does not necessarily lead to core melt. Because these risk values were this low analyzing of the consequences of the RPV rupture was not considered necessary.

In order to minimize the risks of thermal pressurized shocks to the RPV, the temperature of the emergency coolant tanks has been increased and the shut-off heads of high pressure safety injection pumps have been decreased.

It is expected that the lifetime of Loviisa NPS will be determined by the aging of the reactor pressure vessels; until now no reason seems to prevent the operation of the plant up to the design age of 30 years.

BIBLIOGRAPHY

AHLSTRAND, R., KOMSI, M., "Backfitting actions at the Loviisa Nuclear Power Station", Modifications Required for Safety of Nuclear Facilities (Backfitting), Munich, Germany, 1985 (Proc. Sem. Munich, 1985), IAEA-SR-112/31, IAEA, Vienna (1985) 43–44.

AHLSTRAND, R., TÖRRÖNEN, K., VALO, M., BÄRS, B., "Surveillance programs and irradiation embrittlement research of the Loviisa Nuclear Power Plant", Radiation Embrittlement of Nuclear Reactor Pressure Vessel Steels: An International Review Vol. 2 (STEELE, L.E., Ed.), Rep. ASTM STP 909, American Society for Testing and Materials, Philadelphia, PA (1986) 55–69.

HAAPANEN, P., SAMMATTI, P., "Experience from the use of the Loviisa Nuclear Plant simulator for emergency conditions training and research", Experience with Simulator Training for Emergency Conditions (Proc. Sem. Vienna, 1986), IAEA-TECDOC-443, IAEA, Vienna (1986) 165–168.

KOMSI, M., et al., "The probabilistic PTS study of the Loviisa Unit 1 reactor pressure vessel", Top. Mtg on Probabilistic Safety Assessment and Risk Management (PSA '87), Zurich, Sep. 1987.

LAMROTH, H., LAINE, P., Loviisa Nuclear Power Station — An east-west project", Nucl. Eur. **4** (1982).

PALMGREN, A., "Experience of plant performance and methods of improving performance including refueling", ENS/ANS Conf., Geneva, Jun. 1986, Rep. ENC-4, European Nuclear Society, Petit-Lancy, Switzerland (1986).

PALMGREN, A., "Experiences from the Finnish PWR's Loviisa 1&2", Int. Mtg on Further Improvement of LWR Technologies, Apr. 1986, Japan Atomic Industrial Forum, Tokyo.

REGNELL, B., "Ensuring safe operation at the Loviisa Nuclear Power Plant", IAEA Safety Codes and Guides (NUSS) in the Light of Current Safety Issues (Proc. Symp. Vienna, 1984), IAEA, Vienna (1984) 519–532.

REGNELL, B., HELYNEN, S., "Analysis of the effects of hydrogen burning and measures taken for their mitigation at the Loviisa Nuclear Power Plant", Hydrogen Behaviour and Control and Related Containment Loading Aspects (Proc. Spec. Mtg Suzdal, 1983), IAEA-TC-427.6, IAEA, Vienna (1984) 83–89.

SJÖBLOM, K., "Developing utility emergency preparedness exercises", Emergency Planning and Preparedness for Nuclear Facilities (Proc. Symp. Rome, 1985), IAEA, Vienna (1986) 495–499.

SJÖBLOM, K., "The initiating events in the Loviisa Nuclear Power Plant history", Anticipated and Abnormal Transients in Nuclear Power Plants, Top. Mtg Atlanta, GA, Apr. 1987, American Nuclear Society.

SJÖBLOM, K., LAINE, P., Loviisa Nuclear Power Station — An east-west project", Nucl. Eur. **4** (1982).

THE EVOLUTION OF SAFETY MEASURES IN BELGIAN NUCLEAR POWER PLANTS

P.H. GOVAERTS
Vincotte ASBL,
Brussels, Belgium

Abstract

THE EVOLUTION OF SAFETY MEASURES IN BELGIAN NUCLEAR POWER PLANTS.

A brief description is presented of the nuclear power plants operating in Belgium. For the last four plants, the protection against accidents of external origin has led to the duplication of many safety systems. During the safety reassessment of the old plants, which is required every ten years, much has also been done to improve the protection against external accidents, taking into account the limitations owing to existing installations and with the aim to define consistent solutions. For the new plant the protection against external accidents will be integrated from the beginning in the design, as well as features to mitigate 'beyond design' accidents. A probabilistic safety assessment will be made and used as a feedback in the design.

1. Introduction.

The seven nuclear power plants (NPPs) now operating in Belgium have generated 38.7 TW·h in 1986, representing 67% of the electricity production. The other sources of electricity are mainly coal (around 20%), the rest being divided about equally between gas and oil. This is a fundamental change with respect to 1974, when more than 80% of the production came from oil.

Situated at the crossroads of western Europe, with a very high population density, Belgium has always considered safety a prime necessity; the recent evolution of its electricity production emphasizes also reliability.

2. Main characteristics of the plants in operation.

All the plants are pressurized water reactors built by consortia grouping Belgian firms (ACEC, COCKERILL, etc.) and Nuclear Steam Supply System (NSSS) (Framatome or Westinghouse) with Belgian architect-engineers (Tractebel).

Doel 1 and 2 are Westinghouse twin plants of 400 MW(e) each, which began operation in 1974/1975; they are two loop plants with an 8 ft [(1)] core and a rather high average linear power (6,76 kW/ft or 221,8W/cm). These 2 plants share a number of engineered safety features (ESF) systems : 4 high head safety injection pumps, 4 diesels, multiple interconnections between the 4 train electric buses, 4 component cooling pumps, etc. High head safety pumps inject water, up to a maximum pressure of 120 bar, in the cold loops and directly into the reactor vessel.

Tihange 1 is a three-loop Framatome reactor of 900 MW(e), which was critical in 1975. It is similar to Beaver Valley in the USA or to the Fessenheim plants in France, except that it is still fuelled with 15 x 15 type assemblies. It has, like the 900 MW(e) plants of the French programme two safety trains (2 x 100%) for electrical as well as mechanical equipment.

Doel 3 and Tihange 2 (Framatome, 900 MW(e) each) and Doel 4 and Tihange 3 (Westinghouse, 1000 MW(e) each) are three-loop plants designed with an increased separation of functions between independent systems, and with three independent safety trains (mechanical and electrical). Belgian safety authorities have required that the United States Nuclear Regulatory Commission (USNRC) rules be followed and that these plants be protected against accidents of external origin : aircraft crash, explosions, large fires, toxic gases. It has led to the following design :

- A primary containment in prestressed concrete (design pressure around 5 bar corresponding to a large loss of coolant accident (LOCA) or a steam line break)
- A secondary containment in reinforced concrete, to withstand a military aircraft crash (or a civil aircraft of 90 t at a speed of 90 m/s)
- Three safety trains (mechanical and electrical) totally independent, either 3 x 100% or 3 x 50% depending on the systems.

As the control room and the ESF systems are not completely protected against aircraft crashes, a 'second level of

---

1 1 ft = 0.3048 m.

protection' is added and the corresponding equipment is located in bunkerized areas; these systems obey the single failure criterion (mostly 3 x 50%, in some cases 2 x 100%).

This second level comprises:

- A three independent train safety protection (analogic and logic) system independent of the classical protection system from the sensors up to the command of the scram breakers located in a bunkerized area (these breakers and the control rods are common to the first and the second level protection systems);
- Pumps and tanks to inject borated water into the primary circuit (equivalent to the chemical and volume control system (CVCS));
- Pumps and tanks to inject water into the steam generators;
- Pumps to inject borated water in the seals of the primary pumps as soon as these seals are no more fed by the CVCS;
- Diesel groups, the cooling systems associated with the above mentioned systems, the ultimate heat sink (wells in the phreatic zone or artificial ponds);
- Supplementary equipment to protect against toxic and explosive gases, and air intakes far enough from each other so that they both cannot be affected at the same time by explosive gases or a large fire.

The systems and components are similar in the 4 plants, but the start logic of the second level of Framatome differs from the Westinghouse design.

3. Safety level achieved in the latest plants

The protection against external accidents is designed for an automatic operation of the plant in a safe state for at least 3 hours (time period assumed for other operators to come to the plant and to take over the control) whatever the initial state of the plant.

As a consequence a number of components have been classified as safety graded, e.g. discharge valves and heaters of the pressurizer, discharge valves of the steam generators, decay heat removal system, compressed air, reactor coolant pumps trip breakers.

Following USNRC rules implies that the LOCA radiological consequences are calculated with a source term corresponding more or less to a core melt (Regulatory Guide 1.4), independent of the fact that the requirements on emergency core cooling are met (10 CFR 50 Appendix K). However, the doses to the workers and the public considered acceptable in Europe are much smaller than the US limits (10 CFR 100 Reactor Site Criteria). This has led to a particularly sophisticated system to prevent leakage of radioactive products : double containment and steel liner in the primary containment, ventilation of the space between the two containments designed as an ESF system and discharge through filters, ventilation of potentially contaminated auxiliary buildings designed as ESF, containment isolation made by different signals, injection of gas or water between isolation valves (or between the discs of double-disc valves) at a pressure at least equal to 1.1 times the accident pressure.

For the calculation of radiological consequences, all potential leakage paths have been identified and minimized. As a result, a large percentage of the remaining doses is related to the leaks from the safety injection system during the recirculation period, as these leaks bypass both containments.

Containment leakage is monitored through tests made (up to recently) in accordance with 10 CFR 50, Appendix J. Taking into account the experience gained from these tests, the integral test of the containment (type A test) has now been modified, in order to be sure that no unknown path has been left open after refuelling.

The large number of sensors, resulting from 3 independent safety trains in the first and in the second levels of protection, increases prohibitively the number of periodic tests : these are as a consequence performed by an automatic test console and there are overlaps between the tests so that the whole chain, from the sensor to the trip breakers, is tested.

In the design it has been assumed that an external accident does not occur at the same time as one of internal origin. It implies that no external accident may induce an accident of internal origin, for example a small LOCA. Hence the seals of the primary coolant pumps have been protected by the specific equipment quoted above. In case of station black-out (i.e. the loss of the grid, the turbogenerators and the loss of the 3 ESF diesels at the first level of protection), there is an automatic start of the 3 diesels at the second level of protection (diverse with respect to the other ones) and start up of the primary pumps' seals protection system.
This scenario of station black-out, which is an important contributor to core melt in some foreign probabilistic risk assessment studies, is eliminated in the Belgian plants.

Another aspect to which many efforts have been devoted is the qualification of structures, systems and components against seismic effects and against post accident environment (LOCA, steam line break, aircraft crashes,..); the results of these qualification programmes are still under review. It has also been checked that non-seismically qualified equipment may not induce failures of seismic equipment. High energy line breaks have also been postulated and the consequences made acceptable, by supplying restraints by physical separation, by protective screens.

The progressive degradation of steam generator tubes is a worrisome problem in pressurized water reactors (PWRs); roto-peening of the tubes has been performed prior to criticality in Doel 4 and Tihange 3; the first industrial application of shot-peening took place after a few years of operation in Doel 3 and Tihange 2. The improvement attributable to theses processes is monitored at each refuelling, but years are required before results can be assessed.

Other characteristics of Doel 4 and Tihange 3 are an 'integrated' reactor vessel (large reduction of the number of welds) and a 'rapid refuelling' package. These features should lead to a reduction of the doses to the workers during maintenance.

The lessons of the Three Mile Island accident have been incorporated during construction and commissioning of Doel 3 and Doel 4, Tihange 2 and Tihange 3. Very few hardware modifications were necessary except for better monitoring of beyond design conditions (large pressure and radiation in the containment, sampling and measuring of highly radioactive effluents) and better presentation of the status of the plant to the operators and safety engineers (safety parameter display system, on-site technical support centre and other emergency operation facilities).

Accident procedures have also been revised and thoroughly adapted. Training of the operators has been formalized and, due to the peculiarities of these plants with their specific protection systems against external accidents, two full scale simulators are being built and should become operational at the end of 1988. The qualification of these simulators is under way, as it is necessary to know exactly what are the limits of the simulation, which physical phenomena are simulated and with which accuracy, etc. A qualification and commissioning test programme must be established for these simulators, in a similar manner to the commissioning programme of the plant itself.

The safety evaluation of these plants has followed deterministic principles and criteria, and they have not been subjected to a probabilistic safety assessment (PSA) up to now. When compared to projects being assessed now (like the Italian PUN and the British Sizewell B) the Belgian plants incorporate as many or more protection and safety grade systems, with at least the same degree of diversity. Hence they should show as good or better marks from a safety viewpoint; in order to check that a PSA will begin in 1988 on Doel 3 as a preparation for the ten-year safety reassessment to be made for 1992. It is intended to perform a level 1 PSA extended to the loadings on the containment and its failure modes.

4. Safety reassessment of older plants and backfitting

The operating licenses of all Belgian NPPs require that a safety reassessment be made every ten years comparing the plants with the rules and practices from USA and the European countries at the time of the new review. Such a work has been launched in 1983 for Doel 1 and 2 and Tihange 1, taking into account the safety issues dealt with during the safety analysis of the most recent plants, the feedback of operating experience from foreign plants and from the plants under review themselves. Apart from specific improvements (ventilation, gaseous waste treatment, flooding, fire protection,..) most issues concerned the seismic reassessment, the protection against external accidents and against high energy line breaks, an improved physical separation between safety systems, the upgrading of support systems, the classification and qualification of components according to their safety role, the protection against pressurized thermal shock.

The increase of the safe shutdown earthquake (SSE) from 0.1 g to 0.17 g at Tihange has led to the definition and the use of a site-specific spectrum and the development of tridimensional elasto-plastic computer codes. For the qualification of components to a higher SSE, data obtained from industrial installations is used by joining the Seismic Qualification Utility Group initiative in the US.

An original method has been developed by the architect-engineer to study high energy line breaks and to determine if and where restraints should be added.

Studies and experimental verifications are under way to deal with the pressurized thermal shock issue for Doel 1 and 2. The possibility of direct injection of cold water in the reactor vessel makes specific studies mandatory, instead of applying generic conservative criteria.

Accident analyses were updated, and accidents not considered ten years ago have been included, like feedwater line breaks. Due account has been taken in these reevaluations for modifications in systems, like the decrease of the boron concentration in the boron tank of the safety injection system, from 21 000 ppm to 7000 ppm.

To protect an existing plant against external accidents, to improve physical separation between safety related systems, to deal with new issues like ATWS or station black-out requires comprehensive and consistent solutions. It has led to the installation of new systems ('ultimate shutdown') in Doel 1 and 2 and Tihange 1.

Considering Doel 1 and 2 as examples, the design basis for the new system are to cope with the SSE, with the total loss of external and internal electric sources, with a number of aircraft crashes/gas cloud explosions/high energy line breaks and with a fire in the electrical building.

This ultimate shutdown system will consist, per unit, of :
- A system injecting automatically borated water into the seals of the primary pumps as soon as normal injection is lost;
- An additional feedwater system for the steam generators, manually started;
- A diesel group starting automatically upon loss of power on the electric buses.

The corresponding structures and components are seismically qualified.

5. Feedback from operating experience

As one of the lessons learned from Three Mile Island (TMI), the collection and analysis of incidents occurring in foreign plants has been started in 1980, feeding a data bank where more than 1300 incidents are now incorporated. The emphasis is of course on PWR but generic problems (corrosions, diesel generator failures, valves malfunctions, etc..) appearing in other types of reactors are also recorded.

Events described in the various national and international reporting systems are first screened for their safety relevance and interest; when retained they are analysed for their possible causes, their consequences, the corrective actions considered or implemented, their applicability to the Belgian plants. These events are entered in the data bank, from which they can be retrieved according to some classification : plant where it occurred, system or component affected, and type of transient. More extensive analysis is made on a number of topics, selected for their safety significance, due to the frequency of the events or to their seriousness, and a synthesis report is issued on each selected topic, stressing the applicability and the possible corrective measures.

The lessons from TMI, as described in NUREG 0737, have been taken into account for the Belgian plants. Modifications were made mainly with respect to the handling of highly radioactive effluents (sampling, shielding), the emergency operating facilities (safety parameter display system), the accident procedures, and the use of simulators as indicated previously.

On account of its very different technology and its design weaknesses, the Chernobyl accident is not applicable as such to the Belgian plants. It has however reminded us that beyond design accidents may occur through unanalysed scenarios. It has been concluded that core melt accidents, though incredible owing to the various levels of protection against internal and external accidents existing in the Belgian plants, should be considered with the aim of mitigating their consequences. Steps to be implemented are a reassessment of reactivity induced accidents as to their possible causes, an evaluation of the ultimate strength of the containment, a survey of systems mitigating core melt consequences. Emphasis should be laid on diagnosis and mitigation procedures, not on adding hardware. The PSA exercise on Doel 3 to begin in 1988 will also shed light on this.

6. Design criteria for new plants

No decision has been made up to now to build an eighth NPP in Belgium but industrial commitments indicate that it would be a 1400 MW(e) four loop Framatome plant. It is then necessary to combine the French design rules applicable to the NSSS with what has been up to now Belgian practice i.e. application of USNRC rules.

A comparison of French and US design rules has shown that they are not very different and that interfacing them would not create insuperable difficulties, hence the proposal to apply the US rules on a general basis, with an exception for the NSSS where the French rules are applied. For individual components (electrical, mechanical and civil work) either set of rules may be applied.

Siting conditions in Belgium are not similar to those in France and this fact prevents building in Belgium a replica of a French plant. With the renewed interest for severe accidents, no valid reason has been found to deviate from the design rules used for estimating radiological consequences; it means the provisions required to prevent leakages are the same as in the last four Belgian plants.

The necessity to protect the plant against external accidents, in particular aircraft crashes, leads to the bunkerization of the reactor building, the spent fuel building and the auxiliary and electrical buildings, including the main isolation valves on the steam and feedwater circuits. This choice avoids the duplication of safety systems which was characteristic of the two levels of protection present in the last plants.

In order to increase the defence in depth, the design which has been chosen is basically a two train safety system where all active components are duplicated; yielding a four train configuration for active components (4 x 50% or 4 x 100% depending on the systems) and a two train configuration for the passive components like piping. This choice implies an electrical configuration with four independent trains, and four diesels as on-site power.

Additional protection of the reactor coolant pumps seals has been retained, as in the last four plants: moreover the design specifications corresponding to the implementation of the French H and U procedures have also been retained, as they extend the protection of the plant against far beyond design accidents.

As a PSA will be made during construction of the plant, it will be possible to judge the relative merits of these extra precautions and have some feedback about the design in order to optimize it.

7. Conclusions

Siting conditions in Belgium and the application of USNRC rules have led to a rather sophisticated protection against external accidents in the recent plants. Part of it has been fed back to the older plants during the ten year safety reassessment. For the next plant protection against external accidents and features to mitigate the consequences of severe accidents will be integrated from the beginning in the design; a PSA will be made and its results might be used in optimizing the design.

REFERENCES

[1] VERLAEKEN, J., "ARIANE: a novel operational experience feedback system for enhanced power plant safety reliability," Proc. Eur. Nuclear Conf. Geneva, 1986, Vol. 3, European Nuclear Society, Petit-Lancy, Switzerland (1986) 321-327.

[2] GOVAERTS, P., CAVACO, A., "Decay heat removal systems in Belgian nuclear power plants," Proc. Committee on the Safety of Nuclear Installations, OECD Spec. Mtg, Würenlingen, 1983.

[3] ROCH, M. CAVACO, A., "Application of American and French rules for the next Belgian PWR," Proc. 6th Pacific Basin Nuclear Conf., Beijing, 1987.

[4] De BOECK, B., "New Belgian position on containment leakage testing," Proc. 3rd Workshop on Containment Integrity, Washington, DC, 1986.

[5] DECKERS, B., De GELDER, P., "Safety re-evaluation and backfitting of Belgian nuclear power plants", paper presented at Seminar on Modifications Required for Safety of Nuclear Facilities (Backfitting), Munich, 1985.

INHERENT SAFETY FEATURES OF ADVANCED LIGHT WATER REACTORS

W. BÜRKLE, W. BRAUN
Kraftwerk Union AG,
Erlangen, Federal Republic of Germany

Abstract

INHERENT SAFETY FEATURES OF ADVANCED LIGHT WATER REACTORS.

Inherent safety characteristics are of fundamental importance to the design of commercial reactors in order to cope with the whole range of design basis accidents in combination with passive and active safety features. Contrary to popular belief, the term ''inherently'' or ''passively'' safe alone does not guarantee absolute safety, as even inherent safety characteristics are dependent on certain prerequisites, e.g. that the configuration and integrity of the reactor remain constant in the course of an accident. The effectiveness of passive safety features also depends not only on their integrity but also, for example, on their operational readiness and on the availability of an operating medium; all that active safety features require in addition to this is a guaranteed energy supply. Ensuring that all these prerequisites are fulfilled is, in the final analysis, a question of comprehensive quality assurance and quality control with regard to design, construction, materials, fabrication, erection, plant operation and control, maintenance and inservice inspection of the reactor system. Decades of experience in the construction and operation of nuclear power plants with light water reactors worldwide result in this sort of quality standard having been achieved. Exhaustive risk studies have verified this fact and have resulted in the elimination of remaining deficiencies. The safety technology applied to light water reactors has established the degree to which safety can and must be guaranteed, and has thereby set the standards according to which safety technology for other reactor types has successfully been developed. It is, however, futile to compare an individual aspect, such as a specific inherent characteristic or a certain passive subsystem of one type of reactor, with the characteristics and systems of another type, as it is only the combination of all the characteristics, features and quality assurance measures which ensures the safety of a nuclear power plant. These measures have also to cope with the human aspect, which calls for independence from human reliability. Therefore, safety relevant manual actuation at short notice must not be necessary, in order to rule out human maloperation.

Introduction

The Chernobyl reactor accident prompted renewed discussion of the question as to whether our nuclear power plants are sufficiently safe. Since, as far as is known, the accident in the RBMK 1000 was peculiar to this type of construction, and was due primarily to the reactor's lack of inherent stability, the conclusion is frequently drawn that inherent stability alone can ensure reactor safety. It is occasionally even assumed that this very general term **"inherent safety"** is synonymous with a new type of safety of a particularly advanced nature. This is not the case: inherent safety characteristics, supplemented by active and passive safety features, have naturally always been exploited in all reactor types.

The reactor systems used to date in the Federal Republic of Germany have for many years been proving the effectiveness of their inherent safety characteristics and the efficiency of their passive and active safety features. Our safety philosophy is based on the principle of accident prevention by means of quality assurance in the broadest sense and accident mitigation by designing plants to withstand an all-embracing spectrum of design basis accidents. It is thanks not only to the reactor's inherent safety characteristics that these aims are achieved, but also to the passive barriers and active safety systems.

Same safety technology has been developed, tried and tested in the course of the planning, construction and operation of well over 200 light water reactor nuclear power plants in the world outside the centrally planned economy area (WOCA). The project planning, inspection, licensing and safe operation of more than 20 pressurized water and boiling water reactor nuclear power plants in the Federal Republic of Germany have made a vital contribution to this development. At the same time, government-sponsored research programmes for reactor safety have resulted in a constant broadening of experience on which to base safety engineering in light water reactors. The results of the German Risk Study, Phase A of which was completed in 1979 and Phase B of which is to be completed shortly, verify a balanced safety concept for pressurized water reactor power plants. Their safety characteristics are capable of coping with events, the impact of which exceeds that postulated as the design basis. The knowledge and experience which have been gained in this field mean that the safety of light water reactors has become a yardstick for assessing other reactor systems. No other reactor system has been investigated in such depth and with such thoroughness with regard to safety considerations as has the light water reactor.

The role of "inherent safety" in present safety standards

Safe reactor operation in the event of an accident depends, among other factors, on

inherent characteristics, resulting from the plant's mechanical and functional design, which mean that the laws of nature cause the plant to remain in a safe condition or to attain such a condition "of its own accord",

passive safety features, which come into operation in the event of an accident, without the need for switching operations or additional energy supplies, and

active safety features, which differ from the passive ones only in that they require activation and energy to power them.

Examples of **inherent** safety features are, for light water reactors, the negative reactivity coefficient for reactor power and coolant density, which contributes to automatic shutdown in the event of an accident and counteracts unintentional power increases. Other features which can be termed inherent safety characteristics are the fact that the control rods drop under the force of gravity alone, that sudden failure of the pressure-retaining boundary is precluded by the leak-before-break criterion and that heat is removed by natural convection.

The inherent characteristics which can contribute to plant safety differ, however, depending on whether the reactor is of the pressurized water, boiling water, high temperature or breeder type. In the case of high temperature reactors, for example, with their spherical fuel/moderator elements, the inherent features are the negative reactivity coefficient for fuel temperature and the thermal storage capacity of the pebble bed in the event of failure of the heat removal system. The one thing all inherent safety features have in common, however, is the fact that their natural effectiveness is assured only as long as the mechanical and functional design on which they are based remains constant.Furthermore, additional active or passive systems are generally required in order to ensure long-term return of the plant to a stable, safe final condition. Above all, though, the effectiveness of inherent safety principles is, as a rule, restricted to a few special transients and accident sequences.

The dividing line between inherent and passive safety features and systems is not clearly defined. Examples of those considered to be **passive** are the containment in the case of light water reactors, safety injection by means of accumulators, and convective residual heat removal, which is used in many other types of reactor. Structural measures designed to shield and protect the plant can, however, also be classified as passive features. Passive systems come into operation "of their own accord", without activation or drive mechanism; but their effectiveness does depend, for example, on their being in the correct switching status and filled with an operating medium. The operational readiness of passive systems or features is therefore guaranteed only if their function is monitored or if accessibility for the performance of maintenance and inservice inspection is unrestricted. In nuclear power plants equipped with light water reactors or with other types of reactor, the safety systems are designed, wherever possible, to utilize passive principles, such as the use of stored energy or natural convection, even if active elements are involved on account of the necessity for initiation of switching operations or additional drives.

As there is no reactor type in commercial operation in which all conceivable transients and accidents can be handled by inherent safety characteristics and passive safety systems, all reactors, even high temperature reactors as well as other reactor designs recently published are fitted with additional active systems, which are capable of mitigating the consequences of special transients and accidents.

The efficiency of **active** safety systems, whatever their scope and design, depends not only on their structural integrity, their operational switching status and the availability of operating medium, but also on an assured energy supply. It is for this reason that an active system is often looked upon as being inferior from the point of view of its efficiency as a safety feature. However, as great care is taken to ensure that both the signal formation for automatic system activation and the energy supply for the drives are guaranteed, an active safety system is just as efficient as a passive system.

As the effectiveness of any safety feature or system depends on its being of suitable mechanical and functional design, there are no significant differences between the various forms as far as their safety function is concerned. This applies in the same way to the use of inherent safety characteristics, which is also possible only if the system is suitably designed, constructed and planned. Of course, inherent safety characteristics also remain effective only as long as the reactor configuration on which they are based does not change in the course of an accident. Furthermore, in the case of inherent characteristics and passive systems, it is not possible to subject the system to trial operation, inspection and inservice function tests in order to verify the characteristics being exploited: with active systems this is state-of-the-art practice. Consequently, it is immediately and unequivocally apparent whether an active system is functioning correctly or has broken down. Finally, the effect of inherent characteristics frequently does not become determinant until plant conditions far exceed the range which is considered normal.

The generally complex mode of operation of active safety systems is sometimes thought to be a disadvantage. This, however, is only true at first glance, as automation naturally ensures that operator error and other human factors are precluded when the system is called on to operate. As a rule, it is also possible to verify correct functioning of active safety systems without interrupting plant operation. On account of their design, which is aimed primarily at coping with a broad spectrum of accidents and a wide range of operating variables, they function, depending on the accident in question, in combination with passive and inherent safety features, which, in turn, also have a varying scope of application, depending on the type of accident. For example, the inherent feedback effect, which causes a reduction in the reactivity of a reactor as the temperature increases, and which is a prerequisite for every licensable reactor, can serve only to limit reactor power. It is not, however, sufficient to mitigate all design basis accidents without allowing unacceptable amounts of radioactivity to be released. Back-up measures in the form of additional safety features and barriers are essential.

In the final analysis, the evaluation of a safety feature should be based not on collective terms such as "inherent", "active" or "passive", but on the level of knowledge attained, i.e. on the investigations carried out, the experience gained and the resources available, all of which constitute state-of-the-art technology. This type of evaluation reveals that it is only by combining inherent characteristics with active and passive safety features that we have been able to attain such a high safety standard in light water reactors and to mitigate the consequences of the entire accident spectrum.

It is nonetheless common in discussion of this subject for the term "inherently safe" to be applied to advanced reactor designs, because in one type of reactor the materials involved make it impossible for the nuclear fuel to melt, while in another type certain transients can be dealt with passively, without reactor trip being initiated (ATWS). Such sweeping statements are misleading, as the deciding factor where reactor safety is concerned is not the prevention of meltdown or the mitigation of ATWS transients, but the prevention of release of the radioactivity inventory into the environment. To this end, it must be proven that all accidents which could conceivably lead to consequences of this kind can be mitigated. Whether one individual aspect which contributes to this aim can be referred to as "inherent" and another as "passive", or whether active safety features with a high degree of built-in safety are used has little bearing on the degree of safety attained by a nuclear power plant.

Design of safety features

The principle objective of all safety features is to protect the environs of a nuclear power plant. To this end, the Atomic Energy Act of the Federal Republic of Germany specifies that all precautions dictated by current scientific and engineering practice are to be taken, as defined, for example, in the accident guidelines. Risk studies have shown that the proposed safety measures also cope with incidents which go beyond those which constitute the design basis, or at least alleviate the effects of postulated accidents.

In the design of reactor systems, inherent characteristics are exploited wherever possible, and combined with passive and active measures in order to arrive at a suitable concept for a particular type of reactor. In all types of reactor, in the event of a design basis accident, several passive safety barriers combine to prevent the release of radioactivity. For light water reactor the many years of experience have resulted in a well balanced combination of accident preventive measures and mitigation barriers to prevent the release of radioactivity inventory to the environment. In the event of an accident exceeding the design basis, it is ultimately the containment which ensures confinement of radioactivity.

New Safety Concepts

Not only since the Chernobyl accident has it been discussed whether new safety concepts could increase public acceptability of nuclear power and thus stabilize licensing. However, instability of nuclear power plant erection periods and the associated plant cost had only little to do with any real lack of safety features. Therefore, introduction of new and advanced safety principles cannot be expected to shorten construction and licensing times. Moreover, all advanced safety concepts known to date are trying to solve the wrong problem: new results of risk analyses for light water reactors show that not only has the frequency of hypothetical accidents been overestimated by a factor of 10 to 100, but also the effects of the resulting radioactive releases have been overstated by several orders of magnitude. If errors were made in the past, then this was because our attention was focused on serious accidents. Also, certain deficiencies in man-machine interaction may have been recognized too late. Forthcoming progress in safety engineering for light water reactors will thus consist in rounding off and consolidating the present safety concept. This may include careful reconsideration of whether present assumptions and analytical methods could be formulated less over-conservatively, whether system capacities and redundancies could be reduced, and the scope of testing and documentation diminished.

Some specific safety features

There are, in spite of the safety standard already achieved, a few improvements due to light water reactor development for the Siemens/KWU design which are of importance to nuclear safety and are thus worthy of mention.

Traditionally, the reactor pressure vessel has been designed, fabricated and tested to the highest standards in nuclear technology using forged material to a high extent. Today, forged elements are applied also to the entire primary pressure boundary. Sudden unexpected failures of primary piping can therefore be ruled out, thus eliminate pipe whip restraints and facilitating unlimited accessibility for inservice inspection. Besides accessibility, such design meets the following prerequisites set by the licensing authority:

- use of ductile materials with high-grade properties
- conservative stress limitation
- reduction of stress peaks by optimized construction
- use of advanced fabrication and test technologies
- Detection and evaluation of defects by advanced test procedures
- measures for early leakage detection

Availability of the technical equipment can serve as a yardstick for the reliability and safety of the plant. Although an international comparison of unplanned outages (Fig. 1) shows a wide spread in interruptions of operation due to reactor scrams per 1000 hours of criticality, it can be stated that an increasing availability standard has been achieved worldwide.

For low frequencies of unplanned outages, the largest single item of non availability is represented by the annual down-time for refueling and inspection (Fig. 2). Although the position of KWU-built plants is very satisfactory also for this type of non availability, generating power for additional 37 resp. 50 days, efforts are directed towards reduction of maintenance and inspection period. An example of the latter (Fig. 3) at the same time facilitating a major improvement in safety is an inservice inspection unit for the RPV, with ultrasonic heads able to cover the complete inner surface including the field welds of main coolant nozzles.

Safety must not depend on human reliability

It can be observed that the majority of accidents and incidents in the nuclear field as well as in any other technology has been caused not by breakdown of hardware but by human error. Both the Three Mile Island accident in 1979 and the Chernobyl disaster in 1986 were caused mainly by human malbehaviour and maloperation. In less significant nuclear incidents, maloperation played a role as well. Transients could have been stabilized earlier, had the personnel reacted with more promptness and intelligence. With regard to the financial losses caused just by unscheduled outages of large nuclear power plants, not to mention the possible consequences of a radioactivity release accident, the problem of minimizing maloperation and its consequences is of vital importance to nuclear power. The sceptical public, understandably enough , is asking for "error-tolerant" technologies and "forgiving" reactors.

Safe operation and effective mitigation of incidents under the influence of not absolutely failure-proof human beings is not the only aspect of human factors influencing reliability and safety of nuclear installations: the entire safety, including design, fabrication, construction and operation, must not have to depend on human reliability.

Obviously, maloperation by "pressing the wrong button" has to be, and definitely is, precluded by automatic interlocks wherever such errors may be safety-relevant. But also those actuations which are immediately necessary to bring an unscheduled transient under control and to mitigate its consequences must not be left to the control room personnel. Though normally such transients develop not quickly but over a period of minutes, the personnel need some time to analyse the situation, identify the accident, and decide upon suitable procedures. Only then can intelligent measures be taken. As a rule, therefore, no safety-relevant manual actuation must be

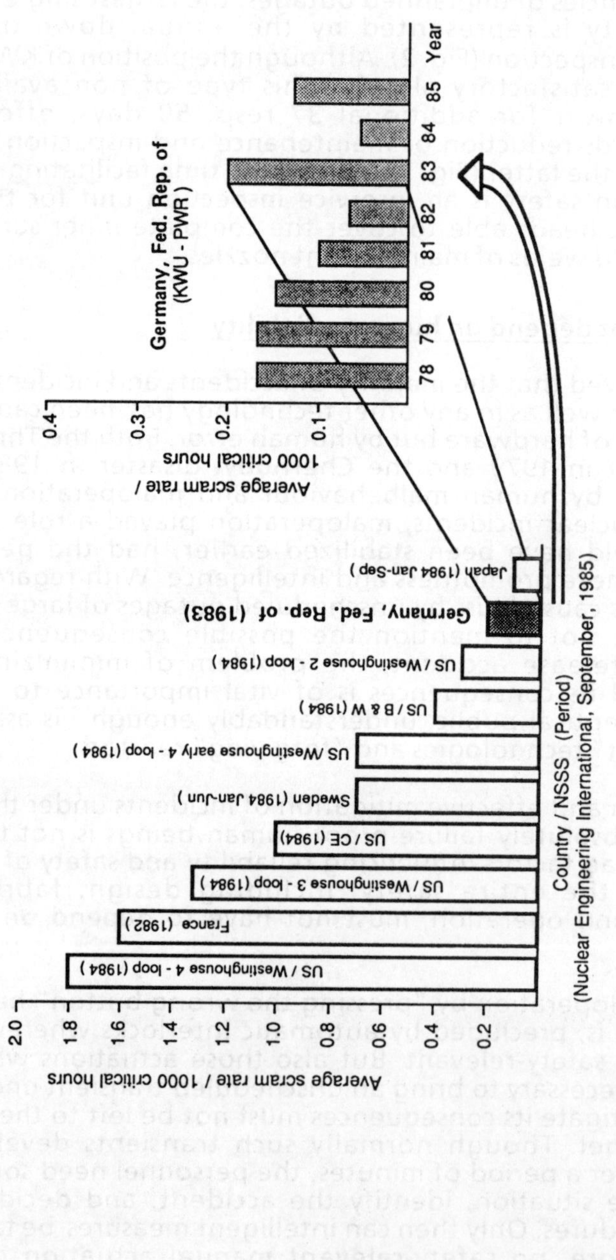

FIG. 1. Scram experience worldwide (PWR).

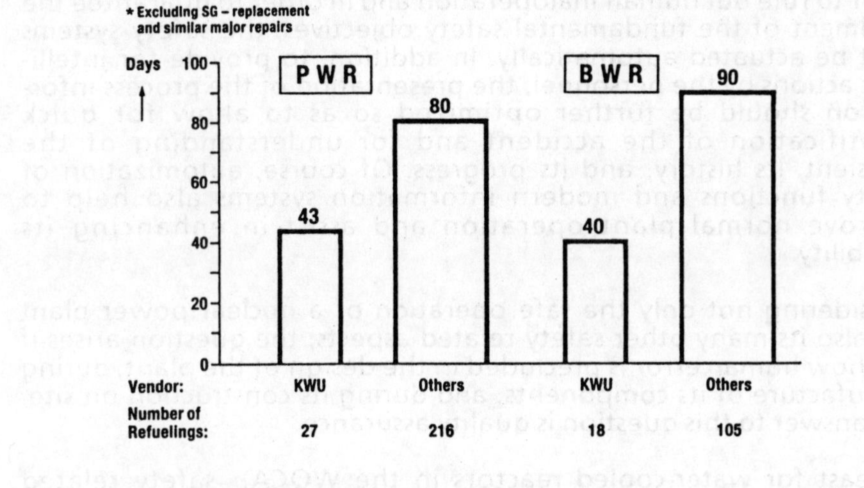

*FIG. 2. Down-time for refueling and inspection 1983–1985 (*mean values in days).*

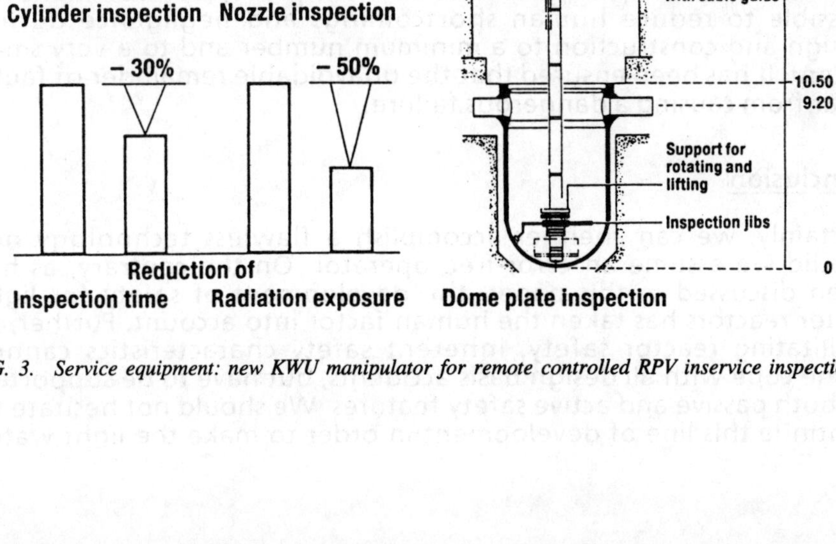

FIG. 3. Service equipment: new KWU manipulator for remote controlled RPV inservice inspection.

necessary in nuclear power plants in the Federal Republic of Germany for at least the first 30 minutes as a minimum. Generally, in order to rule out human maloperation and in order to guarantee the fulfilment of the fundamental safety objectives, the safety systems must be actuated automatically. In addition, to provide for intelligent actions by the personnel, the presentation of the process information should be further optimized so as to allow for quick identification of the accident and for understanding of the transient, its history, and its progress. Of course, automization of safety functions and modern information systems also help to improve normal plant operation and assist in enhancing its reliability.

Considering not only the safe operation of a nuclear power plant but also its many other safety related aspects, the question arises if and how human error is precluded in the design of the plant, during manufacture of its components, and during its construction on site. The answer to this question is quality assurance.

At least for water-cooled reactors in the WOCA, safety related engineering knowledge is well established and so openly exchanged that hitherto undetected basic errors and as yet unnoticed gaps in knowledge can be ruled out. The layout and the design of safety relevant equipment, systems or procedures of a nuclear power plant are not only documented and counter-checked in-house, but fully analysed or even recalculated by an independent inspector organization or the licensing authority. The same holds for specifications, manufacture and the test procedures for safety-related pressure-loaded hardware. The whole process of manufacturing is observed critically and specimens are tested by independent quality control groups and professional inspectors. In particular, the testing of all welds and the evaluation of pressure tests are carried out by up to three inspectors working independently.

This network of quality assurance and quality control makes it possible to reduce human shortcomings and negligence during design and construction to a minimum number and to a very small extent. It has been ensured that the unavoidable remainder of faults is far from causing a dangerous failure.

Conclusion

Certainly, we can neither accomplish a flawless technology nor should we assume an error-free operator. On the contrary, as has been discussed in this paper, the development of safety for light water reactors has taken the human factor into account. Further, in facilitating reactor safety, inherent safety characteristics cannot alone cope with all design basis accidents, but have to be supported by both passive and active safety features. We should not hesitate to continue this line of development in order to make the light water reactor as failure-tolerant as possible and optimize the interface with its user. We are convinced that recent nuclear power plants in the Western World are very close to these goals.

BIBLIOGRAPHY

BRAUN, W., KWU Advanced LWR Principles, Nucl. Eur. **4** (1986) 9.

BRAUN, W., Safety must not depend on human reliability, Kerntechnik **50** (1987) 85.

INHERENT SAFETY DESIGN FEATURES OF THE HIGH TEMPERATURE REACTOR MODULE

I.A. WEISBRODT
Kraftwerk Union AG,
Offenbach am Main

Interatom GmbH,
Bergisch-Gladbach

Federal Republic of Germany

Presented by K. Ebbinghaus

Abstract

INHERENT SAFETY DESIGN FEATURES OF THE HIGH TEMPERATURE REACTOR MODULE.

All over the world, the reactor catastrophe of Chernobyl has again stirred up the discussion about the acceptance of nuclear energy. New reactor concepts such as the modular high temperature reactor, which claim to have a high degree of passive or inherent safety features, are now being increasingly investigated. The modular high temperature reactor concept is characterized by the following inherent safety features: (1) SiC-coated fuel particles, which show no excessive release of fission products at temperatures below 1600°C; (2) Negative temperature coefficient; loss of helium has no effect on the reactivity; (3) Gravity-driven, diverse shutdown systems; (4) Passive decay heat removal to 3-fold redundant surface coolers (surrounding the reactor pressure vessel) or by heating up the environment and the surrounding concrete shielding if heat sink fails; (5) Limitation of peak fuel temperature to values below 1600°C even in all beyond-design accidents. Thus neither core meltdown nor excessive fission product release are possible; (6) High heat capacity of graphitic fuel balls and graphite reflector, thus very slow heat-up in design or beyond-design accidents. Long period available for taking countermeasures. (7) Direct water ingress into core prevented by lowered steam generator; (8) Limited reactivity ramp at steam ingress into core due to low heavy metal content in fuel ball; (9) Air ingress in design accidents very small, as demonstrated by experiments. No possibility of a graphite fire. Even in worst beyond design accidents, e.g. in case of sabotage, air ingress is limited due to high pressure drop in pebble bed. Peak fuel temperature stays below 1600°C. Sufficient time for taking countermeasures at low dose rates; (10) Dose rates for all design and beyond-design accidents very low. Thus access possible for taking countermeasures.

1. Technical features of the modular HTR power plant

The HTR-Module has been under development at KWU/Interatom since 1979. Its technology is based on the technology of the experimental high temperature pebble bed reactor of the AVR consortium in Jülich, which has been operating successfully since 1967,

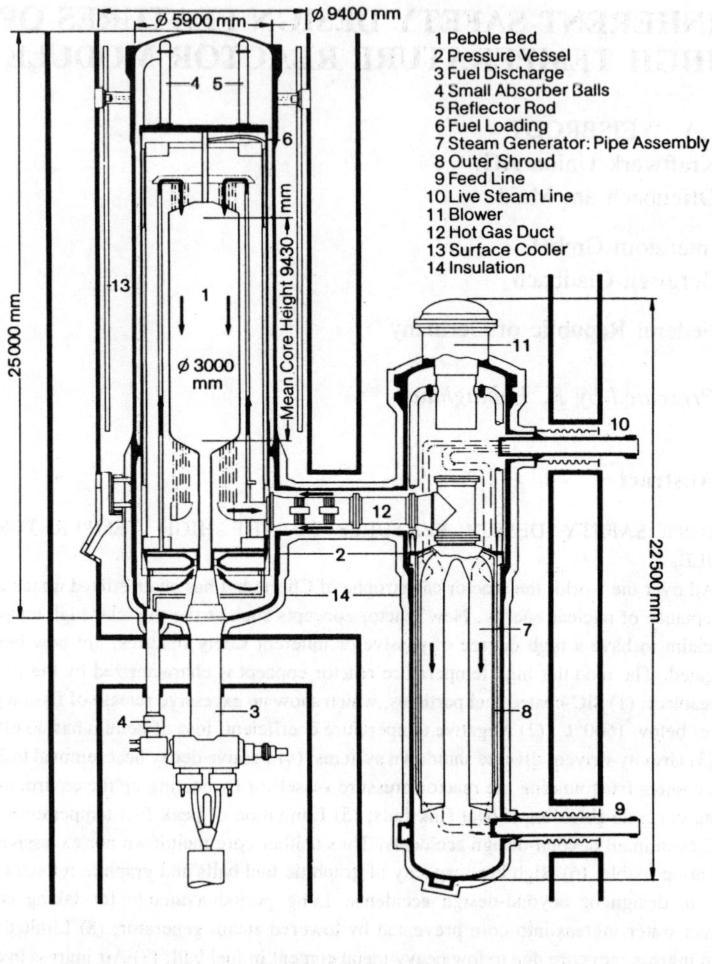

FIG. 1. Cross-section of a modular unit with steam generator (primary circuit).

and, to a very large extent, on the proven technology of KWU's light-water reactors. Moreover, HTR-Module technology has been derived from the comprehensive basic development program for high temperature-reactor technology performed by the Kernforschungsanlage Jülich (KFA) and Interatom (e.g. High Temperature Helium Component Test Loop) and from the comprehensive development program performed by KFA, Hobeg and Sigri for the fuel elements and graphite respectively.

Status

The detailed engineering for the HTR-Module has proceeded so far that the safety report for a power plant with two modules was completed by the end of 1985. In April 1986 an application was filed for a site-independent license for the conceptual design. The respective procedure is now being actively pushed forward, so that the license for the conceptual design is expected by the middle of 1988.

Features

Figure 1 shows a longitudinal section through the HTR-Module with side-by-side arrangement of reactor and steam generator. The main features are

- pebble bed core, fuel balls: 6 cm diameter
- direction of flow through the core from top to bottom
- core diameter 3 m/power density 3 MW/m^3
- core power: 200 MWth
- maximum fuel temperature, even in case of beyond-design accidents: < 1600 °C
- negative temperature coefficient
- arrangement of the gravity-driven control rods and absorber balls in reflector bores
- ferritic steel pressure vessel
- in case the operational decay heat removal systems fail completely, passive decay heat removal via 3-fold redundant surface coolers (with fire hose connections) or ultimately by heating up the concrete and the environment
- lowered steam generator
- primary system consisting of 1 reactor, 1 blower, 1 steam generator

In the light of the Chernobyl catastrophe, some of these features - like limitation of the fuel temperature, negative temperature coefficient, passive decay heat removal, gravity driven absorbers and lowered steam generator - appear to be more or less mandatory for HTR's intended for sites in highly populated areas.

A more detailed description of the technical layout has been given in references 1 to 12.

2. Safety concept especially regarding operational safety

On the one hand, the safety concept is based on the fission product rentention behaviour of the fuel elements respectively the SiC-coated fuel particles, the graphite matrix of the fuel element and the overall primary system of an He-cooled high temperature reactor. On the other hand the passive and inherent safety features of small and, in particular, small high temperature reactors are applied in the layout of the reactor in such a way that an HTR-Module with an optimum degree of passive safety features has been designed. This safety concept has been described in detail in various publications (Ref. 1, 2, 3, 4, 5, 6, 10, 12).

Subsequently special emphasis is given to the aspects of operational safety, in particular, regarding operational errors, malfunctions, disturbed operation, and hypothetical accidents or even sabotage. In the latter case, however, it is taken as being granted that countermeasures can and should be taken after approximately 1 day at the latest.

2.1 Fuel behaviour

Fig. 2 shows the fission product retention from fuel elements with UO_2-Triso particles at various temperatures and in dependence on the heating time. This figure clearly shows that there is great dependence on the temperature and that there is a steep rise in fission product release above 1600 °C. For this reason 1600 °C has been selected as the upper permissible temperature limit for the fuel temperature even for hypothetical accidents.

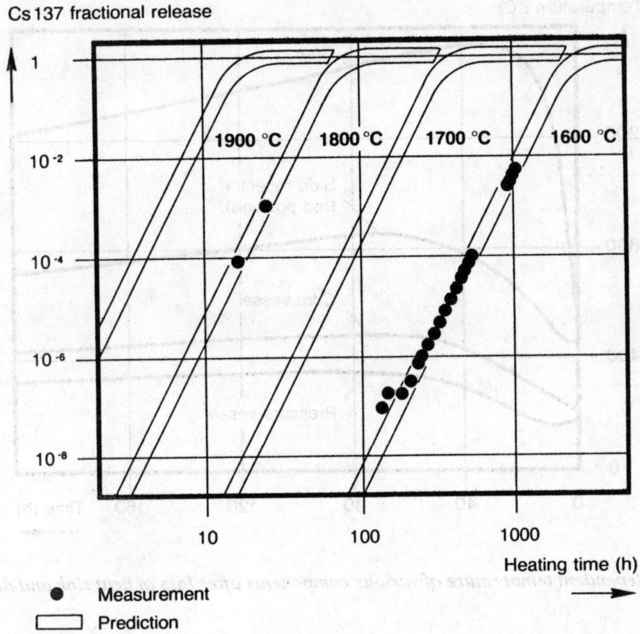

FIG. 2. ^{137}Cs release from fuel elements with TRISO particles (selected experimental results from W. Schenk, KFA; combined with model predictions, H. Nabilek, KFA).

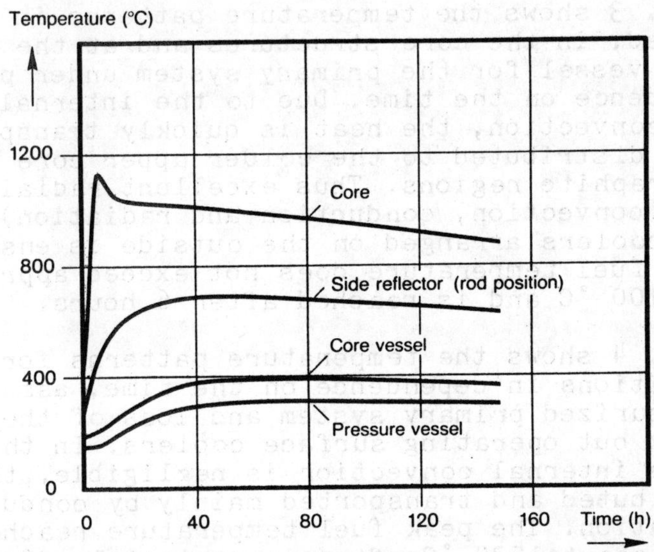

FIG. 3. Time dependent temperature of various components after loss of heat sink (primary pressure: 60 bar).

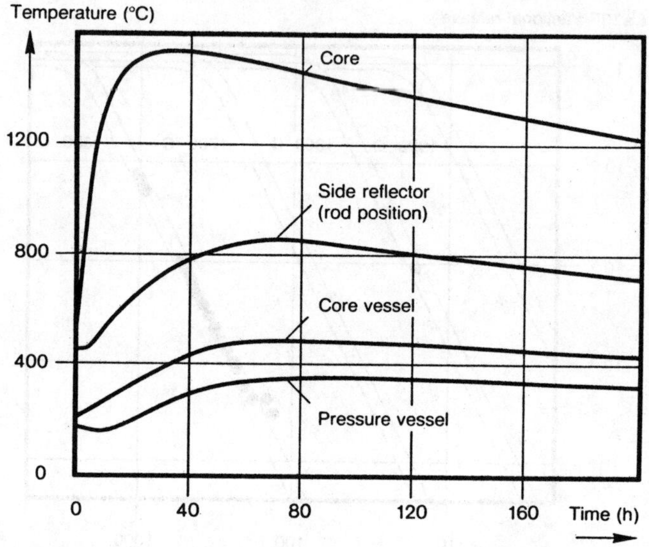

FIG. 4. Time dependent temperature of various components after loss of heat sink and depressurization accident.

2.2 Failure of operational decay heat removal

Fig. 3 shows the temperature patterns in the pebble bed, in the core structures and at the pressure vessel for the primary system under pressure in dependence on the time. Due to the internal natural convection, the heat is quickly transported and well distributed to the colder upper core and radial graphite regions. Thus excellent radial heat removal (convection, conduction and radiation) to the surface coolers arranged on the outside is ensured. The peak fuel temperature does not exceed approximately 1100 °C and is reached after 6 hours.

Fig. 4 shows the temperature patterns for the same positions in dependence on the time, assuming a depressurized primary system and loss of the main heat sink but operating surface coolers. In this case, the internal convection is negligible, the heat is distributed and transported mainly by conduction and radiation. The peak fuel temperature reaches a value of about 1530 °C after approximately 32 hours. However, the permissible temperatures for the fuel, i. e. 1600 °C and for the pressure vessel steel, i.e.

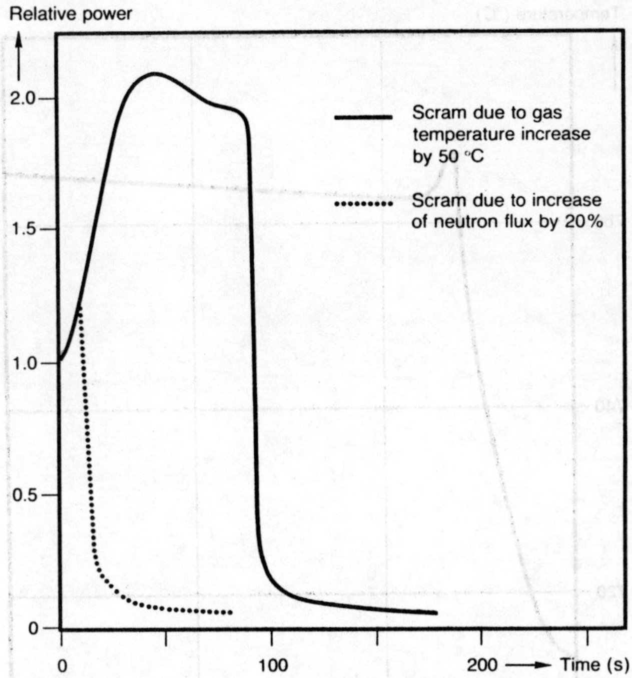

FIG. 5. Core response after withdrawal of all reflector rods with maximum speed ($\Delta\rho$ = 1.1%).

350 °C are not exceeded. The radiation dose rates remain low and the reactor can be restarted immediately after the causes for these disturbances have been eliminated.

2.3 Operational errors with the reflector rods

In the event of inadvertent withdrawal of absorber rods, the reactor scram is automatically initiated by the reactor protection system, with the objective of reinserting the rods. Core shutdown is initiated by the following signals of the reactor protection system:

- Ratio of the neutron flux ϕ/ϕ_0 larger than 1.20

- Hot gas temperature higher than 750 °C.

Figure 5 shows the power response of the reactor as a function of time at simultaneous withdrawal

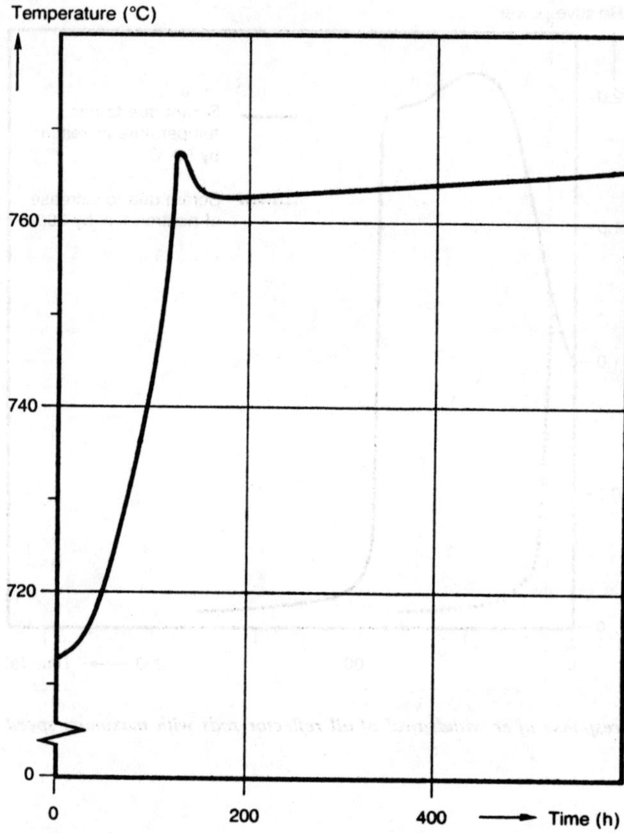

FIG. 6. Development of fuel-element temperature for withdrawal of all reflector rods for the case of "scram due to excessive hot-gas temperature".

of all 6 shutdown rods from the operating position at full power. The reactivity worth of all these rods is 1.2 % or approx. 2 $. Figure 6 shows the response of the maximum fuel temperature as a function of time for the case in which the first scram signal triggered by the neutron flux remains inactive and the reactor scram is initiated by the second scram signal due to excessive hot gas temperature.

This graph shows that only a rather small temperature increase in the core is sufficient to counterbalance the worst hypothetical operational error with the reflector rod drives.

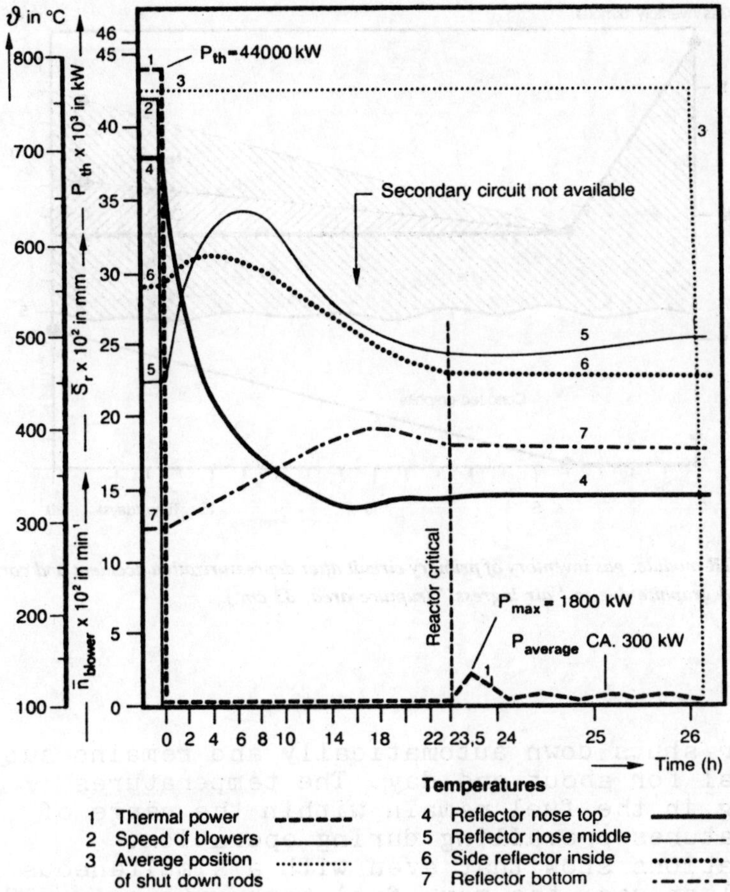

FIG. 7. Simulated failure of shutdown systems and interrupted decay heat removal for that time.

2.4 Operational test at the AVR regarding failure of the control rods

Since the HTR-Module is largely based on AVR technology, it is certainly interesting to consider the following AVR experiment (see ref. 13).

AVR operators have demonstrated by way of experiments - see Fig. 7 - that after failure of the active main heat sink during power operation, with simultaneous failure of the shutdown device, the

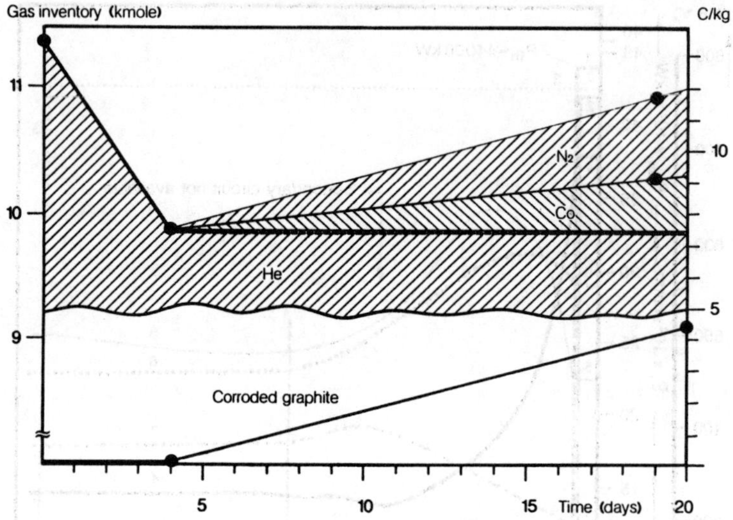

FIG. 8. HTR-module: gas inventory of primary circuit after depressurization accident and corresponding corroded graphite due to "air ingress" (rupture area: 33 cm²).

reactor shuts down automatically and remains subcritical for about one day. The temperatures arising in the fuel remain within the range of temperatures prevailing during operation. Calculations show that even with a simultaneous loss of coolant gas, the max. fuel temperatures would not reach critical values at all.

After this demonstration, the AVR operators have taken this procedure, which was originally considered to be a serious emergency condition, as one of the standard shutdown procedures.

2.5 Air ingress

Following a depressurization accident due to the ND 65 mm design leak, one of the actions to be taken is to close the leak temporarily and to cool down the core within about 10 hours by means of the steam generator. If it is assumed that this action is not possible, the average primary gas temperature rises within approx. 95 hours from approx. 210 °C to approx. 270 °C. Because of the resulting volumetric expansion of the helium, penetration of air through the leak into the primary loop can be ruled out during this period. Subsequently, however, it is possible for air to enter the primary loop by diffusion processes. However, from experiments performed specially on this subject by the Institute for Reactor Safety of the KFA - see Fig. 8 - one can conclude that an air penetration rate of less than 10 litres/h is to be expected by diffusion through the design leak. Even for long periods, therefore, the resulting fuel element corrosion is of negligible importance, since it is only in the range of grams per hour.

2.6 Beyond-design air ingress

Excessive air ingress into the primary system, resulting in a large amount of corroded/burned graphite, can only occur if natural convection through the core is created or enabled. For that to happen, one would have to postulate the simultaneous creation of holes in the lower and upper region of the confinement, and in the lower and upper region of the primary concrete cell - surrounding the module - and of holes/leaks in the lower and upper portion of the steel pressure vessel of the reactor as well as in the upper inner top shield plate. Such simultaneous events could only occur through sabotage. Interim results of our analyses - see Fig. 9 - are as follows in this respect:

Even in the worst case, it would take at least 30 hours for the graphite to become burned/corroded at the hottest spot to such an extent that the first coated fuel particles would be liberated. This is because the maximum air flow is limited by the pressure drop in the pebble bed to 0.6 kg/s. (A double-ended break of the connecting vessel would only result in 0.1 kg air/s). Here, too, the fuel temperature would remain below 1600 °C, so that there would be no excessive release of fission products.

The patterns of maximum fuel temperature in the core for different air flow rates through the core are shown in Figure 9.

Since no inadmissible fission product release occurs, the radiation level outside the primary cell remains so low - and far below dangerous levels - that access is granted to the personnel to perform adequate countermeasures such as closure of the holes

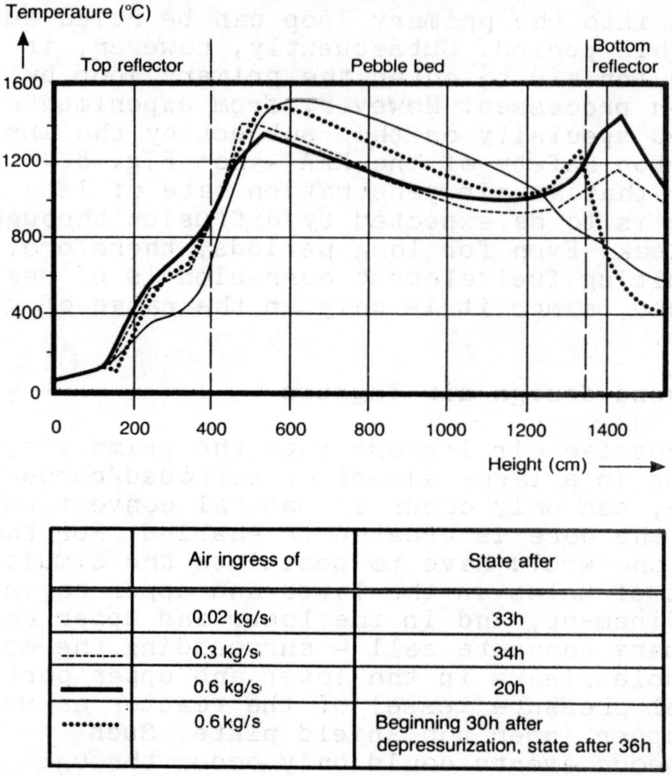

FIG. 9. Maximum temperatures due to graphite corrosion after hypothetical air ingress accidents.

in the confinement, inertization of the confinement, connecting the surface coolers to fire hoses, filling with water etc.

2.7 Water ingress

Water ingress into the primary circuit can be caused by a leaking heat exchanger tube. The complete double-ended break of one heat exchanger tube is detected after approx. 10 s by moisture measurement in the primary loop. Within this time interval, 200 kg of water flow into the primary system. After detection of the accident, the steam generator is isolated by quick-closing valves in the feedwater and main steam line.

The subsequent process of dumping the steam generator via the feedwater line is ended after a primary loop pressure of 60 bar has been reached. The time up to equalization of the pressure was computed to be 15 s. During the drainage process, an average leak rate of 10 kg/s can be estimated because of the pressure drop in the steam generator, and so a total of approx. 350 kg of water penetrates into the primary system. At the end of the emptying procedure, approx. 250 kg of water still remain in the steam generator, therefore, approx. 600 kg enter the primary system in the most unfavourable case.

Two auxiliary systems which can immediately cope with a water ingress accident are installed in the reactor plant. One auxiliary system is installed in the stand-by gas purification system and is capable of removing the moisture from the primary helium by merely passing the helium content of the primary circuit through a water-cooled condenser with a feed-rate of one He-fill per hour. The second auxiliary system is the helium pressurizer of the primary circuit which is part of each module's allocated gas purification.

In relation to the large quantities of graphite in the fuel elements and the reflector, the amount of corroded graphite is negligible, even if the total quantity of 600 kg steam were to react. There are no hazardous consequences on the radiology and there is no formation of explosive gas mixtures, even assuming total reaction of the 600 kg of steam and a subsequent break of a connecting pipe at the primary system with a release into the confinement air. In other words, the released gas mixture $He/CO/H_2$ stays below the ignition limit.

The low-probability event that none of these systems works or that an even larger amount of water is ingressing has been described in detail in ref. 12.

3. Summary

All the analyses of the dynamic and operational behaviour, of the stability of the primary system as well as of operational errors or malfunctions, of design and beyond-design accidents have shown that the HTR-Module is easy to operate and to control. It has only a small number of control and safety signals. The heat capacity of the fuel balls in the core and of the graphite structures is high (about 220 t of graphite per module). Therefore, the time constants for the control or the inertia of the system are large. The primary system is extraordinarily stable regarding the dynamics. In case of operational errors, malfunctions or disturbances, the heating-up time of the core until critical fuel temperatures are reached is very slow, i. e. in the order of one day. The number and the scope of required systems or auxiliaries, i. e. emergency power supply, cooling chains, etc., are small. Even in case of the worst accidents or sabotage, no excessive fission product release occurs and, therefore, the dose rates remain so low that countermeasures of an adequate nature can be taken without endangering the personnel or the environment. There is a large time gap in the order of 1 day before countermeasures should be taken. For some cases, i. e. failure of the cooling water supply for the decay heat removal surface coolers, additional connections have already been provided in order to take such countermeasures.

Therefore, this reactor can be considered as a real Chernobyl-proof "walk-away-reactor".

REFERENCES

[1] SCHULTEN, R., "High temperature gas-cooled reactors", Technical committee meeting, IAEA, Jülich (Oct. 1986)

[2] KRÖGER, W., "Safety characteristics of current HTR", Technical committee meeting, IAEA, Jülich (Oct. 1986)

[3] REUTLER, H., LOHNERT, G.H., "Advantages of going modular in HTRs", Nuclear Engineering and Design, Vol. 78, No. 2 (April 1984)

[4] HÜBEL, H.J., LOHNERT, G.H., "The safety related properties of future HTR-Module power plants", Atomkernenergie-Kerntechnik, Vol. 47, No. 3 (Nov. 1985)

[5] FREWER, H., KELLER, W., "The modular high-temperature reactor", Nuclear Science and Engineering, 90, (1985)

[6] REUTLER, H., "Konzept des modularen HTR in Zweibehälterbauweise", Atomwirtschaft (Aug./Sept. 1985)

[7] JÄGER, W., STEINWARZ, W., "HTR-Modul-Reaktor", Chemieanlagen + Verfahren (Dec. 1984)

[8] ANDLER, M., JÄGER, W., "Anwendung und Aussichten des HTR-Moduls", Atomwirtschaft (Aug./Sept. 1985)

[9] WEISBRODT, I., "Von der Kommune bis zur Dritten Welt", Energiewirtschaftliche Tagesfragen, 36 Jg., Vol. 3 (1986)

[10] SCHMIDT, P., LOHNERT, G., "Das Schutzkonzept des HTR-Moduls gegen Einwirkungen von außen", Jahrestagung Kerntechnik (1986), Proceedings ISSN 0720-9207

[11] WEISBRODT, I., STEINWARZ, W., KLEIN, W., "Status of HTR-Module plant design", Technical Committee Meeting, IAEA, Jülich (1986)

[12] WEISBRODT, I., "Survey on the design status of Interatom/KWU's HTR-Module", San Diego (Sept. 1986)

[13] ZIERMANN, E., AVR Personal Communication

THE DEVELOPMENT OF SAFE ON-LOAD REFUELLING PRACTICES IN ADVANCED GAS COOLED REACTORS

I. GLENDENNING, J. THORN
Generation Development and Construction Division,
Barnwood, Gloucester

P.M. BOOCOCK
Generation Design and Construction Department,
South of Scotland Electricity Generating Board,
Glasgow

G.M. EMSLEY
Mechanical Systems Department,
National Nuclear Corporation Ltd,
Knutsford, Cheshire

United Kingdom

Abstract

THE DEVELOPMENT OF SAFE ON-LOAD REFUELLING PRACTICES IN ADVANCED GAS-COOLED REACTORS.

The advanced gas cooled reactor (AGR) was designed for on-load refuelling at close to full reactor power. Constraints identified during the design and safety assessment process have resulted in all operational AGRs being limited to off-load batch refuelling in early operations. The economic penalties of this operating mode have led to a major development programme aimed at better understanding all aspects of the refuelling process to enable on-load refuelling to be re-established while, at the same time, demonstrating improved safety margins in line with recent trends. The design of the AGR fuel assembly and fuel route is briefly described together with the fundamental safety objectives for handling fuel. The principal hazards are identified as fuel sleeve damage during handling, fuel tie bar failure causing a dropped fuel stringer and, ultimately, the possibility of dropping a complete fuel assembly, made worse in each case if either air can be present in the coolant or if the containment boundary is simultaneously breached. The development of safety arguments to allow on-load refuelling is described for each of these hazards with the aim of showing a low risk of occurrence and limited consequences for sleeve damage and dropped fuel stringer events, and such a remote risk of occurrence for more extreme faults that the more significant potential consequences can be tolerated. Particular emphasis is placed on design changes and operating procedures which have been introduced to improve safety margins and the further improvements which are presently under development, including those with potential for limiting the consequences of some very unlikely faults. With the safety improvements achieved and in prospect it is concluded that high load refuelling is still a realisable economic goal for the AGR without compromising modern safety standards.

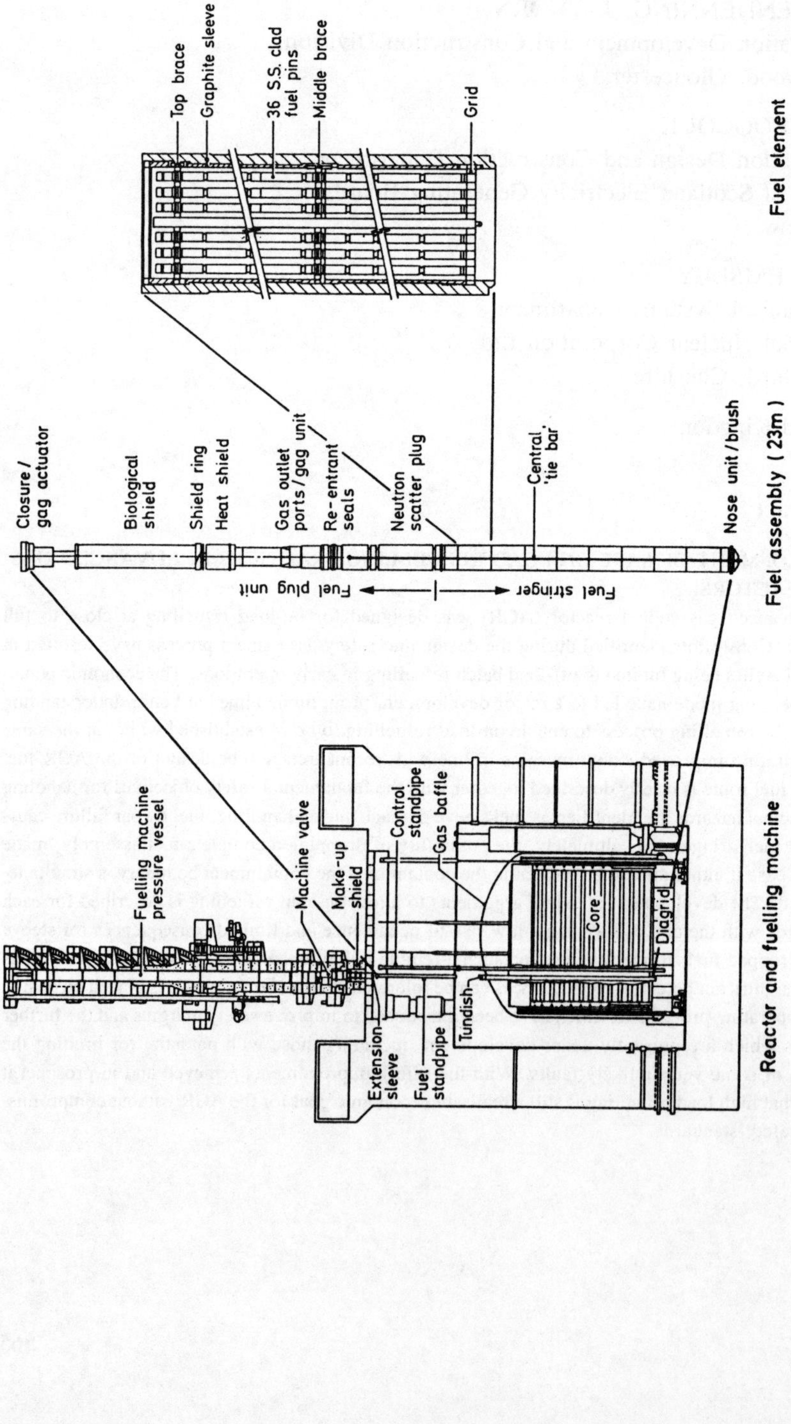

FIG. 1. AGR fuel assembly and reactor handling.

1. INTRODUCTION

The Central Electricity Generating Board (CEGB) and South of Scotland Electricity Generating Board (SSEB) own and operate five Advanced Gas Cooled Reactors (AGRs) in the United Kingdom. Two new AGRs are at an advanced stage of commissioning at Heysham and Torness. All were designed to be refuelled at or near full power in order to maximise reactor availability.

A number of constraints have been identified in the design, development and Safety Assessment processes which prevent the AGR from achieving high power refuelling. Indeed all the operational AGRs have initially been restricted to off-load refuelling until revised safety arguments and some modifications to plant and procedures have permitted on-load refuelling at substantially reduced power. The two new AGRs are expected to be able to refuel on-load from early in their operational lives.

Refuelling off-load or at reduced power adds substantially to the number of thermal cycles experienced by the reactor and associated plant and is potentially life limiting. The loss of generation during refuelling also incurs a very large financial penalty. Both are minimised by batch refuelling up to eight channels in a single shutdown or load reduction and, where possible, increasing reactor output in between fuel handling operations within a batch. Nevertheless there is a very big incentive to achieve high load refuelling at the highest possible reactor power.

This paper described the AGR refuelling system and the safety principles applied in its assessment. The key design features and principal hazards are outlined followed by a description of the main assessment findings and the changes which have been made or are in prospect to achieve improved safety standards whilst also permitting the desired improvement in on-load refuelling power.

2. THE AGR REFUELLING SYSTEM

2.1 General Description

An AGR core has upwards of 308 fuel channels each containing a 23m long 'fuel assembly' comprising a reusable 'plug unit' and a 'fuel stringer' (Fig. 1). This assembly is normally locked and sealed in at the top of its own access standpipe. The reactor is refuelled by replacing each complete assembly after a 4-6 year dwell, at a uniform rate of about 50 assemblies per year.

All fuel assembly handling on an AGR power station is carried out by a single fuelling machine. The machine contains a high integrity hoist and grab and provides all the shielding and pressure and activity containment necessary for safely lifting and transporting irradiated fuel assemblies through the fuel route (Fig. 2).

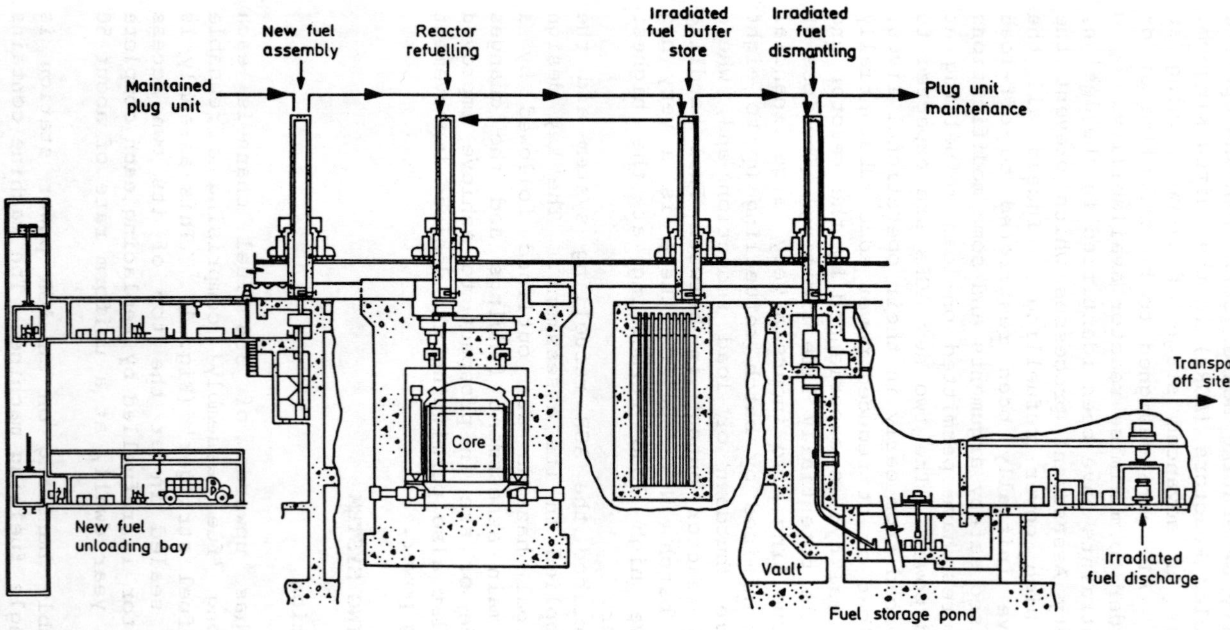

Facility	Reactor				Buffer store	IFD	
						Air	CO_2
Refuelling regime	On-load	Off-load pressurised	Off-load depressurised	Air filled	All	All	All
Power limit	30-50%	30-50kW	10-15kW	Not permitted	30-50kW	~16kW	~20kW
Source of limit	Sleeves / dropped stringer trip	Cooling in infrequent faults	Stringer cooling	UO_2 Oxidation in dropped stringer event	Dropped stringer (pessimistic)	Dropped stringer (with operator action)	Dropped stringer debris

FIG. 2. AGR refuelling system and typical limits.

2.2. Safety Principles

Refuelling safety principles are consistent with the overall safety objectives for operating nuclear plant, to prevent harm to the public and to minimise the exposure of station operators to radiation or other hazards. Plant design and operating procedures are arranged to achieve the lowest practical levels of risk/exposure and safety assessments are required to demonstrate compliance with agreed maximum permissible frequency targets which reduce with increasing hazard.

In practice the principles as applied to AGR fuel handling both at the reactor and at the other fuel route facilities may be interpreted simply as:

i) Measures should be taken to ensure that damage to fuel causing the release of fission products within the facility containment is a remote event.

ii) Should such a release of activity occur, the leakage from the containment should not produce an unacceptable hazard.

These fundamental principles were applied to AGR fuel handling when the systems were originally designed but their detailed interpretation has developed considerably in the intervening years. There is now a greater emphasis placed upon the systematic numerical analysis of fault frequencies and consequences and a shift towards safety cases based upon rigorous Probabilistic Risk Assessments.

This change has greatly increased the complexity and extent of the analysis required to prepare a safety case, often highlighting deficiencies in knowledge and methods in relation to the required level of detail. The result has been an identified need for enhanced protection in a number of areas and the imposition of additional constraints on refuelling although some of these are intended to be relaxed once the necessary development work has been completed.

3. PRINCIPAL HAZARDS

The more obvious requirements to provide adequate shielding, the avoidance of criticality hazards and the proper control of contamination in normal handling operations, had all been fully addressed at the design stage. Potential for a major hazard only arises following damage to fuel.

The AGR fuel element comprises a cluster of 36 fuel pins supported in three concentric rings by a 'grid' and two 'braces' within a 1.0m long x 0.24m diameter graphite sleeve (Fig. 1). Each pin is a sealed stainless steel canister containing a stack of sintered hollow pellets of enriched uranium dioxide. A fuel stringer has eight elements (only seven at Dungeness 'B') with the graphite sleeves spigotted together to form a continuous channel for the CO_2 coolant. The coolant enters this channel at

300°C and leaves at up to 675°C. During handling, the stringer is supported from the plug unit by a central 10mm diameter Nimonic (PE16) 'tie bar'.

Fuel damage can result from fuel handling operations in a number of ways as follows:

3.1. Fuel Sleeve Damage

Fuel sleeve damage can have several consequences. The risk of snagging and tie bar overloading in subsequent fuel movements is greatly increased; fuel below the damage site will be less effectively cooled leading either to early fuel failure or an increased risk of multiple pin failure in unrelated reactor faults such as reactivity or depressurisation events; abnormal stringer cooling during subsequent fuel handling may result in excessive tie bar temperatures and an increased risk of failure.

During early operations at Hinkley Point 'B', a fuel sleeve disrupted during on-load charging of channel 4K05 at 80% power. The loading and subsequent operations had appeared normal until the fault was revealed by the burst can detection system. Similar effects can also result from a gap in the stringer which could occur in the event of a snag on the stringer or lower plug unit late in insertion.

3.2 Dropped Fuel Stringer

Failure of the PE16 tie bar will lead to the dropping of all eight fuel elements, and could occur anywhere within the fuel route with a drop height of up to 30m. The result would be severe damage to the sleeves and pins of the fuel elements at the bottom of the stringer and, where flow blockage also occurs, excessive fuel and clad temperatures within the debris. This would cause a substantial release of fission products within the facility where the drop occurred particularly if heated fuel can be exposed to an air atmosphere. The key design requirement is that a dropped stringer shall not cause a loss of containment. Consideration is also given to the possibility that a circuit breach may cause or coincide with a dropped stringer for which case it is necessary to demonstrate an extremely low frequency unless the damage and activity release are very small.

3.3. Dropped Fuel Assembly

Failure of a plug unit component, the grab or the fuelling machine hoist or chain, would allow up to the complete fuel assembly to be dropped. In this case not only would fuel damage be more extensive than for a dropped fuel stringer but there is a possibility of impact damage to the standpipe and a simultaneous loss of containment. This is potentially the most serious hazard and the design intent has been to ensure that such an event cannot occur.

4. PROTECTION PROVISIONS

In this section, the basic safety arguments are outlined for each of the main hazards, paying particular attention to areas where improvements to overall safety have been achieved or where more detailed assessment make possible the demonstration of increased safety margins.

4.1. Fuel Sleeve Integrity

The investigations which followed the 4K05 incident showed that graphite sleeves are subjected to two loading mechanisms during on-load handling. As each element crosses the above dome region it is subjected to an internal pressure approaching the core pressure drop. This was the cause of the 4K05 failure. Fuel sleeves are also subjected to lateral impacting caused by channel flow induced vibrations of the suspended asssembly. Both problems have been tackled successfully.

Failure resulting from internal pressurisation alone only occurs with significantly defective sleeves. Such defects are not apparent from visual inspection or available non-destructive testing techniques so, since early in 1980, all fuel elements have been subjected to a pneumatic overpressure test to 6.35bar. Fuel which passes this test has a large margin to pressure burst failure under any refuelling conditions. Handling of non-pressure tested fuel already in Hinkley Point 'B' and Hunterston 'B' was restricted to a low core differential pressure and has all been safely discharged. The risk of impact damage is also dependent upon reactor power during fuel handling, vibration levels and impact velocities increasing with core differential pressure. The present design of AGR fuel (Stage 1) which has a double graphite sleeve, with a gas gap to minimise heat leakage, has been found to be unsuitable for handling at full power. Assessed against criteria of $<10^{-7}$ failures/charge and $<10^{-5}$ failures/discharge, where 'failure' is the extension of the largest defect not revealed by the pressure test, it has been shown that Stage 1 fuel can be safely handled at 40-60% reactor power depending on the vibrational behaviour in the particular AGR as established in full scale experiments and confirmed by reactor measurements.

For future use a replacement has been developed, known as Stage 2 fuel. It differs from the Stage 1 design in that it has a single thicker walled graphite sleeve which is significantly more resistant to impact damage and pressure bursting. It also has a machined rib profile on the sleeve outer surface which is very effective in stabilising the flow induced vibrations which occur in charge-discharge. This fuel will also be pressure tested, but to a significantly higher pressure, and has already been shown to be capable of full power refuelling in all the AGR designs with large margins to the target failure criteria. The initial fuel charges of Heysham 2 and Torness are of the Stage 2 design and it will be introduced as feed fuel in other AGRs starting later this year improving both the potential refuelling power level and safety margins. Stage 2 fuel will also be designed for increased discharge irradiation which brings the additional benefit of reducing the overall fuel handling requirement, itself an aid to safety.

4.2 Stringer Snags and Gapping

In normal operation, the fuel tie bar is not tensioned and the fuel stack is held in place under the dead weight of the plug unit below a telescopic joint, enhanced in the case of the Hartlepool/Heysham I design by a powerful anti-gapping spring. If the stringer or this lower part of the plug unit should snag very close to full insertion this may be mistaken for the completion of a normal loading operation. In this event a small gap can be produced in the stringer and lead to overheating of fuel. Considerable effort has been put into the development and installation of new fuelling machine interlocks and revised operating procedures to maximise the probability of such a snag being revealed at the time of fuel loading. Further protection is provided by making careful checks on the performance parameters for each newly fuelled channel looking in particular for anomolous channel gas outlet temperatures/gag settings or assessed channel power.

4.3 Dropped Fuel Stringer

Failure of the fuel tie bar has always been regarded as a credible but low frequency fault. The safety aims, therefore, are to ensure that the risk of failure is understood and kept to the lowest practical and economic level and to take such additional measures as are necessary to achieve an acceptably low radiological hazard arising from the fault.

4.3.1 Tie Bar Failure

The duty of the tie bar is unusual in that after fuel assembly charging, it rests unloaded in guide tubes through the centre of the stringer suffering vibration and wear and the effects of irradiation and thermal ageing for a period of 4-6 years. In this process the tie bar material undergoes complex changes in properties which are not uniform along the bar. It is then required to support the weight of the fuel stringer and lower plug unit during discharge from the reactor and in subsequent fuel movements through the disposal route.

The tie bar material, Nimonic alloy (PE16) was selected for its good strength at elevated temperatures. It is sized to provide large factors of safety to normal handling loads and a factor of at least 2 over the worst snag load after taking account of stress raising features such as wear scars. The main quantifiable cause of possible failure is high temperature creep. The assurance of high tie bar integrity comes from measures to limit the temperature, load and time at risk in all normal and faulted fuel handling sequences.

The tie bar duty is sensitive to refuelling power level but is always well controlled in normal handling. More onerous conditions arise in stopped hoist situations, particularly where a snag is involved. Improvements in the form of reliable cooling supplies from the fuelling machine, including stop initiated emergency supplies where necessary, and the provision of high reliability hoist load sensing, tripping and braking systems have minimised these hazards.

The risk of tie bar failure is quantified by employing a sophisticated stress rupture model for irradiated PE16 developed by CEGB's Berkeley Nuclear Laboratories. The model is based on theory and fitted to a large body of data obtained in tests on irradiated tie bar samples under representative and extreme conditions of load, temperature and wear defects and can predict the failure probability for any normal or fault combination of these parameters which are separately derived.

The target risk adopted is $<5 \times 10^{-4}$ failures/reactor year for cases with an intact pressure boundary, selected to be below the cut off applied to take account of uncertainties arising from dependence on a single component. At present tie bar reliability is not limiting on refuelling power and strategies have been identified which would permit on-load discharge at 60-70% power.

This work has demonstrated that continuous improvements made to quality control and the manufacturing processes involved in tie bar production have achieved a more uniform product which consistently displays a reduced failure risk in Post Irradiation Examination (PIE) testing. This improvement is being monitored to optimise the material specification. An alternative dual tie bar design is also under development. These measures are intended to further reduce limits on refuelling power and the overall risk of failure. The possible use of a dual tie bar, in particular, will greatly increase confidence in tie bar reliability in fuel route facilities where temperature effects are not significant and the quantifiable failure risk is already small.

4.3.2 Dropped Stringer Consequences at the Reactor

In the absence of direct experimental evidence, it was initially argued that a stringer dropped at the reactor might result in some local fuel melting but that this would not hazard the reactor pressure boundary. This concept was soon abandoned and a criterion of no fuel melting was adopted. In the assumed absence of coolant flow and with pessimistically defined fuel and graphite debris this criterion was satisfied only if the reactor was tripped immediately after the dropped stringer event. No such protection had been included in the original AGR design so a new automatic dropped stringer reactor trip (DSRT) was introduced.

A number of systems were investigated but the only one which was both dependable and could be engineered on a short time-scale was based on sensing the load on the fuelling machine grab and initiating a trip if this reduced by an amount equivalent to the loss of the fuel stringer. The measured fuel assembly weight is reduced in the lower 15m of lift by gas forces. These increase with power and impose a refuelling power limit of 40-50% reactor power beyond which the risk of spurious tripping becomes unacceptable. Weight sensing based DSRT systems are being introduced in all AGRs, the system at Hinkley Point 'B' and Hunterston 'B' having been installed before on-load refuelling was restarted in 1982. The single system has been judged adequate in view of the low frequency of tie bar failure but a second, acoustic based system has been installed at Heysham 2 and Torness and systems sensing a second trip parameter are being actively developed for backfitting to the other AGRs.

A number of full size, full height stringer drop tests have now provided a clearer picture of likely fuel debris geometries. These tests use unirradiated fuel but tests are also being conducted on irradiated materials to enable a reliable prediction of irradiated stringer drop damage to be made. The debris is found to be open and significant damage limited to the lower 3 or 4 elements. The key criterion now applied to dropped fuel debris assessments is that it should be stable throughout the post drop period to avoid any risk of fault escalation. The resultant peak debris temperature criterion of 1200°C or below is taken to avoid significant oxidation and local creep collapse of cladding. In the absence of coolant flow through the debris, these criteria can only be achieved if reactor power is kept low before and during refuelling so that post drop decay power levels are kept low.

In a tripped reactor only very small quantities of flow are required to substantially relax the refuelling power constraint. The likelihood of forced convection flow is strongly influenced by the design of stringer bottom components below the first fuel element. With steel components substantial flow is assured, but with graphite features, flow is less reliable and generally not claimed. In future all stringer bottom components will be designed to give an assurance of flow and may incorporate a shock absorber to further limit damage.

Even where forced convection cooling is not available, natural convection in pressurised CO_2 can be very beneficial and major developments are in hand to model and experimentally verify these benefits.

The effect of tighter debris temperature limits is a marked reduction in the release of fission products to the reactor circuit and assessments of the radiological consequences of a dropped stringer at the reactor show that the external hazard is a small proportion of that permitted and is being further reduced.

To refuel at greater than 50% power either of two things must be achieved. Alternative dropped stringer reactor trip systems must be developed for use at high power or it must be shown that in high load refuelling a prompt reactor trip is not necessary. Progress with both is encouraging. The gas forces which effect grab load also greatly reduce impact velocities in the event of a dropped stringer. Four full scale drop tests of Stage 2 fuel stringers simulating conditions for 50% power refuelling and above have caused very little fuel damage. Heat transfer models of the observed geometries and flow suggest that even lower fuel and clad temperatures will result before and after a reactor trip in this situation than in a low power drop. This work is continuing.

4.3.3 Dropped Stringer at Fuel Route Facilities

The channel diameter in fuel route facilities is larger than at the reactor so more severe fuel damage with element overlapping is predicted. Temperature assessments which ignore coolant flow lead to severe restrictions on stringer decay power which can be operationally inconvenient.

Typically a 30-50 kW limit is currently derived for handling at a buffer store and this precludes direct transfer of irradiated fuel from the reactor during high power refuelling. In low power batch refuelling this constraint can be accommodated by rescheduling channels within a batch and at worst by imposing a few hours decay time in the fuelling machine prior to loading into the buffer store.

However, work in hand is demonstrating that natural convection cooling is extremely effective in a pressurised buffer storage tube and there is every prospect that a decay power limit of >70kW can be established with margins. This would not be operationally limiting.

Irradiated Fuel Disposal Facility cells with CO_2 cooling can readily accept stringers of around 20kW which will rarely restrict fuel movements from buffer storage. Where the cell is air cooled, a much tighter limit of around 250°C has to be applied to avoid UO_2 oxidation and a greatly increased radiological hazard. However, by providing CO_2 flooding for the cell and boronated water flooding of the debris this risk is eliminated and similar ratings can be justified. At the present level of about 16kW, the rate of rise of temperature in the debris is sufficiently slow to permit the water flooding provision to be under operator control. Clearly, no such measure is possible at the reactor and, for this reason, refuelling in an air filled reactor is not permitted.

4.4 **Dropped Fuel Assembly**

4.4.1 Fuel Assembly Lifting Integrity

Fuel assembly lifting integrity arguments are based on demonstrating that the design and protection provisions are of such a high standard that a dropped fuel assembly event need not be considered as a design basis fault i.e. the emphasis is on prevention. Where such high reliabilities are required it is normally necessary to have diverse protection systems, and additional hardware has been added to achieve this objective.

Potential causes of a dropped assembly fall into three main categories:

a) Excessive loads applied to the lifting system as a result of snag events.

b) Excessive loads resulting from a free drop of the assembly and snatch loading of the lifting system.

c) Component failure as a result of wrong assembly or wear or deterioration in service from another cause.

The excessive loads, a), will normally only arise if either the stringer or the plug unit become snagged during withdrawal or if the hoist should overrun at the top of its travel. There is protection in depth against such faults at each AGR. Typically, two diverse load sensing systems rapidly detect an increase over the normal assembly weight, trip the hoist, and by the application of brakes and shutting off of motor supplies, bring the assembly quickly to rest. Further limitation of the forces which could be generated in such a fault is provided by slipping clutch drives and limits on the hoist motor torque capability.

The diverse load sensing provisions also detect hoist underload which would result from a snag during insertion, again stopping the hoist sufficiently rapidly that, at worst, only a very small amount of slack chain can be generated before the hoist is brought to rest. This fault is more onerous than overloads since it is assumed that the snag could suddenly clear and the assembly could drop freely until the slack is taken up producing a high snatch load, b). This fault has been the subject of dynamic analysis and experiment and the safety margins in all parts of the hoist system and plug unit are assessed against this fault in particular. Large factors of safety are demonstrated through the lifting system which, wherever possible, is arranged so that the load is shared between a number of components any one of which is capable of supporting the maximum fault load. Elsewhere, as in the case of the grab body and the chains at Hartlepool/Heysham I and Dungeness B, both very large factors of safety and in service inspection provide the assurance of high integrity.

A more severe snatch load fault could arise if leakage from the fuelling machine caused levitation of the fuel assembly which could then drop through a large distance and almost certainly fail the chain. This fault is prevented by ensuring that any credible leak from the fuelling machine is limited to below the level at which levitation could occur. To achieve these low credible leak rates additional analysis and inspection of fuelling machine pressure vessel and pipework has been necessary together with modifications to include flow limiting orifices and valves which are either interlocked closed or close rapidly in the event of a breach. Coincident faults, such as reactor pressure transient resulting from a boiler tube failure, which could possibly cause assembly levitation have also been considered and shown not to threaten lifting system integrity.

The third category, wrong assembly or in service deterioration, is protected against by a very wide range of provisions. These include tight administrative control over plug unit maintenance, provisions for in service inspection or replacement of, in particular, hoist, chain and grab components and careful attention to off reactor operations, to ensure that no damaging loads can be generated nor deterioration go undetected.

These considerations of dropped assembly causes apply to fuel assembly handling at any pressurised facility and are not made significantly more onerous by on-load refuelling. Protection systems provided to detect a dropped stringer will initiate a prompt reactor trip in the event of an assembly drop during on-load refuelling at the reactor, regardless of refuelling power.

The very comprehensive and detailed nature of these assessments and the large margins which have been demonstrated give full confidence that a dropped assembly need not be considered as a design basis fault. However, reviews will be carried out throughout station life to confirm that this position can be sustained or, if necessary, to define requirements for revising any part of the design or operating practices.

4.4.2 Dropped Assembly Consequences

Notwithstanding this confidence in fuel assembly lifting integrity, the Boards are investigating the potential consequences of hypothetical drops of a fuel assembly whilst handling at reactor, decay store or Irradiated Fuel Dismantling Cell.

These studies are a mixture of generic and station specific investigations covering fuel damage, containment impact damage and radiological release assessments with support from drop and impact testing. The expectation is that only small breaches will result from dropped assembly impacts and radiological consequences will be limited to levels acceptable for an event of such low frequency. Consideration will, however, be given to any practical measure to further improve this position.

5. THE PRESENT POSITION AND PROSPECTS

On-load refuelling has been carried out successfully at Hinkley Point 'B' and Hunterston 'B' at 30% power with cycling to 60-70% power between fuel exchange operations in a 6-8 stringer batch. All the safety modifications to permit this had been completed by 1982 and have worked well.

The more recently completed AGRs at Hartlepool, Heysham 1 and Dungeness 'B' are currently restricted to off-load refuelling at a pressurised or depressurised CO_2 filled reactor. The modifications and safety assessments to permit on-load refuelling operation equivalent to those at Hinkley Point 'B' will have been completed within the next two years.

Developments, particularly in the areas of dropped stringer risk and consequences limitation, including additional reactor trip parameters will allow increased refuelling power but only to the limit imposed by the integrity of the present fuel design.

Heysham 2 and Torness will be the first stations to benefit from these developments since they already have Stage 2 fuel as standard. Stage 2 fuel will be introduced in the other AGRs, but the duration of each fuel cycle is such that improvements in refuelling power, possibly to the 70% target level could not be contemplated before 1993-1994. This achievement will be heavily dependent upon the successful completion of the arguments on limited dropped stringer damage and/or additional reactor trips.

On-load refuelling, even perhaps at high power, remains a realisable goal in AGRs and the continuing development programme involving the nuclear industry throughout the UK is ensuring that it is achieved whilst also improving the overall level of understanding and safety of the refuelling process.

L'APPROCHE PAR ETAT: CONCEPTION DE LA SECONDE GENERATION DE PROCEDURES POST-ACCIDENTELLES

J. MESNAGE, E. OLIVIER
Framatome,
Paris – La Défense

J.-P. BERGER, R. SOLDERMANN
Service Etudes et projets thermiques et nucléaires,
Electricité de France,
Villeurbanne
France

Abstract–Résumé

THE SYMPTOM-BASED APPROACH: CONCEPT OF THE SECOND GENERATION OF POST-ACCIDENT PROCEDURES.

In order to deal with the weaknesses of an approach based purely on events, which was the basis of previous procedures for post-accident action, a new methodology known as the symptom-based approach (APE) has been developed. Authorizing a type of operation connected directly with the state of the boiler and the systems used, it provides a means of dealing exhaustively with combinations of material and/or human failure in accident situations. Already introduced in the post-accident action taken in units now in operation, this generalized utilization (complete replacement of procedures based on type of event) is scheduled for 1989 in the units which will be placed in operation as from that date. The latter will possess the instrumentation (vessel level instrumentation) necessary to permit the gradual adaptation of operation as a function of degradation of the state of the boiler. In view of this deadline, the symptom-based operational procedures are now being worked out. The paper gives a more detailed presentation of the role of simulators in the method of preparing documentation on symptom-based accidental operation.

L'APPROCHE PAR ETAT: CONCEPTION DE LA SECONDE GENERATION DE PROCEDURES POST-ACCIDENTELLES.

Afin de pallier les faiblesses d'une approche purement événementielle sur laquelle se basaient auparavant les procédures de conduite post-accidentelle, une nouvelle méthodologie dite «approche par état» (APE) a été élaborée. Autorisant une conduite directement liée à l'état de la chaudière et des systèmes utilisés, elle permet de couvrir de façon exhaustive des cumuls de défaillances matérielles et/ou humaines en situation accidentelle. Déjà introduite dans la conduite post-accidentelle sur les tranches en exploitation, sa mise en œuvre généralisée (remplacement complet des procédures de type événementiel) est prévue pour 1989 sur les tranches qui seront mises en exploitation à partir de cette date. Celles-ci posséderont en effet l'instrumentation requise (instrumentation niveau cuve) pour permettre une adaptation progressive de la conduite en fonction de la dégradation de l'état de la chaudière. En vue de cette échéance, les procédures de conduite par état sont en cours d'élaboration. Ce mémoire présente de manière plus précise le rôle des simulateurs dans la méthode d'élaboration des documents de conduite post-accidentelle par état.

1. EVOLUTION DE LA CONDUITE POST-ACCIDENTELLE

Les procédures de conduite post-accidentelle élaborées jusqu'à présent sont fondées essentiellement sur l'approche événementielle: l'opérateur diagnostique initialement la cause de l'accident à l'aide d'une procédure de diagnostic puis s'engage dans une séquence d'actions prédéterminées afin de ramener de manière optimale la chaudière dans un état plus sûr.

Ce principe présente quelques faiblesses, notamment la nécessité d'établir un diagnostic fiable, et n'est pas adapté pour couvrir des cumuls de défaillances matérielles et/ou humaines.

Une méthodologie originale, dite «approche par état» (APE), fondée sur une conduite directement liée à l'état de la chaudière et des systèmes utilisés, pallie les faiblesses précédentes grâce à l'introduction d'un diagnostic permanent d'état et à l'exhaustivité des situations qu'elle couvre. Cette méthodologie ne peut être toutefois mise en oeuvre de façon pleinement efficace en remplacement total de l'approche événementielle que si les tranches sont équipées d'une mesure de niveau cuve.

1.1. APE sans instrumentation niveau cuve

Dans un premier temps, l'APE a été introduite dans la conduite post-accidentelle à deux niveaux:

1) A l'intérieur des procédures événementielles existantes (I, A, H): la conduite des systèmes les plus importants pour la sûreté, basée sur des paramètres d'états, est indépendante du diagnostic initial (conduite de l'injection de sécurité, de l'aspersion dans l'enceinte, des pompes primaires).

2) En dehors des procédures I, A, H, une surveillance par état (SPI) est effectuée en permanence, permettant de confirmer les actions principales liées à la sûreté, de demander des actions complémentaires ou de décider d'appliquer la procédure U1.

La procédure U1, fondée sur le principe de l'APE, couvre les situations les plus dégradées de la chaudière. Elle vient prendre le relais des procédures de type événementiel lorsque celles-ci s'avèrent inappropriées à résoudre le problème.

Cependant, l'instrumentation requise pour connaître en permanence avec précision l'état du circuit primaire (et en particulier le niveau cuve) n'étant pas encore installée sur les tranches en exploitation, une conduite progressive en fonction de l'état de la chaudière n'est pas possible. Il apparaît donc un domaine mal couvert par cette solution entre le domaine d'application des règles événementielles (accidents de dimensionnement) et celui de U1 (situations ultimes). Dans ce domaine (voir la figure 1), l'opérateur continue d'effectuer les actions qui lui sont demandées dans la règle événementielle en cours, même si elles ne sont plus totalement adaptées et, si la situation se dégrade, il effectuera les actions ultimes envisagées dans U1.

1.2. APE avec instrumentation niveau cuve

Afin de permettre un meilleur recouvrement du domaine ci-dessus, et sachant qu'une mesure de niveau cuve sera installée sur les tranches mises en exploitation à partir de 1989, un groupe d'étude EDF/Framatome/CEA a entrepris une analyse de faisabilité sur la généralisation de la conduite APE de façon à permettre une adaptation progressive de la conduite en fonction de la dégradation de l'état de la chaudière.

Le travail de l'équipe s'est déroulé en deux temps:

– une réflexion théorique sur l'analyse des états physiques de la chaudière (inventaire des informations requises, identification des états, interprétation des données en termes de dégradation de la chaudière, regroupement des états du point de vue des actions requises, objectifs de conduite pour chaque état);

– une ébauche de mise en forme opérationnelle de cette méthode sous forme de logigrammes de conduite utilisables par l'opérateur sans bouleversement de la structure préexistante (notamment au niveau de l'organisation des équipes de conduite).

Les travaux de ce groupe d'étude ayant permis de conclure à la faisabilité de la généralisation de la conduite APE, le processus d'élaboration des documents de conduite a pu être enclenché.

2. REGLES DE CONDUITE PAR ETAT

Les documents d'exploitation en situation incidentelle ou accidentelle mis à la disposition des opérateurs en salle de commande sont établis par l'exploitant (EDF) à partir des règles de conduite rédigées par le concepteur (Framatome ou EDF).

Une règle de conduite décrit le principe de la conduite à tenir, les justifications des actions et des seuils retenus et sert de dossier pédagogique à l'usage de l'exploitant en centrale.

La conduite par état s'organise autour de trois types de documents:

– un document de vérification des actions immédiates (AU – IS – EAS – isolement enceinte): EV;
– un document de diagnostic permanent d'état permettant d'identifier l'état de la chaudière et conduisant au choix des règles de conduite primaire et secondaire: ED;
– des documents de conduite (7 documents de conduite du primaire: ECP_i; 3 documents de conduite du secondaire: ECS; 1 document de conduite de l'enceinte: ECE; 1 document de surveillance des systèmes).

Le domaine couvert par les règles de conduite ECP et ECS est représenté en figure 1.

L'organisation en salle de commande (présentée en figure 2) reste analogue à l'organisation actuelle. La conduite prévue dans une ECP permet de rallier

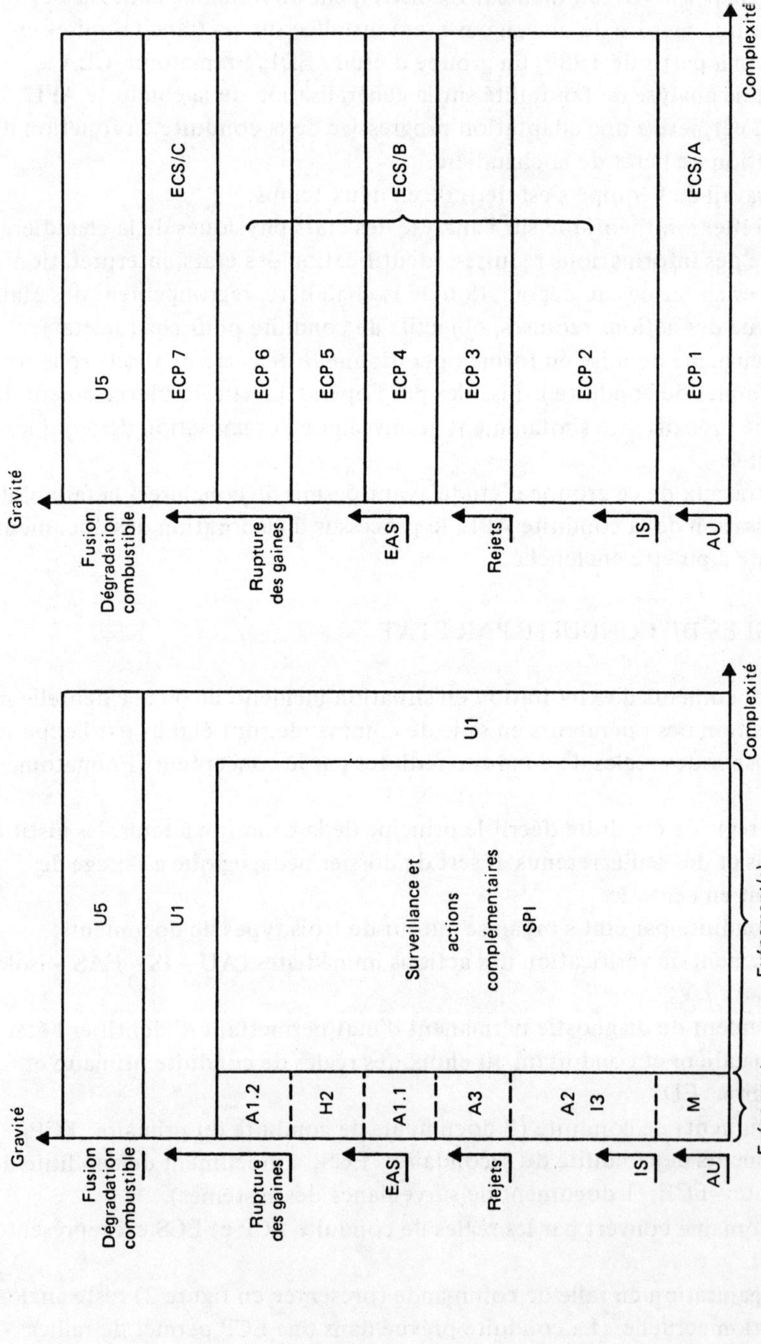

FIG. 1. Domaines couverts par les procédures accidentelles (A: procédures actuelles, B: procédures APE).

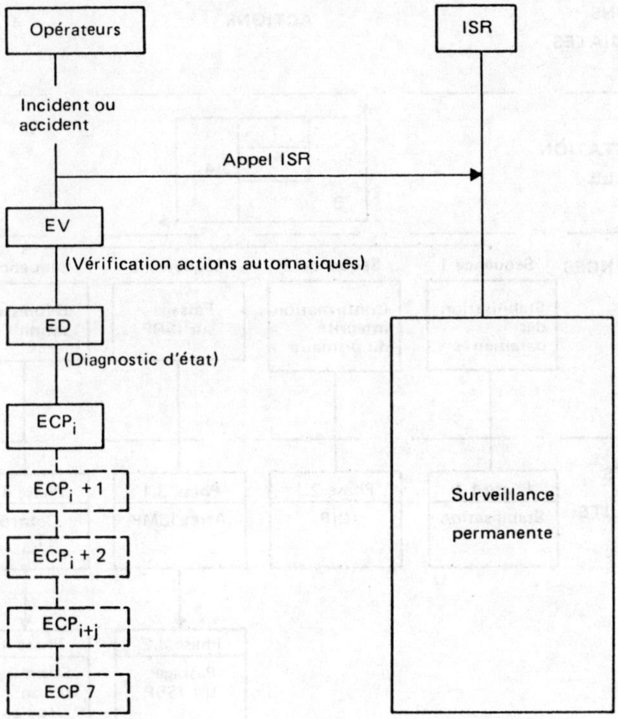

FIG. 2. Organisation des documents de conduite APE.

l'état de repli sans changement de règle de conduite dans le cas d'un événement unique. Le passage d'une ECP à une autre est prévu uniquement en cas de dégradation des paramètres d'états de la chaudière conduisant à une zone d'états non couverte par l'ECP en cours. Ainsi l'opérateur ne pourra s'orienter que vers une ECP de rang supérieur.

2.1. Structure interne d'une règle de conduite

Chaque règle de conduite du primaire et du secondaire (ECP–ECS) a la structure suivante:

– un logigramme général (fig. 3) avec:
  · des actions immédiates,
  · des critères d'orientation vers les différentes séquences de conduite,
  · une présentation succincte des séquences de conduite, ces séquences pouvant être composées de plusieurs phases de conduite;
– un logigramme par phase de conduite (fig. 4) avec:
  · des actions immédiates,
  · les actions spécifiques à chaque objectif avec des priorités propres à chaque séquence,

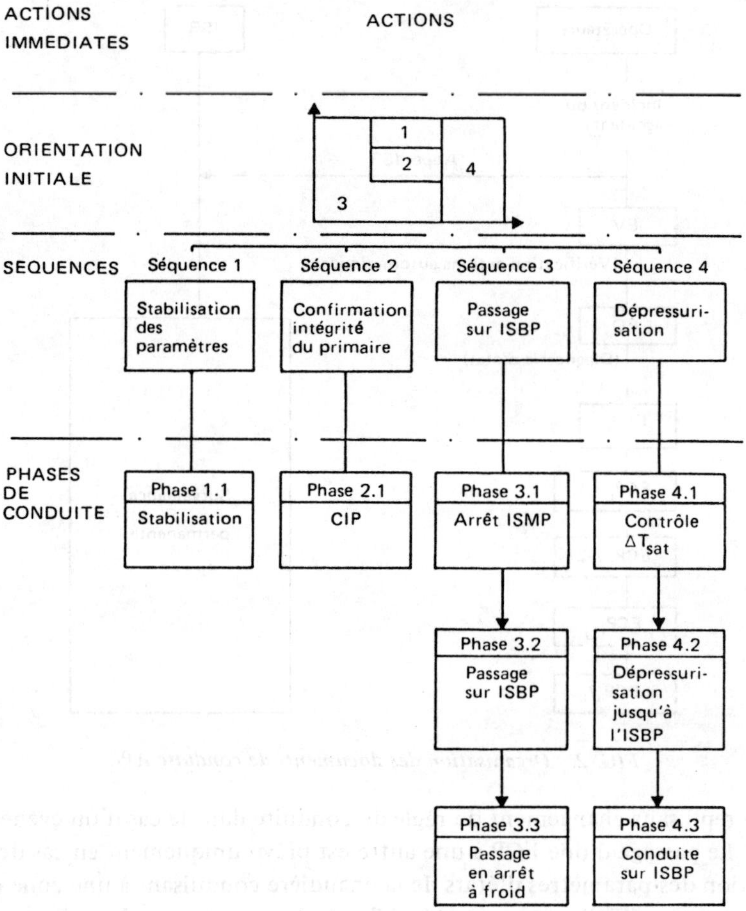

FIG. 3. Logigramme général de la règle de conduite ECP2.

- une surveillance permanente des principaux systèmes utilisés (avec appel éventuel à des restaurations),
- une grille de réorientation qui conduit, selon l'évolution de l'état, soit à rester dans la phase en cours, soit à changer de phase ou de séquence de conduite.

Chaque phase de conduite comprend une conduite de base (en cas de disponibilité des systèmes) et une conduite de substitution lorsqu'un système normalement utilisé dans la conduite est ou devient indisponible. Cette notion de substitution permet d'apporter une aide fondamentale à l'opérateur face à une situation imprévue liée à une défaillance ponctuelle d'un système ou de l'ensemble de plusieurs systèmes.

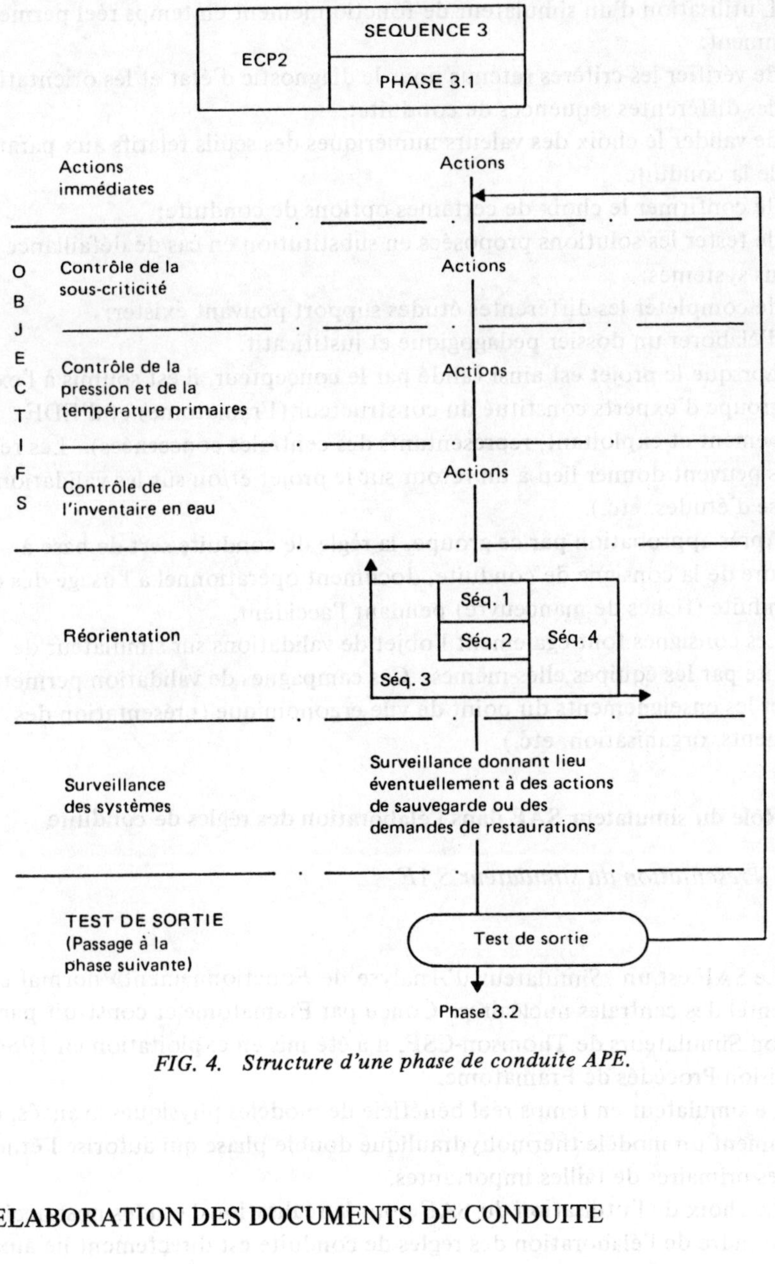

FIG. 4. Structure d'une phase de conduite APE.

3. ELABORATION DES DOCUMENTS DE CONDUITE

3.1. Méthodologie

Le projet de règle de conduite, établi sur les bases présentées ci-dessus, est soumis de manière interactive à diverses validations: soit des études post-accidentelles, soit des tests sur simulateur d'étude.

L'utilisation d'un simulateur de fonctionnement en temps réel permet notamment:

– de vérifier les critères retenus pour le diagnostic d'état et les orientations entre les différentes séquences de conduite;

– de valider le choix des valeurs numériques des seuils relatifs aux paramètres clefs de la conduite;

– de confirmer le choix de certaines options de conduite;

– de tester les solutions proposées en substitution en cas de défaillance de certains systèmes;

– de compléter les différentes études support pouvant exister;

– d'élaborer un dossier pédagogique et justificatif.

Lorsque le projet est ainsi validé par le concepteur, il est soumis à l'examen d'un groupe d'experts constitué du constructeur (Framatome) et d'EDF (équipement et exploitant, représentants des centrales concernées). Les remarques émises peuvent donner lieu à un retour sur le projet et/ou sur les validations (reprise d'études, etc.).

Après approbation par ce groupe, la règle de conduite sert de base à l'écriture de la consigne de conduite, document opérationnel à l'usage des équipes de conduite (fiches de manoeuvre) pendant l'accident.

Les consignes font également l'objet de validations sur simulateur de conduite par les équipes elles-mêmes. Ces campagnes de validation permettent de tirer les enseignements du point de vue ergonomique (présentation des documents, organisation, etc.).

3.2. Rôle du simulateur SAF dans l'élaboration des règles de conduite

3.2.1. Présentation du simulateur SAF

Le SAF est un «*S*imulateur d'*A*nalyse de *F*onctionnement» normal et accidentel des centrales nucléaires. Conçu par Framatome et construit par la Division Simulateurs de Thomson-CSF, il a été mis en exploitation en 1984 dans la Division Procédés de Framatome.

Ce simulateur en temps réel bénéficie de modèles physiques avancés, dont notamment un modèle thermohydraulique double phase qui autorise l'étude de brèches primaires de tailles importantes.

Le choix de l'utilisation du SAF pour la réalisation d'études post-accidentelles dans le cadre de l'élaboration des règles de conduite est directement lié aux avantages de cet outil par rapport aux codes de calcul classiques:

– reproduction d'une conduite rigoureusement équivalente à celle d'un opérateur;

– possibilité de faire évoluer le scénario à volonté en cours de simulation en introduisant par exemple des défaillances successives;

— dépouillement aisé des résultats en raison de la visualisation en temps réel des phénomènes physiques et des possibilités graphiques du SAF (en particulier, l'historique de l'ensemble des paramètres intervenant dans le calcul peut être sauvegardé et consulté a posteriori);

— possibilité d'effectuer des reprises à n'importe quel moment du transitoire si plusieurs variantes de conduite sont envisagées (gain de temps et de coût calcul).

3.2.2. Exemple de validation d'une règle de conduite sur le SAF (ECP2)

3.2.2.1. Présentation de la règle ECP2

Le diagnostic d'entrée dans cette règle est établi à partir de la constatation de l'état de la tranche suivant:

- inventaire en eau peu ou pas dégradé,
- injection de sécurité en service.

ECP2 couvre, outre les événements simples du type:

- injection de sécurité surabondante,
- aspersion intempestive,
- brèches secondaires,
- petites brèches primaires (⩽ 1 pouce),

la combinaison de ces différents types d'incidents cumulés ou non avec la défaillance de systèmes.

Selon l'état du circuit primaire, l'état de repli visé est le passage à l'arrêt à froid ou la stabilisation en arrêt à chaud.

La règle de conduite s'organise en quatre séquences en fonction de l'état de repli visé et de la conduite de l'injection de sécurité associée (voir fig. 3).

Le choix de la séquence de conduite à utiliser est effectuée par l'opérateur à partir des paramètres d'états suivants:

- niveau cuve,
- marge à la saturation,
- pression primaire,
- pression enceinte.

Une surveillance permanente de ces paramètres lui permet de se réorienter d'une séquence vers une autre séquence en cas d'évolution contrôlée ou non contrôlée de l'état de la chaudière.

3.2.2.2. Validation de ECP2 sur le simulateur SAF

Les études menées sur le simulateur avaient pour but d'explorer les domaines suivants:

1) Validation de la substitution des règles événementielles par les règles de conduite par état dans le cas d'événement unique.

TABLEAU I. EXEMPLES DE VALIDATIONS

Type d'accident envisagé	Scénario simulé	Séquences parcourures
IS surabondant	Ouverture intempestive d'une soupape secondaire en aval VVP	Stabilisation
Dépressurisation intempestive du pressuriseur	Blocage mécanique d'une vanne d'aspersion normale avec défaillance fins de course	– Confirmation intégrité du – primaire – Stabilisation
Brèche secondaire non isolable extérieur enceinte	Blocage ouvert d'une soupape VVP en arrêt à chaud	– Confirmation intégrité du primaire – Stabilisation
Brèche secondaire intérieur enceinte	Rupture tuyauterie secondaire en arrêt intermédiaire	– Dépressurisation – Passage sur ISBP
Brèche primaire de petite taille (0,75″)	Fuite primaire non compensable par RCV	– Confirmation intégrité du primaire – Passage sur ISBP

2) Validation exhaustive des séquences de conduite de la règle.

3) Vérification de la validité de la règle dans tout le domaine de fonctionnement défini.

4) Validation des moyens de substitution préconisés.

5) Validation pour des combinaisons d'accidents.

Des exemples de validations effectuées pour ECP2 sont donnés dans le tableau I.

3.2.2.3. Enseignements tirés des validations

Les validations effectuées sur le SAF ont contribué à l'amélioration de la règle de conduite:

– mise en évidence de la nécessité de simplification de la séquence «confirmation de l'intégrité du primaire»,

– révision de certains critères (critère de redémarrage de l'injection de sécurité, seuil de niveau pressuriseur nécessitant l'arrêt des pompes primaires associées à l'aspersion continue),

et ont permis:

– une validation exhaustive des séquences de conduite,
– une validation d'une grande partie des substitutions préconisées,
– la vérification que ECP2 se substitute sans réserve aux règles de conduite événementielles correspondantes.

Plusieurs scénarios envisageant des combinaisons d'accidents et des cumuls de pertes de sources ont été simulés par ailleurs avec EDF et ont démontré l'adéquation de ECP2 à traiter ce genre de situations.

4. CONCLUSION

La méthode décrite précédemment a été appliquée pour chaque règle de conduite, et une grande partie de ces règles a déjà été soumise à l'approbation du groupe d'experts.

Les consignes, écrites à partir des règles de conduite, sont en cours d'élaboration.

Les consignes ECP2 et ECP3 ont déjà fait l'objet de validations sur un simulateur de conduite EDF avec la participation d'équipes d'opérateurs qui se sont adaptés très facilement à cette nouvelle approche.

L'ensemble des documents de conduite devra être disponible en salle de commande en 1989 pour le démarrage de Penly, première tranche sur laquelle sera mise en oeuvre l'approche par état généralisée.

SAFETY IMPROVEMENTS IN DESIGN AND OPERATION OF CONTROL AND INSTRUMENTATION SYSTEMS OF NUCLEAR POWER PLANTS IN INDIA

S.N. AHMAD, U.N. PANDEY, K. NATARAJAN
Department of Atomic Energy,
Nuclear Power Board,
Bombay, India

Abstract

SAFETY IMPROVEMENTS IN DESIGN AND OPERATION OF CONTROL AND INSTRUMENTATION SYSTEMS OF NUCLEAR POWER PLANTS IN INDIA.

The control and instrumentation (C&I) system of an NPP is required to perform not only plant control functions but also safety related and shutdown functions. The C&I safety system has to be operative even when the plant is permitted to be shut down. This requires that the C&I system be ultrareliable. With increasing complexity of NPPs, operator information systems now have a direct bearing on safe operation of the plants. The paper discusses the evolution of the safety related C&I systems in NPPs, of the pressurized heavy water reactor (PHWR) type, in India. Beginning with the Rajasthan atomic power plant, a description is given of the safety and operator information systems backfitted and work done in C&I systems to improve the safety and reliability of the plant. Also, on the basis of experience gained in Rajasthan, improvements have been made in the Madras atomic power plant. These improvements and the ways in which they have contributed to safer and better performance of the plant are reviewed. For the Narora atomic power plant the extensive changes made by updating of technology, increased computerization, and improved operator information systems are discussed. The quality requirements have been made far more demanding for Narora, and the equipment has to meet more stringent operating, environmental and seismic requirements. The problems faced in building equipment and control systems to meet these requirements are outlined. C&I is a field undergoing rapid technological change. How this feature affects standardization in design and retrofitting particularly with respect to safety implications is discussed. The paper concludes with a discussion of the options available to the C&I engineer for the design of systems that will be able to meet the safety and reliability requirements of the NPPs of the 1990s.

Over the years the control and instrumentation (C&I) systems of Indian pressurized heavy water reactors (PHWRs) have undergone many changes on the basis of operational experience as well as owing to new safety requirements laid down from time to time. Even in the first PHWR station, the Rajasthan atomic power station (RAPS), considerable retrofitting has been done. RAPS uses analog instrumentation with discrete components for most control loops. The control room is centralized and a large number of switches, meters, recorders, etc., have been used for operator information and control of the plant. As the station operation continued many C&I systems were improved upon. In particular,

TABLE I. MODIFICATION IN TRIP AND SET-BACK SYSTEM IN THE RAJASTHAN ATOMIC POWER STATION

Trip		*Changes made*
(1)	Adjuster rod cooling flow low	Delay increased in trip
(2)	Adjuster rod cooling flow high	Trip converted to alarm
(3)	Primary coolant channel flow low	New trip parameter introduced
(4)	De-aerator level very low	New trip parameter introduced
(5)	Channel temperature deviation very high	Trip deleted (changed to set-back)
(6)	Primary coolant channel reverse flow	Trip deleted
(7)	Zone mean temperature very high	Deleted
Set-back		
(1)	Primary coolant channel outlet temperature very high	New addition
(2)	Primary coolant channel temperature deviation from zone mean high	Deleted
(3)	Zone mean temperature high	Deleted
(4)	Low condenser vacuum	Deleted

additional aids for analysis of data and quick diagnosis of transient and upset conditions have been provided.

Rapid shutdown of reactor in RAPS is achieved by dumping of the D_2O moderator into a dump tank. Triplicated protective channels have been provided and the relay based logic system requires two out of three coincidence for tripping. This permits the reactor protective system to be extremely reliable, while at the same time providing for high availability. Although the general protective scheme has been retained, on the basis of operating experience the overall system has been simplified and several trips have either been deleted or converted to set-back or trip parameters changed. These changes have been implemented after theoretical and experimental studies. Table I lists the changes to RAPS in this area. The modifications have been carried out so as to improve the plant availability without compromising safety. Some of the changes made are discussed in detail in subsequent paragraphs.

The neutron signal for high neutron trip, obtained from the out of core neutron chambers, was originally corrected with the help of the N-16 signal as the former is affected by the amount of boron poison in the moderator. The N-16 signal was however found to drift and caused spurious trips. As the boron concentration in the moderator changes very slowly, the auto correction based on N-16 has been replaced by manual correction. The neutron signal is now manually adjusted to match the calculated thermal power. The gain adjustment is made every time there is boron addition or removal from the system.

The channel temperature monitoring (CTM) system of RAPS originally used two analog installations to provide redundancy. The system measured the outlet temperature of the 306 coolant channels. The measurements were used to calculate the mean temperature of the eight zones into which the reactor is divided and the deviation of the temperature of each channel from the zone mean. This measurement is used for determining a hot channel. A reactor trip was originally provided on very high channel temperature deviation. An extensive review of the system enabled elimination of all trips connected with this parameter. Only a set-back is now provided on very high outlet temperature. A computer has subsequently been installed to calculate the outlet temperature in RAPS-1 from the zone mean temperature and deviations.

The failed fuel detection (FFD) system has also undergone several changes. The system monitors the delayed neutrons in the samples of the coolant from individual reactor channels. To optimize the design, eight channels are connected through a bulk header to each measuring station. Forty measuring stations are provided to cover all the coolant channels. The selected coolant sample is passed through the measuring station by opening appropriate valves. Initially solenoid valve banks were operated manually and readings taken. Hand calculations were made with data collected to establish the presence of delayed neutrons and the channel containing failed fuel was identified. Subsequently, a CAMAC based FFD system was retrofitted in RAPS. This system automatically scans all the channels in a group and prints the data from which the channel containing failed fuel can be identified.

Another area where retrofitting has been done on a large scale in RAPS is data analysis. In the original system trip/set-back analysis was difficult and time consuming owing to the poor resolution in the recorder charts and determination of cause–effect relationships. This also at times delayed reactor startup. A window annunciator logger, an event sequence recorder and a disturbance analyser have been retrofitted in RAPS. These have greatly assisted the operator in analysis of events and have permitted early plant startup.

All the systems retrofitted in RAPS have ben incorporated in the Madras atomic power station (MAPS), the second PHWR station. The instrumentation systems in RAPS and MAPS are very similar except that the indigenous contribution is much higher in MAPS. Also, in MAPS the quality control norms were defined in a more systematic manner and the equipment was subjected to a better organized quality control programme.

Significant improvements have taken place in the C&I area of the Narora atomic power plant (NAPP) as a result of the experience gained in RAPS and

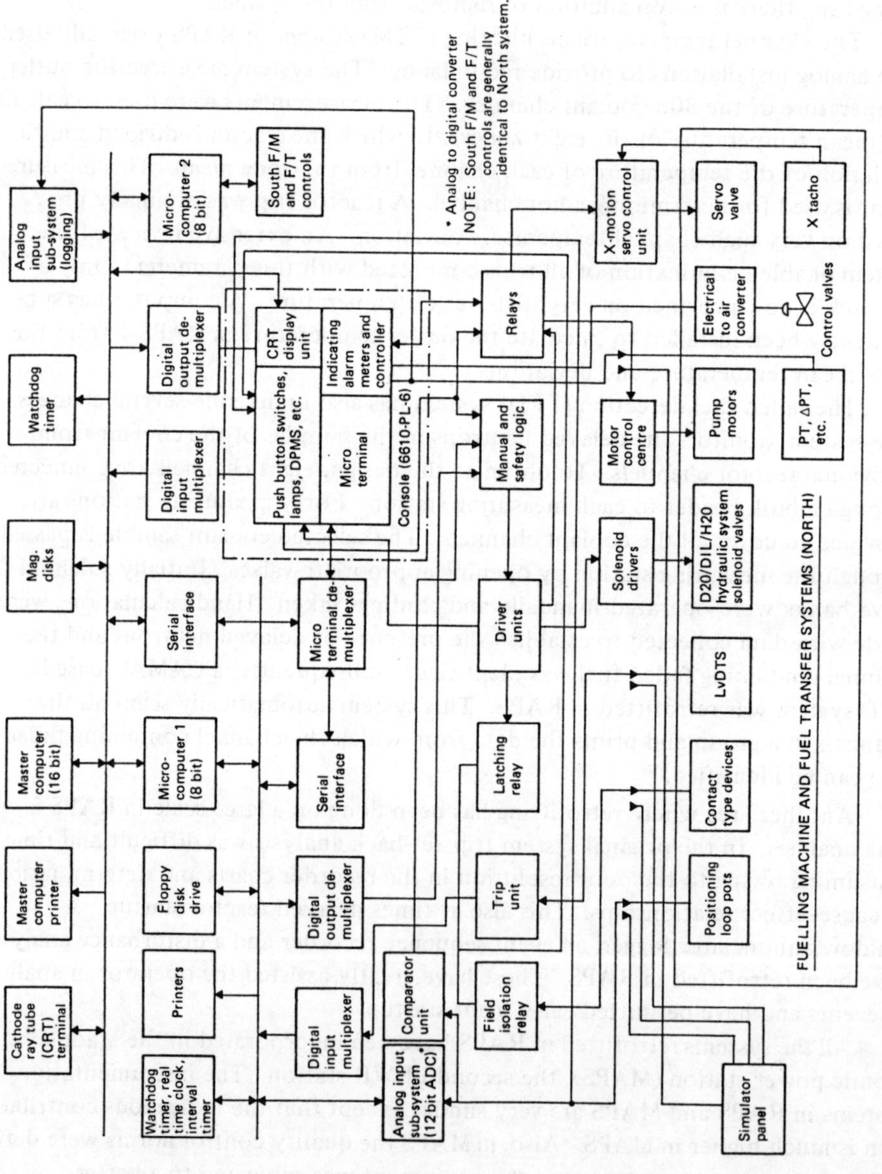

FIG. 1. Block diagram of fuel handling system electrical controls.

MAPS and incorporation of new technologies, particularly microprocessors and computer based systems. In RAPS and MAPS the use of a computer/microprocessor system has been limited to the field of data acquisition. All critical control loops were built around a hardwired analog electronics system. In NAPP, however, computer based control systems have been used in several major areas, some of which are discussed in subsequent sections.

The reactor regulating system (RRS) in NAPP uses a microprocessor system for processing of the reactor power signal and for control of the reactivity mechanisms. This is the first safety related area where a computerized system is being used. The extensive test plans include testing in an operating reactor channel before installation in the reactor system. The simulated tests made have indicated satisfactory system performance. The control strategy in NAPP has been maintained the same as in earlier reactors. This has not only enabled easy comparison of test data but will also permit retrofitting of the new system in existing reactors.

The NAPP reactor regulating system uses three 8085 microprocessors, one for each triplicated control channel. All the inputs/outputs are memory mapped. All address, data and control buses are buffered using tri-state buffers. The analog input signals use pulse modulated isolation amplifiers that provide three way isolation between input, output and supply. The analog to digital convertors (ADC) are used in continuous conversion mode for better system performance.

The regulating signal uses three channels to provide better serviceability and reliability. The median of the three signals is used for control. The median signals are selected through software.

In NAPP a channel is rejected under the following conditions:

- Random access memory (RAM) or erasable programmable read-only memory (EPROM) failure.
- Any two channels declaring the third as faulty.
- Any digital to analog convertor (DAC) failing.
- Any failure of the central processing unit or peripheral detected through ‘watch dog timer’.

In case a channel is rejected control is automatically transferred to the remaining two channels.

Another major area where computerization has been installed is the on-power fuel handling control system for NAPP. The control system is large, complex and is organized in a hierarchical manner. It consists of computer, electronic/electrical, fluid power (using several fluids), process control and mechanical systems (see Fig. 1).

The computer control system is a distributed system consisting of a master computer and Intel 8085 based microcomputers as slaves. The master computer acts as the supervisor and performs system scheduling and maintenance tasks. Each microcomputer controls one fuelling machine and the associated fuel transfer system. The master computer schedules and transfers required codes,

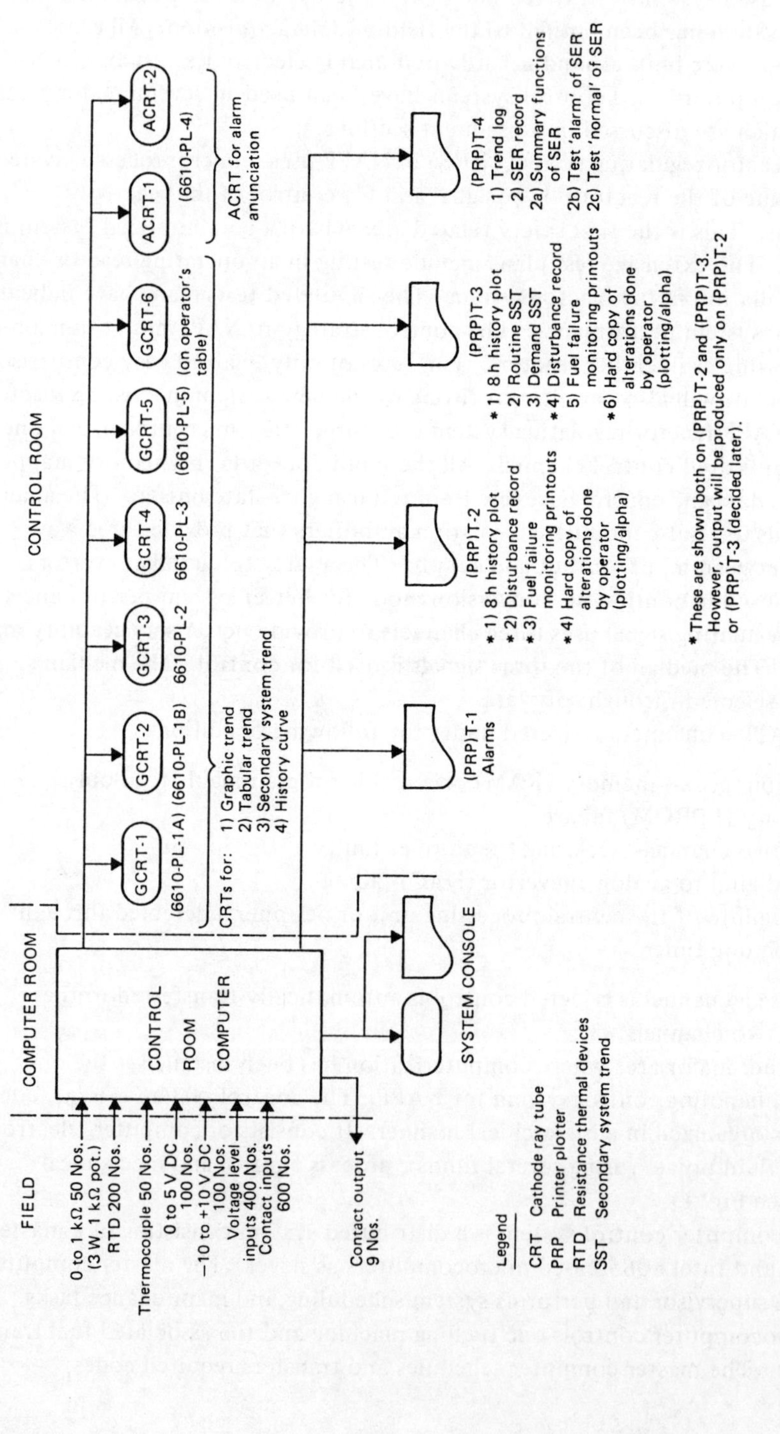

FIG. 2. Control room computer system block diagram.

including sequential steps, to the microcomputers. Sequential steps and all process tasks, including checking of safety logic, are executed in the microcomputers.

The commands from the microcomputers pass through the hardwired manuel and safety logic, where the essential safety logic is rechecked and, if satisfied, commands are issued to the fluid power control system. The fluid power control system then moves the device at the correct speed and force to perform the desired operation. The feedback devices are read by the microcomputers and appropriate control action is initiated. The operation and alarm messages are displayed on video display units and logged by the printers.

Fuel handling operations are done sequentially step by step. The system has been designed so that the fuel handling system can be stopped at any step for an indefinite period. The operator can also intervene at any point and take over control. Facilities have been provided for three modes of operation, namely, auto (full system available), semi-auto (master computer not available) and manual (neither master nor slave computers available). The system can be easily started after power failure or abnormal shutdown.

The NAPP CTM system uses strap-on type resistance temperature sensors in place of the thermowell type. This avoids the large number of penetrations that needed to be made in the high pressure primary heat transport system for the thermowell type. On the instrumentation side solid state comparators have been used eliminating the indicating alarm meters which were used in RAPS/MAPS. Both installations now use microprocessor based data acquisition systems.

In NAPP considerable improvement has also been made in the area of data acquisition and operator information systems to minimize the stress on the operator and to permit easy analysis of all plant conditions. Incorporation of colour cathode ray tubes (CRTs) and hard copy facilities has resulted in deletion of many of the indicators, recorders, annunciator windows, etc. However, parameters connected with safety and those requiring frequent monitoring are still provided with conventional displays. This is to enable plant operation for some reasonable time even when the control room computer is under maintenance. Facilities are available for data logging, event sequence recording and disturbance analysis.

The NAPP control room computer system is shown in Fig. 2. The system accepts about 7000 digital inputs and 800 analog inputs. The scanning time has been optimized on the basis of the various process requirements and varies between 5 to 10 seconds, which includes the time needed for alarm checking, display and printing/plotting. Two CRTs have been dedicated to the display of alarm messages. Another five provide bar graphs, trend tables, etc., either on demand or automatically, depending upon their importance. The graphic trend enables the operator to see the trend of the variables in graphic form with a resolution of 5 or 10 seconds. Audio-visual alarms are also available on the control room computer (CRC).

The event sequence recorder is operated through an 'interrupted programme'. The printer or plotter gives the sequence of occurrence of the events. It is also possible to get a summary printout. The pre-and post-disturbance record can be obtained on the printer for important plant variables. Some of the monitoring functions, for which RAPS/MAPS have separate hardwired systems, are now performed by the CRC. It is expected that in future reactors many more functions will be integrated with the CRC.

Another area planned for computerization is radiation monitoring, which involves a large number and variety of items of equipment. Signals have to be processed and presented suitably. Since there are numerous items of equipment, it has been possible to provide only essential data in the control room. This has posed certain problems in earlier plants. A microprocessor based data acquisition system, different from the main control room system, is planned for NAPP. This system will process the analog signals as well as the alarm relay contacts status from the various pieces of radiation monitoring equipment. There are some 53 analog and almost as many contact inputs. The analog inputs will be processed to display the parameters in engineering units. The system will also compare these values against a set value to generate an alarm, if necessary. The contact inputs which come from various monitors indicate their health as well as alarm status. Visual and audio alarms will be generated for off-normal conditions. All the features available for the main CRC mentioned earlier will be provided, including CRT display with trend, plotter and printer. The system will also have the facility to communicate with the main control room computer and to make any data required available to the operator. Use of microprocessors for the radiation system will also enable integration of certain other systems such as heavy water leak detection and failed fuel monitoring systems to this configuration.

From NAPP onwards the dumping of moderator as a shutdown method has been eliminated. Instead two shutdown systems have been provided. Each shutdown system is individually capable of meeting the fastest design basis reactivity transient. The primary shutdown system (SDS-1) shuts down the plant with the help of 14 mechanical shut-off rods. The secondary shutdown system (SDS-2) shuts down the plant with the help of 12 liquid poison shut-off rods. To provide prolonged shutdown a poison addition system is also provided. This system adds the requisite amount of poison automatically when the need for prolonged shutdown is known.

The philosophy underlying the operation/actuation of the two shutdown systems is that normally SDS-1 will operate, with SDS-2 as stand-by Thus all the trip parameters act on SDS-1. If SDS-1 fails to operate, then SDS-2 will operate. The set points for the actuation of the two systems are different, with SDS-2 set points being higher than those of SDS-1. The selection of the set points is such that both systems act almost simultaneously in case of fast reactivity transients (such as LOCA). For slower transients, the actual failure of SDS-1 is detected by monitoring the travel time of shut-off rods. If more than the specified number of rods are found to be slow the second system is actuated. The timer circuits use microprocessors and the sensor signal processors use analog systems, whereas the logic circuits use relays. Triplicated systems have been used with two out of three logic for actuation of each system. Except for the fact that the reactivity worth of the shutdown systems is lower in NAPP compared to RAPS/MAPS, where dumping is used, this system offers distinct advantages in reliability, speed and reduced startup times.

An altogether new area which has required considerable efforts in NAPP is the field of seismic instrumentation and seismic qualification. Considerable expertise has been generated in this field and most of the critical panels and instrument installations have been seismically qualified. Several designs had to be evolved before an acceptable design could be qualified seismically. Considerable efforts have also been expended in procurement of seismically qualified instruments and components. One major problem has been that though the response spectra are predominant from a frequency of 1 Hz onwards most of the commercially available equipment has a qualification level from 5 Hz onwards only. Special test facilities had to be developed to qualify such equipment.

In NAPP and all future plants an emergency control room will be provided. This will have all essential controls and indications to enable monitoring of the plant status and for safely shutting it down even in the event of the non-availability of the main control room. The emergency control room is physically separated from the main control room.

The incidents at Three Mile Island and Chernobyl have made the safety requirements of nuclear power plants more stringent. Safety related C&I systems are not only becoming more complex but are also required to meet progressively better reliability and availability criteria. At the same time the evolution of technology will involve more extensive use of advanced systems, such as computers, digital transmission, etc. The reactors of the 1990s will thus rely more on programmable systems. The major challenge facing the C&I designer will be the generation of software that is as reliable as the best analog systems. With the rapid change in technology, and the high rate of obsolescence, C&I system design will have to be flexible and capable of continuous upgrading.

NEW SAFETY APPROACHES IN THE DESIGN OF THE ARGENTINE PRESSURIZED HEAVY WATER REACTOR

A.J. GONZALEZ*, K. FRISCHENGRUBER, J.A. RECALDE,
R.B. SOLANILLA, R.C. VANZULLI
Comisión Nacional de Energía Atómica,
Buenos Aires, Argentina

Abstract

NEW SAFETY APPROACHES IN THE DESIGN OF THE ARGENTINE PRESSURIZED HEAVY WATER REACTOR.

The Argentine nuclear power plant ARGOS PHWR 380 — or ARGentine Option for a Safe Pressurized Heavy-Water Reactor of 380 MW(e) — is briefly summarized and the new safety approaches used in its design are described. The paper presents the safety features for normal operation as well as the design criteria applied for potential accident situations. Great emphasis is given to the application of the demanding probabilistic safety requirements from the Argentinian regulatory authority. These include proabilistic safety criteria as well as specific requisites for performing the relevant probabilistic safety analyses. The description also includes the application of deterministic safety criteria both domestic and international, including the safety standards, guides and recommendations issued by the International Atomic Energy Agency. The paper also presents the special safety features incorporated in the design as an ultimate effort toward safety, namely a redundant high-pressure heat sink, improvements in auxiliary and emergency power supply, early and efficient detection of leakages from the primary boundary, and a vented containment. Finally, a short description of the possibilities of back-fitting is included.

1. INTRODUCTION

This paper describes briefly the technical characteristics of the latest design of an Argentinian nuclear power plant (NPP) and discusses the new safety approaches applied to such design. The new NPP is known as ARGOS PHWR 380, an acronym for ARGentine Option for a Safe Pressurized Heavy-Water Reactor of 380 MW(e) and has recently been described in the literature [1]. In its current version, the plant is equipped with a pressurized heavy-water reactor of the pressure vessel type (PHWR), although - conceptually - it might also hold a nuclear steam supply system (NSSS) with a pressure-tube heavy-water reactor (PTHWR) type. Therefore, both the flow diagram and main technical characteristics (see Table 1) of the current version are practically the same as those of the MZFR reactor type [2], which was used as the basis for the PHWR Atucha NPPs [3] [4] (see Figure 1 for simplified flow diagram). Thus, and

* Now at the International Atomic Energy Agency.

TABLE 1. MAIN TECHNICAL DATA (COMPARATIVE TABLE)

	CNA I	ARGOS PHWR 380	CNA II
Reactor Type		P H W R	
Gross Generator Output (MW)	367	375	745
Thermal Reactor Output (MW)	1179	1123	2160
REACTOR CORE			
Type of Fuel		Sintered Pelletized Natural Uranium Dioxide, 37 Rods	
Refuelling		On Load	
Number of Fuel Assemblies	253	244	451
Active Length (mm)	5300	5300	5300
Burnup (MW·d/mg)	6000	6600	7500
Mean Fuel-Rod Power (W/cm)	232	223	232
Number of Control Rods	29	60	18
MAIN CIRCUITS			
Number/Main Coolant Loops	2	2	2
Number/Moderator Coolant Loops	2	3	4
Coolant Flow Rate per Loop (kg/s)	3080	2573	5150
Moderator Flow Rate per Loop (kg/s)	222	150	222
Operating Pressure (bar)	113	115	115
Coolant Temperature (^{o}C)	262/296	277/314	278/312
Average Moderator Temperature (^{o}C)	140/210	165/220	170/220
REACTOR PRESSURE VESSEL			
Internal Diameter (mm)	5360	5366	7368
Weight of Bottom Portion (mg)	320	320	670

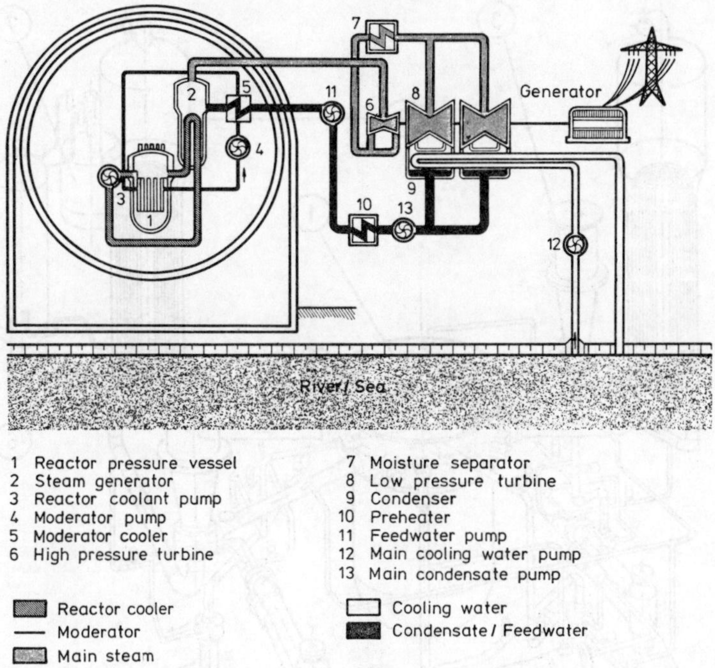

FIG. 1. Simplified flow diagram.

given the excellent operating performance attained by the Atucha I NPP (CNA-I) [3], ARGOS PHWR 380 is comparable with a perfect reference plant of medium power [5]; moreover, the design experience of the Atucha II NPP (CNA-II) [4] - which is currently under construction - has also been fed back into the ARGOS PHWR 380 design.

The new plant includes new boundary design conditions and a number of special features to assure safety. It complies with the demanding safety requirements of the Argentinian national regulatory authority, particularly with its probabilistic safety criteria. Automatically, the plant follows all relevant international safety criteria, notably the standards, guides and recommendations issued by the International Atomic Energy Agency (IAEA). The design has also incorporated special safety features which reflect ultimate lessons learned in the nuclear safety field.

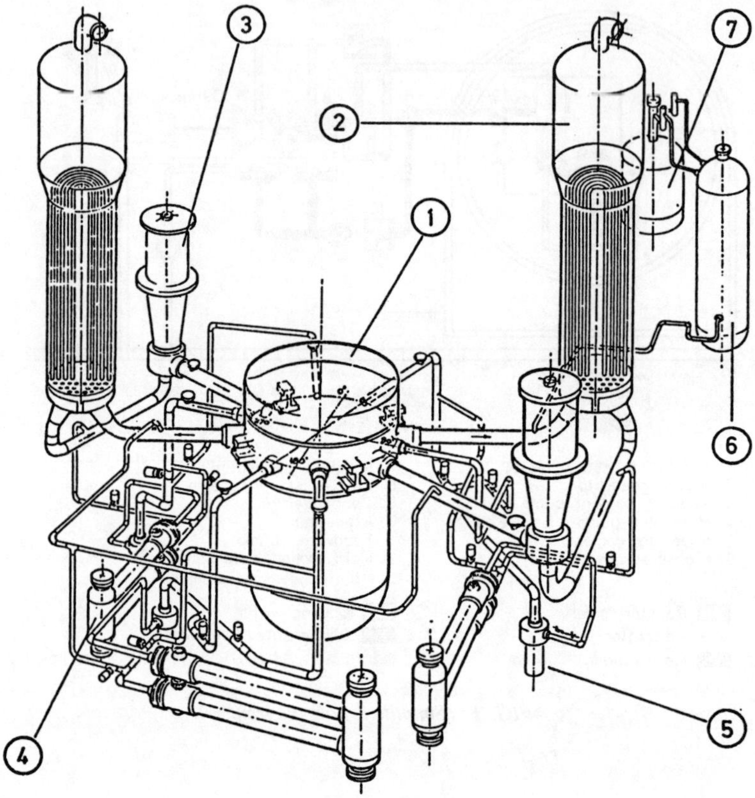

FIG. 2. Reactor coolant and moderator circuit.

2. BRIEF DESCRIPTION OF ARGOS PHWR 380

2.1 The nuclear steam supply system

The current version of the NSSS of ARGOS PHWR 380 bears a close resemblance to a standard two-loop PWR. In fact, the reactor coolant system of ARGOS PHWR 380 is fully comparable in design and arrangement with those of PWRs (see Figure 2). It consists of the reactor pressure vessel (RPV), in the center, and two identical coolant loops, each comprising one coolant pump and one steam generator. In addition, as a typical HWR feature, there is a moderator cooling system, which is subdivided into three identical loops. The moderator system is connected with the moderator tank (similar to a calandria tank) within the RPV, and includes three moderator loops comprising one moderator pump and cooler per loop. Moderator and coolant are held at essentially the same pressure, enabling the use of quite thin-walled coolant channels. As the moderator and coolant systems are interconnected, common auxiliary systems can be used to maintain water quality. The RPV and core design is similar to that of the Atucha PHWRs (see Figure 3). The moderator tank contains 244 vertical fuel channels producing 1067 MW of thermal power, arranged in a triangular lattice array with a pitch of 27.5 cm. The fuel assemblies (one per fuel channel) are identical to those used in the Atucha design (37 pins having an active length of 530 cm). Hydraulically actuated control and shutdown absorber rods are vertically inserted from the upper part of the vessel through the moderator tank. A total of 60 rods is used. Forty-five of them are shutdown rods - allowing for quick shutdown and maintenance of safe long-term subcriticality - and the rest are power and power density control rods. A suitable selection of these vertical control rods provides flattening of the power distribution from the very beginning of the reactor operation, allowing the reactor to be operated at 100% of rated power with a nearly fresh core. This system provides outstanding load-following capability of the reactor.

The operational advantage of the Atucha PHWR, namely that it is designed for automated on-load refuelling, was maintained in the ARGOS PHWR 380 design. The refuelling system of ARGOS PHWR 380 is designed for the automatic refuelling of the reactor during full power operation. New fuel is inserted into the transfer vessel, where a fluid change from H_2O to D_2O takes place. There is a single refuelling machine. The fuel assembly transport system is located within the reactor building and includes a fuel pool with a capacity which can be tailored up to the design lifetime of the plant. The refuelling machine takes over the fuel assembly and performs reloading and reshuffling operations on top of the reactor pressure vessel. The spent fuel is carried to the spent fuel pool into the opposite direction. The refuelling procedure is monitored from the control room. A special feature is the possibility of inserting irradiated fuel assemblies into the core in the same way as fresh ones, as well as specially designed fuel assemblies containing rods for ^{60}Co generation.

2.2 The balance of plant

The plant includes the reactor building, the auxiliary building and other conventional buildings. The balance of plant is designed to ensure economy in construction and operation. The following aspects are considered in the arrangement of the building: clear energy flows, short piping and cable runs; good access for construction, erection, operation and maintenance; adaptability to future extension;

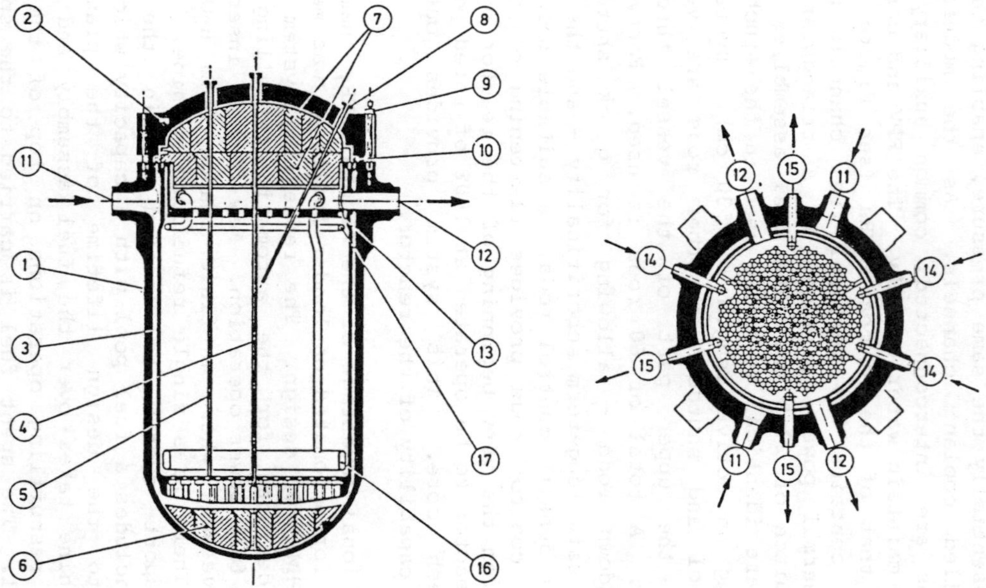

FIG. 3. Reactor pressure vessel and internals.

1 RPV (reactor pressure vessel)
2 Closure head of RPV
3 Moderator tank
4 Coolant channel
5 Guide tube of control rod shutdown rod
6 Lower filler pieces
7 Upper filler pieces
8 Boric acid injection-line
9 Stud
10 Closing joint
11 Coolant inlet
12 Coolant outlet
13 Closure head of moderator tank
14 Moderator inlet
15 Moderator outlet
16 Moderator piping lines
17 Moderator piping lines

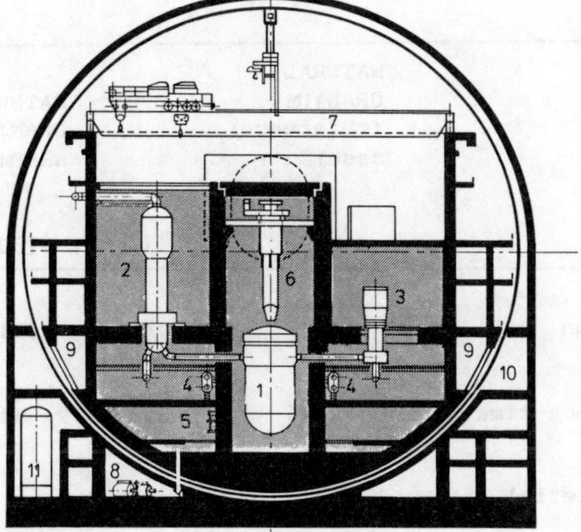

Equipment compartments — Operating compartments

1. Reactor pressure vessel
2. Steam generator
3. Reactor coolant pump
4. Moderator cooler
5. Moderator pump
6. Refueling machine
7. Reactor building crane
8. Safety injection pump
9. Pipe duct
10. Cable spreading area
11. D_2O Storage tank

FIG. 4. Reactor building cross-section.

and, clear separation of the controlled area for radiation protection purposes. The reactor building (see Figure 4) is of the Atucha type and similar to the building of same type PWRs of triple power. It consists of two concentric spheres and its main function is a high performance containment of radioactive materials. The good volume power relation simplifies the enhancement of the outstanding containment capability of the building (see Section 3.2.3 vented containment). The internal sphere is metallic and has been designed to withstand the maximum pressure which might result from any conceivable loss of coolant or moderator accident. The spherical form was chosen as optimal, taking into account stress distribution parameters. The steel sphere is, in turn, surrounded by an external spherical structure of concrete, which is intended to act as a secondary containment and as a radiation shielding, as well as to resist all relevant external events which might damage the inner containment. The space between the metallic and the concrete structures forms the reactor building annulus where, in its lower part, the most important auxiliary systems are placed. Thus, they are located close to the primary system

TABLE 2. FUEL CYCLE ALTERNATIVES (MAIN DATA)

	NATURAL URANIUM (reference case)	L.E.U.	Pu SPIKED NATURAL URANIUM ASSEMBLIES (80%)		Pu SPIKED PLUTONIUM ASSEMBLIES (20%)
Fuel consumption	0.99	0.41	0.48		0.06
(f.a./f.p.d.)				0.54	
Average residence time	246	594	407		814
(f.p.d.)				453	
Fissionable material (%)					
- Fresh U-235	0.71	1.00	0.71		0.68
Pu	0	0	0		2.10
- Discharged U-235	0.29	0.15	0.18		0.18
Pu	0.24	0.29	0.28		0.81
Average burnup	6 000	14 500	9 200		26 000
(MW·d/mg H.M.)				11 040	
Average thermal power per channel (MW)	4.6	4.6	4.3		6.0
				4.6	
Pu production (kg fiss. Pu/GW(e) f.p.y.)	430	215			

L.E.U. .... Low enriched uranium
f.a. .... fuel assembly
f.p.d. .... full power day
H.M. .... heavy metal
f.p.y. .... full power year
fiss.Pu .... fissionable plutonium

and to the reactor itself and well protected against external events. The building houses not only the NSSS but also the refuelling equipment and the 12-year capacity fuel storage pool.

Other systems are contained in the reactor auxiliary building. They include: the coolant and moderator cleaning and upgrading systems, the nuclear ventilation system and all the equipment necessary for handling liquid and gaseous radioactive wastes. All the other buildings are in the so-called conventional part; no special requirements are needed for their design, except that the switchgear and emergency power supply building - reflecting the importance to safety of the equipment it houses - is protected against external events and internally subdivided into three redundant modules, which are physically separated.

2.3 Fuel cycle

The ARGOS PHWR 380 is designed to operate with the natural uranium fuel cycle as well as with advanced fuels without significant systems or component changes. Two important alternatives to the use of pure natural uranium are the optimized low enriched uranium (LEU) fuel cycle and the plutonium-spiking fuel cycle (see Table 2 for the fuel cycle alternatives of ARGOS PHWR 380) [6] [7].

Using the natural uranium fuel cycle, a fuel economy of 20% (comparing with standard PWRs with a high burnup cycle) is already attainable. This is done by using 0.99 fuel assemblies per full power day (corresponding to 6000 MW·d/mgU discharge burnup), compatible with a load cycle of 100-80-100% of rated power. In the equilibrium burnup core condition, the fuel loading scheme is 3 zone-1 way: that is, fresh fuel is introduced into a channel at an intermediate radial position. From there, partially depleted fuel is shuffled to the core centre and finally to the reactor periphery, from where the fuel assemblies with the highest burnup are discharged. By adopting this radial shuffling scheme and by flattening the radial power distribution, an average discharge burnup of 6600 MW·d/mg can be obtained in case of a minimum reactivity reserve. The average discharge burnup can be further increased through the use of advanced fuel cycles.

3. SAFETY FEATURES

The basic safety criteria applied to ARGOS PHWR 380 involve design features for normal operation and for potential accidental situations. Compliance with these criteria - coupled with the well known intrinsic safety advantages of HWRs - makes ARGOS PHWR 380 a safe NPP option built on the basis of the existing experience in nuclear engineering and industry, particularly on the extended experience achieved by very similar class of reactors, the PWRs.

3.1 Design features for normal operation

The basic design criterion for the normal operation of ARGOS PHWR 380 is that it should comply strictly with the system of dose limitation recommended by the International Commission on Radiological Protection (ICRP) [8] and adopted - inter alia - by the IAEA, the World Health Organization, the International Labour Organization and the Nuclear Energy Agency of the OECD [9]. The Argentinian regulatory authority has implemented the ICRP system by issuing regulations on occupational exposure [10] and on limitation of releases [11]. The system's requirement for the optimization of radiation protection - which is still being implemented by the nuclear industry - has been applied to the design of ARGOS PHWR 380.

Radiation protection features for the normal operation of ARGOS PHWR 380 are optimized to keep doses as low as reasonably achievable (the ALARA principle). Optimization is achieved by using the internationally recommended decision-aiding techniques of cost-benefit analysis [12]. Optimization is carried out under the constraint that the design should ensure - under any relevant circumstances - that the annual dose to exposed individuals be lower than established dose upper bounds. During the design process, protection by technical means was preferred to that achievable by operational procedures.

The following design limits (NB: limits, rather than objectives or goals) have been, in every case, respected:

- For occupational exposures: Access to any area where the dose index may exceed 0.5 mSv/h is prevented by physical barriers. In areas without restrictions, the concentration of radionuclides in air must be lower than 0.1 of the Derived Air Concentration (DAC) (or the concentration of radioactive materials in air which would expose workers to the recommended limits). No individual may be exposed to concentrations higher than 0.1 DAC without individual protection devices. Maintenance and in-service inspection can be performed with dose rates lower than 0.5 mSv/h, normal repair with dose rates lower than 3 mSv/h and unfrequent repair with dose rates lower than 12 mSv/h.

- For exposure of members of the public: Design is constrained by an annual dose upper bound in the critical group of 0.3 mSv. The collective dose commitment must not exceed 0.015 man Sv/MW·a of electrical energy generated.

Adherence to such limiting conditions implies, in practice, that ARGOS PHWR 380 includes particular radiation protection systems that are not commonly used. For instance, the design incorporates inter alia systems for the retention of ^{14}C releases and for on-line monitoring of tritium in local environments.

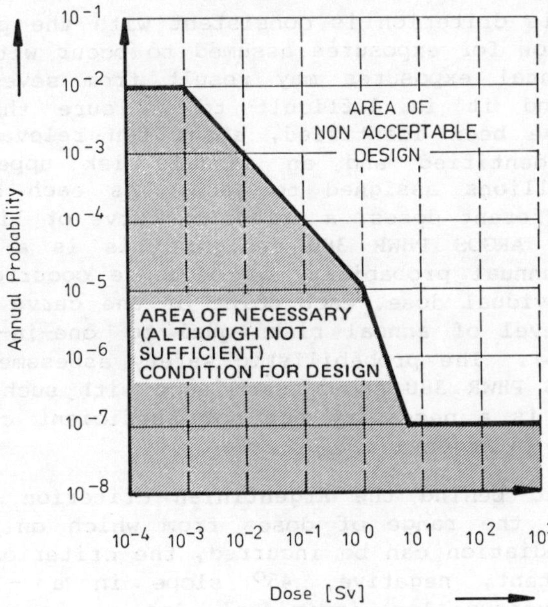

FIG. 5. Liquid criterion curve.

3.2 Design criteria for potential accidental situations

3.2.1 Probabilistic safety criteria

A unique feature of ARGOS PHWR 380 is that its safety design is mainly based on probabilistic safety criteria (PSC), which is a requirement of the Argentinian regulatory authority [13]. A priori probabilistic safety analyses (PSA) were carried out at the design stage and their results fed back into the reactor design taking into account the most recent reliability data and design criteria. Such analyses are not purely theoretical: they are substantiated by the experience gained with the CNA II safety design, which was also partially based on PSA [14]. The PSA results are checked against the quantitative PSC issued by the Argentinian regulatory authority, which are in line with the most recent international developments in PSC [15],[16]. These requirements were used as an <u>a priori</u> comprehensive condition for design, rather than as an <u>a posteriori</u> confirmation of design compliance.

The limiting criterion is that an annual risk upper bound of one-in-a-million must be respected for any individual who might hypothetically be subject to accidental exposures from the NPP. This criterion is consistent with the philosophy of dose limitation for exposures assumed to occur with certainty. Since accidental exposures may result from several accident sequences, and it is difficult to be sure that all such sequences have been identified, about ten relevant sequences are being identified and an annual risk upper bound of one-in-ten-millions assigned to each. As each sequence may result in different doses, a criterion curve or limit line was used for the ARGOS PHWR 380 design: this is a relationship between the annual probability of sequence occurrence and the expected individual dose, each point of the curve representing a constant level of annual risk equal to one-in-ten-millions (see Figure 5). The probabilistic safety assessment performed for the ARGOS PHWR 380 shows compliance with such a criterion curve which is a necessary but not sufficient condition for NPP licensing in Argentina.

The logic behind the Argentinian criterion curve is as follows. For the range of doses from which only stochastic effects of radiation can be incurred, the criterion curve must show a constant, negative, 45° slope in a - log annual probability versus log individual dose - coordinate axis plane. This would ensure that the risk, i.e., the annual probability of incurring the dose multiplied by the probability of serious deleterious effects given the dose (the latter being in the order of 10^{-2} Sv^{-1}), will be kept constant. One of the coordinate points in this part of the curve would obviously be the following: [Annual probability = 10^{-5}; Individual dose = 1 Sv], because the product 10^{-5} a^{-1} x 1 Sv x 10^{-2} Sv^{-1} results in an annual risk of 10^{-7} which is the risk upper bound established by the Authority for any scenario from the postulated initiating events. In the dose range where non-stochastic effects of radiation may occur (i.e., for individual doses higher than approximately 1 Sv), the slope of the curve should increase, in order to take into account the higher risks of death at these levels of dose. For doses higher than approximately 6 Sv, the probability of death approaches unity. From this level to higher doses, the criterion curve should remain constant at an annual probability of 10^{-7} (because the exposed individual would inevitably die regardless of the level of the dose). Between the coordinate points defined by [Annual probability = 10^{-5}; Individual dose = 1 Sv] and [Annual probability 10^{-7}; Individual dose = 6 Sv], the criterion curve should show a shape inverse to the dose-response relationship which, at that range, is approximately S-shaped; however, for the sake of simplification, the regulatory authority has decided to approximate these two points by means of a linear-shaped relationship. Finally, the criterion curve has been truncated at an annual probability level of 10^{-2}, because the occurrence of incidents having a higher annual probability

TABLE 3. MAIN FEATURES OF THE REGULATIONS ON FAILURE ANALYSIS APPLIED TO THE ARGOS PHWR 380 DESIGN

- The probability of occurrence of each identified failure sequence, as well as the corresponding activity of released radionuclides, are assessed by using event and fault trees, while taking into account the following criteria:

  - The failure analysis must systematically encompass all foreseeable failures and failure sequences, considering the common-mode failures, the failure combinations and - most important - the situations exceeding the design basis. (Failure in this context means an alleatory event preventing a component from performing its safety function, as well as any other event which may additionally occur as a necessary consequence of such deficiency. Failure sequence, on the other hand, means a sequential series of possible failures which can, although not necessarily, occur after an initiating event).

  - A failure or a failure sequence is selected as representative of a group of failures or of failure sequences. The failure or failure sequence that is selected from the group is that delivering the worst consequences and the analysis takes into account the sum of the probabilities of the failure or failure sequences in the group.

  - The analysis considers that a protection function may have lost operativeness, either before the occurrence of the failure or of the failure sequence or as a result of such occurrence.

  - The analysis of failures, of failure sequences or of any part thereof is based on experimental data as far as it is possible. If this cannot be done, the valuation methods are validated through appropriate tests.

  - Failure rates assigned to safety-related components for evaluating the probability of system failure must be justified. In case that justifiable values were not available for some of the components, levels of failure rate prescribed by the competent authorities are used.

TABLE 3. (cont.)

- Failure analyses take into account maintenance and testing procedures, and the time interval between successive maintenance and testing actions.
- The failure rates postulated for human actions are justified taking into account the complexity of the task, the psychological stress involved and any other factors which might influence that failure rate, balanced with the level of automatization for each interaction in concern.

- The doses on the critical group, that would result from the release of radionuclides due to a failure or failure sequence, must be assessed by accepted methods. (The critical group is defined as a group of people, near to the nuclear power plant, sufficiently homogeneous with regard to the doses expected to be incurred, and representative of the most exposed individuals in case of an accident). The assessment should take into account the meteorological conditions of dispersion at the site and their probabilities. The assessment should not take into account the eventual application of countermeasures, even if they are forecasted in the emergency planning.
- The annual probability of occurrence of any failure sequence, if plotted as a function of the resulting effective dose equivalent assessed as indicated above, must result - as a necessary but not sufficient design condition - in a point located outside the non-acceptable area of the limit criterion curve. Otherwise, the design must be adjusted accordingly.

(regardless of the dose) is unacceptable for the regulatory authority. The ARGOS PHWR 380 design also complies with applicable regulations on failure analysis for PSA [17] (see Table 3).

However, ARGOS PHWR 380 compliance with the individual-related PSC described in the previous paragraph is a necessary but not sufficient condition for its safe design.

Ensuring that no single individual will incur an unduly high probability of harm because of potential radiation exposure might not be sufficient to ensure the appropriateness of the safety level and it may need to be improved by taking into account, for instance, that from a high number of individuals incurring an acceptably low probability of harm may still result in an unacceptable high expectation of harm. But, for exposures which have a very low probability of occurrence, the use of the concept of expectation to reduce risks further, below the limits, is not straightforward [18]. The engineering options, then, should be compared, taking into account the full mathematical distributions of individual risks they can produce. There is also the additional problem of including - in the comparison process - quantities or preferences which may not be translated into commensurate units, such as the degree of risk aversion for higher consequence accidents, social costs for restrictions or inconveniences, and the degree of relative importance of the various possible manifestations of radiation health effects.

The problem in comparing safety options that already comply with the individual PSC has been solved in the process for optimizing the ARGOS PHWR 380 safety design by using utility functions and decision theory [19]. Preferences for quantities of differing types are expressed using a utility function which prescribes how the different types of quantities are to be combined for the purposes of comparison. The resulting utility functions are then processed by a decision mechanism so as to arrive at a "best-under-the-circumstances" (i.e., optimized) safety option. The result of this process is a design for the ARGOS PHWR 380 that can be considered as safe as is reasonably achievable. Again, it is more than a theoretical excercise: if ARGOS PHWR 380 had been designed with two moderator loops, it would have complied with the criterion curve of the individual-related PSC; however, a third loop has been added following a decision-making process aimed at safety optimization. The optimization process may be further refined when the full spectrum of the site related information is available.

3.2.2. Deterministic safety criteria

ARGOS PHWR 380 also complies with deterministic criteria which supplement the probabilistic requirements described above, as follows:

<u>Compliance with deterministic requirements from the Argentinian authority</u>

These requirements include: general safety criteria in the design [20] (minimizing the consequences of any eventual failure); reactor core design [21] (ensuring safe operation during the whole lifetime of the reactor); residual heat removal systems [22] (ensuring that fuel elements shall not suffer damage); pressurized primary boundary [23] (preserving integrity under any operational, testing or failure conditions); fuel behaviour in the reactor [24] (minimizing the possibilities of activity releases); safety-related protection and instrumentation system [25] (considering all the tentative situations under operation and failure conditions); shutdown system [26] (ensuring reliable shutdown, under any conceivable conditions); containment system [27] (ensuring a proper activity confinement function); main electrical supply [28] (ensuring the necessary power supply for the protection, instrumentation and safety-related systems); and, a quality assurance system [29] (assuring the adequate quality of the NPP safety systems).

Compliance with international safety standards, guides and recommendations

The ARGOS PHWR 380 complies - as a necessary but not sufficient condition for a safe design - with all applicable safety codes, standards, guides and recommendations issued by the IAEA and, in particular, with those of the Nuclear Safety Standards programme [30].

Compliance with other deterministic criteria based on practical experience and current engineering judgement

Obviously, ARGOS PHWR 380 complies with the conventional safety criteria intended to ensure safe reactor trip and long-term holding of subcriticality, as well as reliable residual heat removal, limiting the release of radioactive materials into the environment. Great emphasis was given in the ARGOS PHWR 380 design to the achievement of these basic objectives in the most reliable manner, by means of the following measures, among others:

- Passive engineered safeguards: To control the release of radioactive materials, several passive engineered barriers are provided: the fuel matrix, the fuel cladding; the closed and seal welded high-pressure boundary of the reactor coolant system; the spherical full pressure steel containment; the secondary concrete containment; and the annulus between the steel and the concrete spherical structures, which is exhausted by a specially designed system. But, even for the most unlikely accident sequence resulting in uncontrolled core meltdown, a special venting system for the reactor building is provided allowing for a controlled release through a passive filtering system (see Section 3.2.3). Increase of pressure within the steel containment beyond its design pressure is avoided, and thus its eventual disruption is prevented.

- Active engineered safeguards: A wide spectrum of accidents and incidents was considered in the design. To keep the plant in a controlled state under these accidental conditions, active engineered safeguards are automatically actuated and controlled by the reactor protection system. They are designed to meet high reliability targets by means of: conservative and careful design; quality assurance and control; examination and in-service inspections; and inherently safe operating characteristics. The active engineered safeguards are provided in a diverse manner and each diverse system is redundantly three-branched. Any one of these three branches is designed to cope with a given postulated accident sequence. This principle was also observed in the physical separation of sub-systems, thus avoiding consequent ("knock-on") failures.

- Defence-in-depth concept: In addition to the instrumentation and control required for normal operation, a condition limitation system is provided, which acts between the normal feedback controls and the reactor protection system, keeping the plant variables within the range specified and enhancing not only the safety but also the availability of the plant. This condition limitation system is also redundant. The reactor protection system also triggers and controls the active engineered features. This is a self-controlling three-channel dynamic system. Reflecting the state of the art, the reactor protection system detects all dangerous deviations from the parameters and triggers all necessary countermeasures automatically. No manual action is required during the first 30 minutes after any conceivable incident. This provides additional protection against any tentative improper human action.

3.2.3 Special safety features

Finally, as an ultimate effort towards safety, ARGOS PHWR 380 offers some special safety features, as follows:

Additional high pressure heat sink

The reactor moderator system can be used to remove heat in a high-pressure mode. Under normal shutdown conditions, the residual heat can be removed via the steam generators - as in PWRs, maintaining coolant recirculation either by operating the main coolant pump or by simple natural convection. ARGOS PHWR 380 includes the additional possibility of using the moderator system as a high pressure heat sink. For this operation mode, the moderator is pumped from the bottom of the moderator tank, cooled in the moderator coolers and injected into the main coolant system. In an emergency core cooling condition, the moderator system serves also as a high pressure injection system. The necessary commutations for the different operation modes are performed automatically according to the already mentioned general design philosophy requiring that no operator action should be necessary within the first 30 minutes after any conceivable incident.

The high pressure residual heat removal system is designed for high pressure and temperature. In all conceivable incidents, it can keep the reactor in a hot condition after shutdown as long as it is required or, if convenient, it can cool it down following a predetermined temperature gradient. All branches of the RHR chain are redundant and physically separated. Since the steam generators are also available for heat transfer, the plant has two diverse, high pressure, highly available heat sinks for the different accident sequences which may need to be considered.

Reliable auxiliary and emergency power

The PSA of ARGOS PHWR 380 indicated that the power supply is in the critical pathway for risk. Accordingly, the auxiliary power supply has been designed to assure adequate reliability levels and a high degree of protection against interruptions. The system is divided into two diverse, redundant and independent systems, which are located in separate sections of the switchgear building. In case of a common mode failure of the normal power supply, an emergency system takes over the feed of safety related loads. This emergency power supply system is divided into three redundant systems, that are physically separated. Each of these systems comprises: one diesel motor generator, a non-interruptible AC bus bar and a DC supply with batteries, rectifiers and converters. The design also provides that, in case of a prolonged loss of power, the installation can switch an external transportable generator to the emergency power net.

Early detection of leakages from the primary boundary

One of the most important advantages of the ARGOS PHWR 380 design is the possibility of early detection and location of potential leakages in the coolant and moderator systems. This is achieved by tritium detection, the most sensitive method for this purpose, which can be effectively implemented in this reactor type. Therefore, even the smallest leaks can be detected long before they can threaten the integrity of the primary boundary of the reactor. In case of leakage, the localization and further repair can be achieved with a minimum effort and at a very early stage.

Vented containment

The ultimate lesson learned in the field of nuclear safety is the need to ensure the confinement of radioactive material also in cases of severe hypothetical accidents involving core melt-down. For that purpose, ARGOS PHWR 380 was equipped with a venting system whose objective is preventing the disruption of the steel containment and the consequent uncontrolled release of radioactive materials into the environment that could occur in such an extreme case. Should the pressure increase unexpectedly within the containment, the venting system is designed to stabilize the pressure at a safe value by regulating the release of excess gases and steam into the atmosphere.

The intended functioning of the vented containment of the ARGOS PHWR 380 can be exemplified by following a postulated accident scenario, the so-called "emergency power" case, in which the highly reliable emergency power supply would be unavailable and non-recoverable. In such a case, after reactor shutdown, the temperature of the reactor coolant would increase as a result of residual heating in the core. The primary pressure would reach a level that would actuate the safety valve. Primary coolant steam would then flow into the pressurizer relief tank, where the pressure would eventually lead to bursting of a rupture disk. Radioactive coolant steam would then be released into the containment environment, where it would be retained. The design of this extremely important boundary includes provision for sufficiently reliable isolation of pipes and ducts penetrating the steel sphere by redundant valves. Otherwise, there would be an immediate release of radioactive materials into the reactor building annulus, followed by an uncontrolled release into the environment and a subsequent unacceptable high public exposure and environmental contamination.

After the containment isolation, which can be ensured either by automatic or by human action, all consequences would be limited within the containment environment. The core would dry out and eventually melt. The cladding of the fuel rods would burst and its temperature would increase until a zirconium water reaction would take place. Hydrogen, along with fission products, would be released into the containment environment. There, another important system of the containment would become effective: the early ignition of the hydrogen by appropriate catalysts which will operate without auxiliary power. This effect will ensure the integrity of the containment, by avoiding a later hydrogen deflagration with possible rupture of the steel sphere.

Following with this catastrophic scenario and supposing that the electrical power supply continues to be absent, the melting core is assumed to penetrate the bottom of the reactor pressure vessel and come into contact with the floor of the reactor vault. There might then be an exothermic reaction between the melted core material and the concrete. The released energy would increase the pressure inside the containment significantly (that pressure would already amount some bars, because of the blowdown of the primary coolant). One of the most significant features of the ARGOS PHWR 380 design would now become effective. The containment design pressure of about five bar would be reached only after some 10-14 days, when all the short-lived fission products would have already decayed and most of the radioactive material would have been deposited. This is due to the very favourable relation between the very large containment volume and the comparatively small energy content of the primary system. Without any countermeasures at this stage, pressure would increase inside the containment up to its rupture, which would take place some 20 days after the initiation of the described hypothetical scenario. In such a case, the consequences would be extremely modest compared with other accident scenarios, where such containment function does not exist. However, a containment rupture would lead to an uncontrolled release of the radioactive material still remaining in the containment into the environment. The ARGOS PHWR 380 design therefore incorporates venting, to permit the controlled release of gases and aerosols from the containment into the environment through a filtering system, relieving the pressure and thus avoiding the containment failure.

The design criterion for the vented containment is that the result of this hypothetical and extremely unlikely situation will be such that even the critical group of the population would not be exposed to projected doses higher than 0.1 Sv. This level of projected dose would not usually justify radiological intervention or countermeasures. In fact, the International Atomic Energy Agency has recommended (for radiological emergencies) that "the level of projected dose liable to be received in the short term, below which evacuation is unlikely to be justified, will usually be about ..." [31] in the order of 50 mSv. An integrated dose rate - over the long term - of the order of 0.1 Sv should therefore not usually trigger radiological countermeasures. This is consistent with recent developments in the optimization of radiation protection concerning emergency measures [32].

3.3 Possibilities of back-fitting

Certain new safety approaches in the ARGOS PHWR 380 design can be back-fitted to the CNA-I, which is presently in operation in Argentina, as well as to the CNA-II, currently under construction. Possible back-fittings to the PTHWR Embalse NPP should be evaluated on a case by case basis.

Safety systems that can be easily incorporated to the Atucha type NPPs include some of the special safety features described above; namely, the high-pressure heat sink, the improvement in auxiliary and emergency power supply and the vented containment (including control of explosive hydrogen). Although the leakage detection, via tritium measurement, can also be used in the Atucha type NPPs, the current design is not necessarily adaptable.

As far as the PSC are concerned, they cannot obviously be back-fitted to CNA-I. The individual related probabilistic criterion has already been used in the design of CNA-II; however, no probabilistic criterion for safety optimization was used in this design and, now, back-fitting results are impracticable.

In relation to normal operation features, the main difference between the ARGOS PHWR 380 and the Atucha NPPs designs is in the full compliance with the requirement to optimize radiation protection. This can eventually be back-fitted by assessing the optimization of operational radiation protection at the Atucha NPPs on a case by case basis.

4. CONCLUSIONS

ARGOS PHWR 380 is an NPP option that presents several non-safety advantages, such as: being a medium power reactor, with lower financial and indirect costs, and with a simpler adaptation to electrical distribution networks; a design featured to ensure installation and operational economy; efficient generation of by-products; fuelling with the independent natural-uranium fuel cycle, as well as with optimized LEU assemblies and even with the plutonium it generates.

But, since confidence on nuclear safety has evolved and is now a relevant aspect to be considered in the development of nuclear power, ARGOS PHWR 380 does mainly offer the usual outstanding safety features of heavy water reactors furthermore upgraded by following the ultimate developments in the nuclear safety field. This enhancement has not been done starting from zero but using all the past experience of an already economic and extremely safe industry, particularly reflecting the worldwide experience achieved through the PHWRs and their very similar class of reactors, the PWRs.

REFERENCES

[1] González, A.J.; Frischengruber, K.; Recalde, J.A.; Solanilla, R.B.; and Vanzulli, R.C.; "Argentina offers a 380 MWe PHWR with enhanced safety features"; Nuclear Engineering International; Vol. 32, No.394; May (1987).

[2] Der Mehrzwecksforschungsreaktor, Atomkernenergie, Kerntechnik, Vol. 46; (June 1985).

[3] Herzog, G. and Sauerwald, K.; "La Central Nuclear de Atucha"; Atom und Strom; Year 15, No. 4; (April 1969).

[4] "ATUCHA II, Building a 745 MWe pressure-vessel PHWR in the Argentine"; Nuclear Engineering International, Vol. 27, No. 9 (Sep. 1982).

[5] Small and Medium Power Reactors: Project Initiation Study, Phase I; Report prepared by the International Atomic Energy Agency and the OECD Nuclear Energy Agency; IAEA-TEC DOC-347. IAEA; Vienna; (1985).

[6] Frischengruber, K. and Dusch, F; in IAEA Technical Committee on Advanced Light and Heavy Water Reactor Technology Development; Potential and Advanced Fuel Cycles in KWU Type PHWRs; IAEA; Vienna; (1985).

[7] International Nuclear Fuel Cycle Evaluation; Working Group 8 on Advanced Reactor Systems and Fuel Cycle Concepts, INFCE/PC/2/8; IAEA; Vienna; (1980).

[8] International Commission on Radiological Protection; "Recommendations of the International Commission on Radiological Protection"; ICRP Publication No. 26; Oxford, Pergamon Press; (1977); Annals of the ICRP, vol. 1, No. 3; (1977)

[9] International Atomic Energy Agency; "Basic Safety Standards for Radiation Protection"; IAEA Safety Series 9; Vienna, Austria; (1982).

[10] Comisión Nacional de Energía Atómica; Buenos Aires, Argentina. Consejo Asesor para el Licenciamiento de Instalaciones Nucleares, "Exposición ocupacional"; CNEA, Buenos Aires (1979); Norma CALIN 3.1.1.;

[11] Comisión Nacional de Energía Atómica; Buenos Aires, Argentina. Consejo Asesor para el Licenciamiento de Instalaciones Nucleares, "Limitación de efluentes radiactivos"; CNEA; Buenos Aires (1979); Norma CALIN No. 3.1.2.

[12] International Commission on Radiological Protection; "Cost Benefit Analysis in the Optimization of Radiation Protection"; in Annals of ICRP, ICRP Publication 37; Volume 10 No. 2/3 (1983).

[13] González, A.J.; "The Regulatory Use of Probabilistic Safety Analysis in Argentina"; in Proceedings of the International Meeting on Thermal Nuclear Reactor Safety; Chicago, USA; NUREG/CP-0027;(1982).

[14] Fabian, H. and Frischengruber, K.; Safety concept and evaluation for the pressurized heavy water reactor ATUCHA II; Atomenergie, Kerntechnik, Vol. 46.
[15] International Atomic Energy Agency, "Status, Experience, and Future Prospects for the Development of Probabilistic Safety Criteria"; Report of the Technical Committee Meeting; IAEA; TEC DOC (in preparation).
[16] International Atomic Energy Agency; "The application of radiation protection principles to sources of potential exposure: towards a unified approach to radiation safety". IAEA Consultative Document (in preparation).
[17] Comisión Nacional de Energía Atómica; Buenos Aires, Argentina. Consejo Asesor para el Licenciamiento de Instalaciones Nucleares, "Criterios radiológicos relativos a accidentes"; CNEA, Buenos Aires (1979). 2 p Norma CALIN 1.3.1.;
[18] Beninson, D. and Lindell, B.; "Critical views on the applications of some methods for evaluating accident probabilities and consequences"; International Conference on Current Nuclear Power Safety Uses; proceedings series, STI/PUB/566, Vol.2, pgs. 325-341; Stockholm; 20-24 October, 1980.
[19] Beninson, D.; "Optimization of radiation protection as a special case of decision theory" (IAEA-SM-285/38). Proceedings of the International Symposium on the Optimization of Radiation Protection; Vienna; (March 1985).
[20] Comisión Nacional de Energía Atómica; Buenos Aires, Argentina. Consejo Asesor para el Licenciamiento de Instalaciones Nucleares. "Criterios generales de seguridad en el diseño"; CNEA; Buenos Aires; (1980); Norma CALIN no. 3.2.1.
[21] Comisión Nacional de Energía Atómica; Buenos Aires, Argentina. Consejo Asesor para el Licenciamiento de Instalaciones Nucleares. "Núcleo del reactor" CNEA: Buenos Aires; (1979); Norma CALIN no. 3.3.1.
[22] Comisión Nacional de Energía Atómica; Buenos Aires, Argentina. Consejo Asesor para el Licenciamiento de Instalaciones Nucleares. "Sistemas de remoción de calor"; CNEA; Buenos Aires; (1979); Norma CALIN no. 3.3.2.
[23] Comisión Nacional de Energía Atómica; Buenos Aires, Argentina. Consejo Asesor para el Licenciamiento de Instalaciones Nucleares. "Circuito primario de presión"; CNEA; Buenos Aires; (1980); Norma CALIN no. 3.3.3.
[24] Comisión Nacional de Energía Atómica; Buenos Aires, Argentina. Consejo Asesor para el Licenciamiento de Instalaciones Nucleares. "Comportamiento del combustible en el reactor"; CNEA; Buenos Aires; (1980); Norma CALIN no. 3.3.4.
[25] Comisión Nacional de Energía Atómica; Buenos Aires, Argentina. Consejo Asesor para el Licenciamiento de

Instalaciones Nucleares. "Sistema de protección e instrumentación relacionada con la seguridad"; CNEA; Buenos Aires; (1980); Norma CALIN no. 3.4.1.

[26] Comisión Nacional de Energía Atómica; Buenos Aires, Argentina. Consejo Asesor para el Licenciamiento de Instalaciones Nucleares. "Sistemas de extinción"; CNEA; Buenos Aires; (1981); Norma CALIN no. 3.4.2.

[27] Comisión Nacional de Energía Atómica; Buenos Aires, Argentina. Consejo Asesor para el Licenciamiento de Instalaciones Nucleares. "Sistemas de confinamiento"; CNEA; Buenos Aires; (1981); Norma CALIN no. 3.4.3.

[28] Comisión Nacional de Energía Atómica; Buenos Aires, Argentina. Consejo Asesor para el Licenciamiento de Instalaciones Nucleares. "Alimentación eléctrica esencial"; CNEA; Buenos Aires; (1980); Norma CALIN no. 3.5.1.

[29] Comisión Nacional de Energía Atómica; Buenos Aires, Argentina. Consejo Asesor para el Licenciamiento de Instalaciones Nucleares. "Garantías de calidad"; CNEA; Buenos Aires; (1980); Norma CALIN no. 3.6.1.

[30] International Atomic Energy Agency; Nuclear Safety Standards for Nuclear Power Plants; IAEA Safety Series No. 50; IAEA, Vienna (1980).

[31] International Atomic Energy Agency; "Principles for Establishing Intervention Levels for the Protection of the Public in the Event of a Nuclear Accident or Radiological Emergency"; IAEA Safety Series 72; IAEA; Vienna (1985).

[32] Beninson D. and González A.; "Optimization in Relocation Decisions" (IAEA-SM-285/37). Proceedings of the International Symposium on the Optimization of Radiation Protection; Vienna; (1985).

POSTER PRESENTATIONS

POSTER PRESENTATIONS

IAEA-CN-48/133P

SIMULATION DES SEQUENCES DE RELESTAGE DES AUXILIAIRES SECOURUS PAR LES GROUPES ELECTROGENES DES CENTRALES NUCLEAIRES

F. IFFENECKER, C. MORZELLE, C. DEGRAVE
Service Etudes et projets thermiques et nucléaires (SEPTEN),
Electricité de France,
Paris, France

Electrité de France (EDF) a développé un code de calcul pour simuler les séquences de relestage des auxiliaires secourus de ses centrales nucléaires par des groupes électrogènes à diesel en cas de perte des sources électriques externes.

Ce code modélise les éléments suivants:

– le diesel suralimenté avec sa régulation de vitesse;
– l'alternateur et sa régulation de tension (équations de Park);
– les auxiliaires secourus (moteurs asynchrones et composants entraînés).

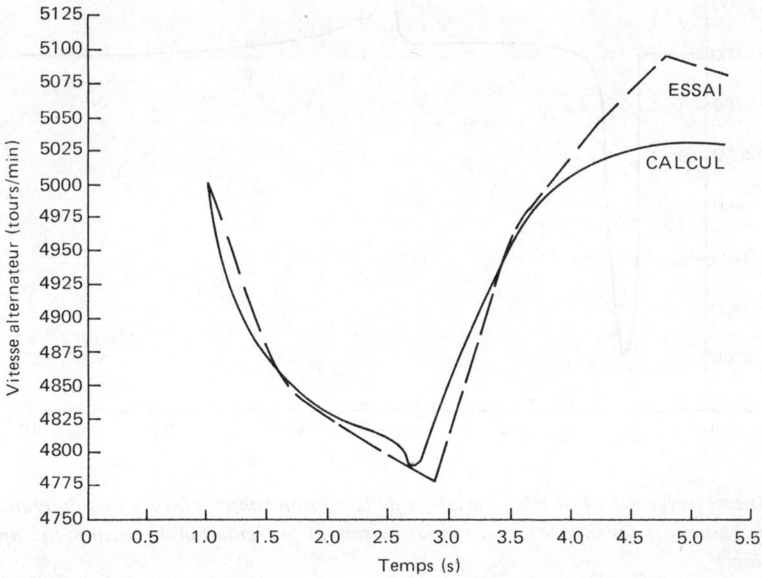

FIG. 1. Diesel palier 1300 MW: variation de la fréquence lors de la mise sous tension d'un groupe d'injection de sécurité. Comparaison calcul–essai.

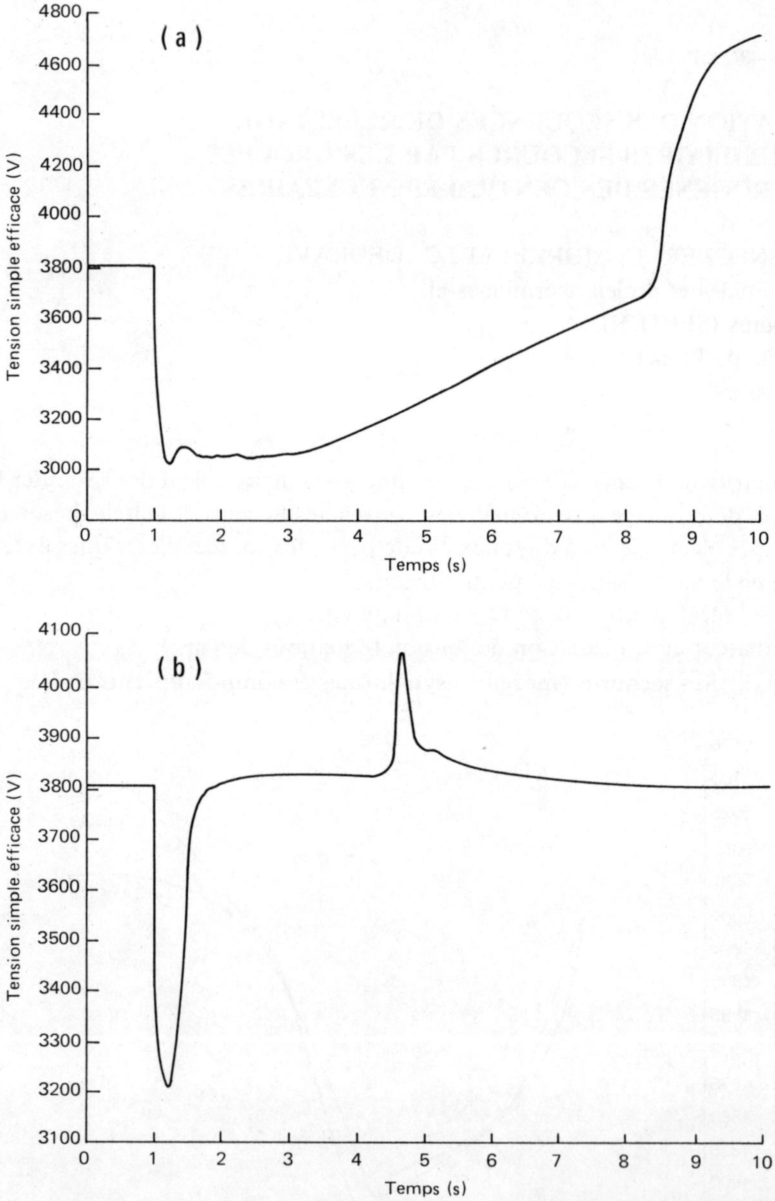

FIG. 2. Diesel palier N4 1400 MW: variation de la tension simple efficace lors du cran de relestage le plus contraignant. (a: avant modification de la régulation de tension; b: après modification.)

Pour les centrales du palier 1300 MW, le code a été calé et qualifié à l'aide d'essais réalisés durant la procédure de mise en service de la centrale de Paluel.

Les mesures durant ces essais furent effectuées par la Direction de études et recherches d'EDF en utilisant une chaîne d'acquisition synchrone à haute fréquence d'échantillonnage (Note EDF/DER HP/126/85-07, 1985), ce qui a permis d'obtenir des données très précises (puissances active et réactive).

Cet outil a été utilisé pour vérifier que les groupes électrogènes à diesel des centrales nucléaires du palier 1300 MW sont conformes à la réglementation française en matière de chute de fréquence et de tension. La figure 1 montre l'évolution calculée et l'évolution mesurée de la fréquence lors de la mise sous tension de la pompe d'injection de sécurité moyenne pression (1,4 MW).

Avant construction des groupes électrogènes à diesel du nouveau palier français N4 (1400 MW), le code a été utilisé pour vérifier leur capacité à respecter les critères de sûreté.

Les résultats ont montré qu'il était nécessaire de modifier la structure de la régulation de tension. Les figures 2a et 2b montrent l'évolution de la tension avant et après modification lors du cran de relestage le plus sévère.

IAEA-CN-48/143P

FRAMATOME'S 'SAF' ENGINEERING SIMULATOR

A first step towards defining the engineer's simulator tool of the year 2000

T. CONSTANTIEUX, M. DENOISEUX
Real Time Simulation Centre,
Framatome,
Paris-La Défense, France

Among the techniques available to engineers today, computerized simulation is taking on an ever growing importance. The SAF simulator, designed by Framatome for the use of its own engineers, has been in service since 1985. The SAF simulator provides continuous assistance to the engineer; from the preliminary design stage to the precise definition of operating procedures, including safety analysis and sizing computations.

The SAF represents a first step towards a comprehensive simulation tool for the engineer in the year 2000 who will be used to 'dialoguing' with the computer from a very young age. Interactive and thus 'alive', SAF combines both extensive programming and data processing capabilities. Its simulation domain can still be considerably extended.

The SAF simulator is available for use by Framatome's engineers but also by our customers and the safety authorities. It enables modelling all the nuclear units built by Framatome in their entirety. Highly modular and equipped with easy to use compilers, SAF can be readily modified by the user, to enable testing new models or new systems, in the complex and detailed environment of the nuclear unit being analysed.

Employing the advanced computer programs used in project design, the SAF simulator is a particularly high performance tool for simulating and analysing complex accident scenarios, including multiple equipment failures and possible operator errors, which may extend to complete draining of the reactor vessel and the release of radioactive fission products within the containment structure.

IAEA-CN-48/249 P

КОМПЛЕКСНОЕ УПРАВЛЕНИЕ КАЧЕСТВОМ ЭКСПЛУАТАЦИИ АЭС

Е.П. ЛАРИН, С.А. ЛЕСНОЙ, В.В. ТАРАТУНИН
Министерство атомной энергетики СССР,
Москва,
Союз Советских Социалистических Республик

Накопленный опыт эксплуатации АЭС свидетельствует о важности и необходимости комплексного подхода к качеству эксплуатации АЭС, т. е. с учетом безопасности, надежности и эффективности их использования. Научная обоснованность, системность и оптимальность принимаемых решений должна основываться на количественных показателях.

Сегодня проблема безопасности должна разрабатываться в единстве с проблемой обеспечения качества эксплуатации.

В качестве основных положений единой концепции повышения качества эксплуатации энергоблоков можно сформулировать следующее:

1. Качество эксплуатации энергоблока – совокупность характеристик процесса эксплуатации, которые обеспечивают его техническую и оперативную готовность, экономичность при безусловном выполнении требований безопасности.
2. Ключевыми, управляемыми звеньями в системе обеспечения качества эксплуатации энергоблоков являются вопросы повышения надежности и вопросы подготовки персонала.
3. Принципиальная возможность управления надежностью определяется возможностью целенаправленного воздействия на качество проектирования, изготовления, транспортировки, хранения, монтажа, эксплуатации и ремонта оборудования и энергоблоков в целом.

При этом выделяются два уровня управления:

– внешний уровень управления реализуется при поставке высококачественного оборудования и при качественном его монтаже;
– внутренний уровень управления реализуется воздействием на качество эксплуатации.

4. Принципиальная возможность управления качеством подготовки персонала определяется возможностью создания эффективной системы обучения, переподготовки, системы оценки эффективности подготовки.

Цели комплексного управления качеством эксплуатации могут быть достигнуты только на основе обобщения опыта эксплуатации АЭС.

Накопленный за более чем 500 реакторо-лет опыт эксплуатации АЭС в СССР показывает, что источники исходных событий любых аварийных ситуаций разделяются на две основные группы:

– первая – аномалии технологических процессов;
– вторая – отказы оборудования.

Возрастающие потоки информации о работе АЭС потребовали создания ряда баз данных различного назначения. К таким базам относятся:
– базы данных по надежности оборудования;
– базы данных по событиям, важным с точки зрения безопасности;
– базы данных по технико-экономическим показателям АЭС.

Информационной основой баз данных служат специально разработанные карты отказов и отчеты об инцидентах, поступающие с АЭС.

С целью эффективного использования информации, вводимые в указанные базы данных, во ВНИИАЭС Минатомэнерго СССР разработан пакет прикладных программ, унифицированный для машин серии ЕС.

Для оптимального использования опыта эксплуатации должны быть разработаны специальные базы для данных и комплексы прикладных программ:

– по сбору, первичной обработке, накоплению и хранению необходимой информации (программные комплексы ДИАКРИЗ, BRISSO);
– по определению достигнутого уровня надежности и эффективности оборудования АЭС (программные комплексы RACORS, GRAF);
– по оптимальному планированию планово-предупредительных ремонтов (программный комплекс MARR).

Использование этих пакетов прикладных программ позволит с более высоким уровнем научной и технической обоснованности, основанной на опыте эксплуатации, решить ряд задач, связанных с повышением надежности и безопасности АЭС.

IAEA-CN-48/164P

PERFORMANCE OF AN INTEGRATED THERMOHYDRAULICS AND AEROSOL BEHAVIOUR CODE (CONTAIN) IN A BENCHMARK ON LACE LA-4 SIMULATING LATE CONTAINMENT FAILURE

F. BEONIO-BROCCHIERI, E. CAGLIOTI,
A. MARKOVINA, R. RICCHENA
Joint Research Centre, Ispra Establishment,
Commission of the European Communities,
Ispra

In the last few years the problem of assessing the thermohydraulics and aerosol behaviour in LWR containments in the event of severe accidents has been extensively analysed by many groups in Europe and overseas. This activity has included a number of comparative exercises on computer codes treating the problem in order to verify their consistency with the general state of knowledge. Validation against experimental results was considered a more advanced objective and was only recently undertaken. A first step in this direction was the benchmark exercise on the DEMONA-B3 experiment (Kernforschungszentrum Karlsruhe GmbH and Battelle Frankfurt Laboratory) organized by the Commission of the European Communities (CEC).

The present work is in the framework of the second step in this direction and addresses a more complex problem. The set of LACE experiments represents a significant advance in the database available on large experiments over a range of parameters. In particular LACE-LA4 simulates the late failure of the containment with consequent depressurization. As this sequence has been theoretically analysed in the past and is also very important from a safety point of view, it was chosen by the CEC as a case study to analyse in depth the differences in modelling and numerical methods amongst the various codes in a benchmark exercise.

This poster presentation describes the analytical work performed by the team at the Joint Research Centre, Ispra, utilizing the code CONTAIN 1.03 in the framework of the CEC benchmark on LACE LA-4. The special features of this code are analysed and commented upon and in particular the possibility of an integrated treatment of thermohydraulics and aerosol behaviour.

The main phenomena which characterize LA-4 are discussed in the light of the present status of knowledge and modelling of specific effects.

Of interest are the results obtained with CORRADI (a modified version of CORRAL-2 which we have implemented to describe also the diffusiophoretic effects). Calculations are based on the best fit of the experimental data produced in the experiment A5 of containment system experiment which is in many aspects very similar to LACE LA-4.

IAEA-CN-48/167P

ASSESSMENT OF LWR SAFETY CODES USING LOBI EXPERIMENTAL DATA

H. STÄDTKE, W. KOLAR, L. PIPLIES
Joint Research Centre, Ispra Establishment,
Commission of the European Communities,
Ispra

During the last two decades a great effort has been made to develop complex system codes which are capable of predicting off-normal and accident conditions in LWR power plants. Some of these codes, e.g. TRAC and RELAP, have received worldwide recognition and are presently used in many countries; others like DRUFAN and CATHARE are used mainly in the originating countries, the Federal Republic of Germany and France. All these codes are directed towards describing the governing phenomena expected to occur in postulated accident scenarios including loss of coolant accidents (LOCAs) and special transients, and to provide best estimate calculations for all safety relevant system parameters.

To demonstrate the capabilities of such safety codes in describing reactor safety relevant phenomena, various national and international code assessment programmes have been established where code predictions are compared with experimental data from differently scaled separate-effects and integral test facilities. The final goal of the code assessment programmes is to quantify the uncertainties in predicting the reactor behaviour during accident conditions as a basis for the evaluation of realistic safety margins. A substantial contribution to the overall database for the assessment of LWR safety codes is provided by the LOBI test facility which has been in continuous operation since 1979 at the Joint Research Centre, Ispra Establishment, Commission of the European Communities. In this facility, which is a 1:700 scaled model of a PWR with inverted U-tube steam generators, about 50 experiments simulating the thermohydraulic conditions during large and small break LOCA and special transients have so far been performed.

For most of the LOBI experiments pre- and post-test calculations have been performed with the system code RELAPS. On the basis of a detailed comparison of measured and predicted data various model improvements have been implemented which have resulted in a considerable improvement of the overall prediction capability of the RELAPS code. The outcome of this activity is part of the multinational effort for the establishment of standard thermohydraulic codes with international acceptance for LWR safety analyses and licensing calculations.

IAEA-CN-48/185P

REVALUATION OF OPERATING LIMITS DUE TO LOSS OF COOLANT ACCIDENTS (SMALL AND INTERMEDIATE BREAKS) IN PWRs

G. PETIT, J.M. MIRACOURT
Service Etudes et projets thermiques et nucléaires,
Electricité de France,
Villeurbanne, France

Improved thermohydraulics models (with separate two phase flow description) have been implemented in advanced codes. These models must be qualified through comparison with numerous analytical experiments spreading over the whole spectrum of thermohydraulic phenomena which could occur during accidents. The results must then be compared to the results of integral experiments in order to check the global coherence of the models put together.

The CATHARE code, which has been developed and assessed by Electricité de France, Commissariat à l'Energie Atomique and Framatome following the above described methodology, ensures a reliable transposition to reactor calculations. Now the code is used to assess the consequences of several accidents taken into account within and beyond the design basis and especially for small and intermediate break LOCA studies.

The models used in the first generation codes led to core uncovering for break sizes around 3 in and therefore to high clad temperatures. This was mainly due to rough representation of stratification and countercurrent flow phenomena leading to retention of water in the upper parts of the primary circuit.

These results limit the realization of core loading and the plant power capability under normal operation.

The advanced code CATHARE is now able to provide a more accurate calculation of small and intermediate breaks. These results show core uncovering. Furthermore, the code is able to give a good representation of post-accident monitoring, for example: stopping of primary coolant pumps and cooling of the primary circuit using secondary discharge valves.

These more pertinent studies give the opportunity to remove previous operation limits imposed by small and intermediate breaks, the consequences of which are always smaller than those of large breaks.

The advanced codes, also assessed for the large break area, will provide (in the near future for the CATHARE code) a better evaluation of these accidents in order to give a new quantitative analysis of the related safety limits.

---

[1] 1 in = 25.4 mm.

IAEA-CN-48/287P

RECENT REVIEW OF REACTIVITY TRANSIENTS AND ACCIDENTS FOR ITALIAN NUCLEAR POWER PLANTS

E. LO PRATO,
Direzione Sicurezza Nucleare
e Protezione Sanitaria (ENEA-DISP),
Comitato Nazionale per la Ricerca e per lo Sviluppo
dell'Energia Nucleare e delle Energie Alternative,
Rome, Italy

INTRODUCTION

A special safety review of plants under construction or operating in Italy, concerning reactivity events, was performed recently by Direzione Sicurezza Nucleare e Protezione Sanitaria (ENEA-DISP), Comitato Nazionale per la Ricerca e per lo Sviluppo dell'Energia Nucleare e delle Energie Alternative [1], the Italian control authority. This review was part of planned activities concerning the Chernobyl accident analysis and evaluation, and was given the following goals:

(a) Verification that the Italian plants cannot undergo degraded reactivity events, according to 'Chernobyl type' scenarios;
(b) Individuation (for the plants mentioned) of the effects of possible reactivity event degradation produced by 'influence factors' of the Chernobyl accident, and evaluation of the connected safety margins allowed by the present design/construction features of the plants.

The plants investigated are very different in type and age, and are described as follows:

— Three operating units, installed at Latina (a gas graphite reactor plant, in operation since 1962), at Trino Vercellese (a Westinghouse PWR plant, in operation since 1965) and at Caorso (a General Electric BWR-4 plant, in operation since 1981);
— Three units under construction, at Montalto di Castro (Alto Lazio Power Station, equipped with two General Electric BWR-6 plants) and at Latina (the CIRENE prototype plant, equipped with a D_2O moderated, boiling water cooled, channel reactor);
— A PWR plant in the design stage (Projetto Unifacto Nucleare), derived from the Westinghouse model 312A, which will be installed in future Italian stations, according to the present Italian Energy Plan.

METHODOLOGY OF THE REVIEW

The review activities accounted for some results of the studies mentioned on the Chernobyl accident analysis and evaluation. These studies provided a set of 'influence factors', which assumed outstanding importance in determining the start and the evolution of the accident sequence in the Soviet reactor. The influence factors are connected, in part, with the RBMK reactor concept (unstable core dynamics, inadequacy of the scram and containment systems, lack of interlocks, etc.), in part with human factors and operation management problems. The importance of these factors was acknowledged by the Soviet participants in the post-accident meeting at Vienna and was emphasized by studies performed in several countries.

Accounting only for the 'hardware' influence factors, the investigations related to goal (a) above were carried out by comparing the design features of the Italian plants with the corresponding features of the damaged Chernobyl plant (a comparison on the subject of human factors and operation control was performed, too, but was not accounted for, because it was considered to be outside the scope of the conceived specific investigation).

The review activities given in (b) above were performed, for each Italian plant, as follows:

- Re-examination of the reactivity event analysis presented in the plant safety report and of the effectiveness/reliability of the installed preventive and mitigating defences;
- Individuation of realistic degraded conditions which can determine the development of event sequences more severe than the design sequences (concurrence of faults, operator errors, bypass of protective devices, lack of conformity to operation rules, etc.);
- Preliminary evaluation of the possible implications of these sequences and of the safety margins provided by plant safeguards and protective devices, including features installed in compliance with special regulatory requirements (anticipated transient without scram, hydrogen control, etc.);
- Individuation of safety topics which require further investigation or development effort (i.e. development or validation of models/computation codes).

RESULTS/CONCLUSIONS

The main results of the review mentioned in (a) is that a significant set of influence factors of the Chernobyl accident cannot be applied to any Italian plant. In particular:

- None of these factors is applicable to LWR plants, operating or under construction;

— The existence of a positive void coefficient in the CIRENE reactor is largely offset by safety margins provided by other design features (high core compactness, sophisticated supervision and automatic control systems, double and fast acting scram system, etc.);
— In the Latina plant, the lack of a containment external to the reactor 'sphere' is counterbalanced by other features such as: very low reactivity and good dynamics of the core, low activation level of the coolant gas, etc.

Then, as far as the hardware features are concerned, the possibility that any Italian reactor may undergo a degraded reactivity accident as severe as the Chernobyl accident, must be excluded. Incidentally, this conclusion might be reinforced if the comparison concerning human factors and management were not disregarded.

Of course, severe accidents of the reactivity type exceeding design limits can be conceived of Italian reactors, with the possible exception of the Latina plant. However, preliminary results of the analysis performed in the framework of (b) show that:

— Such events must be connected with very improbable concurrences of faults and/or operator errors;
— There is no evidence that the considered events are more severe — in terms of consequences — than events of different kinds, but of the same probability class;
— Even considering reactivity events of very low probability, reasonable credit may be given for the effectiveness of the safeguards installed in the Italian plants.

Further investigation may include typical subjects such as: special event sequences of extremely low probability, dynamic performance of some reactors in limited regions of their operation range and containment performance (and effectiveness of their controlled venting systems) in reactivity events of very low probability.

REFERENCE

[1] DIREZIONE SICUREZZA NUCLEARE E PROTEZIONE SANITARIA (ENEA-DISP), COMITATO NAZIONALE PER LA RICERCA E PER LO SVILUPPO DELL'ENERGIA NUCLEARE E DELLE ENERGIE ALTERNATIVE, Studi Impiantistici e Gestionali Post Chernobyl, Doc. ISP(86)16, ENEA/DISP, Rome (1986).

IAEA-CN-48/288P

ARIES: A COMPUTER BASED SYSTEM FOR THE REAL TIME MONITORING OF ATMOSPHERIC DISPERSION IN A NUCLEAR EMERGENCY

R. CARACCIOLO
Direzione Sicurezza Nucleare
e Protezione Sanitaria (ENEA-DISP),
Comitato Nazionale per la Ricerca per lo Sviluppo
dell'Energia Nucleare e delle Energie Alternative,
Rome, Italy

As part of the programme to revise emergency procedures in case of nuclear accidents, Direzione Sicurezza Nucleare e Protezione Sanitaria (ENEA-DISP), Comitato Nazionale per la Ricerca e per lo Sviluppo dell'Energia Nucleare e delle Energie Alternative, since 1980, has been working to develop and implement ARIES (Accidental Release Impact Evaluation System).

ARIES [1] was designed to be a centralized resource of a trained and specialized staff devoted to all aspects of emergency response and to reduce duplication of capabilities, software development and maintenance. This concept was not intended to replace local functions or responsibilities but to complement and enhance local emergency capabilities.

The main services offered by such a system are:

— Real time description of radioactive contaminant dispersion in the atmosphere and deposition on the soil;
— Assessment of short term effects of contaminants on environment and population;
— Co-ordination and direction of radiation monitoring teams;
— Provision of a suitable set of countermeasures.

To provide these services, ARIES comprises the following components:

— ARIES Central Facility (ACF) to manage the system and co-ordinate technical and operational phases;
— Telecommunication network (voice and data) with ACF as central focal point and links with SMA (Italian Air Force Weather Service), associated sites, operation rooms, etc.;
— Computer codes to describe contaminant transport, assess its impact on environment and population, and handle emergency countermeasures;
— Databanks, where data related to the sites and nuclear plants are stored;
— Dedicated computer, hardware instrumentation and system procedures to manage the above mentioned codes according to event sequence.

As far as computer codes are concerned, ARIES has operational models that estimate consequences of atmospheric releases of radioactivity on local and regional bases. SPADE [2], a sequential puff model, based on site meteorology, is used for close-in local estimates (to distances of 10 km) during the first ten minutes after the notification of the release. Regional calculations are performed with three-dimensional numerical transport and diffusion codes MATHEW/ADPIC [3].

These codes, developed by Lawrence Livermore National Laboratory (LLNL) and utilized by ARIES on the basis of co-operation between ENEA and LLNL [4], estimate air concentration and ground deposition from continuous or instantaneous point sources. They are mass conservative and take into account the orography.

Database and management systems that allow for the rapid extraction of pertinent information needed during an emergency are an essential part of ARIES. This component was developed for the construction of plant siting maps [5] that were used to select the most suitable areas in Italy for nuclear power plants. Moreover, several integrations and improvements have been made regarding quality, quantity and spatial resolution of territorial parameters.

The main phases that are activated at ACF during a run of the system are:

— The filling-in of an on-line questionnaire for entry of accident information (plant, type of accident, time of accident, etc.);
— The rapid acquisition of meteorological (local and regional) and source term (if available) data and terrain information;
— Checking of the data by means of computer generated graphics;
— Running of the codes (intermediate and/or complex ones);
— Graphical display of calculated dose contours and surface contamination levels overlaid on a site geographical map;
— Acquisition of radiometric data from monitoring teams and comparison between measured and calculated data;
— Evaluation of possible countermeasures;
— Communication of the analysis to the involved authorities.

An initial qualitative assessment is available within 15 minutes after the start of the system and a complete assessment within 45 minutes. Thereafter ARIES provides continuing support on an hourly basis until termination of the emergency.

ARIES has been in an operative phase since 1985 and during the Chernobyl emergency its main components were successfully used to assess the consequences on Italian territory, evaluate the source term, manage a large amount of radiometric data ($\sim 2 \times 10^5$), etc.

REFERENCES

[1] BROFFERIO, C., CARACCIOLO, R., PANDOLFI, G., ''Methods and models adopted in Italy in the evaluation of consequences of accidental releases'', Radioactive Releases and their Dispersion in the Atmosphere following a Hypothetical Reactor Accident (Proc. Sem. Risø, 1980), CEC, Luxembourg (1980) 483–497.

[2] DESIATO, F., ''SPADE: A real-time langrangian puff model for local scale atmospheric diffusion evaluation'', Workshop on Real-Time Computing of the Environmental Consequences of an Accidental Release to Atmosphere, CEC, Luxembourg (1985).

[3] DICKERSON, M.H., ARAC Status Report: 1985, Rep. UCRL-53641, Lawrence Livermore Natl Lab., CA (1985).

[4] DICKERSON, M.H., CARACCIOLO, R., Joint Research and Development and Exchange of Technology on Toxic Material Emergency Response between Lawrence Livermore National Laboratory and Ente Nazionale per l'Energia Elettrica, 1985 Progress Report, Rep. UCRL-53696 (1986).

[5] CARACCIOLO, R., et al., Land Survey Methodology, Note CNEN-10, Comitato Nazionale per l'Energia Nucleare, Rome (1980).

IAEA-CN-48/145P

AN ADVANCED DESIGN FOR SNR-2

K. EBBINGHAUS
Interatom GmbH,
Bergisch-Gladbach

M. KÖHLER
Kraftwerk Union AG,
Offenbach am Main

Federal Republic of Germany

Based on experience in design and construction of the SNR-300 and SPX-1 fast breeder NPPs, an advanced design has been developed for the follow-up project SNR-2.

The design simplifies the necessary safety measures of the nuclear island and will lead to investment costs which are comparable to those of LWRs in the Federal Republic of Germany.

The cost goal is achieved mainly by designing the nuclear island in such a way that it is self-sufficient for all safety related operation thus allowing a design of the balance of plant according to conventional rules.

The main design features are as follows:

— The preventive safety concept against core disruptive accidents combines inherent reactivity features, enhanced by the design, and a combination of passive and active means to ensure an ultimate shutdown of the core in any event and before critical structural loadings have been reached.

— Realistic failure assumption and an adequate core monitoring system to protect the core against fault propagation and melting of fuel elements.

— The choice of structural material, design measures, and stress criteria to assure the system integrity in such a way that large failures of the coolant boundary are excluded.
— The removal of the decay heat and the cooling of hot components mainly by passive means in order to ensure the necessary reliability of heat removal and the safety related independence of the nuclear island from the balance of plant.
— The fuel handling chain, including a novel design of an internal storage and the transfer of spent fuel elements under sodium inside the plant and possibly also during transport to the reprocessing plant, thus assuring decay heat removal by natural circulation.

The final arrangement of the secondary system is still under consideration, in particular with respect to decay heat removal capability during the first period of the shutdown transient. Eight loops have been selected as reference to allow high availability even in the case of a steam generator failure (operation near 100% power with seven out of eight secondary loops).

The steam generator design is based on straight tube design and various alternatives including a J-type solution were investigated.

The design features described above lead to substantial simplifications for the plant arrangement when compared with a typical Federal German LWR, in particular regarding protection against external events of the relevant system of the balance of plant.

REFERENCES

[1] KÖHLER, M., BÜRKLE, W., "Design and safety aspects for the nuclear island of SNR 2", International Conference on Fast Breeder Systems, Richland, WA (1987).

[2] EITZ, A.W., WOLFF, U., RIETHMÜLLER, R., "Power generation cost of the SNR 2 in comparison to German LWR ", ibid.

IAEA-CN-48/251/P

КОНЦЕПЦИЯ БЕЗОПАСНОСТИ РЕАКТОРНОЙ УСТАНОВКИ АТОМНОЙ СТАНЦИИ ТЕПЛОСНАБЖЕНИЯ

Ф.М.МИТЕНКОВ, В.В. ЕГОРОВ, В.С. КУУЛЬ,
В.А. МАЛАМУД, Ю.Г. НИКИПОРЕЦ, О.Б. САМОЙЛОВ,
И.Н. СОКОЛОВ
Государственный комитет по использованию атомной энергии СССР,
Москва,
Союз Советских Социалистических Республик

Рассмотрены результаты исследований по безопасности реакторной атомной станции теплоснабжения АСТ-500.

1. ПРОЕКТНЫЕ АВАРИЙНЫЕ РЕЖИМЫ

Аварии разгерметизации первого контура в связи с наличием страховочного корпуса, локализующей арматуры на трубопроводах первого контура на КС проходят с сохранением активной зоны под уровнем теплоносителя.

2. СВЕРХПРОЕКТНЫЕ АВАРИИ

Реактивностные аварии, связанные с захолаживанием реактора и отказом аварийной защиты, характеризуются в АСТ медленной динамикой и небольшими отклонениями параметров от номинальных значений (10–30%).

Теплоотводная авария с отказом системы аварийного расхолаживания проходит без существенного повышения давления в реакторе. При этом в качестве страховочной системы аварийного теплоотвода выступает система теплоотвода через ИПУ на ПКД второго контура, функционирующая на пассивном принципе за счет выпаривания воды второго контура.

Разгерметизация корпуса реактора без срабатывания аварийной защиты. В данной аварии интенсивное парообразование в активной зоне приводит к самоглушению реактора за счет отрицательного парового коэффициента реактивности.

В авариях с разгерметизацией корпуса реактора при негерметичном страховочном корпусе и в аварии с разрывом трубопровода систем очистки при незакрытии двойной запорной арматуры активная зона сохраняется под уровнем теплоносителя в течение 12 часов без принятия мер.

3. ГИПОТЕТИЧЕСКИЕ АВАРИИ

Перечень рассмотренных гипотетических аварийных режимов охватывал все наиболее тяжелые аварии из класса реактивностных, теплоотводных аварий и аварий разгерметизации, возможным исходом которых могло быть плавление активной зоны.

Прекращение теплоотвода в сеть с отказом АЗ реактора

Вскрытие разъемов корпуса реактора ограничивает рост давления в реакторе и приводит к самоглушению реактора за счет интенсивного кипения теплоносителя в активной зоне. Максимальное давление в реакторе составляет 6,4 МПа.

Значение разрушающего давления, определенного с учетом постулированного стандарта дефекта, составляет для корпуса реактора 8,4 МПа, для корпуса страховочного – 2,8 МПа. При давлении выше 5,0 МПа происходит потеря герметичности разъемов корпуса реактора, обусловленная упругой вытяжкой шпилек. Давление раскрытия разъемов страховочного корпуса составляет 1,5 МПа.

Разгерметизация корпуса реактора в нижней части при крупной негерметичности перегородки между полостями КС

Авария проходит с кратковременным частичным осушением активной зоны, неприводящим к недопустимому повышению температуры твэлов.

Принятые в реакторной установке АСТ-500 технические решения, такие, как естественная циркуляция теплоносителя первого контура, системы аварийного теплоотвода, функционирующие на пассивном принципе действия, использование страхоснижение параметров и энергонапряженности активной зоны, определяют внутреннюю безопасность и самозащищенность реакторной установки, относя ее к классу систем повышенной безопасности.

Анализ широкого набора сверхпроектных и гипотетических аварий с постулированием отказов высоконадежных систем безопасности позволил установить, что с учетом физико-технических особенностей установки, наличия резерва времени, достаточного для дистанционного дублирования автоматических действий по вводу в работу систем безопасности, использования страховочных защитных систем плавление топлива является технически нереализуемым событием (вероятность менее 10^{-7}). Радиационное воздействие АСТ на среду незначительно.

IAEA-CN-48/87P

VARIOUS REPOSITORY DESIGNS FOR THE DISPOSAL OF HEAT GENERATING RADIOACTIVE WASTES (REPROCESSING WASTES AND SPENT FUEL)

B. HARTJE, H.J. ENGELMANN, C. SCHRIMPF
Deutsche Gesellschaft zum Bau und Betrieb
von Endlagern für Abfallstoffe mbH (DBE),
Hannover, Federal Republic of Germany

In recent years in the Federal Republic of Germany work has been done on two concepts for the disposal of high level radioactive wastes. The reference concept foresaw a repository for vitrified wastes in small canisters from a reprocessing plant. The plan is for these canisters to be lowered from a gallery into a vertical 300 m deep borehole. In the framework of the 'Systems Study Alternative Entsorgung' the DBE developed a disposal technique for 60 Mg heavy casks with spent fuel. Figure 1 shows the transport and emplacement concept for these casks in the mine and the final disposal in a tunnel. The borehole technique is also planned for heat-generating MAW drums, which contain sludge suspension, cladding, core parts from the reprocessing plant and HTR fuel elements.

In January 1985 the Federal Government decided that it complied best with the legal requirements by adhering to reprocessing and recycling as the preferred version of Entsorgung. The Cabinet stated also that direct disposal of spent fuel was permissible for some types of fuel elements. So the Federal Ministry for Research and Technology (BMFT) is financing a project for designing various repository layouts for the disposal of all types of heat generating wastes. These various designs will be analysed considering the boundary conditions of the geological formation (salt dome) [1]. The result of this comparative analysis will be the selection of a repository design catering for both possible concepts.

The different designs take into consideration the variation of

— The relationship between spent fuel and vitrified wastes from reprocessing;
— The design of the disposal packages and techniques;
— The time between discharge from the reactors and disposal.

The report will illustrate six principal designs of the repository layout, e.g.

— Vertical borehole emplacement only
— Dual purpose repository (four variations) (Fig. 2)
— Emplacement in disposal tunnels only.

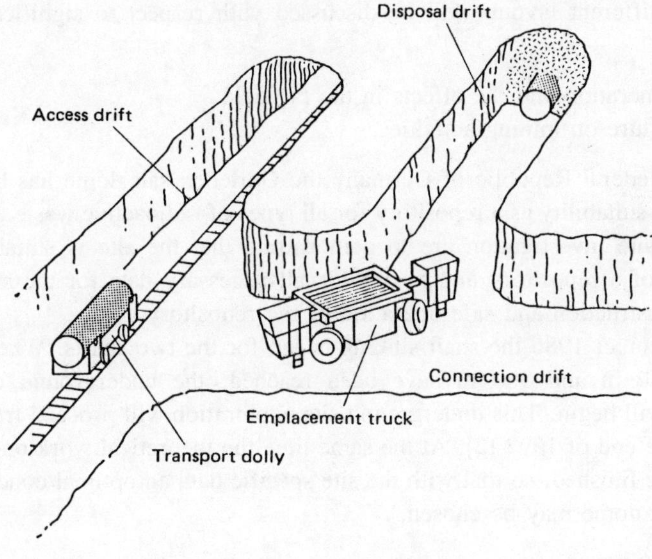

FIG. 1. Horizontal emplacement technique.

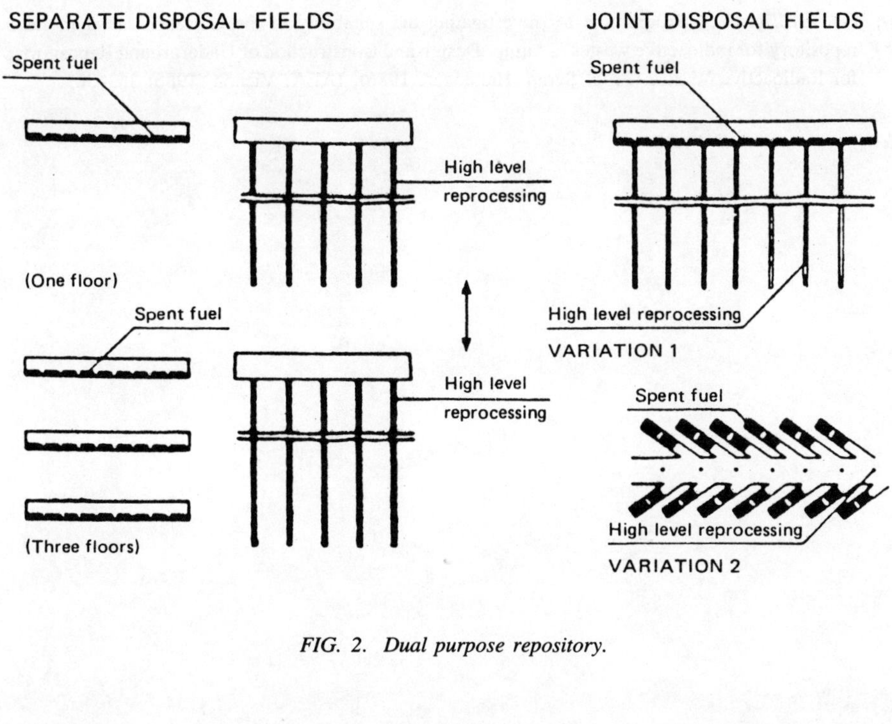

FIG. 2. Dual purpose repository.

These different layouts will be discussed with respect to significant aspects such as

— Heat generation and the effects in the far field
— Expenditure on mining activities.

In the Federal Republic of Germany the Gorleben salt dome has been investigated for its suitability as a repository for all types of radioactive waste. The objectives of the site investigation are to demonstrate that the site is suitable for the construction of a repository and to provide all necessary data for the site specific planning, construction and safe operation of the repository.

In September 1986 the shaft sinking began for the two shafts. When the final depths of 840 m and 960 m have been reached, the underground exploration programme will begin. This underground site exploration will proceed from the end of 1989 to the end of 1992 [2]. At the same time the theoretical work on repository design will be finished, so that with the site specific data an optimal concept for the Gorleben salt dome may be chosen.

REFERENCES

[1] PAPP, R., CLOSS, K.D., ''Results of the German alternative fuel cycle evaluation and further efforts geared toward demonstration of direct disposal'', Proc. Waste Management 1986, Vol. 2, Tucson (1986).

[2] GRÜBLER, G., ''Programme for investigating the suitability of the Gorleben salt dome as a repository for radioactive wastes'', Siting, Design and Construction of Underground Repositories for Radioactive Wastes (Proc. Symp. Hannover, 1986), IAEA, Vienna (1986) 363–369.

IAEA-CN-48/184P

DISPOSAL OF RADIOACTIVE WASTES IN RELATION TO HEALTH PROTECTION FOR EMPLOYEES AND THE PUBLIC

V. PICCARDO, A. CANDELORA,
L. CRESCIMANNO, G. MONACELLI
Direzione Generale Servizi Igiene Pubblica,
Ministerio della Sanità,
Rome, Italy

The disposal of radioactive wastes represents one of the greatest problems not yet extensively dealt with, after the production and the peaceful uses of nuclear energy.

Solutions to this problem constitute a very important guarantee for the protection of the employees and the public, both now and in the future, against the risks, derived from the environment, of solid, liquid or gaseous wastes interacting with the human organism.

A great difficulty arises from the diversity of the radioactive wastes derived from the different origins, physical forms, chemical composition, nature and concentration of the radionuclides contained in them.

Italian legislation foresees that the radioactive waste disposal should be under the local authority's control. The central bodies of the State, and the Health Ministry in particular, have the functions of guidance and co-ordination.

In such a decentralized system there is a need for relevant norms and general criteria to which the peripheral organs can comply.

Close technical control at national level is maintained by the Direzione Sicurezza Nucleare e Protezione Sanitaria (DISP) of Comitato Nazionale per la Ricerca e per lo Sviluppo dell'Energia Nuleare e delle l'Energie Alternative (ENEA) following the IAEA classification which subdivides radioactive wastes into three categories based on the radionuclide identity and concentrations. Each category involves different ways of disposal.

The first category comprises the wastes characterized by relatively short half-lives and permits their disposal as normal wastes after a period of some months or, at the most, a few years. They derive principally from the employment of radioactive materials in fields of medical and scientific research. The principal problem presented by this category of wastes is that of having the waste points spread over the national territory which could create a difficulty in evaluating the impact on health.

The largest volumes of radioactive wastes come under the second category, which comprises above all wastes produced during the operation of the nuclear plants and from the fuel treatment plants. For these wastes to reach low concentrations of radioactivity requires a period varying from a few decades to a few centuries.

The disposal criterion which, at the current stage of studies, seems to offer greater guarantees from the radioprotection point of view, is that of the isolation of these types of radioactive wastes, eliminating any possibility of dispersion and return to mankind until the radioactivity has decreased to levels comparable with the natural background.

The creation of suitable engineering techniques or the identification of sites for disposal requires time and the introduction of such new techniques is of great urgency because of the increasing accumulation of wastes in the second category. At present volume reduction provides a temporary remedy to the problem and gains time for studies to be carried out to determine the storage characteristics necessary to minimize risk to the public.

The need is particularly felt for an international dialogue about the stage of such studies and the methodologies adopted for the disposal of the high activity radioactive wastes, belonging to the third category. At present the most suitable solution seems to be the disposal of the solidified wastes into deep geological formations. The aim is to create natural geological barriers that stop the movement of the radionuclides, in order to guarantee that future generations are not exposed to radioactivity greater than that which the earth naturally contains.

IAEA-CN-48/242 P

БИТУМИРОВАНИЕ ЖИДКИХ РАДИОАКТИВНЫХ ОТХОДОВ НА ЛЕНИНГРАДСКОЙ АЭС

Н.И. АРЖАНОВ, А.И. БОРЗУНОВ, А.Ф. ВОТИНЦЕВ,
В.И. ДАВЫДОВ, Б.В. КЛЮЧНИКОВ, С.Г. МАРЯРОВ,
А.С. НИКИФОРОВ, А.С. ПОЛЯКОВ, К.Д. РОГОВ,
М.В. СТРАХОВ, С.Н. ФИЛИППОВ
Государственный комитет по использованию
атомной энергии СССР,
Москва,
Союз Советских Социалистических Республик

Установка битумирования предназначена для отверждения жидких радиоактивных отходов (ЖРО) АЭС. Промышленная эксплуатация установки на кубовом остатке в непрерывном режиме осуществляется на Ленинградской АЭС с 1985 года.

Установка включает: оборудование приема и дозирования отходов и чистого битума. Роторно-пленочный битуматор РБ-1000-14, аппараты конденсации и очистки парогазовой фазы, систему транспортировки компаунда, хранилище отвержденных отходов, систему автоматического пожаротушения.

Оборудование установки размещено в защитных каньонах. Управление осуществляется с центрального пульта.

Длина компаундопровода 50–150 м (в зависимости от заполняемого отсека хранилища). Хранилище состоит из 12-ти отсеков и имеет объем 27 000 м^3.

А. Кубовой остаток	
Концентрация, г/дм^3	200–300
PH	11–12
плотность, кг/дм^3	1,13–1,16
Б. Связующие	
битум БНД 90/130, БНД 40/60	
температура размягчения, °C	45–55
пенетрация, 0,1 мм	40–130
температура вспышки, °C	200
влажность, %	1
плотность, кг/см^3	0,96–1,02
Результаты эксплуатации	
переработано кубового остатка, м^3	6000
поступило на захоронение бит. комп., м^3	300
производительность установки, л/ч	400
давление греющего пара, МПа	0,5–0,6
поверхность нагрева битуматора, м^2	10
частота вращения ротора, мин$^{-1}$	49
степень наполнения компаунда, %	40
влажность, %	2
скорость выщелачивания радионуклидов, г/см^2 · сут	10^{-4}–10^{-5}
Удельная активность, Бк/л	
вторичного конденсата	$4{,}1 \times 10^1$–$1{,}4 \times 10$
отходящих газов	10^{-4}–10^{-5}

CHAIRMEN OF SESSIONS

Volume 4

Technical Session 3.1A	H. RABOLD	German Democratic Republic
Technical Session 3.1B	Wan-Yong CHON	Republic of Korea
Technical Session 3.2A	F.P. MARTINEZ	Spain
Technical Session 3.2B	W. LEPECKI	Brazil
Technical Session 3.3A	C. SENNIS	Italy
Technical Session 3.3B	M. COPIC	Yugoslavia

SECRETARIAT OF THE CONFERENCE

Scientific Secretariat

Chief: L. BENNETT

Members:

F. CALORI	E. MUELLER-KAHLE
B. CSIK	A. NECHAEV
P. DE	V. ONUFRIEV
J. FISCHER	R. PONTIER
J. HIRLING	K. SCHENK
A. KARBASSIOUN	B. THOMAS
G. LINSLEY	F. WALD
J. MARQUES DE SOUZA	

Administrative Secretariat

Administrative Secretary: E. PILLER

Editors

Proceedings Editors: R. PENISTON-BIRD, E. KATZ, E. REILLY

French Editor: J.-N. AQUISTAPACE

Russian Editor: O. MELNIK

Spanish Editor: L. HERRERO

HOW TO ORDER IAEA PUBLICATIONS

An exclusive sales agent for IAEA publications, to whom all orders and inquiries should be addressed, has been appointed in the following country:

UNITED STATES OF AMERICA	UNIPUB, 4611-F Assembly Drive, Lanham, MD 20706-4391

In the following countries IAEA publications may be purchased from the sales agents or booksellers listed or through major local booksellers. Payment can be made in local currency or with UNESCO coupons.

ARGENTINA	Comisión Nacional de Energía Atómica, Avenida del Libertador 8250, RA-1429 Buenos Aires
AUSTRALIA	Hunter Publications, 58 A Gipps Street, Collingwood, Victoria 3066
BELGIUM	Service Courrier UNESCO, 202, Avenue du Roi, B-1060 Brussels
CHILE	Comisión Chilena de Energía Nuclear, Venta de Publicaciones, Amunategui 95, Casilla 188-D, Santiago
CHINA	IAEA Publications in Chinese: China Nuclear Energy Industry Corporation, Translation Section, P.O. Box 2103, Beijing IAEA Publications other than in Chinese: China National Publications Import & Export Corporation, Deutsche Abteilung, P.O. Box 88, Beijing
CZECHOSLOVAKIA	S.N.T.L., Mikulandska 4, CS-116 86 Prague 1 Alfa, Publishers, Hurbanovo námestie 3, CS-815 89 Bratislava
FRANCE	Office International de Documentation et Librairie, 48, rue Gay-Lussac, F-75240 Paris Cedex 05
HUNGARY	Kultura, Hungarian Foreign Trading Company, P.O. Box 149, H-1389 Budapest 62
INDIA	Oxford Book and Stationery Co., 17, Park Street, Calcutta-700 016 Oxford Book and Stationery Co., Scindia House, New Delhi-110 001
ISRAEL	Heiliger & Co. Ltd. 23 Keren Hayesod Street, Jerusalem 94188
ITALY	Libreria Scientifica, Dott. Lucio de Biasio "aeiou", Via Meravigli 16, I-20123 Milan
JAPAN	Maruzen Company, Ltd, P.O. Box 5050, 100-31 Tokyo International
PAKISTAN	Mirza Book Agency, 65, Shahrah Quaid-e-Azam, P.O. Box 729, Lahore 3
POLAND	Ars Polona-Ruch, Centrala Handlu Zagranicznego, Krakowskie Przedmiescie 7, PL-00-068 Warsaw
ROMANIA	Ilexim, P O. Box 136-137, Bucharest
SOUTH AFRICA	Van Schaik Bookstore (Pty) Ltd, P.O. Box 724, Pretoria 0001
SPAIN	Díaz de Santos, Lagasca 95, E-28006 Madrid Díaz de Santos, Balmes 417, E-08022 Barcelona
SWEDEN	AB Fritzes Kungl. Hovbokhandel, Fredsgatan 2, P.O. Box 16356, S-103 27 Stockholm
UNITED KINGDOM	Her Majesty's Stationery Office, Publications Centre, Agency Section, 51 Nine Elms Lane, London SW8 5DR
USSR	Mezhdunarodnaya Kniga, Smolenskaya-Sennaya 32-34, Moscow G-200
YUGOSLAVIA	Jugoslovenska Knjiga, Terazije 27, P.O. Box 36, YU-11001 Belgrade

Orders from countries where sales agents have not yet been appointed and requests for information should be addressed directly to:

Division of Publications
International Atomic Energy Agency
Wagramerstrasse 5, P.O. Box 100, A-1400 Vienna, Austria

88-02291